STUDIES IN INTERFACE SCIENCE

Proteins at Liquid Interfaces

STUDIES IN INTERFACE SCIENCE

SERIES EDITORS
D. Möbius and R. Miller

Proteins at Liquid Interfaces

Edited by

DIETMAR MÖBIUS
Max-Planck-Institut für Biophysikalische Chemie
P.O. Box 2841
Göttingen
Germany

REINHARD MILLER
Max-Planck-Institut für Kolloid- und Grenzflächenforschung
Rudower Chaussee 5
Berlin-Adlershof
Germany

1998
ELSEVIER
Amsterdam – Lausanne – New York – Oxford – Shannon – Singapore – Tokyo

ELSEVIER SCIENCE B.V.
Sara Burgerhartstraat 25
P.O. Box 211, 1000 AE Amsterdam, The Netherlands

Library of Congress Cataloging in Publication Data
A catalog record from the Library of Congress has been applied for.

ISBN: 0 444 82944 X

♾ The paper used in this publication meets the requirements of ANSI/NISO Z39.48-1992 (Permanence of Paper).

Transferred to digital printing 2005

Printed and bound by Antony Rowe Ltd, Eastbourne

Foreword

The interfacial behaviour of surfactants and proteins, and their mixtures, is of importance in a wide range of areas such as food technology, detergency, cosmetics, coating processes, biomedicine, pharmacy, biotechnology. Surface and interfacial tension measurements and interfacial dilation and shear rheology are methods to characterise such layers at liquid interfaces. Relations between these properties and the complex behaviour of foams and emulsions become more and more established and gain greater importance. Emulsification processes and emulsion properties are determined by drop break-up and coalescence processes. An analogous situation exists in foam formation and its stabilisation. Besides hydrodynamic conditions these local events are controlled by dynamic interfacial properties, modified by adsorbed surfactants and/or polymers. Quantitative correlations between adsorption layer properties and the behaviour of a foam or emulsion films or even of disperse ensembles like foams and emulsions have to be developed to improve production efficiency, product quality and stability.

Extensive knowledge of the properties of adsorption layers at interfaces is required to predict or control the behaviour of practically important disperse liquid systems. For this reason, numerous systematic experimental studies of proteins and their mixtures with surfactants have been performed using various complementary methods: dynamic and equilibrium interfacial tensions, ellipsometry, surface light scattering, shear and dilational rheology. The interaction of proteins and surfactants in bulk, and its relevance for interfacial properties, and the role of contaminants are vital to understand interfacial layers. New phenomena, for example the formation of skins at the surface of protein layers or adsorption rates of proteins which are orders of magnitude faster than expected from a diffusional model are features of high actuality.

New or recently developed experimental techniques broaden the sources of information, such as FRAP which enables one to measure molecular mobility in adsorption layers, or the Brewster Angle Microscopy and Atomic Force Microscopy which allow the visualisation of macroscopic structures of adsorption layers. IR-ellipsometry, FTIR and CD at liquid interfaces, SHG and other new optical principles have been tested and proven to be applicable to monolayers at different liquid interfaces also for studies of protein layers.

Further, the development of theories to describe the thermodynamic surface state or the exchange of matter for proteins and protein/surfactant mixtures necessary to interpret relaxation experiments have reached a level which at least provides qualitative understanding of many phenomena.

Summarising studies by leading groups in Europe, Australia and North America on interfacial layers of proteins and protein/surfactant mixtures have yielded remarkable results. Therefore, it seems very reasonable to review them in a book. The book will have 11 chapters. As kind of introduction chapter 1 is dedicated to a systematisation of proteins from a general point of view with emphasis is made on those proteins which are mainly used in model investigations or in the subsequent chapters of the book. Chapter 2 deals with the basic problem of adsorption isotherms for proteins. Besides a historical overview about isotherms of polymers in general it will give the very recent developments of adsorption isotherms especially derived for the purpose of describing the equilibrium state of adsorption layers of proteins at liquid interfaces. These theoretical models do not only provide access to explicit functions of surface pressure on surface concentration but also on bulk concentration, and relationships for the adsorption layer thickness. Special emphasis was laid in chapter 4 on the problem of reversibility of protein adsorption.

Chapters 7 and 8 are dedicated to several new techniques developed or modified in order to study specific properties of protein layers. One of these recently developed techniques, the "fluorescence recovery after photobleaching" (FRAP) is discussed in chapter 7. This technique allows to learn about the mobility of molecules in an adsorption layer and hence give information about its structure. Consequently this mobility should be somehow related to the rheological parameters of this layer. The axisymmetric drop shape analysis as a technique to measure dynamic surface tensions without touching or deforming the interface is described in chapter 8, and a large number of applications is added. Also chapter 6 is dedicated to a variety of applications of protein and mixed protein/surfactant layers. Thus systems like beer or champaign are explained from the point of view of interfacial science.

Half of the book is dedicated to systems containing both proteins and classical surfactants, in particular lipids. This is evident as in most application proteins act in connection with surfactants and all interfacial properties are controlled by the interplay between these components. Correlations are given between the surface rheological properties and the stability of emulsions and foams. The particular importance of lipid/protein mixtures for living systems is discussed in chapter 10. This type of mixed layers mimics membrane layers. Particular interest in such systems comes from medical research on pulmonary surfactants.

In chapters 3, 5 and 9 results are summarised comparing protein layers at water/oil and water/air interfaces. Adsorption mechanisms, and surface rheology are discussed and a large variety of data about different proteins is given.

In most of the chapters practical applications of specific systems are discussed. In particular, the mixed gelatine/surfactant systems relevant in photographic industries is described from different points of view in the final chapter 11. It is shown that this highly sophisticated technology is based on non-equilibrium properties of such mixed adsorption layers.

In summary, the present book reflects the state of the art of research and application of protein interfacial layers rather than a snapshot of only some recent developments in this field. Emphasis is placed on experimental details as well as theoretical developments. Thus, the presented results should be of interest for a broad audience working in fundamental research and in the applications in food technology, pharmacy, coating, biotechnology, medicine.

The authors and editors are very grateful to Sabine Siegmund and Oliver Senkel for technical support during the preparation of the manuscript.

Contents

Proteins at Liquid Interfaces
D. Möbius and R. Miller (Editors)

PROTEINS: SOME PRINCIPLES OF CLASSIFICATION AND STRUCTURE

K. D. Schwenke

Forschungsgruppe Pflanzenproteinchemie, Universität Potsdam/Förderverein Proteinchemie e.V. c/o Biologische Bundesanstalt, Stahnsdorfer Damm 81, D-14532 Kleinmachnow

Contents:

1. CLASSIFICATION OF PROTEINS - USEFULNESS AND LIMITATION

Scientists confronted with the diversity in types of proteins tried to determine typical properties which could be chosen as criteria for classification purposes. A pioneer in this field was Thomas B. Osborne, Research Scientist in the Connecticut Agricultural Experiment Station, New Haven. Working with plant proteins he proposed a classification on the basis of the solubility of proteins [1] which is still used recently by many researches. Osborne made a distinction between two types of protein: simple and conjugated. The former consists of albumins which are soluble in water at neutral or slightly acid reaction, coagulable by heat and precipitable by saturating their neutral solutions with salts such as sodium chloride or magnesium sulphate (recently mostly ammonium sulphate at 70-100% saturation); globulins which are insoluble in water but soluble in salt solutions; prolamins, which are soluble in 70-80 % ethanol; glutelins, which are not dissolved by neutral aqueous solutions but partially dissolved by alkaline solutions, by saline solutions or by alcohol. The conjugated proteins group was divided into nucleoproteins, glycoproteins, phosphoproteins, hämoproteins and lecithoproteins (recently generally lipoproteins).

Osborne also pointed out that differentation between albumins and globulins may not be distinct in every case. The presence of small amounts of salts or other subsidiary compounds in the protein-containing material can cause the solubilisation of globulins by water, a fact which is often ignored nowadays. Nevertheless, water solubility is an useful criterion also for classification of typical albumins of animal origine such as serum albumins, α-lactalbumin, β-lactoglobulin and ovalbumin. On the other hand, prolamins and glutelins are typical plant proteins and attempts to classify water and salt insoluble animal proteins such as casein into the group of glutelins, as done so by some authors, might be correct from a rather formalistic point of view but they are far from the reality when considering the protein structure (see section 6).

A commonly used principle for protein classification on a structural basis, distinguishes between water and salt soluble globular proteins of a rather sphere-like molecular shape (ellipsoids of revolution with an axial ratio not greater than about 4) and fibrillar proteins insoluble in water and salt solutions, e.g. collagen and keratins [2]. The latter are structural proteins which have a stabilising structure forming function (collagen, elastins, keratins) or take place in the muscle contraction (actomyosin complex).

A subdivision of the various proteins according to their origin was also used for distinguishing between animal and plant proteins. Animal proteins were subdivided in this manner into those from blood plasma, tissue liquides, muscles, connective tissue etc. as well as proteins from milk, egg and other sources [3].

Protein function have also been used as a criterion for classification. Accordingly, one distinguishes between the above mentioned structural proteins and those having an active role in the metabolism such as enzymes and enzyme inhibitors, hormones, carrier proteins (e.g. serum albumin, haemoglobin), contractile proteins (myosin, actin), storage proteins (ovalbumin, wheat gliadin, plant 11S globulins) and others. This classification may be useful from a pure biochemical or physiological point of view, but is less helpful when the relationships between structure and function are not considered. Such relationships should be very useful in subdividing some protein families e.g. groups of enzymes and enzyme inhibitors, plant storage proteins, contractile proteins and others.

A logical and systematic classification of the various protein groups comprising several thousand species should be based on the primary and spatial structure. Though a number of protein structures has been elucidated, information is still limited to allow a sufficient realisation of classification. An approach to use secondary structure data for protein classification is mentioned in section 2.

2. STRUCTURAL ORGANISATION IN PROTEINS

2.1. The amino acid sequence determines the spatial structure

Three levels of structural organisation in proteins were proposed by Linderström-Lang [4]: primary structure referring to the amino acid sequence, secondary structure denoting the regular arrangement of the polypeptide backbone, and tertiary structure as the three-dimensional organisation of globular proteins. Quarternary structure, referring to the arrangement of aggregates of the globular proteins, has been added as the fourth level by Bernal [5]. Actually, two more structural levels can be distinguished: supersecondary structure, referring to physically preferred aggregates of secondary structure and domains denoting those parts of the protein which form well-separated globular-regions [6]. Renaturation experiments have shown that the amino acid sequence contains the entire structural information of a protein. Accordingly, the relationship between the six structural levels are interdependent, with elements of a lower level determining the elements of higher levels in a structural hierarchy: amino acid sequence→ secondary structure→supersecondary structure→domains→tertiary structure (globular protein) →quaternary structure.

To understand the regular arrangement of the backbone of the protein polypeptide chain in the secondary structure, one has to take into consideration the following structural restrictions (for details see references 6, 8, 9, 10].

1. The C-N bonds in the peptide amide groups have a partial double bond character due to resonance between the structures given in Fig. 1a which is reflected by a shortening of the C′-N-bond length from 1.46-1.50 Å in aliphatic amines to 1.33 Å in simple peptides, and a prolongation of the C-O bond length from 1.215 Å to 1.24 Å. Thus the peptide bond is fixed in a plane and free rotation about the C′-N bond is inhibited.

2. The formation of secondary structure is strictly influenced by the limitation of free rotation about the C′-N-bond, the C_α-C′ bond and N- C_α bond characterised by the dihedral angles ω, ψ and ϕ respectively (Fig. 1b). ω amounts to 180° in the most frequent trans-peptide bonds and to 0° in cis-peptide bonds. The free rotation about the C_α -C′ and the N- C_α bonds is limited by steric contact non-bonded interactions. These can be calculated on the basis of van der Waals atom radii. There are three rotational regions which are clearly not allowed. These are (i) with $\psi = 180°$, $\phi = 0°$ the carboxyl oxygens of the peptide group overlap; (ii) with $\psi=0°$, $\phi=180°$ there is a potential overlap of hydrogen atoms; and (iii) with $\psi=0°$; $\phi=0°$; there is an overlap of 0-and H-atoms. It is obvious that the residue with the smallest side chain, e.g. glycine, restricts the ψ, ϕ rotation least, whereas bulky amino acid side chains contribute markedly to steric restrictions and affect the formation of secondary structure.

The polypeptide backbone forms a linear group, if successive peptide units assume identical relative orientations or if all (ψ, ϕ)-angles are the same. This linear group is a helix, which is conveniently described by the rise per element d, the number of elements per turn n, and the distance r of a marker point on each element (here the C_α-atom) from the helix axis. d is taken as positive and the helix chirality can be read from the sign of n [6].

a)

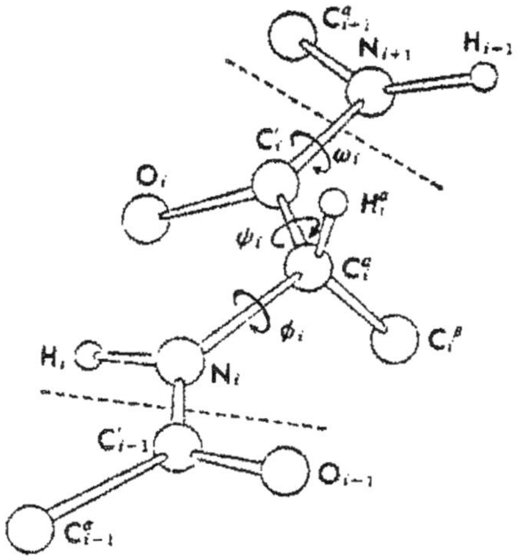

b)

Fig. 1 a. Resonance between two structures of the peptide bond, b. Dihedral angles in a polypeptide. When the peptide unit rotates clockwise (viewed from C_α) around the C-N and C-C` bonds while the C_α is held fixed, the angles are taken to be positive.

The parameters for linear groups formed by polypeptides are summarised in Table 1. The dependencies of favoured secondary structures on ψ and ϕ are commonly presented in the Ramachandran plot (see ref. 10). Secondary structures are stabilised by hydrogen bonds between peptide amide and carbonyl groups. In the α-helix, the C=0 bond is parallel to the helix axis and a straight hydrogen bond is formed with the N-H group, which is the most stable geometrical arrangement. It is also the interaction of all constituent atoms of the main chain, which are packed closely together, which allows for van der Waals attraction across the helical axis and contributes to the stabilisation of the helix. Therefore, the α-helix is the most abundant secondary structure in proteins.

If the stabilisation by hydrogen bonds is not ensured in a single linear group, inter-hydrogen bonds can be formed between helices. This is the case in the polyproline and collagen helices, where the C=0 bond is directed perpendicular to the helix axis and the triple-helices are formed [10].

The situation is similar in β-structures, where the β-strands form intrachain or interchain hydrogen bonds, resulting in the formation of β-pleated sheets. The interchain β-structure is found in fibrous proteins, whereas the intrachain β-structure is present in many globular proteins. The existence of reverse turns allows the reversion of the direction of the polypeptide chain by 180° and in the case of β-strands the stabilisation by intrachain hydrogen bonds. In the reverse turn conformation, a hydrogen bond is formed between the carbonyl oxyygen of amino acid residue and N-H group of the residue (n+3). Reverse turns are very abundant, their different types comprise about one quarter of all residues in globular proteins [11-14].

Table 1. Structural parameters in basic conformations formed by polypeptides

Linear group	Φ (°)	Ψ (°)	ω (°)	residue per turn **n** and chirality	rise per residue ***d*** (Å)	radius of helix ***r*** (Å)
α-helix (right handed) (commonly found in globular proteins and coiled in fibrous proteins)	-57	-47	+180	+3,6	+1,5	+2,3
α-helix (left-handed) (hypothetical)	+57	+47	+180	-3,6	+1,5	+2,3
3_{10} helix (only small pieces)	-49	-26	+180	+3	+2,0	+1,9
π-helix (hypothetical)	-57	-70	+180	+4,3	+1,1	+2,8
β-sheet (parallel) (occasionally in adjacent chain segments of globular proteins)	-119	+113	+180	± 2	+3,2	+1,1
β-sheet (antiparallel) (commonly found in proteins and synthetic polypeptides)	-139	+135	-178	± 2	+3,4	+0,9
collagen-helix (in fibers)	-51 -76 -46	+153 +127 +148	+180	-3,3	+2,9	+1,6

Corresponding to the bulky character of the amino acid side chains the different amino acid residues in the primary structure have a propensity to favour the formation of α-helix, β-sheets or a reverse turn or to break them. The result of an analysis of the frequency of occurence of each of the 20 amino acid residues in a number of more than 50 different proteins is summarised in Table 2.

Accordingly, each of the amino residues, except for Arg, favours one of the three possible types of secondary structure. β-sheet is favoured by amino acid residues chains branched at C_β (Val, Ile, Thr) and aromatic residues (Phe, Tyr, Trp). Reverse turns are favoured by amino acids with short polar side chains (Ser, Asp, Asn), Gly and Pro. Histidine (His) has a low β-sheet propensity and it has a low frequency to occur in any of the secondary structures. The formation of α-helix is favoured by the other amino acid residues.

An approach for the prediction of secondary structures was elaborated by Chou and Fasman [16] on the basis of the primary structure and the frequence of occurence of each of 20 amino acids in the α-helix, β-sheet and reverse turn. Athough many other methods for predicting the secondary structure have been proposed [6, 17], the Chou-Fasmann method has become the most popular.

Table 2. Capacities of amino acid residues to form α-helix, β-sheet and reverse turn according to [15]

Residue	Helix	β-sheet	Reverse turn	Residue	Helix	β-sheet	Reverse turn
Glu	h	B	i	Ile	i	h	b
Ala	h	I	b	Asp	i	b	h
Leu	h	I	b	Thr	b	h	i
His	h	I	b	Ser	b	i	h
Met	h	I	b	Arg	i	i	(i) (b)
Gln	h	B	i	Cys	(h)	b	(b)
Trp	i	(h)	(b)	Asn	i	b	h
Val	i	H	b	Tyr	b	h	i
Phe	i	H	b	Pro	b	b	h
Lys	h	B	i	Gly	b	i	h

h, former; i, indifferent; b, breaker

2.2. Protein classification on the basis of secondary structure

A first classification of proteins on the basis of secondary structure has been attempted by Jirgensons [18]. He divided globular proteins into three classes. Class I comprised proteins with high α-helix content (ex. myoglobin), class II those with low α-helix content + β-structures (ex. lysozyme), class III united nonhelical proteins. The latter were subdivided into group A with a high content of antiparallel β-structure (ex. concanavalin A, chymotrypsin), group B_1 with low β-structure content (ex. immunoglobulins), group B_2 as rigid nonhelical proteins (ex. soybean trypsin inhibitor) and group C representing proteins with flexible chains (ex. phosvitin). A more recent classification considers five structural classes: α-proteins, β-proteins , α+β-proteins, α/β-proteins and coil-proteins [6, 8] (Table 3). Since large proteins usually consist of two or more smaller, structurally independent domains, these classifications are based on domain structures rather than the whole protein molecules.

The secondary structures of many kinds of proteins have been compiled by Richardson [19].

Table 3. Classification of proteins on the basis of secondary structures according to[8]

Classification	Secondary Structures	Examples
α-Proteins	with α-helix only	Myoglobin, haemoglobin, ferritin, phospholipase C
β-Proteins	mainly with β-sheets	Immunoglobulin, concanavalin A, Chymotrypsin
α+β-Proteins	with α-helix region and β-sheet region that exist apart in the sequence	Insulin, pancreatic trypsin inhibitor, ribonuclease A, lysozyme
α/β-Proteins	with alternating segments of α-helix and β-sheet	Subtilisin, alcohol dehydrogenase, hexokinase, phosphorylase, Taka-amylase A
Coil proteins	with no regular secondary structures	Bowman-Birk-type protease inhibitor

2.3. Methods of investigating secondary structures

The most powerful method for determining secondary structures is X-ray crystallography. It allows to elucidate which segment of the polypeptide chain of a protein has an α-helical, β-sheet or reverse-turn structure. However, application is limited to proteins which crystallise.

The development of NMR method for investigating proteins in solution is an efficient alternative for the determination of secondary and tertiary structure [20].

The most abundant method of determination of protein secondary structure in solution is circular dichroism (CD) spectroscopy in the far ultraviolet region. The CD-spectrum of the α-helix has a negative maximum at 222 nm due to the n-π^* transition of the peptide group and another at 208 nm due to the π-π^* transition, and also a positive maximum at 190 nm due to π-π^* transition. The CD spectrum of the β-sheet is characterised by a negative maximum at 218 nm due to n-π^* transition and a positive maximum at 195 nm due to the π-π^* transition.

The evaluation of the CD- spectroscopic data was previously performed using CD spectra of α-helix, β-sheet and random coil of synthetic polypeptides such as poly-L-glutamic acid and poly-L-lysine as reference standards. Later, basis spectra taken from CD spectra of proteins with known secondary structures were used [21, 22, 23]. Notwithstanding, deviations of the secondary structure data derived from CD-spectra and the real secondary structure data determined by X-ray crystallography were frequently found. Recently, the improved CD instrumentation and the methods of evaluation of CD spectra [24-26] allow a reliable estimation of the content of α-helix, β-conformations, turns and remainder.

Predictions of α-helix are excellent, the absolute error for predictions of β-turn is generally small and the correlation coefficient for total β-sheet is 0,91. An excellent review on secondary structure evaluation in proteins through CD spectroscopy is given by Johnson [27].

A very useful and sensitive method of secondary structure evaluation in proteins is infrared spectroscopy, especially the Fourier transform infrared (FT-IR) technique, which allows investigations in aqueous solution [28-30]. The determination of secondary structure is based principally on the analysis of amide band (1600-1700 cm^{-1}) representing the C=0 stretching vibrations of the amide groups coupled to the in-plane NH-bending and CN-stretching modes. α-helical segments give a bond centred between approx. 1650 and 1658 cm^{-1}, whereas bands between 1620 and 1640 cm^{-1} and between 1670 and 1695 cm^{-1} are assigned to β-strands. Turns were reported to be associated with bands around 1665, 1670, 1683, 1688 and 1694 cm^{-1} [28].

2.4. Supersecondary structures

As postulated by Crick [31] α-helices often appear as coiled-coil helices. Their most regular form occurs in fibrous proteins such as α-keratin [32] and tropomyosin [33], where two α-helices are wound around each other, forming a left-handed superhelix. In globular proteins, coiled-coil α-helices are observed in short pieces. The formation of this kind of supersecondary structure is energetically favoured. It allows side chain meshing, intimate contact between α-helices and appreciable van der Waals binding energy [6].

Other supersecondary structures in globular proteins are the formation of hairpins of two α-helices (α-hairpin) or β-strands (β-hairpins) and β-X-β-units, where the connecting unit β-X-β-can be an α-helix (βαβ) or a β-strand (βββ) [6, 8, 34, 35). A combination of two βαβ units is found frequently and is called "Rossmann fold" [36]. The arrangement of β-sheets in three adjacent antiparallel strands and rather short connections (reverse turns) is defined as β-meander. It is the most frequent pattern of three consecutive strands in globular proteins [36].

2.5. Structural domains - a constituent part of the tertiary structure

Globular proteins of large molecular weight are usually composed of smaller globular folded units, which are connected only losely with each other. These structural domains are outlined by overall clefts in the electron density distribution calculated from X-ray analysis. Most structural domains contain 100-150 residues which correpond to a globule of diameter 25 Å [6]. The formations of domains is a result of the chain folding process [38]. They can be regarded as those pieces of a long polypeptide chain which fold independently of each other due to a high degree of "neighbourhood correlation". In this way, structural domains can be defined as the folding units of a protein.

An example of globular proteins with an exceptionally obvious domain structure are the immunoglobulins, where the domains are ordered along the polypeptide chain like pearls on a string [6, 39, 40].

2.6. Surface and internal structure of globular proteins

As early as the 1920s I. Langmuir and E.K. Rideal, two prominent pioneers of surface chemistry, predicted the distribution of hydrophilic and hydrophobic residues within the protein molecules, which was described as "oil drops with a polar coat". This distribution, which is characterised by the location of hydrophilic and ionisable residues on the surface of the molecules and hydrophobic residues mainly buried in the interior, is also called the "nonpolar-in, polar-out" rule. X-ray crystallographic data of investigated globular proteins have confirmed this prediction. The formation of a hydrophobic core in the globule can be assumed as one of the essential driving forces in protein folding [42,43].

The hydrophobicity of amino acids was evaluated by Kauzmann [41] and later by Tanford and Nozaki [44], who measured the free energy of transfer of amino acids from water to ethanol or dioxane. These solvents are believed to resemble the protein interior. The free energy difference ΔG_t for the side chain was obtained by subtracting the value for Gly. The results are listed in Table 4.

Table 4. Hydrophobicities ΔG_t and accessible surface area (A) of the side chains of amino acid residues

Residue	Original data from Nozaki and Tanford [44] (kcal/mol)	ΔG_t (kJ/mol)	A (Å^2) according to [46]
Trp	-3.4	-14.2	217
Phe	-2.5	-10.5	175
Tyr	-2.3	-9.6	187
Leu	-1.8	-7.5	137
Val	-1.5	-6.3	117
Met	-1.3	-5.4	160
Ala	-0.5	-2.1	67
His	-0.5	-2.1	151
Thr	-0.4	-1.7	102
Ser	+0.3	+1.3	80

All hydrophobic residues have negative values of ΔG_t which indicates that they avoid an aqueous environment. ΔG_t for the glycine residues was found to be around zero. This means that the backbone prefers neither the interior nor the surface of the protein. Some hydrophilic side chains, e.g. those of lysine and arginine, contain hydrophobic -CH_2-groups, whose contribution to the hydrophobicity should be taken into account. [see ref. 42 p. 121].

A linear relationship between the values of -ΔG_t and the water accessible surface area (Tab.4) was found for completely nonpolar side chains (Ala, Val, Leu, Phe) [45]. Ser, Thr, Tyr and Trp, which contain a dipole show a similar relationship but with a decrease of about 1,5 kcal/mol in ΔG_t [6, 46].

The relative accessibility of a residue was defined as the ratio of the accessible surface area of the residue in a native protein to that of the residue in the completely unfolded protein. Miller et al [47] estimated the distribution of the amino acid residues between the interior and the surface of the protein molecule assuming that residues with a relative accessibility below 5 % are buried in the interior and residues with a relative accessibility above 5 % are located in the surface of the molecule. Accordingly, Val, Leu, Ile and Phe occupy about 44 % of the internal residues and only about 14 % of the surface residues. Cys also shows a preference for the interior. While Ala, Gly, Ser and Thr are distributed equally between the surface and interior, the hydrophilic charged residues Asp, Glu, Lys and Arg occupy about 27 % of the surface residues and only 4 % of the internal residues [8, 47]. The packing density of the protein interior, defined as the ratio of the volume enclosed by van der Waals envelope to the volume actually occupied in a crystal or liquid, was found to be 0.75 for lysozyme and ribonuclease A [48]. Correspondingly, 75 % of the total volume of the protein interior is occupied by protein atoms. Calculated values of the mean volume occupied by amino acid residues in the interior of proteins were found to be the same as that found in crystals of the corresponding amino acids [45, 49]. One can therefore conclude that a polypeptide chain with a secondary structure will be folded into a crystal-like tertiary structure and that the description of a protein molecule should be more approximated to a crystal rather than an oil drop [8].

2.7. Quarternary structure

Larger proteins are mostly built up from non-covalently linked subordinate entities, called subunits. The smallest subunit that can be released from a quarternary structure without cleavage of covalent bonds is called a monomer. A monomer may be composed of a single polypeptide chain or may contain two or more chains linked covalently (e.g. by disulphide bonds). These constituent polypeptide chains are also called subunits by some authors, which may lead to some confusion. Therefore, in the interest of clarity, the above definition of quarternary structure as an ensemble of non-covalently interacting subunits, which was introduced by Bernal [5] and used by Klotz et al [50] in a competent review article, should be preferred.

A protein composed of subunits is called oligomeric protein. It may consist of one type or several types of monomers. The minimum-size subunit that, on association with an integral number of identical subunits, will generate the quarternary structure, has been defined as a protomer [51]. In many cases, the protomer is identical with the monomer. In some other cases, such as oligomeric enzyme proteins containing a catalytic (C) and a regulatory (R) subunit, the protomer is an associate of more than one subunit. The "classical" example is aspartate transcarbamylase, where the protomer is composed of two subunits (C and R) and therefore called a dimer. In this case six C•R protomers generate the quarternary structure [52].

Due to the non-covalent interaction of the subunits in the quarternary structure, dissociation of the oligomeric protein into subunits can be achieved by denaturants such as urea, guanidin hydrochloride or sodium dodecylsulphate.

Dissociation of oligomers can also be caused by binding of small inorganic or organic ions. The opposite effect, association of subunits is also possible. Other factors causing dissociation are metabolits or cofactors in the case of enzymes. A corresponding survey is given by Klotz et al [50], who also listed the subunit stoichiometry. Accordingly, by far the majority of proteins possessing a quarternary structure have two (dimers) or four (tetramers) subunits and most of them are identical subunits. Their assembly may posses one of several different geometries. The number of possible arrangements increases with the number of subunits but can be limited by the following restrictions:

1. All subunits in a oligomeric protein must be in equivalent (or pseudoequivalent) environments, which has been fulfilled in the majority of proteins studied by X-ray diffraction.

2. The bonding potential or binding regions of the subunits must be saturated otherwise higher aggregates would be encountered. In this way the arrangements of subunits in an oligomeric ensemble are restricted to closed sets.

From these two restrictions it follows that a regular packing of the subunits around a central point in the oligomer allows only a point group symmetry. For small protein oligomers this point groups may be categorised as (1) cyclic, (2) dihedral, (3) cubic (for detailed information see ref. 50, 51, 53-55).

Examples of oligomers with a subunit arrangement in a cyclic symmetry are the dimers of β-lactoglobulin and insulin with a single twofold rotational axis (C_2-symmetry). Oligomers with dihedral symmetry are e.g. haemoglobulin, and the hexameric 11S storage proteins from plant seeds (see section.6). The former's four subunits are arranged as a tetrahedron with three twofold axes (D_2 - symmetry), and the latter has six subunits which are proposed to be arranged as a trigonal antiprism with one threefold and three twofold rotational axes (D_3 symmetry). To classify haemoglobin in this way, one must ignore the differences between the haemoglobulin α and β subunits. Cubic symmetry is of importance for proteins with a great number (≥ 12) of subunits, e.g. multisubunit spherical viruses [56].

The subunit association in an oligomer may be isologous or heterologous. In the first case the contact sites between two subunits are identical. This is valid for dimers of equivalent subunits. In oligomers with an odd number of subunits only heterologous subunit interactions with non identical contact sites are possible, whereas the subunit association in an oligomer with an even number of subunits may be either isologous or heterologous [50].

The formation of protein associates is favourable both from structural and functional reasons. Subunit association in 11S plant storage proteins provides a high packing density and reduces the osmotic pressure. It also results in the reduction of the accessible surface area of globular proteins [46]. Aggregation makes the formation of special geometric structures possible e.g. long tubes in virus coat proteins [56, 57] and a fibrous structure in F-actin, one of the components of the small filaments in muscle sarcomeres [58]. Oligomerisation enlarges the number of possible enzyme characteristics by introducing cooperativity between subunits and various types of regulation (e.g. associates of catalytic and regulatoric subunits, formation of multi-enzyme complexes).

Useful methods for investigating the geometric arrangement of subunits in oligomeric proteins besides X-ray crystallography are electron microscopy [59] and small angle X-ray scattering which allows investigation in solution [60]. To study association-dissociation equilibria in oligomeric proteins, analytical ultracentrifugation is a very suitable method [61, 62]. The most frequently used and very simple method of estimating the number of subunits in an oligomeric protein is the polyacrylamide gel electrophoresis in the presence of sodium dodecylsulphate (SDS-PAGE) [63].

3. FACTORS DETERMINING PROTEIN STABILITY

Protein structure is stabilised both by covalent disulphide bonds and a complexity of non-convalent forces (Table 5). These are electrostatic interactions, hydrogen bonds, hydrophobic interactions and van der Waals potential. Really, hydrogen bonds are predominantly electrostatic interactions and van der Waal potentials comprise three terms, i.e. electron shell repulsion, dispersion forces and also electrostatic interactions.

3.1. Disulphide bonds

Disulphide bonds have a key role in stabilising protein structures by intra- and interchain crosslinking. The maximum number of disulphide bonds found in proteins are 17 and the average number is 3. There are also proteins without disulphide bridges. Extracellular proteins have a greater number of disulphide bonds than intercellular ones. The mechanical properties of extracellular proteins such as keratins (wool, hair) are determined by interchain S-S-crosslinks. Similarly the cohesive-elastic properties of wheat dough is highly influenced by the interchain S-S-bonds of glutenin [64].

In intracellular proteins, thiol-disulphide exchange may play an active role in regulating enzyme activity. In immunoglobulins, the disulphide bridge between a light and a heavy chain maintains the specifity of a binding site [50]. Disulphide bridges are an integral part of structural motifs in proteins. A frequently found structural element is the sequence Cys-Cys, which forms the basis for linking three chain segments in close proximity [6]. The Cys-Cys sequence has been shown to be a highly conservative structural motif in 2 S albumins from different plant seeds which also contains a Cys-X-Cys triplet [65-67]. In several proteins (e.g. pepsin, silk fibroin), cysteine residues in sequences Cys-$X_{2\text{-}4}$-Cys are bridged and have tight interchain loops which tend to be flat and rigid.

Tab 5. Different types of interaction and bonds in proteins

Typ	Example	Binding energy (kJ/mol)	Free energy change water-ethanol (kJ/mol)	Bond length (Å)	Factors of weaking or disrupting	Factors of strengthening
Covalent bond	-S-S-	about -250		1-2	reducing agents	
Electrostatic interactions	$-COO^{-} \cdots H_3N^{+}-$ $-\overset{\delta+}{C}=\overset{\delta-}{O} \cdots \overset{\delta-}{O}=\overset{\delta+}{C}-$	-21 +1.3	- 4.2	2-3	salts, high or low pH	
Hydrogen bonds	O-H•••O N-H•••O=	-17 -13		+2.8 about 3	urea guanidin HCl, detergents heating	cooling
Hydrophobic forces	- Ala•••Ala- - Val•••Val- - Phe•••Phe- - Trp•••Trp-		-3 -8 -13 -19		Detergents Organic Solvents	increasing temperature
Van der Waals interactions	permanent and induced dipoles	about 4		3-4		

3.2. Non-covalent interactions

From Table 5 it can be seen that the binding energy of non-covalent forces is one to two orders of magnitude lower than that of covalent bonds. The stabilising effect by non-covalent interactions results from their great number and cooperativity in proteins.

Van der Waals attraction results from dispersion forces between any pairs of atoms due to dipole-dipole interactions. The attraction forces, which are proportional to the sixth power of the inverse distance between atomic nuclei [68], are maximal at an atomic distance of 3-4 Å corresponding to the van der Waals contact distances [6]. The van der Waals energy decreases rapidly when the atomic distance becomes only 1 Å greater than the contact distance. At shorter distances, electron shell repulsion increases in proportion to the 12th power of the inverse distance between the atoms [69].

The energy of van der Waals bonding for one pair of atoms (about 4 kJ/mol) is only slightly higher than that of the average thermal energy of molecules at room temperature (2,5 kJ/mol). Therefore van der Waals forces become only important if a great number of atoms of a molecule simultaneously approach a corresponding great number of a second molecule. This can be realised only when steric complementarity exists. This may cause specifity, though a single van der Waals bond is non-specific. The repulsion forces at shorter contact distances are just as important for the realisation of specifity of the attraction forces.

Due to the high number of charged amino acid side chains belonging to Asp, Glu, Lys, His, Arg, Tyr and Cys residues, proteins should be considered as polyelectrolytes. Internal neutralisation of positive and negative charges results in a zero net charge at the isoelectric point (pI) of proteins. It is generally assumed that proteins are most stable at its pI and the stability decreases with increasing shift of pH to lower (acidic range) or higher (basic range) values due to electrostatic repulsion. In some cases, highest stabilisation has been observed at pH‘s far from pI (e.g. T4 phage lysozyme) and contributions other than the total net charge should be involved in determining the protein stability [70].

Ionisation is commonly studied by acid-base titration of proteins. However, titration curves are very complex because of the great number of ionisable groups. Moreover, the dissociation of a charged group is influenced by the neighbouring ionised groups, hydrophobic residues and hydrogen bonds. Thus, the apparent pK-values of ionisable groups of amino acid side chains may vary by one unit.

One has to take into consideration that most covalent bonded atoms, even neutral molecules, carry partial charges. This is due to the asymmetric bond electron distribution. Thus, the partial charges of atoms in the polypeptide backbone and neutral side chains also contribute to electrostatic interactions.

In Table 5 data for a salt bridge between Lys and Asp and the interaction of two carbonyl groups at van der Waals contact distance are given. According to Coulombs` law the electrostatic interaction energy depends on the dielectric constant ε of the surrounding medium which is difficult to calculate for microscopic dimensions. ε is therefore assumed to be 4, which is the macroscopic constant of amide polymers (see ref.6).

Salt bridges stabilising the protein structure can also be found between negatively charged groups of proteins by interaction with multivalent cations. The bridging of polypeptide chains by Ca^{2+} and bound or free phosphate residues occurs in casein micelles (see section 6). Stabilisation of enzyme proteins by chelation with multivalent metal ions also occurs. Calcium and zinc ions are for example essential for stabilising α-lactalbumin [71, 72]. A detailed discussion of electrochemical properties of proteins is found in ref. 6, 8 and 74.

Hydrogen bonds have an essential stabilising effect in protein secondary and tertiary structures. They are formed between a H-atom and a contact partner with a large negative partial charge, e.g. between amide and carbonyl groups, hydroxyl and carbonyl groups, two hydroxyl groups (ice; Tyr-phenols), amide and hydroxyl groups, amide and imidazol nitrogen (His), amide and sulfur. They are characterised by the following particularities: both partners can approach one another to within a distance closer than the van der Waals radii allow. This can occur because the shell repulsion between contact partners becomes small due to an appreciable shift of the electron shell of the hydrogen atom to the atom to which hydrogen is covalently bound.

The interaction energy of hydrogen bonds resulting from a high attractive Coulomb energy and a high dispersion energy is intermediate between the energies of van der Waals contact and covalent bonds (see Table 5). Hydrogen bonds found in proteins were analyzed with regard to the distance between the hydrogen donor, that is the atom to which hydrogen is covalently bound, and the hydrogen acceptor atom. Compared with calculated van der Waals radii, the observed distances were found to be reduced by 25 % in the case of hydroxyl-carbonyl interaction and by 20 % for amide-carbonyl and amide-hydroxyl interactions [75].
A further characteristic of hydrogen bonds is that they are linear. This results from the fact that the positively charged H-atom is located between two negatively charged atoms (O or N) and that it assumes the lowest potential energy when all three charges are aligned.

Protein formation and stability depend both on non-covalent binding energies (van der Waals interaction and hydrogen bonds) and on free energy. Therefore, in Table 5 the free energy change , ΔG_t , of transfer from water to ethanol is given as a stability parameter of hydrophobic forces (see also section 2.6. and Tab.4). From the thermodynamic parameters for dissolution of hydrocarbons in water, it has been concluded that the driving force for the transfer of hydrophobic side chains from water to the interior of the protein molecule is entropic [42].

The hydrophobic forces in proteins is thus to be understood as entropy-driven interactions of hydrophobic residues surrounding by structured water, which leads to a distortion of the water structure and reorientation of water molecules.

The contribution of the various non-covalent factors to the thermodynamic stability of the native structure of the protein may be expressed as:

$$\Delta G_N = \Delta G_{hb} + \Delta G_{ele} + \Delta G_{h\varphi+} + \Delta G_V - T\Delta S_{conf.} \qquad (1)$$

where ΔG_{hb}, ΔG_{ele}, $G_{h\varphi+}$ and G_V are the free energy contribution from hydrogen bonding, electrostatic, hydrophobic and van der Waals interactions respectively. The $T\Delta S_{conf}$ is the unfavourable change in the free energy resulting from the loss of conformational freedom of the polypeptide (conformational entropy). The conformational entropy is a destabilising factor in the native protein, whereas it stabilises the unfolded protein molecule. This destabilising free energy competes with the sum of the stabilising factors in eqn.1. An estimation of individual factors to the protein stability resulted in a value of only 10-20 kcal/mol (42-84 kJ/mol) [8,76]. The structural stability would be therefore profoundly affected by any perturbation causing even a small decrease in the free energy of stabilisation. On the other hand the change in the conformational entropy on unfolding is small in a protein with intact disulphide bonds compared to that after reduction of the latter. This underlines the stabilising effect of disulphide bonds. The problem of protein stability is treated in a number of competent review articles [49, 77-82].

4. HYDROPHOBICITY

Using the free energy of transfer from water to organic solvent (ΔG_t) for the various amino acid residues, Bigelow calculated the average hydrophobicity ($H\varnothing_{av}$) of proteins [83]. According to this author, both the average hydrophobicity and the charge frequency (parameter of hydrophobicity) are the most important molecular features which have the greatest influence on the physical properties, such as solubility, of proteins. Protein solubility can be expressed as the manifestation of the equilibrium between hydrophilic (protein-solvent) and hydrophobic (protein-protein) interactions [84].

Accordingly, proteins with lower average hydrophobicity and higher charge frequency would have a higher solubility. This empirical relationship seems be true for most proteins. However, it does not explain the solubility characteristics of proteins. Thus, two proteins with the same hydrophobicity and charge frequency, can exhibit distinctly different solubility characteristics depending on their amino acid sequence, and consequently differences in the spatial arrangement of the residues in their tertiary structure [84]. For example, myoglobin and serum albumin have almost the same charge frequency (0.34 and 0.33 respectively) but are different in their average hydrophobicity (4.69 kJ/mol residue and 4.56 kJ/mol residue for serum albumin and myoglobin, respectively). From the higher hydrophobicity of serum albumin one may expect a lower solubility of this protein. In fact, serum albumin is extremely soluble at its isoelectric pH, whereas myoglobin is insoluble at its pI [84]. One should conclude, that the

$H\varnothing_{av}$ value derived by Bigelow may be an indicator of the extent of hydrophobic surfaces buried at the interior of the protein and which stabilising the protein structure by hydrophobic interactions. On the other hand, solubility and other solution related physico-chemical properties of the protein, e.g. emulsifying properties, are determined by the extent of the hydrophobic surfaces exposed at the exterior. In most globular proteins, e. g. myoglobin, lysozyme and ribonuclease S, about 40-50 % of the surface is found to be composed of uniformly distributed nonpolar patches [85]. Nevertheless, these proteins exist in the soluble monomeric state, obviously because of a sufficiently great contribution of hydration and intermolecular repulsive forces at the protein surface. However, intermolecular hydrophobic interaction can become relevant, when the fraction of hydrophobic patches on the surface exceeds a critical level. In this case self-association to oligomeric proteins occurs as found in globular plant storage proteins. Another reason for protein self-association can be the asymmetric distribution of the hydrophobic patches as found in caseins (see section 6).

For investigations to determine the "effective hydrophobicity" or "surface hydrophobicity", different empirical methods were introduced, namely partition methods, high performance liquid chromatography (HPLC), hydrophobic ligand binding methods, contact angle measurement and fluorescence probe methods [86]. From these, fluorescence probe techniques have become the most widely applied methods using 1-anilino-naphtalene-8-sulphonate (ANS) and cis-parinaric acid [87, 88] as probes. Both substances exhibit a very low quantum yield of fluorescence in water, but become highly fluorescent on binding to hydrophobic regions in proteins. Using cis-parinaric acid, Kato and Nakai [89] developed a simple procedure for the estimation of relative surface hydrophobicity of proteins. This consists of the measurement of the fluorescence of protein-parinaric acid conjugates at different protein concentrations at 420 nm, with excitation at 325 nm. The initial slope are of the plot of fluorescence intensity vs protein concentration is taken as a measure of the surface hydrophobicity (S_0) of the protein. A highly significant correlation was found between S_0 and the interfacial tension and the emulsifying activity index of native, denatured and surfactant-bound proteins [89]. Increasing surface hydrophobicity resulted in a continuous decrease of interface tension but in an increase of the emulsifying activity index.

5. DENATURATION

Protein denaturation can be defined as each change in the native conformation (i.e. in the region of secondary, tertiary and quarternary structure) which takes place without change of the primary structure, i.e. without splitting of peptide bonds. Complete denaturation may correspond to a totally unfolded protein. In this case, the transition from the native state N to the denatured state D can be approximated by a two-state transition (Eq. 2)

$$N \leftrightarrow D \qquad (2)$$

and the free energy change (ΔG_D) of unfolding may be expressed by eqn.3.

$$\Delta G_D = \Delta H_D - T\Delta S_D \qquad (3)$$

where ΔH_D and $T\Delta S_D$ are the enthalpy and entropy terms. Upon unfolding, the entropy and enthalpy changes are large and positive, whereas the free energy change is relatively small. The large increase in enthalpy upon denaturation indicates a much lower energy level of the native conformation of the proteins, and the change in entropy reflects the anticipated increase in disorder when the protein denatures. A third important thermodynamic parameter is the change in the constant pressure heat capacity, C_p ,which is related to the enthalpy and entropy change as shown in Eq. 4.

$$\Delta C_P = \left(\frac{\delta \Delta H^0}{\delta T}\right)_P = T\left(\frac{\delta \Delta S}{pT}\right) \quad (4)$$

When the unfolding causes aliphatic and aromatic side chains, which are normally buried in the interior of the molecule, to become exposed to the solvent, a most striking increase appears in ΔC_p which can reach a magnitude of 8.4 kJ/deg. mol. This change, and the temperature dependence of ΔH and ΔS, indicate temperature-dependent changes in the aqueous solvent as well as in the mode of solvation. Water is oriented into structures of solvation around nonpolar groups in the denatured state at low temperature, and more random and disorganised solvation interaction occurs at high temperature. It has been shown by Privalov [78] that the plots of the changes in enthalpy of unfolding per gram of protein against temperature intersect at a common temperature of 110 ^{0} C for most proteins and that the slopes of these plots, ΔC_p, are proportional to the content of hydrophobic residues. An intersect at around 110 ^{0} C has been also found for the plots of the changes in entropy of unfolding per gram of protein.

Disulphide bridges may enhance considerably the thermal stability. Thus, the Bowman-Birk trypsin inhibitor from soybeans, a small protein (M_r=7800) with seven disulphide bridges [90], is extremely stable against heat treatment and is responsible for the residual inhibitor activity in toasted soybean meals [91].

The tendency to become unfolded increases with the increase in the pH-distance from the pI due to the rise of electrostatic repulsion. Oligomeric proteins show thus a pH-dependent dissociation into subunits [50].

Protein denaturation can also be caused by salts in high concentration. The denaturing power increases in the order SO_4^{2-}< CH_3COO^- < Cl^- < Br^- < ClO_4^- < CNS^- for anions and $(CH_3)_4N^+$, NH_4^+, K^+, Na^+ < Li^+ < Ca^{2+} for cations corresponding to Hofmeisters` lyotropic series [92].

Urea and guanidin hydrochloride (Gdn-HCl) are powerful denaturing agents which can change at high concentration (about 8 M urea or 6 M Gdn-HCl) the conformation of most proteins to random coil. An example of the denaturation of a helical protein by 8 M urea is given in Fig. 2, which shows the change of the CD spectrum of napin, the 2S storage protein from rapeseed [93]. The helix content in the secondary structure decreased from 45 % in the native protein to 20 % in 8 M urea. This high residual helicity is due to the stabilisation of the protein by disulphide bridges, the reduction of which decreased the helix content to less then 10 % [94]. When the disulphide bridges remained intact, renaturation of the denatured protein could be observed. The stabilisation of disulphide bonds also prevents the destruction of the secondary structure under acidic conditions (pH 3.25) which are far from the pI (around pH 11) of the protein. At pH close to the pI the α-helix content changed only slightly at room temperature but decreased considerably after heating which causes an oxidative splitting of disulphide bonds in alkaline solution.

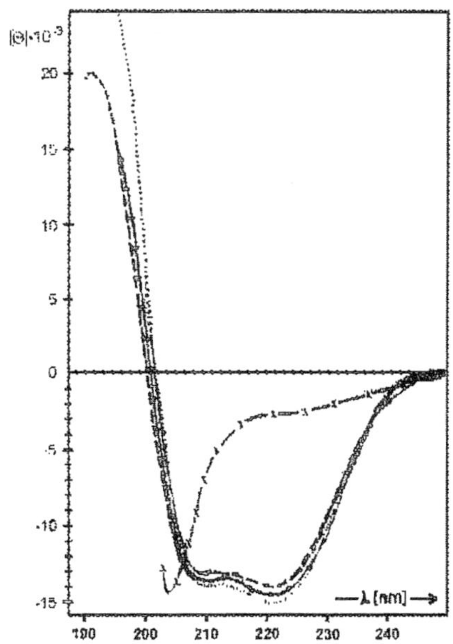

Figure 2. Far ultraviolet CD spectra of native and denatured 2 S protein (napin)from rapeseed. Denaturation by urea shifted the negative double maximum at 222 and 208-210 nm to 204 nm. Solvents:Teorell-Stenhagen buffer at pH 7.25 (–∘–), 3.25 (•••) and 11.8 (– –); 8 M urea (x–x); according to [93].

The effect of 6 M Gdn•HCl and disulphide bond reduction on the unfolding of various proteins has been studied in Tanfords laboratory by means of hydrodynamic methods, acid-base titration and optical rotation measurements [95-98]. The measurement of the intrinsic viscosity, $|\eta|$, which is a function of the effective hydrodynamic volume (ml/g) and is related to the compactness of the protein molecule, gave a clear response to the conformational changes [97]. While $|\eta|$ for rather compact and spherical molecules is about 3 ml/g, it increased considerably in 6M Gdn•HCl but attained maximum values when denaturation took place after reducing the disulphide bonds. Studying more than 10 different proteins, Tanford [95, 97, 99, 100] derived a linear relationship between the values of log $[\eta]$ and the logarithm of the number of residues n for reduced proteins, indicating the existence of randomly coiled high-polymer chains. Recent spectroscopic studies on a number of proteins gave however unambiguous evidence of residual structure in some cases even at very high denaturant concentration [101].

The use of urea or Gdn•HCl as denaturants allows the development of unfolded forms (D) from the native protein (N) to be followed . On the one hand, high concentrations of these denaturants can cause random-coiled structures, on the other hand, the stepwise increase in denaturants concentration from 0 to 6 or 8 M allows detailed investigations of the equilibrium between the native and unfolded forms (Eq. 2) to be investigated and provides evidence of the existence of various intermediate states [77]. Although a two-state transition according to Eq. 2 has been observed for some proteins, there are many proteins (e.g. phosphorylase b and cytochrom c), where the denaturation proceeds via a number of discrete intermediate steps [43, 77]. Accordingly, the denaturation process may be more accurately written as:

$$N \leftrightarrow I_1 \leftrightarrow I_2 \leftrightarrow I_n \leftrightarrow D \qquad (5)$$

where I_i represent intermediates who´s structure are between those of the native and the completely denatured state. As an example, the denaturation of β-lactoglobulin by urea proceeds corresponding to Eq. 6:

$$N_s \leftrightarrow N_u \leftrightarrow D_r \leftrightarrow D_i \qquad (6)$$

where N_s and N_u are the stable and unstable forms of the native protein, respectively, and D_r and D_i are the reversible and irreversible states of the denatured protein [102, 103]. The characterisation of the intermediates depends on the sensitivity of the method of analysis.

Besides viscosimetry, optical methods (CD, UV difference and fluorescence spectroscopy) are frequently used for analysing the denaturation curves.

Acid and thermal denaturation often do not bring about complete unfolding. An important intermediate state of denaturation, which has been realised in acidic or alkaline solutions in the presence of salts at moderate concentrations, is termed "molten globule" [104, 105].

It is characterised by the maintenance of a rather compact molecular state, i.e. only slight increase of the hydrodynamic radius. Moreover, it shows a CD-spectrum which is essentially identical to that of the native protein in the far ultraviolet region, but is similar to that of a completely unfolded protein in the near ultraviolet region corresponding to the changes of aromatic chromophores in the tertiary structure [105]. Accordingly, the molten globule state can be defined as a partially denatured state of a globular protein, which retains the ordered secondary structure but not the tertiary structure of the native protein. It is also discussed to appear as an early intermediate in the refolding process of proteins [105, 106]. Discussing recent data on the structure of proteins at a liquid interface, Dickinson and Matsumura [107] concluded, that the state of an adsorbed globular protein at a liquid interface is close to the molten globule state.

Since the interface adsorption of proteins is a key process in the formation of food emulsions, food proteins such as α-lactalbumin and β-lactoglobulin, have preferentially been investigated with regard to their behaviour at liquid interfaces [107]. The surface denaturation involved in the process of adsorption of lysozyme at the air-water interface has thoroughly been studied by Xu and Damodaran [108]. The importance of a more or less directed process of denaturation for inducing favourable protein functionality in food systems has been underlined by several authors in monographs and text books on food proteins [e.g. 109-113].

A very convenient method for inducing protein denaturation with special regard to improving surface functionality is the change of the charge by chemical modification [114] (see section 7). This can be realised for example by succinylation that changes the positively charged amino groups of lysine residues into negatively charged carboxyl groups, the excess of which can cause conformational changes due to the effect of electrostatic repulsion. Thus, successive succinylation induces stepwise dissociation of oligomeric proteins [50]. As an example, the dissociation of the hexameric 11S storage protein from pea seeds by succinylation is shown in Fig. 3. The dissociation proceeds via a 7S halfmolecule to the monomeric 3S subunits [115]. Using ultracentrifugation, viscosimetry, CD and fluorescence spectroscopy, it has been observed that the succinylated protein, though partially dissociated, retains a globular structure up to a critical degree of modification, where a sudden unfolding takes place. This has been shown to occur in various succinylated 11S plant proteins [115-118].

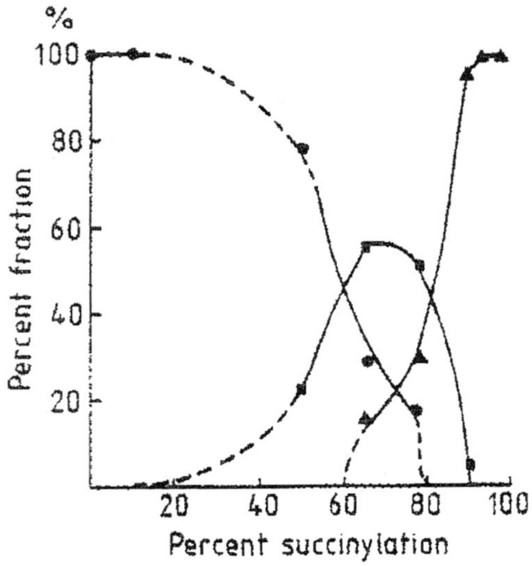

Fig. 3. Ultracentrifuge study of the dissociation of the oligomeric legumin from peas in dependence on the degree of succinylation, according to [115], • 11S, ■ 7S, ▲ 3S.

The results of viscosimetric and CD spectroscopic measurements of variously succinylated 11S globulin from rapeseed is also given in Fig. 4. Both the drastic change of the intrinsic viscosity η, and the mean residue ellipticity at 280 nm, characteristic of changes in the molecular shape and the tertiary structure respectively, indicate protein unfolding at a critical degree between 60% and 70 % modification.

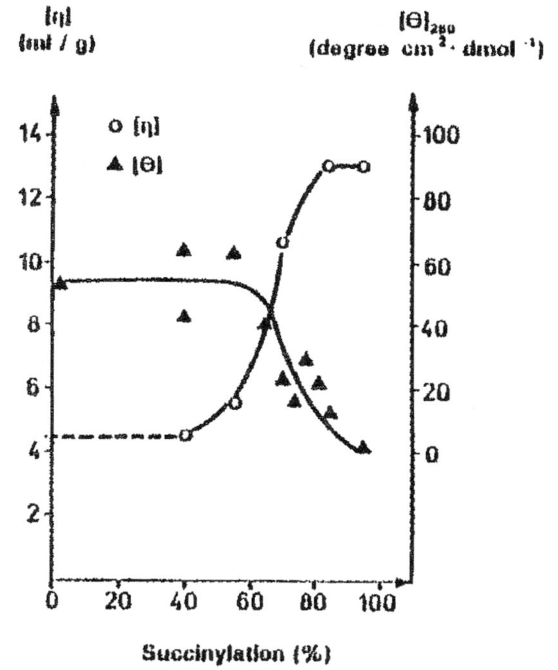

Fig. 4. Conformational changes in an oligomeric 11 S storage protein (cruciferin from rapeseed) as function of the degree of succinylation: /η/intrinsic viscosity; $[\theta]_{280}$ ellipticity at 280 nm from near ultraviolet CD spectrum; according to [116]

6. EXAMPLES OF PROTEINS PREFERABLY USED IN INTERFACE ADSORPTION STUDIES

Proteins used in interface adsorption studies are mainly food proteins. This is because of their natural widespread availability in high concentrations in food systems or raw materials. They are also commercially available due to their ease of isolation and purification. Moreover, adsorption studies at liquid interfaces with proteins are essentially motivated by the requirement to elucidate structure-functionality relationships in emulsifying and foaming food protein systems.

6.1. Small - and medium - size globular proteins

Characteristics of globular protein containing in milk, i.e. β-lactoglobulin, α-lactalbumin and serum albumin, and egg white, i.e. lysozyme and ovalbumin, are shortly outlined. Table 6 summarises essential chemical and physicochemical properties of some of them.

β-lactoglobulin is a component of the whey proteins in bovine milk, amounting to about 90 % of the total milk proteins. It exists in several genetic variants (e.g. A, B, C, D, E) of which A and B are the most abundant ones. The amino acid sequence of β-lactoglobulin B consists of 162 amino acids with a calculated molecular weight of 18277 [119]. It consists mostly of 9 anti-parallel β-sheets wrapped so as to form a flattened cone or calix (Fig. 5). While the core of the molecule is in an eight-stranded β-barrel (strands A-H), strand I is involved in the formation of a dimer by causing anti-parallel interactions with its counterpart.

The oligomerisation of β-lactoglobulin is reversible and depends on the pH according to eqn.7.

$$A \leftrightarrow A_2 \leftrightarrow (A_2)_4 \leftrightarrow A_2 \leftrightarrow A \qquad (7)$$

$$pH < 3.5 \quad 3.7 < pH < 5.1 \qquad pH > 7.5$$

Monomers are only stable at pH <3.5 and pH>7.5. In the pH-range 5.2 - 7.5, the protein exists primarily as a dimer with a molecular dimension given in Table 6. In addition to the monomer-dimer association, octamerisation of β-lactoglobulin A has been observed in the isoelectric region (pH 5.1) [122]. The pH-dependence with maximum octamerisation indicates the involvement of carboxyl groups with the possible formation of hydrogen bonds between the protonated carboxyl-groups. At pH>8.6 an irreversible denaturation occurs. The pH-dependent structural changes of the protein are summarised by Paulsson [123]. β-lactoglobulin has one buried SH-group which becomes exposed under denaturating conditions (e.g. high pH, heating, high pressure) and gives rise to the formation of disulphide-bridged dimers.

α-lactalbumin is the second protein component in bovine milk and amounts to about 4 % of the total milk proteins. It occurs in three genetic variants A, B, C, of which A and B are the most abundant components with known primary structure [124]. The tertiary structure of α-lactalbumin, which could not be determined directly by X-ray crystallography due the difficulty of crystallising the protein has been derived from the three-dimensional structure of lysozyme which appeared to be highly homologous to α-lactalbumin [125, 126] (Fig. 6). Thus, the amino acid sequences of both proteins show 49 identical positions and an identical localisation of the 4 disulphide bridges. CD spectra also indicate a high structural similarity of lysozyme and α-lactalbumin. Small angle X-ray scattering data revealed a compact globular structure of α-lactalbumin [127] (see Table 6).

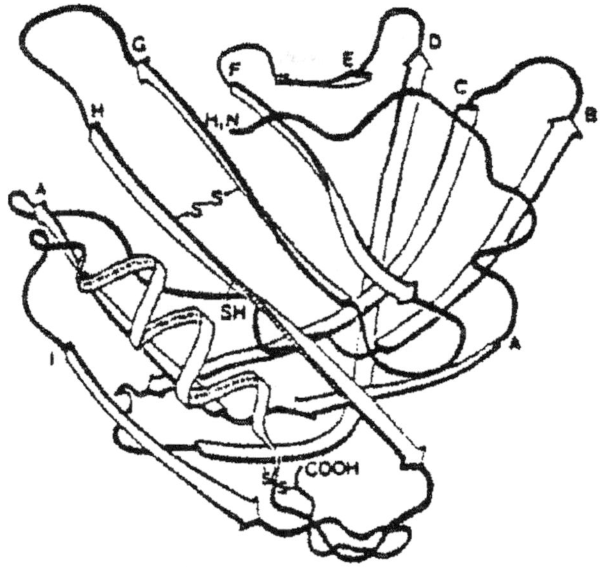

Fig. 5. Tertiary structure of β-lactoglobulin, according to [121]

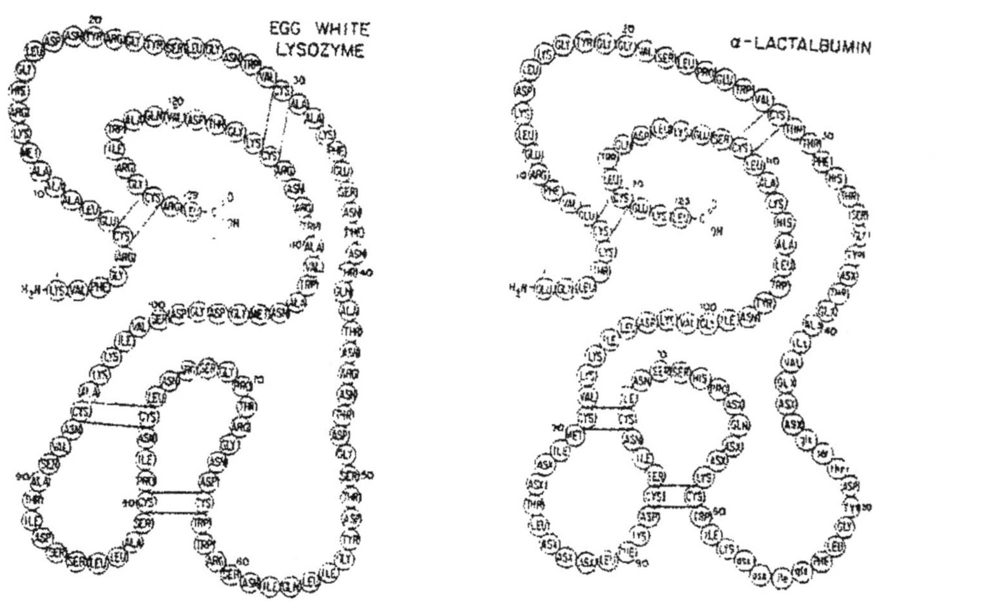

Fig. 6. Diagrammatic comparison of the covalent structure of bovine α-lactalbumin and hens egg-white lysozyme, according to [126]

At pH-values below the pI (4.2-4.5) reversible association to dimers and trimers and aggregation to higher molecular weight products was observed [128, 129]. α-lactalbumin is stabilised by Ca^{2+} against thermal unfolding [130]. Moreover its conformation and stability is affected by binding Zn^{2+} and other cations [123].

Table 6. Properties of some globular food proteins

	α-lactoglobulin	β -lactalbumin	Lysozyme	Serum albumin
Molecular weight (kDa)	18.3	14.2	14.6	66.3
Number of amino acid residues	162	123	129	582
Number of SH groups	1	-	-	1
Number of SS bonds	2	4	4	17
pI	5.35-5.41	4.2-4.5	9.5-10.7	5.13
Secondary structure (%) α-helix	10	26	29	55
β-sheet	50	14	15	16
remainder	40	60	56	29
Molecular dimensions (nm)	3.6x6.95 prolate	2.2x4.4x5.7 oblate	4.5x3.0x3.0 oblate	4.1x14.1 Prolate

Lysozyme was the second protein and the first enzyme whose detailed molecular structure was worked out by X-ray analysis [131, 132]. Like haemoglobulin and myoglobulin, lysozyme conforms the principle of "hydrophobic in, hydrophilic out". All of its charged polar groups are located on the surface, as are its uncharged polar groups with one or two exceptions, while the great majority of its nonpolar, hydrophobic groups are buried in the interior. A cleft is formed which contains the active site, where the substrate (bacterial cell wall polysaccharides) is fixed by hydrogen bonding to hydrophilic β-sheets [10, 132]. It is thought that lysozyme retained its original biological function during the evolution of organisms, whereas α-lactalbumin, which is thought to have the same molecular ancestor as lysozyme, has changed its function to be involved in the synthesis of lactose [125].

Bovine serum albumin (BSA) represents about 5 % of the whey proteins in bovine milk. It is the main component in blood serum. The amino acid sequence [133, 134] is given in Fig. 7, which also shows the distribution of disulphide bonds. In contrast to the above described proteins, BSA is rich in helical structure [135] (Table 6).

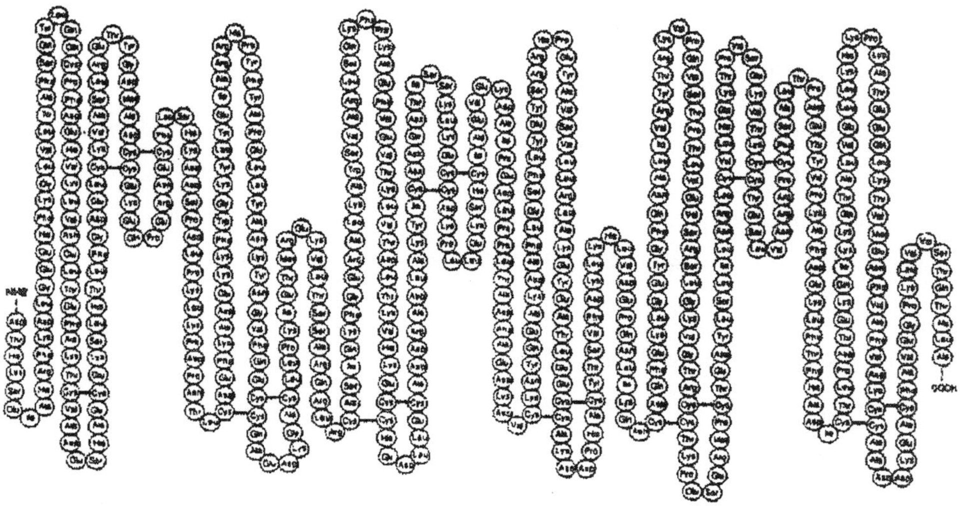

Fig. 7. Covalent structure of bovine serum albumin, according to [134]

A triple domain structure was proposed which includes three very similar structural domains, each consisting of two large double loops and one small double loop [133]. This was confirmed by three-dimensional structure analysis, performed on human serum albumin [136]. Large solvent channels (9 nm by 9 nm) that pass through the crystal parallel to the crystallographic C-axis, have been found by analysis of the electron density in the crystal structure. This coincides with a previously estimated hydration shell of water molecules in the interior [137, 138]. Studies of the pH-dependence of physicochemical properties revealed a marked increase of the intrinsic viscosity [139] and a stepwise decrease of the α-helix content from 55 % to about 35 % when the pH decreased to 2.7 [140]. Below pH 4, the molecule becomes fully uncoiled within the limits of its disulphide bonds. This so called E-form (E = expanded) has a molecular size 2.1x25 nm, two-thirds longer than the model at neutral pH [141, 142]. A continuum of molecular species of BSA has been postulated, which differ in their inherent stability and in the pH-range in which the transition from the native to the acid form occurs [141]. BSA functions as a carrier molecule in blood with the ability to bind different biological materials especially by hydrophobic interactions.

Ovalbumin, the major protein component of egg white, is a monomeric phosphoglycoprotein with a molecular weight of about 43 kDa. It is composed of 385 amino acid residues of known sequence [143]. Ovalbumin contains 4 SH and one S-S group, it is phosphorylated at two serine residues and glycosylated at one asparagine residues. There are three components A_1, A_2, A_3 with the approximate ratio 85:12:3, which differ in the extent of phosphorylation and therefore in the pI which amounts to 4.75, 4.89 and 4.49, respectively [144]. During storage of eggs, even at low temperature, ovalbumin is modified by SH/SS exchange into a variant with greater heat stability, called S-ovalbumin [145]. The denaturation temperature of ovalbumin and S-ovalbumin is 84.5 and 92.5^{0}C respectively, as measured by differential scanning calorimetry at pH 9 [146]. Spectroscopic and viscosimetric studies indicate that S-ovalbumin has a slightly more compact conformation than ovalbumin [147]. Ovalbumin is easily denatured at surfaces which thus makes it especially suitable for the formation of foams. The stability of ovalbumin foams has been mainly related to surface denaturation-induced aggregation and surface gelation and, to a lesser extent, to disulphide cross-linking between the denatured aggregate molecules [148].

6.2. Proteins forming micelles: the caseins

Caseins are the major protein fraction in bovine milk amounting to about 80 % of the total milk proteins. Their components are $\alpha_{s,1}$ and $\alpha_{s,\ 2}$-caseins, β-caseins and κ-caseins besides the γ-casein fraction which have been shown to be a proteolytic breakdown product of β-caseins [71, 113, 123, 149]. As similarly with ovalbumin, their main chemical characteristic is that they are phosphoproteins, i.e. they bear phosphoryl ester groups at some serine residues in the primary structure [150]. Each casein fraction is composed of several genetic variants. Table 7 summarises some main properties of the casein fractions [71, 123, 149, 152].

$\alpha_{s,1}$-casein is the major protein component of the casein micelle amounting to about 34 % of total milk proteins. It consists of 5 genetic variants A, B, C, D with known primary structures [151, 152]. The sequence of the 199 residues of the major variant B shows an unequal distribution of charged polar residues and apolar residues (Table 8).

Table 7. Properties of casein fractions

	$\alpha_{s,1}$B	$\alpha_{s,2}$A	βA^2	κB
Molecular weight (kDa)	23.6	25.2	24	18
Number of amino acid residues	199	207	209	169
P-residues	8	13	5	1
Number of Cys residues	0	2	0	2
pI	4.44-4.76		4.83-5.07	5.45-5.77
Average hydrophobicity (kJ/res.)	4.89	4.64	5.58	5.12
Secondary structure				
α-helix	4-15		1-10	8
β-sheet	18-22		13-25	29
Turns	29-45			33
Remainder	18-40		70	30

Seven of the total eight phospho-serine residues are located in sequence positions 41-80 which contain additionally 12 carboxyl groups. Thus, this region is extremely charged and hydrophilic. In contrast to this, both the N-terminal and the C-terminal regions are characterised by an accumulation of hydrophobic residues. Their high average-hydrophobicity (Table 8) is responsible for the strong tendency of the protein to associate. The association is, however, limited by electrostatic repulsion of the phosphate residues.

The even distribution of proline residues over the whole sequence hinders the formation of extended regular secondary structures [149]. Ono et al. [153] concluded from spectroscopic measurements that only 20 % of the peptide amide groups contributes to formation of a regular structure and that about half of the secondary structure of α_s-casein is α-helix and other half β-structure. In variant A of

$\alpha_{s,1}$-caseins, the residues 14-26 of the sequence are lacking.

$\alpha_{s,2}$-casein has a marked bipolar structure with an accumulation of anionic groups in the

N-terminal region and a concentration of cationic groups in the C-terminal region [71, 149]. Both α_s-caseins have strong Ca-binding properties [154] and are precipitated by addition of Ca^{2+}. (The name „α_s" means calcium-sensitive).

Table 8. Distribution of charged and apolar side chains in $\alpha_{s,1}$- casein and β-casein, adopted from [71], H average hydrophobicities of sequences

α_{s1}-casein			β-casein		
residues	net charge	H (kJ/mol)	residues	net charge	H (kJ/mol)
1-40	+3	5.6	1,43	-16	3.28
41-80	-22.5	2.68	44-92	-3.5	5.98
81-120	0	5.48	93-135	2	4.91
121-160	-1	5.29	136-177	3	6.14
161-199	-25	4.87	178-209	2	7.27

The β-caseins representing about 25 % of the total milk proteins, are comprised of 7 genetic variants (A^1, A^2, A^3, B, C, D, E) [149]. The primary structure of β-casein A^2 [152, 153] shows an accumulation of the 5 phosphate residues and most of ionisable residues in the N-terminal region (sequence positions 1-40). The sequence region 44-209 is extremely hydrophobic which makes β-casein the most hydrophobic of all caseins. This "soap like" structure favours the formation of micellar aggregates in solution. Like in α_s-caseins, the proline residues in β-caseins are even distributed over the primary structure, preventing the formation of extended ordered secondary structures. The percentage of aperiodic structures has been found to amount to more than 70 % [156]. The aperiodic fraction decreased at enhanced temperatures in favour of an increased percentage of β-structures [71]. The association of β-casein is strongly temperature dependent. It occurs as a monomer at low temperatures (0-4°C) and associates to large aggregates with 20-24 monomers at room temperature [157, 158]. The interaction of β-casein with Ca^{2+} is strongly temperature dependent and does not result in precipitation at $T \leq 1^{o}C$ [159]. The Ca-binding capacity corresponds to the number of bound phosphate residues [160].

The third casein-component, κ-casein, represents about 10 % of the total milk proteins and contains two genetic variants A and B with known primary structure [71, 149, 152, 161, 162]. Recently performed secondary structure analysis using FT-IR spectroscopy revealed 8%

α-helix, 29 % β-sheet and 33 %reverse turns [163]. The monomer of κ-casein does exist only under reducing conditions and occurs normally in an oligomeric form, mostly as trimer, obviously due to intermolecular S-S-bridging [71]. The primary structure is characterised by a highly uneven charge and polarity distribution [164]. Thus, the 53-residue C-terminal sequence (positions 116-169] has a excess of negative charge belonging to the only P-Ser residue, Glu carboxyl groups and some sialic acid residues linked to Thr and Ser residues. Another cluster of charge, although both anionic and cationic, occurs in the N-terminal 20-25 residues, which is quite hydrophilic. The central segment of the polypeptide chain is highly hydrophobic. Thus, the primary structure suggests a highly solvated, loosely coiled or unfolded C-terminal segment of more than 50 residues. On the other hand, the central portion of the molecule should be thought to be compactly folded into a globular structure due to its hydrophobicity [164].
Corresponding to its amphiphilic nature, κ-casein forms large spherical micelle-like aggregates in the absence of other caseins. This aggregation to form 600-650 kDa polymers is thought to occur by lateral interaction of the hydrophobic N-terminal perhaps globular portion, thereby placing the highly solvated C-terminal segment on the surface of the sphere [164].

Large spherical casein micelles are formed by association of α_s-, β- and κ-casein in the presence of free phosphate and calcium ions and held together by electrostatic and hydrophobic interactions. α_s-and β-caseins are „surrounded“ by κ-casein molecules, the flexible hydrophilic C-terminal region of which forms the surface layer of the micelle. The high negative charge of this exposed κ-casein-sequence region prevents a collapse of the micelle by electrostatic repulsion [71, 154, 165]. The size and stability of the micelles are determined both by the successful "covering" of α_s-and β-caseins by κ-casein and the concentration of phosphate and calcium. The micelle diameter varies between 50 nm and 300 nm [149]. Medium-size micelles with a diameter of 140 nm contain about 25,000 monomer units [71]. When the micelles are attacked by the "milk clotting enzyme" chymosin, a highly specific proteolytic splitting of the hydrophilic tail of the κ-casein takes place which occurs between the positions 105 Phe – 106 Met. Thus, the hydrophobic core of κ-casein becomes exposed, causing the collapse of the micelle by hydrophobic interactions [71, 166].

6.3. Oligomeric plant storage proteins

Storage proteins in legume and oil seeds are of prominent importance for supplying the world protein requirements [167]. Investigation of seed storage proteins has thus been highly motivated by the requirement to produce plant protein products with an acceptable functionality, such as emulsifying and foaming properties [168]. It has been shown that the storage proteins of dicotyledons, such as oilseeds and grain legumes, are composed of few fractions possessing very similar physicochemical properties. They are classified according to their sedimentation behaviour in the analytical ultracentrifuge and therefore, named 11 S, 7 S and 2 S proteins [169, 170]. The 2 S proteins are normally a minor fraction and contains beside real storage proteins [65-67, 93, 94], biologically active proteins, e.g. proteinase inhibitors [91].

Both the major storage protein fractions, 11 S and 7 S proteins, are oligomeric globular proteins. The 11 S globulins are composed of 6 non-covalently linked subunits, each of which contains a disulphide bridged pair of a rather hydrophilic acidic 30-40 kDa α-polypeptide chain and a more hydrophilic basic 20 kDa β-polypeptide chain [169, 170]. Thus, the molecular weight of the subunit amounts to 50-60 kDa and that of the hexameric protein 300-360 kDa. The sedimentation coefficients range from 11 S to 13 S [169]. The oligomeric structure is stabilised by neutral salts in high ionic strength (I 0.25-0.50). It undergoes a dissociation into monomeric 3 S subunits via a trimeric 7 S component in acidic or alkaline solutions or in the presence of denaturants such as urea [170]. An early structure model based on electron microscopic analysis, considered all the α- and β-polypeptide chains as rather equivalent monomers in the formation of their quaternary structure. Thus, 12 subunits were considered to form a double layer of two hexagons [171]. In a more recent structure model, derived from small-angle-X-ray scattering data, the α-β-pairs were considered as subunits and an arrangement of six subunits in a trigonal antiprism was proposed [60]. Though arguments against both structure models were given [172], the trigonal antiprism model seems to be the better approach to the real subunit arrangement which corresponds more adequately to the dissociation behaviour and the properties of the constituent polypeptide chains [173, 174]. Thus, in the latter model the hydrophobic β-chains were proposed to be situated in the interior of the densely packed oligomer, whereas the very hydrophilic C-terminal strands of the α-chains should be exposed at the surface of the molecule. Indeed, this structural prediction could be experimentally confirmed by studying the course of limited proteolysis (e.g. [175]). The exposure of the hydrophilic C-terminal sequence of α-chains should particularly affect the surface functional properties of the 11 S proteins. Some important physicochemical properties of different 11 S globulins are summarised in Table 9.

Table 9. Physicochemical properties of 11S and 7S plant storage proteins derived from small-angle X-ray scattering and hydrodynamic studies, according to [176].

	Helanthinin (Helianthus annuus)*	11 S Cruciferin (Brassica napus) **	Legumin (Vicia faba)***	7S Phaseolin Phaseolus vulgaris)+
Molar mass $M_{S,D}$ [g/mol]	$(3.0\pm0.1)\times10^5$	$(3.0\pm0.1)\times10^5$	$(3.0\pm0.2)\times10^5$	$(1.45\pm 0.05)\times10^5$
Stokes radius R_S [nm]	5.56±0.12	5.56±0.15	6.3±0.15	4.71±0.1
Radius of gyration RG [nm]	3.96±0.8	4.08±0.8	4.45±0.07	4.05±0.5
Maximum dimension [nm]	11.0±0,5	11.2±0.5	13.0±0.5	13.0±0.5
Moleculare shape	oblate ellipsoid of revolution	oblate ellipsoid of revolution	oblate ellipsoid of revolution	disc-like
Molecular dimensions [nm]	11x11x8.8	11.2x11.2x8.8	12.6x12.6x8.8	12.5x12.5x3.75

*) Sunflower **) Rape ***) Broad bean +) French bean

The group of 7 S globulins constitutes proteins with molecular weights between 150 and 210 kDa [169]. This range of molecular weights results from the presence of two different subunits (50 kDa and 70 kDa) which associate to trimers composed of one type of subunits (3x50 kDa or 3x70 kDa) or to mixed trimers [177, 178]. Using small-angle-X-ray scattering, the quaternary structure of a typical 150 kDa 7 S globulin, phaseolin from french bean (Phaseolus vulgaris), has been shown to consist in an arrangement of y-shaped subunits separated by deep clefts filled with solvent [176]. The molecular dimensions are given in Table 9. This type of quaternary structure has been essentially verified by X-ray diffraction analysis on crystals of phaseolin and another 7 S globulin, canavalin from jack beans (Canavalia ensiformis) [179, 180, 181].

Both the 11 S globulins and 7 S globulins, named also legumin-like proteins and vicilin-like proteins respectively, according to the related proteins in a number of leguminous seed, both have a β-sheet structure and are poor in α-helix content [182, 183]. The refinement to 2.2 Å resolution of the three-dimensional structure of phaseolin was the basis of a postulated canonical model for the structure of 7 S proteins [184]. This model also provides a means for interpreting the structure of the 11 S family. Accordingly, a close relationship between the 11 S and 7 S globulin families has been derived. Although a high degree of homology exists between the 7 S proteins from different plants, there are structural peculiarities in each protein. Thus, characteristic differences in the tertiary structures of the 7 S globulins from french bean (phaseolin), soybean (β-conglycinin) and broad bean (vicilin) has been observed using absorption spectroscopy, fluorescence probe hydrophobicity measurents and Trp-fluorescence quenching studies [185]. The surface hydrophobicity of phaseolin was significantly lower than that of vicilin and β-conglycinin. Apparently, the non-polar regions in phaseolin are mostly buried in the interior of the protein. The exposure of hydrophilic patches in the other proteins favours the association to higher aggregates, which has been observed for β-conglycinin [186] and the 7 S globulin from peanuts, conarachin [187]. The self-association of vicilin has been studied with regard to the formation of micelle-like supermolecular structures [188].

6.4. Proteins of the wheat gluten complex

The proteins of the wheat gluten complex are storage proteins with an unique structure. They are essentially interesting with regard to their ability to form supermolecular network structures as the basis of wheat dough visco-elastic properties [190, 191]. According to its extended structure, which is built up of rather long (40-100 kDa) intermolecular S-S-bridged polypeptide chains containing a high number of reverse turns forming sequence repeats, the major protein fraction glutenin should be most responsible for the elastic properties of the network [71, 191, 192]. Gliadins, the second protein fraction, are the monomeric prolamins of wheat. They constitute a polymorphic group of proteins comprising α-, β-, γ- and ω-types according to the decreased electrophoretic mobility [71]. γ-gliadins are the most hydrophobic components and ω-gliadins the least [193]. The gliadin polypeptides have molecular weights in the range of 30kDa to 60 kDa [191]. They contain intramolecular S-S-bonds but some of them form high-molecular weight (HMW) gliadins by intermolecular S-S-bridging. The gliadins have been shown to contribute markedly to the viscous properties of the gluten complex [71, 194]. The gliadin polypeptide chains are organised in repetitive domains, consisting of manifold repetitions of sequences containing essentially Gln, Pro and Phe/Tyr, and of non-repetitive domains [195]. The relative length of these domains differs in the various gliadin types.

The emulsifying properties of gliadins show marked differences due to structural peculiarities. ω-gliadins, which are highly repetitive prolamins with very low basic amino acid content, exhibit very poor emulsifying properties [196]. The surface properties of gliadins have been related to their sequences, especially to the size of repetitive domains. It has been shown, that non-repetitive domains are primarily responsible for gliadin adsorption at non-polar interfaces [196]. The conformation of gliadins vary from quasi random coil to folded monomer and aggregate depending on the pH and salt concentrations [196, 197]. These variations in conformation are accompanied by a change of accessible surface hydrophobicity.

6.5. Myosin, a globular head–fibrillar tail structure

Myosin is the most abundant myofibrillar component, constituting approximately 43% of myofibrillar proteins in mammalian and avian muscle tissue [198]. It has a molecular weight of about 500 kDa and consists of approximately 4500 amino acid residues. The principal structure of myosin consists in the arrangement of two heavy 220 kDa polypeptide chains in an assymmetric molecule with two pear-shaped globular heads attached to a long α-helical rodlike tail. The globular heads are relatively hydrophobic, are able to bind actin, the second protein component of myofibrills, and exhibits ATPase-activity. The rodlike tail, a coiled-coil double helix, is relatively hydrophilic and contains a high portion of charged side chains such as Arg, Glu and Lys residues. The native myosin molecule is approximately 150 nm long and has a diameter of about 8 nm in the head region and 1.5-2.0 nm in the rod region. Proteolytic treatment of myosin with papain cleaves myosin at the so called hinge region in the helical part yielding two fragments, the 350 kDa heavy meromyosin (HMM) and the 125 kDa light meromyosin (LMM). HMM is about 60 nm long and LMM about 90 nm. The secondary structure of the S-1 subfragment displays a dominating portion of about 40 % α-helices. The tertiary structure of this fraction from chicken pectoralis myosin has been determined using single-crystal X-ray diffraction [201].

Myosin contains 40 Cys residues with a distribution of about 27 residues in the head region and about 13 residues in the rod and without a formation of any S-S bond in the head or in the tail. Myosin is extractable from muscle tissue by high-ionic strengths (e.g. 0.3 M KCl, 0.5 M phosphate pH 6.5).

In meat emulsions, salt soluble proteins are of a first-range importance in forming interfacial films that encapsulate fat particles or oil droplets. The emulsifying capacity of myosin in 0.3 or 0.6 M NaCl solution at low protein concentration (5mg/ml) has been shown to be superior to other muscle proteins, ranging in the order myosin>actomyosin>sarcoplasmic proteins>actin [202]. The outstanding emulsifying properties of myosin can be ostensibly related to two unique structural factors. One the one hand, there is an uneaven distribution of charged polar residues in the molecule, with a prevalence of hydrophobic residues in the head region (or HMM S-1 subfragment) and a preponderance of polar groups in the tail or LMM portion. On the other hand, the myosin structure shows a high length to diameter ratio (roughly 40:1) which is conducive to strong protein-protein interaction and molecular flexibility at the interface.

The general structure-function relationship of myosin and other muscle proteins have been discussed in a recent review article [203].

7. CHEMICAL AND ENZYMATIC MODIFICATION –A TOOL FOR DIRECTED CHANGE OF THE PROTEIN STRUCTURE

A great number of chemical and enzymatic methods has been elaborated for a directed modification of protein structure and correspondingly for a change of surface functional properties [114, 204, 205]. Most popular are various methods of acylation which modify the charge by reaction with the positively charged ε-amino groups of lysine residues. In this way, acetylation results in a charge neutralisation at the amino groups and an increased hydrophobicity, which can be drastically enhanced by introducing long chain fatty acid residues, such as palmitoyl groups. When succinyl residues are attached to the protein instead of acetyl residues, the positively charged amino-groups change into negatively charged N-succinyl groups. Both acetylation and succinylation are not specific to NH_2-groups and results in side-reactions with OH and SH groups as well. Some acyl residues, such as citraconyl and maleyl residues, can be easily removed from the protein by mild acidic hydrolysis, allowing a reversible blocking of functional groups in proteins.

Reductive alkylation is a very useful method of introducing an alkyl residue (e.g. CH_3 group) into protein amino groups by reaction with a carbonyl compound (e.g. formaldehyde) following by reduction of the formed Schiff- base by e.g. cyanoborhydride. When the carbonyl compound is a sugar, glycosylation of the protein can be performed resulting in "neoglycoproteins" with an enhanced hydrophilicity. Chemical modification of carboxyl groups of glutamic and aspartic acid residues by esterification with alcohols increases the positive net charge of the proteins and changes the effective hydrophibicity. On the other hand, hydrolytic splitting of the amide ammonia from glutamine and asparagine residue (deamidation) increases the negative net charge . A powerful method of chemical modification of proteins is by phosphorylation. Phosphorus oxychloride is the most frequently applied reagent, though its reaction with proteins is more difficult to govern than that with acetic anhydride or succinic anhydride. A lot of side-reactions of protein with $POCl_3$ has been observed including phosphorylation of most functional groups and crosslinking of polypeptide chains.

Protein crosslinking can easily be performed by using bifunctional reagents causing acylation, alkylation (also at SH groups), carbonyl-amine reactions and others. There are some enzymes, such as transglutaminase, peroxidase and polyphenol oxydase which are especially suited to catalyse intermolecular crosslinking of proteins. From these, tranglutaminases have recently been intensely studied as enzymes which catalyse the interchain linking of ε-amino groups of lysines and γ-amide groups of glutamic acid residues. It has been also successfully applied for the enzymatic deamidation of proteins.

The availability of some protein-kinases, catalysing protein phosphorylation, made the enzymatic phosphorylation of proteins, even food proteins, highly attractive. Enzymatic phosphorylation proceeds specificly at definite sequence regions („consens sequences") in proteins and is therefore limited in its extent to these target sites.

Specific modification of charge and hydrophobicity which cause distinct conformational changes should be very helpful in deriving relationships between protein structure and surface activity.

8. References

1. Th.B. Osborne, The Vegetables Proteins, 2nd edition, Longmans, Green & Co, London, 1924, pp 16-35
2. H.D. Jakubke and D. Jeschkeit, Brockhaus ABC Biochemie, VEB F.A.Brockhaus Verlag, Leipzig, 1976
3. I.Prendl and K.Felix, In Hoppe-Seyler/Thierfelder, Handbuch der physiologisch- und pathologisch-chemischen Analyse, 10. Auflage, Band IV/1, S.212, Springer Verlag, Berlin, 1960
4. K.U.Linderström-Lang, Lane Medical Lectures: Proteins and Enzymes, Stanford Univ.Press, Stanford Cal., 1952
5. I.D.Bernal, Discuss.Faraday Socc., 25(1958)7
6. G.F.Schulz and R.H. Schirmer, Principles of Protein Structure, Springer Verlag, New York, Heidelberg, Berlin, 1979
7. C.B.Anfinsen, Science, 181(1973)223
8. K.Hamaguchi, The Protein Molecule - Conformation, Stability and Folding, Springer Verlag, Berlin, 1992
9. A.G.Walton, Polypeptides and Protein Structure, Elsevier, New York, 1982
10. R.Dickerson and I.Geis, The Structure and Action of Proteins, Harper & Row, Publ., New York, 1969

11. I.L.Crawford, W.N.Liscomb and C.G.Schellman, Proc.Nat.Acad.Sci. USA, 70(1973)538
12. P.Y.Chou and G.D.Fasman, Biochemistry, 13(1974)211
13. P.N.Lewis, F.A.Momamy and H.A.Scheraga, Biochem., Biophys, Aeta, 303(1073)211
14. I.D.Kuntz, J. Am., Chem.Soc., 94(1972)4009
15. M.Levitt, Biochemistry, 17(1978)4277
16. P.Y.Chon and G.D.Fasman, Adv. Enzymol., 47(1978)45
17. G.D.Fasman, Prediction of Protein Structure and the Principles of Protein Conformation, Plenum Press, New York, 1989
18. B.Jirgensons, Optical Activity of Proteins and other Macromolecules, Springer Verlag, Berlin, 1973
19. J.S.Richardson, Adv.Protein Chem., 34(1981)168
20. K.Wütherich, NMR of Proteins and Nucleic Acids, John Wiley & Sons, New York, 1986
21. V.P.Saxena, D.B.Wetlaufer, ProcNatl.Acad.Sci.USA, 68(1971)969
22. Y.H.Chen, J.T.Chang and K.H.Chang, Biochemistry, 13(1974)3350
23. C.T.Chang, C.-S.Wu and J.T.Yang, Analyt.Biochem., 91(1978)13
24. St.W.Provencher and J.Glöckner, Biochemistry, 20(1981)33
25. S.Brahms and J.Brahms, J.Mol.Biol., 138(1980)149
26. W.C.Johnson jr., Methods Biochem.Anal., 31(1985)61
27. W.C.Johnson jr., Ann.Rev.Biophys.Chem., 17(1988)145
28. W.K.Surewicz and H.H.Mantsch, Biochem.Biophys.Acta, 952(1988)115
29. F.Dousseau, M.Therieu and M.Pezolet, Appl.Spectrosc., 43(1989)538
30. F.Dousseau and M.Pezolet, Biochemistry, 29(1990)8771
31. F.H.C.Crick, Acta Crystallogr., 6(1953)689
32. K.D.B.Fraser and T.P.Mac-Rae, Nature, 233(1971)138
33. D.L.D.Caspar, C.Cohen and W.Longley, J.Mol.Biol., 41(1969)87
34. J.Janin, Bull.Inst.Pateur, 77(1979)337
35. M.J.E.Sternberg and J.M.Thornton, J.Mol.Biol., 105(1976)367
36. S.T.Rao and M.G.Rosmann, J.Mol.Biol., 76(1973)241
37. J.S.Richardson, Nature, 268(1977)495
38. D.E.Wetlaufer, Proc.Nat.Acad.Sci., USA, 70(1973)697
39. B.J.Sutton, Curr.Opinion Immunol, 2(1989)106

40. R.Huber, J.Deisenhofer, P.M.Colman, M.Matsushima and W.Palm, Nature, 264(1976)415
41. W.Kauzmann, Adv.Protein Chem., 14(1959)1
42. Ch.Tanford, The Hydrophobic Effect, Jon Wiley & Sons, New York, 1973
43. R.H.Pain, Mechanism of Protein Folding, IRL Press at Oxford University Press, Oxford, 1994
44. Y.Nozaki and Ch.Tanford, J.Biol.Chem., 246(1971)2211
45. C.Chotia, Nature, 254(1975)304
46. S.Miller, A.M.Lesk, J.Janin and C.Chotia, Nature, 328(1987)834
47. S.Miller, J.Janin, A.M.Lesk and C.Chotia, J. Mol.Biol., 196(1987)641
48. F.M.Richards, J.Mol.Biol., 82(1984)1
49. C.Chotia, Annu.Rev.Biochem., 53(1984)537
50. I.M.Klotz, D.W.Darnall and N.R.Langerman, In The Proteins, Third edition Vol I, H.Neurath and R.L.Hill(Ed.), Academic Press, New York, 1975, p.293
51. J.Monod, J.Wyman and J.P.Changeux, J.Mol.Biol., 12(1965)88
52. H.K.Schachman, In Protein-Protein Interactions, R.Jaenicke and E.Helmreich (Ed.), Springer Verlag, Berlin, 1972, p.17
53. H.H.Jaffé and M.Orchin, Symmetry in Chemistry, Wiley, New York, 1965
54. K.R.Hanson, J.Mol.Biol., 22(1966)405
55. K.R.Hanson, J.Mol.Biol., 38(1968)133
56. A.Klug, Symp. Int. Soc. Cell. Biol., 6(1967)1
57. M.A.Lauffer, Entropy-driven Processes in Biology. Polymerisation of Tabacco Mosaic Virus Protein and Similar Reactions, Springer Verlag, Berlin, 1975
58. C.J.Lusty and H.Fasold, Biochemistry, 8(1969)2933
59. J.T.Finch, In: The Proteins, Third Edition, Vol. I, H.Neurath and R.L.Hill (Ed.), Academic Press, New York, 1975, p.413
60. N.Plietz, G.Damaschun, J.J.Müller and K.D.Schwenke, Eur. J. Biochem., 130(1983)315
61. K.E.van Holde, In:The Proteins, Third Edition, Vol.I, H.Neurath and R.L.Hill (Ed.), Academic Press, New York, 1975, p.225
62. S.Nakai and M.Nonaka, In:Protein Interaction, H.Visser (Ed.), VCH, Weinheim, 1992, p.57
63. K.Weber and M.Osborn, In:The Proteins, Third Edition, Vol.I, H.Neurath and R.L.Hill (Ed.), Academic Press, New York, 1975, p.179
64. J.S.Wall, J.Agric, Food Chem., 19(1971)619

65. F.S.Sharif and S.S.-L.Li, J. Biol.Chem., 257(1982)14753
66. Ch.Ampe, J.van Damme, L.A.B.de Castro, M.J.A.M.Sampaio, M. van Montagu and J.Vandekerckhove, Eur. J. Biochem., 159(1986)597
67. B.P.Salmanowicz and J.K.P.Weder, Z.Lebensm., Unters. Forsch. A, 204(1997)129
68. F.London, Z.Phys.Chem., Abt.B, 11(1930)222
69. J.E.Jones, Proc.Roy.Soc., London, Ser.A 106(1924)441
70. D.E.Anderson, W.J.Becktal and F.W.Dahlquist, Biochemistry, 29(1990)2403
71. H.D.Belitz and W.Grosch, Lehrbuch der Lebensmittelchemie, 3.Auflage, Springer Verlag, 1987
72. K.R. Acharya, D.I.Stuart, N.P.C.Walker, M.Lewis and D.C.Phillips, J. Mol. Biol., 208(1989)99
73. M.Mitani, Y.Harushima, K.Kuwajima, M.Ikeguchi and S.Sugai, J.Biol.Chem., 261(1986)8824
74. J.B.Mathew, Annu.Rev.Biophys.Chem., 14(1985)387
75. G.C.Pimentel and A.L.Mc Clellan, The Hydrogen Bond, Freeman & Co, London, 1960
76. T.E.Creighton, Biopolymers, 22(1983)49
77. C.N.Pace, Crit. Rev.Biochem., 3(1975)1
78. P.L.Privalov, Adv. Protein Chem., 33(1979)167
79. P.L.Privalov and S.J.Gill, Adv.Protein Chem., 39(1988)191
80. W.Pfeil, Mol.Cell.Biochem., 40(1981)3
81. T.Alber, Annu.Rev.Biochem., 58(1989)765
82. A.S.Yang and B.Honig, J.mol.Biol., 231(1993)459
83. C.C.Bigelow, J.Theoret. Biol., 16(1967)187
84. S.Damodaran, In Food Proteins, J.E.Kinsella and W.G. Soucie (Ed.), The American Oil Chemists Soc., Champaign, 1989, p.21
85. B.Lee and F.M.Richards, J.Mol.Biol. 55(1971)379
86. S.Nakai, In. Food Proteins-Properties and Characterisation, S.Nakai and H.W.Modler (Ed.), VCH Publishers, New York, 1996, p.2
87. L.A.Sklar, B.S.Hudson and R.D.Simoni, Biochemistry, 16(1977)5100
88. L.A.Sklar, B.S.Hudson and R.D.Simoni, Proc.Natl.Acad.Sci., USA, 72(1975)1645
89. A.Kato and S.Nakai, Biochem. Biophys.Acta, 624(1980)13
90. S.Odani and T.Ikenaka, J.Biochem., 74(1973)697
91. J.J.Rackis, W.J.Wolf and E.C.Baker, In:Nutritional and Toxicological Significance of Enzyme Inhibitors in Foods, M.Friedman (Ed.), Plenum Press, New York, 1986, p.299

92. P.H.von Hippel and T.Schleich, Acc.Chem.Res., 2(1969)257
93. K.D.Schwenke, B.Raab and M.Schultz, Studia Biophys., 51(1975)161
94. K.D.Schwenke, B.Drescher, D.Zirwer and B.Raab, Biochem.Physiol.Pflanzen, 183(1988)219
95. Y.Nozaki and C.Tanford, J.Am.Chem.Soc., 89(1967)742
96. R.Roxby and C.Tanford, Biochemistry, 10(1971)3348
97. C.Tanford, K.Kawahara and S.Lapanje, J.Am.Chem.Soc.89(1967)729
98. S.Lapanje and C.Tanford, J.Am.Chem.Soc., 89(1967)5030
99. C.Tanford, Adv.Protein Chem., 23(1968)121
100. C.Tanford, Adv.Protein.Chem., 24(1970)1
101. K.A.Dill and D.Shortle, Annu.Rev.Biochem., 60(1991)795
102. M. Joly, A Physico-chemical Approach to the Denaturation of Proteins, Academic Press, New York, 1965
103. J.I.Boye, C.-Y.Ma and V.R.Harwalkar, In: Food Proteins and their Applications, S.Damodaran and A.Paraf (Ed.), Marcel Dekker, New York, 1997, p.25
104. O.B.Ptisyn, J. Protein Chem., 6(1987)272
105. O.B.Ptisyn, In: Protein Folding, Th.E.Creighton (Ed.), W.H.Freeman & Comp., New York, 1992, p.243
106. H.Christensen and R.H.Pain, In: Mechanism of Protein Folding, R.H.Pain (Ed.), IRL Press, Oxford, 1994, p.55
107. E.Dickinson and Y.Matsumura, Colloids & Surfaces, Biointerfaces, 3(1994)1
108. S.Xu and S.Damodaran, In: Food Proteins-Structure and Functionality, K.D.Schwenke and R.Mothes (Ed.), VCH Weinheim, 1993, p.270
109. J.E.Kinsella and W.G.Soucie, Food Proteins, Am. Oil Chemist`s Soc., Champaign, 1989
110. S.Nakai and H.W.Modler, Food Proteins-Properties and Characterisation, VCH, New York, 1996
111. S.Damodaran and A.Paraf, Food Proteins and their Applications, Marcel Dekker, New York, 1997
112. J.F.Zayas, Functionality in Foods, Springer Verlag, Berlin, 1997
113. J.C.Cheftel, J.L.Cuq and D.Lorient, Lebensmittelproteine, Behr`s Verlag, Hamburg, 1992
114. K.D.Schwenke, In: Food Proteins and their Applications, S.Damodaran and A.Paraf (Ed.), Marcel Dekker, New York, 1997, p.393

115. K.D.Schwenke, D.Zirwer, K.Gast, E.Görnitz, K.-J.Linow and J.Gueguen, Eur.J.Biochem., 194(1990)621

116. K.D.Schwenke, K.-J.Linow and D.Zirwer, Nahrung, 30 (1986)263

117. K.D.Schwenke, M.Schultz, K.-J. Linow, K.Gast and D.Zirwer, Int.J.Peptide Protein Res., 16(1980)12

118. K.J.Shetty and M.S.N.Rao, Int.J.Peptide Protein Res., 11(1978)305

119. G.Braunitzer, R.Chen, B.Schrank and A.Stangl, Hoppe-Seylers Z.Physiol.Chem., 353(1972)832, 354(1973)868

120. L.Sawyer, M.Z.Papiz, A.C.T.North and E.E.Eliopoulos, Biochem.Soc.Trans, 13(1985)265

121. M.Z.Papiz, L.Sawyer, E.E.Eliopoulos, A.C.T.Nort, J.B.C.Findlay, R.Sivapradarao, J.B.C.Jones, M.E.Newcomer and P.J.Kraulis, Nature, 324(1986)383

122. L.M.Gilbert and G.A.Gilbert, Methods Enzymol., 27, Part D, (1973)273

123. M.Paulsson, Thermal Denaturation and Gelation of Whey Proteins and their Adsorption at the Air/Water Interface, Doctoral Thesis, Univ.Lund., 1990

124. W.N. Eigel, I.E.Butler, C.A.Ernstrom, H.M.Farrell, V.R.Jenness, V.R. Harwalkar and R.Mc.Whitney, J.Dairy Sci., 67(1984)1599

125. W.J.Brown; A.C.T. North, D.C.Pillips, K.Brew, T.C.Vanemann and R.L.Hill, J.Mol.Biol., 42(1969)65

126. R.L.Hill, K.Brew, Th.C.Vanaman, I.P.Trayer and P.Mattock, Brooknaven Symposia in Biology No.21, Structure, Function and Evolution in Proteins, 1969, p.139

127. W.R.Krigbaum and F.R.Kugler, Biochemistry., 9(1970)1216

128. M.J.Kronman and R.E.Andreotti, Biochemistry, 3(1964)1145

129. M.J.Kronman, R.E.Andreotti and R.Vitals, Biochemistry, 3(1964)1152

130. Y.Hiraoka, T.Segawa, K.Kuwajima, S.Sugai and M.Murai, Biochem.Biophys.Res. Commun, 95(1980)1098

131. C.C.F.Blake, G.A.Mair, A.C.T.North, D.C.Phillips and V.R.Sarma, Nature, 206(1965)757

132. C.C.F.Blake, G.A.Mair, A.C.T.North, D.C.Phillips and V.R.Sarma, Proc.Roy.Soc., London, Ser.B, 167(1967)365

133. J.R.Brown, Proceed, 11th FEBS Meeting, 50(1977)11

134. R.G.Reed, F.W.Putman and T.Peters jr., Biochem J., 191(1980)867

135. I.Sjöholm and I.Ljungstedt, J.Biol.Chem., 248(1973)8434

136. C.D.Carter, X.-M. He, S.H.Munson, P.D.Twigg, K.M.Gernert, B.Broom and T.Y.Miller, Science, 244(1989)1195
137. A.K.Wright and M.R.Thompson, Biophys.J., 15(1975)137
138. P.G.Squire, P.Moser and C.T.O'Konski, Biochemistry, 7(1968)4261
139. T.Raj and W.H.Flygare, Biochemistry, 13(1974)3336
140. M.Sogami and J.F.Forster, Biochemistry, 7(1968)2172
141. J.F.Foster, In: Albumin Structure, Function and Uses, V.M.Rosenoer, M.Oratz and M.A. Rothschild (Ed.), Pergamon Press, Oxford, 1977, p.53
142. W.F.Harrington, P.Johnson and R.H.Ottewill, Biochem. J., 62(1956)569
143. A.J.Nisbet, R.H.Saundry, A.J.G.Moir, L.A.Fothergill and I.E.Fothergill, Eur.J.Biochem., 115(1981)335
144. N.N.Nitabatake, A.Ishida and E.Doi, Agric.Biol.Chem., 54(1988)967
145. M.B.Smith and J.F.Back, Aust.J.Biol.Sci, 18(1965)365
146. M.B.Smith and J.F.Back, Aust.J.Biol.Sci., 21(1968)539
147. R .Nakamura, M.Hirai and Y.Takemori, Agric.Biol.Chem.,44(1980)149
148. E.Doi and N.Kitabatake, In: Food Proteins and their Applications, S.Damodaran and A.Paraf (Ed.), Marcel Dekker, New York, 1997, p.325
149. E.Schlimme and W.Buchheim, Milch und ihre Inhaltsstoffe, Verlag Th.Mann, Gelsenkirchen, 1995
150. G.Taborsky, Adv. Protein Chem., 28(1974)1
151. J.C.Mercier, F.Grosclaude and B.Ribadeau-Dumas, Eur.J.Biochem., 23(1971)41
152. G.Brignon, B.Ribadeau-Dumas and J.C.Mercier, FEBS Lett., 71(1971)111
153. T.Ono, K.Yutani and S.Odagiri, Agr.Biol.Chem., 38(1974)1603
154. P.Walstra and R. Jenness, Dairy Chemistry and Physics, J.Wiley & Sons, New York, 1984
155. B.Ribadeau-Dumas, G.Brignon, F.Grosclaude and J.C.Mercier, Eur.J.Biochem., 25(1972)505
156. A.L.Andrews; D.Atkinson, M.T.A.Evans, E.G.Finer, J.F.Green, M.C.Phillips and R.N.Robertson, Biopolymers, 18(1979)1105
157. R.A.Sullivan, M.M.Fitzpatrick, E.K.Stanton, R.Annino, G.Kissel and F. Palermiti, Arch.Biochem. Biophys., 55(1955)455
158. P.H.von Hippel and D.F.Waugh, J.Amer.Chem.Soc., 77(1955)4311
159. D.F.Waugh, L.K.Creamer, C.W.Slatterly and G.W.Dresdner, Biochemistry, 9(1970)786
160. T.G.Parker and D.G.Dalgleish, J.Dairy Sci., 48(1981)71

161. J.C.Mercier, G.Brignon and B.Ribadeau-Dumas, Eur.J.Biochem., 35(1973)222
162. H.Swaisgood, J.Dairy Sci., 58(1974)583
163. T.K.Kumosinski, E.M.Brown and H.M.Farrel jr., J.Dairy Sci., 74(1991)2879
164. H.E.Swaisgood, CRC Crit.Rev.Food Technol., 3(1973)375
165. T.A.J.Payens, J.Dairy Res., 46(1979)291
166. P.Jollès, Mol.Cell.Biochem.7(1975)73
167. A.M.Altschul and H.L.Wilcke (Ed.), New Protein Foods Vol.5, Seed Storage Proteins, Academic Press, Orlando, 1985
168. A.Pour-El (Ed.), Functionality and Protein Structure, ACS Symposium Ser. 92, Am. Chem.Soc., Washington, 1979
169. E.Derbyshire, D.J.Wright and D.Boulter, Phytochemistry, 15(1976)3
170. V.Prakash and M.S.N.Rao, CRC Crit.Rev.Biochem., 20(1986)265
171. R.A.Badley, D.A.Atkinson, H.Hauser, D.Oldani, J.P.Green and J.M.Stubbs, Biochem.Biophys.Acta, 412(1975)214
172. M.J.Miles, V.J.Morris, V.Carroll, D.J.Wright and V.Newby, Int.J.Biol.Macromol., 7(1985)125
173. P.Plietz, B.Drescher and G.Damaschun, Int.J.Biol.Macromol., 9(1987)161
174. P.Plietz and G.Damaschun, Studia Biophys. 116(1986)153
175. S.Dudek, C.Horstmann and K.D.Schwenke, Nahrung, 40(1996)171
176. P.Plietz, G.Damaschun, D.Zirwer, G.Gast, B.Schlesier and K.D.Schwenke, Kulturpflanze, 32(1984)159
177. R.Casey and C.Domoney, Kulturpflanze, 32(1984)99
178. B.Schlesier, Kulturpflanze, 32(1984) S 223
179. S.Johnson, G.Grayson, L.Robinson, R.Chahade and A.McPherson, Biochemistry, 21(1982)4839
180. A.Mc Pherson, J.Biol.Chem., 255(1980)10472
181. M.C.Lawrence, E.Suzuki, J.N.Varghese, P.C.Davis, A.van Donkelaar, P.A.Tullocn and P.M.Colman, EMBO J., 9(1990)9
182. D.Zirwer, K.Gast, H.Welfle, B.Schlesier and K.D.Schwenke, Int.J.Biol.Mocromol., 7(1985)105
183. M.Subirade, M.Pezolet and J.Gueguen, In: Food Proteins-Structure and Functionality, K.D.Schwenke and R.Mothes (Ed.), VCH, Weinheim, 1993, p.158
184. M.C.Lawrence, T.Izard, M.Beuchat, R.J.Blagrove and P.M. Colman, J. Mol.Biol., 238(1994)748

185. S.S.Desphande and S.Damodaran, Int.J.Peptide Protein Res., 35(1990)25
186. I.Koshiyama, Cereal Chem., 45(1968)394
187. P.Johnson and W.E.F.Naiswith, Disc.Faraday Soc., 13(1953)98
188. M.A.H.Ismond, E.D. Murray and S.D.Arntfield, Food Chem., 20(1986)305
189. R.Lasztity, Nahrung, 30(1986)235
190. R.Lasztity and F. Békés (Ed.), Gluten Proteins, World Scientific, Singapore, 1987
191. P.Y.Payne, P.A.Harris, C.N.Law, L.M.Holt and I.A.Blackman, Ann.Techn.Agric., 29(1980)309
192. Y.Popineau and B.Godon, Cereal Chem., 59(1982)55
193. H.D.Belitz, J.-J.Kim, R.Kieffer, W.Seilmeier, U.Werbeck and H.Wieser, In: Gluten Proteins, R.Lasztity and F.Békés, World Scientific, Singapore, 1987, p.189
194. P.R.Shewry and A.S.Tathan, Biochem.J., 267(1990)1
195. Y.Popineau and F.Pineau, In: Food Proteins-Structure and Functionality, K.D.Schwenke and R. Mothes (Ed.), VCH, Weinheim, 1993, p.290
196. Y.Popineau and F.Pinot, Lebensmittel Wiss. u. Techn., 21(1988)113
197. L.D.Yates and M.L.Greaser, J.Mol.Biol., 168(1983)123
198. A.M.Pearson and R.B.Young, Muscle and Meat Biochem., Acad.Press Inc., San Diego, CA, 1989
199. I.Rayment, W.R.Rypniewski, K.Schmidt-Base, R.Smith, D.R.Tomchick, G.Wesenberg and H.M. Holden, Science, 261(1993)50
200. S.J.Galluzo and J.M.Regenstein, J.Food Sci., 34(1978)1761
201. Y.L.Xiong, In: Food Proteins and their Applications, S.Damodaran and A.Paraf (Ed.), Marcel Dekker, New York, 1997, p.341
202. G.E. Means and R.E. Feeney, Chemical Modification of Proteins, Holden Day, San Francisco, 1971
203. S. Magdassi (Ed.), Surface Activity of Proteins - Chemical and Physicochemical Modifications, Marcel Dekker, New York, 1996

Proteins at Liquid Interfaces
D. Möbius and R. Miller (Editors)

ADSORPTION AND INTERFACIAL TENSION ISOTHERMS FOR PROTEINS

Valentin B. Fainerman[1] and Reinhard Miller[2]

[1] Institute of Technical Ecology, Blvd. Shevchenko 25, 340017 Donetsk, Ukraine

[2] Max-Planck-Institute für Kolloid- und Grenzflächenforschung, Rudower Chaussee 5, D-12489 Berlin-Adlershof, Germany

CONTENTS

1. INTRODUCTION

The great practical significance of the adsorption of polyelectrolytes, and, in particular, proteins at fluid interfaces has stimulated the development of various theoretical models which describe the equilibrium and dynamic behaviour of adsorption layers in these systems. The properties of protein adsorption layers differ in a number of aspects from those characteristic to monolayers of usual surfactants. First, for proteins surface denaturation may take place, leading to the unfolding of protein molecule within the surface layer at least at low surface pressures. The value of partial molar surface area for proteins, in contrast to surfactants, is large and can vary. This property, and also the fact that the number of configurations for adsorbed protein molecule exceeds significantly that of molecules in the bulk, leads to a significant increase of the non-ideality of the surface layer entropy. This also makes it impossible to apply the most simple models (Henry, Langmuir) for the description of protein adsorption layers. The principle question about the existence of an equilibrium adsorption layer of proteins, or the reversibility of protein adsorption at liquid interfaces, remains unsolved. Thus, a systematic experimental verification of theoretical models for the description of protein adsorption is complicated significantly.

From the analysis of conformational variations for adsorbed macromolecule, Frisch, Simha and Eirich [1-3] had calculated the adsorption layer thickness and presented a first adsorption isotherm for polymers. During the subsequent 40 years, a number of theoretical results had been published concerning the adsorption of uncharged polymers and polyelectrolytes. Essentially, one can distinguish between three main directions in these studies, thermodynamic, scaling and statistical mechanics approaches.

In the present chapter main attention is paid to thermodynamic models and the statistical and scaling models will be discussed only briefly, referring the reader to more comprehensive analysis in form of reviews [4-8] and original papers (see below). A thermodynamic approach developed only recently for surfactant mixtures occupying different partial molar surface areas and to for individual surfactants capable of varying their molar surface area due to reorientations within the surface layer [9-12], led to new results also for the analysis of protein adsorption [9,13]. These results and their further development are the subject of this chapter.

2. MAIN THEORETICAL APPROACHES

2.1. STATISTICAL MECHANICS MODELS

The statistical methods enable one to consider a detailed structure for various conformations of macromolecules adsorbed at a surface. For the molecules possessing mobile chains, loop-train-tail conformations are common, with varying combinations of trains and loops. Rodlike macromolecules can either adsorb by one end only, or remain entirely localised (similarly to flexible molecules) within the surface layer. Mobile macromolecules can also be fixed to the surface (usually to solids) in a single point, either an end point or an intermediate point. Starting from the early studies of Frisch, Simha and Eirich [1-3], the statistical models were developed further in order to better consider an increasing number of possible configurations of adsorbed macromolecule and to present a more realistic description of the intermolecular interactions both within the bulk and at the surface. However, the analytical treatment of the resulting equations for the adsorption and interface tension, which was possible in early studies, became more cumbersome.

Singer [14] had calculated the total number of various configurations Φ for a macromolecule located within the surface (two-dimensional quasi-crystalline model) neglecting the intermolecular interactions polymer-surface and polymer-solvent molecules, that is, the contribution of the entropy to the free energy was considered only. With $\Delta S = k \ln \Phi$, $\Delta F = -T\Delta S$ and $\Pi = -T(d\Delta S/dA)$, Singer had derived an equation of state for the surface layer in the following

$$\Pi = \frac{RT}{\omega_0}\left[\frac{z}{2}\left(1-\frac{\omega_0}{\omega_1}\right)\ln\left(1-\frac{2}{z}\theta_1\right)-\ln\left(1-\theta_1\right)\right] \tag{1}$$

where Π is the surface pressure, R is the gas law constant, T is the temperature, ΔS is the change in surface entropy, k is the Boltzmann constant, A is the surface area, $\theta_1 = \omega_1\Gamma_1$ is the degree of surface coverage, Γ_1 is the adsorption, ω_1 is the surface area of the macromolecule unfolded at the surface, ω_0 is the quasi-crystal cell area (water molecule or macromolecule segment), and z is the co-ordination number. The first term in the square brackets on the right hand side of Eq. (1) is added to the ordinary Szyszkowski-Langmuir equation to describe the non-ideality of the surface layer entropy. A more complicated expression for Π was obtained by Frisch and Simha in [15] by considering the ability of macromolecules to form loops. However, for long chains ($\omega_1 >> \omega_0$) their equation transforms into Eq. (1).

A number of studies dealt with the conformation of isolated polymer molecules for various adsorption energies [16-22]. For this case the configuration entropy was defined by loops and tails, while the enthalpy of polymer adsorption depend on the interaction energy of the trains with the surface. In the model developed by Hoeve [23] the adsorption layer was divided into two parts. The molecules situated within the first part are assumed to be in direct contact with the surface, while the remaining part consists of the loops only [24]. Interaction between adsorbed linear trains was taken into consideration. In subsequent studies, Hoeve had also accounted for the interaction between the loops. Silberberg [25] applied a two-dimensional quasi-crystalline model of the surface, assuming however that the macromolecule is capable of adsorbing in trains and loops of various lengths. Both segment-segment and segment-solvent interactions were considered. Possibly the most detailed statistic theory was developed by Scheutjens and Fleer [5,26]. Here a quasi-crystalline model was also employed; however, the solution adjacent to the surface was divided into parallel layers, each represented by a two-dimensional quasi-crystal. The features included into the consideration were: all possible conformations of the adsorbed molecule, the adsorption energy for the trains being in contact with the surface, the entropy of mixing of the chains with the solvent, the attraction or repulsion between the segments and the solvent molecules. In their model, Scheutjens and Fleer accounted for the contribution of various conformations of the chains into the concentration profile of the polymer in the vicinity of the surface, and calculated the adsorption as a function of the polymer chain length, estimated the relative portions of trains, loops and tails, the adsorption layer thickness etc. The theoretic values coincide satisfactorily with experimental data for the adsorption of polymers possessing various degree of polymerisation. Model calculations for the adsorption of polyelectrolytes, i.e. charged polymers possessing mobile chains, were performed by Hesselink [27] and Leermakers et al. [28].

In spite of an obvious success of the statistical models for the description of polymer adsorption layers, their extreme complexity has also to be mentioned: usually the relevant calculations can be performed only numerically, with a number of parameters involved, which are often unknown. Therefore, a scaling theory although providing only qualitative estimates, is rather simple as compared to statistic models, and can be successfully applied for the description of adsorption layers of polyelectrolytes.

2.2. SCALING ANALYSIS

The structure of polymers in so-called semi-dilute solutions with overlapping coils can be represented in the form of a grid possessing the characteristic dimension ξ. This characteristic dimension, called the correlation length, is the most important concept of the scaling analysis [29], because various physico-chemical properties of the polymers can be expressed via ξ and g, which is the number of monomers per subunit characterised by this correlation length. The scaling theory expresses ξ and g by a power law

$$\xi \approx d \cdot x^{-3/4} \tag{2}$$

$$g \approx x^{-5/4} \tag{3}$$

Here d is the size of the statistical unit of the polymer chain (the monomer), and x is the volume fraction of the polymer. For example, the osmotic pressure can be expressed by

$$\Pi_{os} \approx kT/\xi^3 \tag{4}$$

It follows from Eqs. (2)-(4) that $\Pi_{os} \approx x^{9/4}$, that means the osmotic pressure is approximately quadratically proportional to the concentration of the polymer. The concepts of the scaling analysis introduced by de Gennes [29] were further developed and generalised in [30-37]. It was shown in [37] that the correlation length ξ for semi-dilute solutions coincide to within a scalar factor with the electrostatic screening length, which in turn exceeds the Debye length. This means that in semi-dilute solution mobile chains are essentially the random walks of electrostatic blobs. Within these blobs comprised of a few monomers, the Coulomb interaction is insufficient to produce a deformation of the polyelectrolyte molecule, that is, the chain can be bent only in the places where the blobs join each other. If an inorganic electrolyte is present (which is the common case in protein studies) the situation becomes somewhat different. For example, if the concentration of the salt is high, the correlation length is similar to that characteristic for an uncharged polymer ($\xi \approx x^{-3/4}$), while for low concentration of inorganic electrolyte this dependence is $\xi \approx x^{-1/2}$ [37]. The polymer chain length L also depends on the concentration of the salt: $L \approx x^{-1/4}$ for low, and $L \approx x^{-1/8}$ for high concentrations of inorganic electrolyte, respectively. For high polymer concentration, which is common for example in adsorption layers of proteins, the electrostatic screening length in presence of inorganic electrolyte becomes lower than the dimension of the electrostatic blob. It means that in this case blobs are destroyed, which increases the flexibility of the chain.

The scaling theory was applied to both qualitative analysis of the adsorption from the polyelectrolyte solutions, and to quantitative estimation of corresponding characteristics [7, 38-43]. The surface pressure of adsorbed or spread monolayers Π is essentially a parameter analogous to the two-dimensional osmotic pressure; therefore as a first approximation from Eqs. (2) and (4) it follows that $\Pi \approx \Gamma^{9/4}$. A detailed analysis performed in [43] led to a more precise expression for the surface pressure, into which the contributions of both polymer chains and counterions are included additively. An expression for the adsorption layer thickness was also derived. Similarly to all other relationships valid in the scaling theory, in the equations derived in [38-43] scalar factors are disregarded; however the general dependencies can be readily understood from the physical point of view, and conform with experimental data. One can agree with de Gennes opinion [7], that the statistical and scaling models are supplementary. Moreover, some concepts of the scaling analysis, for example those concerning electrostatic blobs, the prevailing contribution of ions into the surface pressure of adsorption layer etc., can be successfully applied also to thermodynamic models for the adsorption of proteins at liquid interfaces

2.3. THERMODYNAMIC MODELS

The thermodynamic analysis of the adsorption of proteins at liquid interfaces is based on an equation derived by Butler [44] for the chemical potential of a component in the surface

$$\mu_i^S = \mu_i^{0S} + RT \ln f_i^S x_i^S - \gamma \omega_i \tag{5}$$

and the corresponding equation for the chemical potential of the same component in the solution bulk

$$\mu_i^\alpha = \mu_i^{0\alpha} + RT\ln f_i^\alpha x_i^\alpha \tag{6}$$

where μ_i^{0s} and $\mu_i^{0\alpha}$ are the standard chemical potentials, γ the surface tension, ω_i the partial molar surface, f_i the activity coefficient, $x_i = N_i / \Sigma N_i$ the molar portion, and N_i the number of moles of ith component. Here the superscripts 's' and 'a' refer to the surface (interface) and the bulk.

The pure solvent is usually taken as the standard state (i = 0), that is, $x_0^\alpha = x_0^s =$, $f_0^\alpha = f_0^s = 1$ and $\gamma = \gamma_0$, while for the dissolved components infinite dilution is taken as the standard state, that is, $x_i^\alpha \to 0$, $f_i^\alpha = f_i^s = 1$ and $\gamma = \gamma_0$. For this choice the equilibrium condition can be derived from Eqs. (5) and (6) as [12, 13]

$$\ln \frac{f_i^s x_i^s}{K_i f_i^\alpha x_i^\alpha} = -\frac{\Pi\omega_i}{RT} \tag{7}$$

where $K_0 = 1$ and $K_i = \left(x_i^s / x_i^\alpha\right)_{x_i^\alpha \to 0}$ for $i \geq 1$ are the distribution coefficients at infinite dilution, $\Pi = \gamma_0 - \gamma$ is the surface pressure, γ_0 is the surface tension of the solvent. If one assumes ideal behaviour of the dissolved species in the bulk, $f_i^\alpha = 1$, then from pairs of equations (7) for the solvent and a dissolved species it follow the equations of state for the surface layer [45-47]

$$\Pi = -\frac{RT}{\omega_0}\left(\ln x_0^s + \ln f_0^s\right) \tag{8}$$

and the adsorption isotherm [47]

$$\ln \frac{f_i^s x_i^s}{K_i x_i^\alpha} = \frac{\omega_i}{\omega_0}\left(\ln x_0^s + \ln f_0^s\right) \tag{9}$$

Introducing the surface coverage instead of the molar fraction of the components, that is, setting $x_i^s \approx \theta_i = \Gamma_i \omega_i$, with $\theta_0 = 1 - \sum_{i \geq 1} \theta_i$, one can transform Eqs. (8) and (9) into

$$\Pi = -\frac{RT}{\omega_0}\left[\ln\left(1 - \sum_{i \geq 1} \theta_i\right) + \ln f_0^s\right] \tag{10}$$

$$K_i c_i = \frac{\theta_i f_i^s}{\left(1 - \sum_{i \geq 1} \theta_i\right)^{n_i} \left(f_0^s\right)^{n_i}} \tag{11}$$

where $n_i = \omega_i/\omega_0$, c_i are the bulk concentrations. The equation of state (10) was used in [45, 46] to analyse the adsorption of proteins at the solution/air interface. For this case Ter-Minassian-Saraga [46] showed the existence of an interrelation between protein denaturation processes within the surface and the activity of the solvent molecules (water), while Joos [45] demonstrated that the degree of surface denaturation decreases with the increase in surface pressure of the adsorption layer. Note that ω_0 was assumed in [45, 46] to be equal to the area of one water molecule, i.e. approximately 0.1 nm^2. If the values of f_i^s are independent of θ_i, then the adsorption isotherm Eq. (11) can be simplified by introducing a dimensionless concentration. Introducing the concentration $c_{1/2}$ which corresponds to 50 % saturation of the adsorption layer, Lucassen-Reynders [48] obtained an adsorption isotherm for single dissolved species

$$\frac{c}{c_{1/2}} = \frac{2\theta_1}{\left[2\left(1-\theta_1\right)\right]^{\omega_1/\omega_0}} \tag{12}$$

Eq. (12) follows from the general relation (11) not only for f_i^s = const., but also for the case of non-ideal entropy of the surface layer [48] which is due to the difference in the geometry of the protein and the solvent molecules. The calculations of the surface coverage as a function of c for various ratios ω_1/ω_0 performed in [48] showed that, in contrast to the Langmuir isotherm ($\omega_1 = \omega_0$), Eq. (12) predicts a dependence which is often observed for protein solutions - a sharp increase of the adsorption within a very narrow concentration range. If $\omega_1 \neq \omega_0$, then $x_1 \neq \theta_1$, and therefore Eq. (11) follows from Eq. (9) as an approximation only. One can however identify the molar fraction of the component within the surface, which enters Eqs. (5) and (7), with the portion of the surface area occupied by the molecules of different ω_i. This method, which is similar to that employed by Flory [49] and Huggins [50] for the analysis of the behaviour of polymers in the solution bulk, was first applied to surface layers by Singer [14]. Lucassen-Reynders [48] also used this concept in the analysis of the dependence of the surface layer entropy of polymer solution on the size of molecules which are mixed together. The following expression for the activity coefficient of a component in a mixture was obtained

$$\ln f_k^{sE} = 1 - n_k \sum_i (\theta_i / n_i) \qquad (13)$$

Here the superscript 'E' denotes that the non-ideality is due to the entropy of mixing. For the case of only one dissolved species, using Eq. (13) one can transform the equation of state (10) into (cf. [48])

$$\Pi = -\frac{RT}{\omega_0}\left[\ln(1-\theta_1) + \left(1 - \frac{\omega_0}{\omega_1}\right)\theta_1\right] \qquad (14)$$

Comparison of Eq. (1) and (14) shows that if only the first term of the expansion of $\ln(1-2\theta_1/z)$ is retained in Eq. (1), (this is valid as the co-ordination number z usually ranges between 4 and 6 [51]), then Eq. (1) transforms directly into Eq. (14). It was noted above that the formation of loops results in a small term in Eq. (4). Therefore, to give an approximate estimate for the non-ideality of the surface layer, it is sufficient to account only for the size of protein molecules and the solvent.

One general remark has to be made with respect to the models derived in [45, 46, 48]. It was assumed there that ω_0 is close to the area of a water molecule, and therefore the adsorption of a protein molecule leads to the desorption of $n_0 = \omega_1/\omega_0$ water molecules. This is however true only when the adsorption layer is comprised of water molecules, thus the adsorption layer thickness is ≈0.3 nm. Real protein adsorption layers are much thicker. Moreover, their thickness increases with the adsorption of proteins. Thus, from the theoretical point of view, the procedure employed in [43], where the real thickness of the protein layer was taken into account, seems to be more reliable. In this case the portion of water molecules within the surface increases, while the number of desorbed water molecules per one protein molecule becomes significantly larger than ω_1/ω_0. Clearly one has to introduce the corrections into Eqs. (10) and (11) for the case $\omega_1 \neq \omega_0$ to account for the finite thickness of the adsorption layer. In [12, 13, 47] it was proposed to choose the dividing surface and to define the surface coverage θ_i in such a way that the mean molar surface area for all dissolved components becomes equal to the mean surface area of the solvent. This choice leads to the exclusion of the adsorption layer thickness from further consideration and Eqs. (10) and (11) become rigorous for any size of dissolved molecules.

In a thermodynamic study by Joos and Serrien [9] it was shown that protein denaturation within the surface is more complete at low surface pressure. If the protein molecule possesses say two modifications, 1 and 2, with different partial molar surface areas ω_i, then the ratio of the molar fractions of these modifications within the surface layer obeys the equation (cf. [9])

$$\frac{x_2^s}{x_1^s} = K_{12} \exp\left[\frac{\Pi(\omega_1 - \omega_2)}{RT}\right] \tag{15}$$

where K_{12} is a constant. If $\omega_1 > \omega_2$, with increasing Π the concentration of modification 2 within the surface layer increases. Equation (15) is the analytical expression for the general physico-chemical principle of Braun-Le Châtelier, applied to the adsorption layer of proteins. It shows, in complete agreement with experimental data, that the surface pressure acts as a self-regulation mechanism of the thickness of a protein adsorption layer. Eq. (15) follows from a pair of equations (7) when two different modifications of the same protein are taken instead of two components. The concept developed by Joos and Serrien was generalised for the case of arbitrary number of different modifications of a protein molecule in the surface layer [13] (discrete model), and for the case of continuous variations of the states of adsorbed molecules (continuity of states). In the following paragraphs the results of these studies [12, 13, 47] will be presented in more details.

3. THEORY OF EQUILIBRIUM ADSORPTION OF PROTEINS

3.1. SURFACE LAYER MODEL

The relations (8) and (9), and also (10) and (11) are the most general form for the equation of state of a surface layer and the adsorption isotherm at liquid/fluid interfaces. To particularise to any special case, one has to define the molar surface area of the solvent ω_0 and the activity coefficients for the solvent f_0^s and dissolved species f_i^s within the surface. It was shown above that this can be done in the framework of various models of the surface layer.

Let us consider first that only one dissolved species exists in one state, assuming that the surface layer is ideal. For this case and at $\omega_0 = \omega_1$ the Eqs. (10) and (11) can be transformed into the well-known equation of von Szyszkowski

$$\Pi = \frac{RT}{\omega_1}\ln(1+bc) \tag{16}$$

Here b is an adsorption equilibrium constant. To comply with experimental data, one has to employ such surface layer model, for which the molar surface area of the solvent in Eqs. (10) and (11) is equal to the molar surface area of the surfactant. It was shown by Lucassen-Reynders [52, 53] that this requirement can be satisfied if one chooses the position of the dividing surface such that the total adsorption of the solvent and surfactant are equal to $1/\omega_1$

$$\Gamma_0 + \Gamma_1 = 1/\omega_1 \tag{17}$$

For $\Gamma_1 = 0$ the Lucassen-Reynders' dividing surface is shifted towards the solution bulk by the distance $\Delta = (\omega_0/\omega_1)\cdot d_{H_2O}$ as compared to the Gibbs' dividing surface. d_{H_2O} is the diameter of a water molecule. For a saturated monolayer ($\Gamma_1 = 1/\omega_1$) however, these two surfaces coincide with each other. Note that for proteins ($\omega_1 >> \omega_0$) the Δ becomes negligibly small, and therefore for any adsorption the Lucassen-Reynders' dividing surface coincides with the Gibbs' dividing surface.

For surfactant mixtures or single molecules having several adsorption states within the surface the corresponding values of ω_i differ and the definition of the dividing surface transforms into the more general relation

$$\sum_{i=0}^{n}\Gamma_i = 1/\omega_\Sigma \tag{18}$$

where ω_Σ is the mean partial molar area of all surfactants and/or all states at the interface. To define ω_Σ the following relation can be used

$$\omega_\Sigma = \left(\sum_{i\geq 1}\Gamma_i\omega_i\right) \Big/ \left(\sum_{i\geq 1}\Gamma_i\right) \tag{19}$$

The form of equation (19) differs from analogous relations proposed elsewhere for mixed monolayers [53-55], however the general principle in all cases is the same - the contribution of various ω_i to ω_Σ depends on the adsorptions for individual components or states. The advantage of Eq. (19) over other relationships is the invariance of resulting the Eqs. (10) and (11) on the definition of the coverage θ_i. As the total adsorption observed in the experiments is

$$\Gamma_\Sigma = \sum_{i=1}^{n} \Gamma_i \tag{20}$$

the value of $\sum_{i\geq 1}^{n} \theta_i$ in Eqs. (10) and (11) can be substituted by $\Gamma_\Sigma \omega_\Sigma$, because

$$\omega_\Sigma = \frac{\sum_{i=1}^{n} \omega_i \Gamma_i}{\Gamma_\Sigma} \tag{21}$$

This substitution means that also for each component (or state) the relation $\theta_i = \Gamma_i \omega_\Sigma$ holds. As the value of ω_Σ for all components or states is the same, $\Gamma_i \omega_\Sigma$ is the molar portion of the component (state) within the surface layer. Therefore the transformation from Eqs. (8) and (9) to Eqs. (10) and (11) by replacing x_i^s through θ_i is rigorous. Another important advantage following from the location of the dividing surface according to Eq. (18) and ω_Σ according to Eq. (19) is that there is no contribution of the non-ideality of entropy of mixing, Eq. (13), into the solvent activity coefficient. When the relation $\omega_0 = \omega_\Sigma$ holds, then from Eq. (13) it follows that $f_0^{sE} = 0$. And finally, using the dividing surface definition of Lucassen-Reynders the adsorption layer thickness can be excluded from the consideration. In this approach the actual number of water molecules displaced from the adsorption layer during the adsorption of protein molecule needs not to be accounted for.

3.2. IDEAL ADSORPTION LAYER

Let us consider first that a protein molecule can exist in the surface layer in i different states with different ω_i. The adsorption layer is assumed to be ideal so that $f_i^s = 1$ ($i \geq 0$) can be used in Eqs. (10) and (11) and Eq. (10) transforms into an equation of state for the surface layer

$$\Pi = -\frac{RT}{\omega_\Sigma} \ln(1 - \Gamma_\Sigma \omega_\Sigma) \tag{22}$$

The corresponding surface tension isotherm, let us call it the generalised Joos equation [56], can be obtained by the summation over all equations (7) for all states and components (i.e. from i = 0 to i = n)

$$c\sum_{i=1}^{n} b_i \exp\left(-\frac{\Pi\omega_i}{RT}\right) = 1 - \exp\left(-\frac{\Pi\omega_\Sigma}{RT}\right) \tag{23}$$

Here b_i are the adsorption equilibrium constants for ith state of the protein molecule. The adsorption isotherm for ith state can be derived from Eq. (11)

$$b_i c = \frac{\Gamma_i \omega_\Sigma}{\left(1 - \Gamma_\Sigma \omega_\Sigma\right)^{\omega_i/\omega_\Sigma}} \tag{24}$$

The analytical expression for the ratio of the adsorptions can be derived from the Eq. (7) for the ith and j^{th} state

$$\frac{\Gamma_i}{\Gamma_j} = \frac{b_i}{b_j} \exp\left[\frac{\Pi\left(\omega_j - \omega_i\right)}{RT}\right] \tag{25}$$

The Eqs. (23) to (25) contain the adsorption equilibrium constants for all states of the protein molecule within the surface layer. In the first approximation one can assume that $b_i = b = \text{const}$. If the states with larger ω_i values possess larger surface activity, then it would be reasonable from the physical point of view to include this dependence into Eqs. (23) - (25). Here it is convenient to express all b_i via the adsorption equilibrium constant corresponding to one of the states, say b_1, by a power law

$$\frac{b_i}{b_1} = \left(\frac{\omega_i}{\omega_1}\right)^\alpha \tag{26}$$

where α is a constant. Eq. (26) resembles the dependence of the adsorption equilibrium constant in a homologous series of a surfactant type, for which an exponential dependence of b_i on the chain length exists

$$\frac{b_k}{b_1} = e^{\beta(k-1)} \tag{27}$$

where k and l are the numbers of CH_2 groups in the surfactant molecule, β is a constant for the homologous series.

Now the ω_i in all states with $i > 1$ can be expressed via the molar surface area of the molecule in the first state (assuming that this surface area in state 1 is the minimum) and an increment $\Delta\omega$

$$\omega_i = \omega_1 + \Delta\omega(i-1) \tag{28}$$

The resulting equations can be simplified for the case when they are expressed in terms of ω_1 and the increment Δi, that is:

$$\omega_i = i\omega_1 \tag{29}$$

where i can be either integer or fractional with $\Delta i = \Delta\omega/\omega_1$. Therefore the value of i in Eqs. (23) - (24) can vary from 1 to $n = \omega_{max}/\omega_1$, with $\omega_1 = \omega_{min}$,. Using Eq. (28), Eq. (26) transforms into

$$b_i = b_1 i^{\alpha} \tag{30}$$

For $\alpha = 0$ one obtains $b_i = b_1 = const$, while for $\alpha > 0$ the values of b_i increase with increasing ω_i. Eqs. (25), (29) and (30) allow to express the mean molar surface area in Eq. (21) via ω_1

$$\omega_\Sigma = \omega_1 \frac{\sum_{i=1}^{n} i^{(\alpha+1)} \exp\left(-\frac{i\Pi\omega_1}{RT}\right)}{\sum_{i=1}^{n} i^{\alpha} \exp\left(-\frac{i\Pi\omega_1}{RT}\right)} \tag{31}$$

The total adsorption of Eq. (20) can be expressed via Γ_1

$$\Gamma_\Sigma = \Gamma_1 \sum_{i=1}^{n} i^{\alpha} \exp\left[-\frac{(i-1)\Pi\omega_1}{RT}\right] \tag{32}$$

and the adsorption in the ith state via the total adsorption

$$\Gamma_i = \Gamma_\Sigma \frac{i^{\alpha} \exp\left[-\frac{(i-1)\Pi\omega_1}{RT}\right]}{\sum_{i=1}^{n} i^{\alpha} \exp\left[-\frac{(i-1)\Pi\omega_1}{RT}\right]} \tag{33}$$

From Eqs. (26) and (19) and the equations for the surface tension (23) and adsorption isotherm (24) we obtain

$$b_1c = \frac{1-\exp\left(-\frac{\Pi\omega_\Sigma}{RT}\right)}{\sum_{i=1}^{n} i^\alpha \exp\left(-\frac{i\Pi\omega_1}{RT}\right)} \tag{34}$$

$$b_1c = \frac{\Gamma_i\omega_\Sigma}{i^\alpha\left(1-\Gamma_\Sigma\omega_\Sigma\right)^{i\omega_1/\omega_\Sigma}} \tag{35}$$

Equations (22), (31) - (35) constitute the complete description of an ideal protein adsorption layer. Some conclusions can be drawn from these equations. For large Π from Eq. (33) it follows that, the portions of the states with $i > 1$ decrease, and the main state is state 1. In that case $\omega_\Sigma \rightarrow \omega_1 = \omega_{min}$ and the adsorption layer thickness δ attains its maximum value

$$\delta_{max} \cong \frac{V}{\omega_1} \tag{36}$$

with V - molar volume of a protein molecule. In the opposite case, when Π is very small ($\Pi \rightarrow 0$) it follows from Eq. (31) that

$$\omega_{\Sigma 0} = \omega_\Sigma\Big|_{\Pi\rightarrow 0} = \omega_1\left(\sum_{i=1}^{n} i^{\alpha+1} \Big/ \sum_{i=1}^{n} i^\alpha\right) \tag{37}$$

leading for $\alpha = 1$ to

$$\omega_{\Sigma 0} = \omega_1\left(\frac{2n+1}{3}\right) \tag{38}$$

Thus, at very low surface pressures the mean partial molar area for all possible states of the protein molecule is approximately equal to ω_{max}, which can correspond to the surface area of a completely unfolded (denatured) protein molecule. In this case the adsorption layer thickness is minimal

$$\delta_{min} = \frac{3}{2}\frac{V}{\omega_{max}} \tag{39}$$

For $\Pi \to 0$ the surface layer is composed mainly of molecules in states with large ω_i. For example, from Eq. (33) and $\alpha = 1$ it follows that

$$\Gamma_{i0} = \Gamma_\Sigma \frac{2i}{n(n+1)} \tag{40}$$

and the adsorption in the state i = n is approximately n times larger than that in state 1.

The general case of an ideal adsorption layer formed by i different surfactants, each capable to possess j states, was considered in [12].

3.3. NON-IDEAL ADSORPTION LAYER

Equation (13) expresses the activity coefficient for an adsorbed protein molecule, assuming that the entropy of mixing is non-ideal, or, more precisely, that the molar surface areas of the solvent and protein are different. When the enthalpy of mixing is non-ideal, i.e. the intermolecular interaction plays a role, the activity coefficient can be calculated from the regular solution theory [57 - 59]

$$RT \ln f_k^{sH} = \sum_i \sum_j \left(A_{ik}^s - \frac{1}{2} A_{ij}^s \right) \theta_i \theta_j \tag{41}$$

where $A_{ij}^s = U_{ii}^s + U_{jj}^s - 2U_{ij}^s$, U_{ii}^s and U_{ij}^s are the energies of interaction between the molecules of the same species and different species, respectively. From additivity of the contributions it follows that

$$\ln f_i^s = \ln f_i^{sH} + \ln f_i^{sE} \tag{42}$$

For the solution of two surfactants or one surfactant possessing two different states, from Eqs. (10), (11) (13), (41) and (42) one obtains

$$\Pi = -\frac{RT}{\omega_0}\left[\ln(1-\theta_1-\theta_2) + \theta_1\left(1-\frac{1}{n_1}\right) + \theta_2\left(1-\frac{1}{n_2}\right) + a_1\theta_1^2 + a_2\theta_2^2 + a_{12}\theta_1\theta_2 \right] \tag{43}$$

$$b_i c_i = -\frac{\theta_i}{(1-\theta_1-\theta_2)^{n_i}} \exp(-2a_i\theta_i - 2a_{12}\theta_j) \cdot \exp[(1-n_i)(a_1\theta_1^2 + a_2\theta_2^2 + a_{12}\theta_1\theta_2)] \tag{44}$$

where $a_1 = A_{01}$; $a_2 = A_{02}$; $a_{12} = (A_{01} + A_{02} - A_{12})/2$; $b_i = K_i \exp(n_i - a_i - 1)$; $i = 1,2$; $j = 1,2$ $(j \neq i)$. Clearly, if Eq. (44) describes the case of two states, then $c_1 = c_2 = c$. Note that the equation of state (43) coincides with that derived by Damaskin [60]. Moreover, many of the known equations of state and adsorption isotherms (cf. review [61]) are particular cases of Eqs. (43) and (44). However these equations are of minor practical importance, because even for the most simple case of two surfactants or two states they depend on too many parameters (n_1, n_2, b_1, b_2, a_1, a_2 and a_{12}).

To simplify the problem, we assume that the non-ideality of enthalpy of the surface layer is independent of the adsorption state, and is defined by total adsorption only. Then for the present surface layer model of Eq. (41) relations can be derived which describe the contribution of the non-ideality of enthalpy of mixing

$$\ln f_0^{sH} = a\Gamma_\Sigma^2 \omega_\Sigma^2, \tag{45}$$

$$\ln f_i^{sH} = a\left(1 - \Gamma_\Sigma \omega_\Sigma\right)^2, \qquad i \geq 1 \tag{46}$$

where a is a intermolecular interaction constant. From (13) the contribution of the entropy of mixing can be estimated

$$\ln f_i^{sE} = 1 - \omega_i \sum_{j\geq 0} \Gamma_j = 1 - n_i, \qquad i \geq 1 \tag{47}$$

$$\ln f_0^{sE} = 1 - \omega_0 \sum_{j\geq 0} \Gamma_j = 0 \tag{48}$$

Assuming that the difference in the values of b_i is due to the non-ideality of entropy of mixing only, i.e. setting $K_i = K =$ const, and using (45)-(48) one can transform (10) and (11) into

$$\Pi = -\frac{RT}{\omega_\Sigma}\left[\ln\left(1 - \Gamma_\Sigma \omega_\Sigma\right) + a\left(\Gamma_\Sigma \omega_\Sigma\right)^2\right] \tag{49}$$

$$bc = \frac{\Gamma_i \omega_\Sigma}{\left(1 - \Gamma_\Sigma \omega_\Sigma\right)^{n_i}} \exp\left(-n_i\right) \exp\left[-2a\Gamma_\Sigma \omega_\Sigma + a\left(1 - n_i\right)\left(\Gamma_\Sigma \omega_\Sigma\right)^2\right] \tag{50}$$

with b = K exp(-a - 1). Eq. (25) now reads

$$\frac{\Gamma_i}{\Gamma_j} = \exp\left(\frac{\omega_i - \omega_j}{\omega_\Sigma}\right) \cdot \exp\left(\frac{(\omega_j - \omega_i)\Pi}{RT}\right) \tag{51}$$

Comparing Eqs. (25) and (51), one can see that the pre-exponential term b_i/b_j in Eq. (25) is replaced by the factor $\exp\left[(\omega_i - \omega_j)/\omega_\Sigma\right]$ in Eq. (51). Depending on ω_i, this factor corresponds approximately to that introduced earlier in form of the coefficient α in Eq. (26) which varies between 0 and 1. Therefore the consideration of the non-ideality of entropy of mixing compensates in part the effect introduced by the differences in the partial molar area of different states on the adsorption activity. However, the approach to define the ratio b_i/b_j by relation (30) as the pre-exponential factor is more general. In addition, the presence of ω_Σ in the exponential factor makes the adsorption equation and respective equations more comprehensive.

3.4. INFLUENCE OF THE ELECTRIC CHARGE

Proteins are typically polyelectrolytes, i.e. they contain ionised groups. At the isoelectric point both hydroxyl groups and amino groups possess equal degree of ionisation, and thus the whole molecule is electroneutral. In strong acidic medium the hydroxyl groups become neutral and the molecule acquires an excess of positive charges, while the neutralisation of the amino groups in strong alkaline media results in a transformation of the protein molecule into an anion. Therefore the maximum total charge of a protein molecule in acidic or alkaline media can be equal to the number of amino-acid residues, while at the isoelectric point the charge can be equal to the total number of all groups. The charges in the protein molecule are more or less bound by counterions. A polyelectrolyte molecule in a semi-dilute solution can be regarded as a random walk of electrostatic blobs [37]. For polyelectrolytes the blob charge not bound by counterions can usually amount to several units. It can be assumed that at the isoelectric point the charge of different blobs possess opposite signs. The total number of blobs can be rather high so that the entire protein molecule appears electroneutral. As the degree of counterion bounding both in separate blobs and in the whole protein molecule amounts to 90 %, (which is close to the corresponding value characteristic for ionic micelles), thus the number of unbounded charge units of a protein molecule remains sufficiently large, tens or hundreds units.

The interaction between unbounded charges has to result in strong repulsion between polyelectrolyte chains. This effect was confirmed experimentally [62].

Davies [63] derived an adsorption isotherm and a surface equation of state for charged surfactant molecules based on the Gouy-Chapman theory. The same electric double layer (DEL) model was used by Borwankar and Wasan [64], where non-ideality of the surface layer was taken into account. Combining the results obtained in [64] with Eqs. (45), (47), and the condition $\omega_0 = \omega_\Sigma$, one can transform the surface equation of state Eq. (10) into

$$\Pi = -\frac{RT}{\omega_\Sigma}\left[\ln\left(1-\Gamma_\Sigma\omega_\Sigma\right) + a\left(\Gamma_\Sigma\omega_\Sigma\right)^2\right] + \frac{4RT}{F}\left(2\varepsilon RTc_\Sigma\right)^{1/2}\left[\mathrm{ch}\varphi - 1\right] \tag{52}$$

where F is the Faraday constant, ε is the dielectric permittivity of the medium, c_Σ the total concentration of ions within the solution, $\varphi = zF\psi/2RT$, z is the number of non-bound unit charges in the protein molecule, ψ is the electric potential. Substituting the chemical potential by the electrochemical potential [64], the following expression can be obtained instead of the adsorption isotherm (11)

$$K_i c = \frac{\theta_i f_i^s}{\left(1-\sum_{i\geq 1}\theta_i\right)^{n_i}\left(f_0^s\right)^{n_i}}\exp(2\varphi) \tag{53}$$

The electrical potential is given by surface charge density

$$\mathrm{sh}\varphi = \frac{z\Gamma_\Sigma F}{\left(8\varepsilon RTc_\Sigma\right)^{1/2}} \tag{54}$$

Analysis of Eq. (52) shows that for 1:1 ionic surfactants at low ion bulk concentration the approximate relation $\varphi >> 1$ is valid [65]. This approximation leads to the linear dependence of Π on Γ_Σ in the electrostatic term of Eq. (52). For protein solution the situation is quite different. If the concentration of ions is high, the Debye length $\lambda = (\varepsilon RT/F^2 c_\Sigma)^{1/2}$ is small. For example, for $c_\Sigma = 0.1$ mol/l a value of $\lambda = 1.3$ nm results. This means that for protein solutions the DEL thickness can be smaller than the adsorption layer thickness. Therefore the concentration of ions in Eqs. (52) and (54) is just their concentration within the adsorption layer, which can exceed a concentration of 1 mol/l due to the ionisation of hydroxyl and amino groups, and the contribution of counterions. It follows from Eqs. (52) and (54) that for large c_Σ the approximation $\varphi \ll 1$ can be used. Thus, using the series expansions

$$\mathrm{sh}\varphi = \varphi + \varphi^3/3! + \dots \tag{55}$$

$$\mathrm{ch}\varphi = 1 + \varphi^2/2! + \dots, \tag{56}$$

one obtains an equation of state for non-ideal charged protein surface layers

$$\Pi = -\frac{RT}{\omega_\Sigma}\left[\ln(1-\Gamma_\Sigma\omega_\Sigma) + (a - a_{el})(\Gamma_\Sigma\omega_\Sigma)^2\right] \tag{57}$$

where $a_{el} = z^2F/\omega_\Sigma(8\varepsilon RTc_\Sigma)^{1/2}$. From (53), (55), (56), and the relation

$$\exp(2\varphi) = (\mathrm{ch}\varphi + \mathrm{sh}\varphi)^2 \tag{58}$$

one obtains a protein adsorption isotherm for any ith state of the molecule

$$b_1 c = \frac{\Gamma_i\omega_\Sigma \exp\left[-a\Gamma_\Sigma^2\omega_\Sigma^2\left(i\frac{\omega_1}{\omega_\Sigma}-1\right) - i\frac{\omega_1}{\omega_\Sigma} - 2a\Gamma_\Sigma\omega_\Sigma + 2\frac{a_{el}}{z}\Gamma_\Sigma\omega_\Sigma + \left(\frac{a_{el}}{z}\Gamma_\Sigma\omega_\Sigma\right)^2\right]}{i^\alpha(1-\Gamma_\Sigma\omega_\Sigma)^{i\omega_i/\omega_\Sigma}} \tag{59}$$

To analyse its effect an estimation of the electrostatic constant a_{el} is necessary. Assuming $\omega_\Sigma = 6\cdot 10^6$ m^2/mol (i.e., 10 nm^2 per protein molecule) and $z = 50$, for $c_\Sigma = 10$ mol/l one obtains $a_{el} = 100$. In the Flory-Huggins-theory of amorphous polymer solutions [49, 50] the constant a (Flory's parameter χ) is of the order of unity. Thus one can neglect $a = \chi$ in Eq. (57) as compared to a_{el}. Moreover, for high a_{el} values the logarithmic term in Eq. (57) can also be neglected. This leads to the approximate relation $\Pi \approx \Gamma_\Sigma^2$ which agrees well with the scaling theory, resulting to $\Pi \approx \Gamma_\Sigma^{9/4}$ [43]. These simplifications enable one to transform the equation of state for proteins (57) into

$$\Pi = -\frac{RT}{\omega_\Sigma}\left[\ln(1-\Gamma_\Sigma\omega_\Sigma) - a\Gamma_\Sigma^2\omega_\Sigma^2\right] \tag{60}$$

The constant a in Eq. (60) and in the following is given $a_{el} - \chi \cong a_{el}$. Formally Eq. (60) differs from Eq. (49) only by the sign of a.

The consideration of the molecular charge leads to a significant simplification of adsorption isotherm (59). First, as for all states of protein molecule the constant b_1 is the same, it is convenient to assume $i = 1$ in Eq. (59). Moreover, let us recall that $\omega_1 \leq \omega_\Sigma$. To simplify Eq. (59) is important to note that the constants $a = \chi$ and a_{el}/z are of the same order with opposite signs. It will be shown below that for large a_{el} the surface layer coverage $\Gamma_\Sigma\omega_\Sigma$ remains low even for large Π. These considerations also show that the exponential term in Eq. (59) is close to unity and is independent of $\Gamma_\Sigma\omega_\Sigma$. Therefore the exponential term can be omitted so that additionally two insufficiently defined parameters (z and a) disappear as well. Eq. (59) transforms into Eq. (35), where the non-ideality of entropy of mixing is given by the constant α. The Eqs. (31)-(33) remain their importance also in the case of non-ideal ionised proteins in the surface layer.

3.5. CONTINUITY OF STATES

The adsorption model presented above assumes discrete states of the protein molecule in the surface layer. Neighbouring states differ from each other by the molar area increment $\Delta\omega$, given by Eq. (28). From the viewpoint of scaling theory, the value of $(\Delta\omega)^{1/2}$ has to be close to the size of the electrostatic blob. As it was mentioned in paragraph 2.2, in a protein adsorption layers the flexibility of chains increases at high concentrations of both protein and inorganic electrolyte. This enables one in some cases to consider, instead of a model with discrete states, that neighbouring states differ only by an infinitesimal value $d\omega$. One of the advantages of this continuity model is that $\Delta\omega$ can be excluded from the list of parameters which describe the adsorption of the protein. Using the increment in the form $\Delta\omega = \omega_1$ leads also to the elimination of $\Delta\omega$. To perform the transformation from the discrete to the continuous model to the summation in Eqs. (31)-(33) has to be replaced formally by integration. For example, the summations in Eq. (31):

$$\sum_{i=1}^{n} i^{\alpha} \exp\left(-\frac{i\Pi\omega_1}{RT}\right) \qquad (61)$$

and

$$\sum_{i=1}^{n} i^{(\alpha+1)} \exp\left(-\frac{i\Pi\omega_1}{RT}\right) \tag{62}$$

transform into a corresponding integral

$$\frac{1}{\omega_{max}-\omega_1}\int_{\omega_1}^{\omega_{max}}\left(\frac{\omega}{\omega_1}\right)^{\alpha}\exp\left(-\frac{\Pi\omega}{RT}\right)d\omega \tag{63}$$

and

$$\frac{1}{\omega_{max}-\omega_1}\int_{\omega_1}^{\omega_{max}}\left(\frac{\omega}{\omega_1}\right)^{\alpha+1}\exp\left(-\frac{\Pi\omega}{RT}\right)d\omega \tag{64}$$

The sum in Eqs. (32) and (33) becomes the integral

$$\frac{1}{\omega_{max}-\omega_1}\int_{\omega_1}^{\omega_{max}}\left(\frac{\omega}{\omega_1}\right)^{\alpha}\exp\left[-\frac{\Pi(\omega-\omega_1)}{RT}\right]d\omega \tag{65}$$

In some cases simple analytical expressions can be obtained instead of cumbersome Eqs. (31)-(33). For example, for $\alpha = 0$ and $\omega_{max} >> \omega_1$ (in fact, $\omega_{max} \rightarrow \infty$), from (63) and (64) instead of Eq. (31) one obtains

$$\omega_{\Sigma} = \omega_1\left(1+\frac{RT}{\Pi\omega_1}\right) \tag{66}$$

For non-zero Π the results obtained from Eq. (66) agree well with values calculated from Eq. (31). For the continuous model for $\alpha = 0$ instead of Eq. (32) one obtains

$$\Gamma_{\Sigma} = \frac{RT\Gamma_1}{\Pi(\omega_{max}-\omega_1)}\left[1-\exp\left(-\frac{\Pi(\omega_{max}-\omega_1)}{RT}\right)\right] \tag{67}$$

To illustrate the proposed theory of protein adsorption, and to compare the results with experimental data, numerical calculations of Eqs. (31)-(32), (35), (60) and (59) have been performed. The transition from the discrete to the continuous model was arranged by decreasing the increment $\Delta\omega$ until the results become independent of $\Delta\omega$. As an example, the dependence of the surface pressure of a protein solution (molecular mass $M = 24000$ g/mol, $\omega_1 = 2$ nm^2, $\omega_{max} = 60$ nm^2) on the adsorption layer coverage $\theta = \Gamma_\Sigma \cdot \omega_\Sigma$ is shown in Fig. 1.

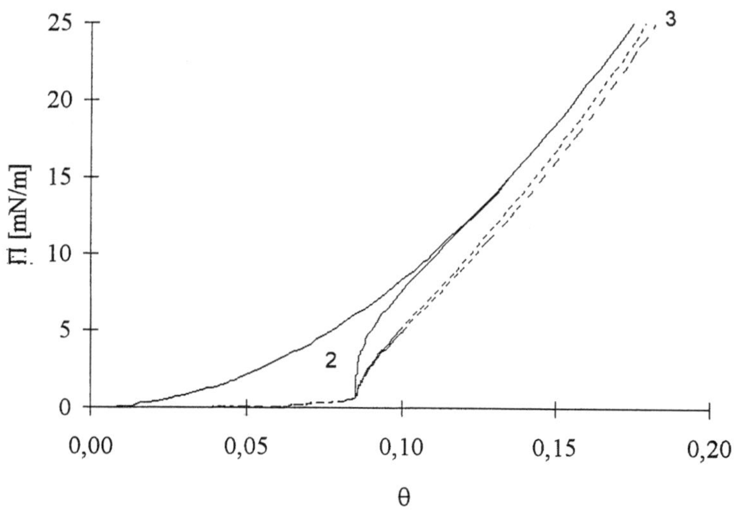

Fig. 1. Dependence of surface pressure on the adsorption layer coverage, calculated for a protein solution with $M = 24000$, $\omega_1 = 2$ nm^2 and $\omega_{max} = 60$ nm^2, $\alpha = 2$ and $a = 400$. Curve 1 calculated for $n = 1$, curve 2 - for $\Delta\omega = \omega_1$, curve 3 - for $\Delta\omega = 0.1\omega_1$, curve 4 - continuous model ($\Delta\omega < 0.02\omega_1$).

Unless otherwise explicitly specified, in the following numerical values of the partial surface area ω refer to one molecule. One can see that for $\Delta\omega = \omega_1$ the curve differs from that calculated for the continuous model. However, already for $\Delta\omega = 0.1\omega_1$ the results for both models are close to each other. If $\Delta\omega = \omega_1$, then from Eq. (33) it follows that for $\Pi > 10$ mN/m all states with $i > 1$ disappear and the curve $\Pi = \Pi(\theta)$ coincides with that calculated for $n = 1$, that is, when only one state with $\omega_i = \omega_1$ exists. For this case Eq. (60) transforms into the equation of state formulated by Frumkin [60].

3.6. GENERALISATION OF LUCASSEN-REYNDERS' THEORY

Lucassen-Reynders [48] considered the difference between molar areas of solvent and protein molecules; however only one state of the protein molecule within the adsorption layer was assumed. This model can be generalised for the case of i different states of a protein molecule in the adsorption layer. For the case of non-ideal surface layer entropy, Eq. (13), the equation of state (10) becomes

$$\Pi = -\frac{RT}{\omega_0}\left[\ln\left(1-\sum_{i\geq 1}\Gamma_i\omega_i\right)+\sum_{i\geq 1}\Gamma_i\omega_i\left(1-\frac{\omega_0}{\omega_1}\right)\right] \tag{68}$$

As shown in [48], ω_0 is close to the area of a monomer or a water molecule. Using Eq. (13), one can transform the adsorption isotherm of the state j Eq. (11) into

$$b_j c = \frac{\Gamma_j\omega_j}{\left(1-\sum_{i\geq 1}\Gamma_i\omega_i\right)^{\omega_j/\omega_0}} \tag{69}$$

where $b_j = K_j \exp(n_j - 1)$, $n_j = \omega_j/\omega_0$. If K_j = const for all states, then the ratio of the adsorptions in different states can be expressed by

$$\frac{\Gamma_i}{\Gamma_j} = \exp\left(\frac{\omega_i-\omega_j}{\omega_0}\right)\cdot\exp\left(-\frac{(\omega_j-\omega_1)\Pi}{RT}\right) \tag{70}$$

And finally, using expressions given above, one obtains the total degree of surface coverage

$$\theta = \sum_{i}^{n}\Gamma_i\omega_i = \sum_{i=1}^{n}\omega_i\exp\left(\frac{\omega_i-\omega_1}{\omega_0}\right)\cdot\exp\left(\frac{(\omega_1-\omega_i)\Pi}{RT}\right) \tag{71}$$

As the values of K_j were assumed to be equal for all states, the adsorption equilibrium constants for different states can be expressed by one constant, say, that for state 1 possessing minimum surface area ω_1

$$b_j = b_1\exp(n_j - n_1) \tag{72}$$

The protein adsorption can be described by one isotherm only, instead of the set of j adsorption equations. For example, in state 1

$$b_1 c = \frac{\theta\omega_1}{(1-\theta)^{\omega_1/\omega_0}\sum_{i=1}^{n}\omega_i \exp\left(\frac{\omega_i-\omega_1}{\omega_0}\right)\exp\left(\frac{(\omega_1-\omega_i)\Pi}{RT}\right)} \tag{73}$$

In agreement with the adsorption isotherm originally derived by Lucassen-Reynders [48], Eq. (73) also predicts a steep dependence of θ on c for small Π. With increasing Π, however, the slope of this dependence decreases, because the states with larger ω_i are excluded from the adsorption layer.

3.7. ANALYSIS OF THE EFFECT OF MAIN PARAMETERS

Let us consider now the influence of the different thermodynamic parameters of the equation of state (60) and adsorption isotherms (35) or (59) on the equilibrium protein adsorption and the surface pressure. Fig. 2 shows the dependence of surface pressure on adsorption calculated from Eq. (60) for the case of a single molecular surface state, i.e., for n = 1, for the discrete and continuous states models of the protein adsorption.

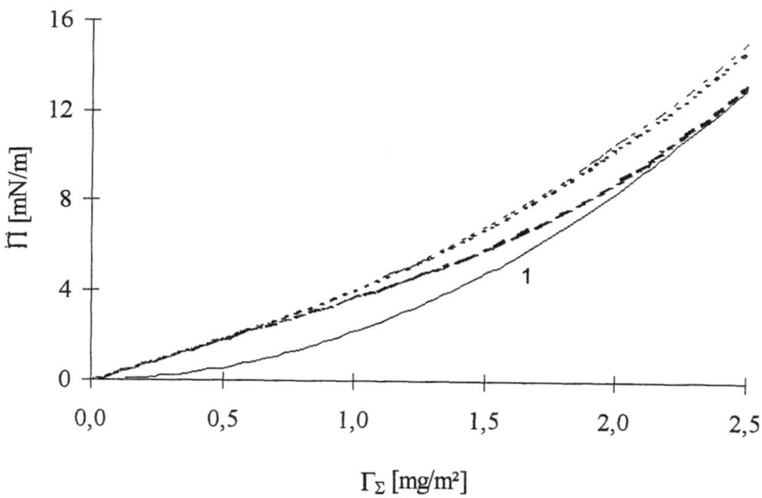

Fig. 2. Dependence of surface pressure on the protein adsorption. The values are the same as in Fig. 1.

One can see that for small Γ_Σ the continuous state model predict higher values of Π due to protein denaturation at the surface. High values of the increment $\Delta\omega = \omega_1$, which is characteristic to inflexible chains, do not affect the dependence of Π on Γ_Σ at small Γ_Σ, but for $\Gamma_\Sigma > 2$ mg/m^2 molecules in states with $i > 1$ disappear and curves 1 and 2 become indistinguishable. In further graphs the continuous state model will be used.

The effect of the coefficient α which accounts for the adsorption activities of different states and the non-ideality of entropy of the surface layer, is illustrated in Figs. 3 and 4. Here all curves are normalised with respect to b_1c such that for $\Pi = 25$ mN/m the value of b_1c is the same for each curve.

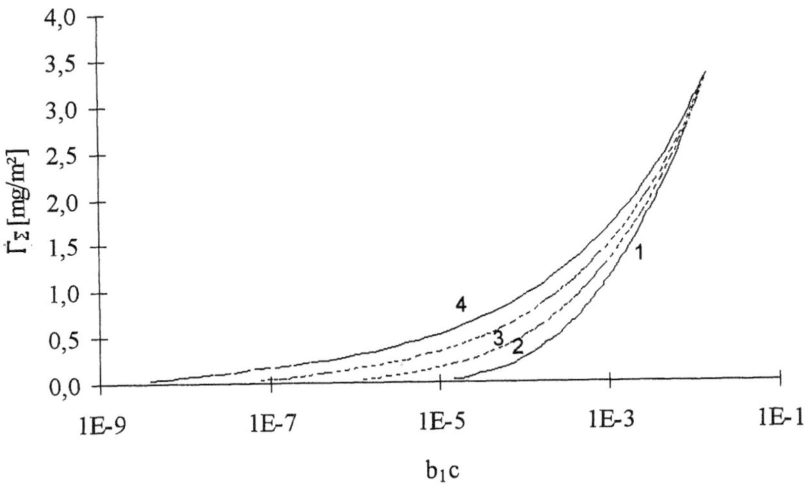

Fig. 3. Dependence of surface pressure on the reduced protein concentration b_1c ($M = 24000$, $\omega_1 = 2$ nm^2, $\omega_{max} = 60$ nm^2 and $a = 400$) for $\alpha = 3$ (curve 4), 2 (curve 3), 1 (curve 2) and 0 (curve 1).

An increase in α results in a remarkable adsorption and in an increase in surface pressure at very low protein concentrations. This is the consequence of a high adsorption activity of the states possessing high ω_i-values. In this case, however, there is no steep increase in Γ_Σ and Π within a narrow concentration range, as it is characteristic to proteins. It seems that the value of the coefficient α should not exceed 1. In other words, it should not exceed $\alpha \approx 0.5$, which corresponds to the contribution of the non-ideality of entropy. As the curves for $\alpha = 0$ and $\alpha = 1$ in Figs. 3 and 4 are hardly distinguishable from each another, one can assume a value of $\alpha = 0$ for proteins. This leads to a reduction in the number of parameters involved in Eqs. (60) and (35) or (59). Thus the adsorption of proteins in the framework of the continuous states model can be described by a set of four parameters: ω_1, ω_{max}, a and b_1. Note that in the framework of the discrete model for $\omega_1 = \Delta\omega$ the number of parameters also four.

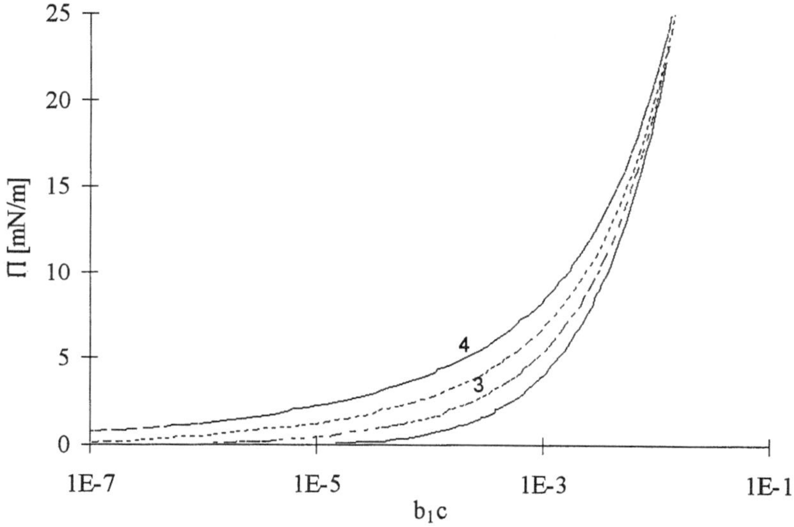

Fig. 4. Dependence of adsorption on b_1c. The values are the same as in Fig. 3.

The effect of the intermolecular repulsion constant is illustrated by Figs. 5 and 6. With decreasing values of a the dependence of Γ_Σ on b_1c becomes steeper, while the Π dependence on Γ_Σ, in contrast, becomes more pronounced with increasing a. This last effect can be explained also by the fact that the parameter a effects only the equation of state (60). For large values of a only minor effect on the dependence of Γ_Σ on b_1c occur when the isotherm (35) is used. At constant surface pressure the increase of a leads to a decrease of the adsorption due to stronger interion repulsion. As the adsorption of proteins in a fully covered adsorption layer is of the order of 3 to 8 mg/m^2, values of a in the range of 100 to 600 are quite realistic. Such values agrees with results obtained for ionic surfactants. For example, for sodium dodecyl sulphate in the presence of electrolyte (NaCl) the value of a (note the sign inversion made above) is 1.5 to 1.8 [65]. It is clear that for proteins with a number of charged groups tens or hundreds times larger, the value of a should increase in the same order of magnitude. The intermolecular repulsion of chains leads to a decrease in the surface layer coverage so that the layer remains loosely packed even at high surface pressures. The dependence shown in Fig. 7 indicates that for a = 600 the adsorption layer coverage does not exceed 15 %. Further increase in the adsorption with the formation of a densely packed adsorption layers is possible only for lower values of the intermolecular repulsion constant a. An approximate theoretical model which describes the behaviour of the adsorption layers of proteins in concentrated solutions and predicts an increase in the adsorption without almost any increase in surface pressure will be considered below.

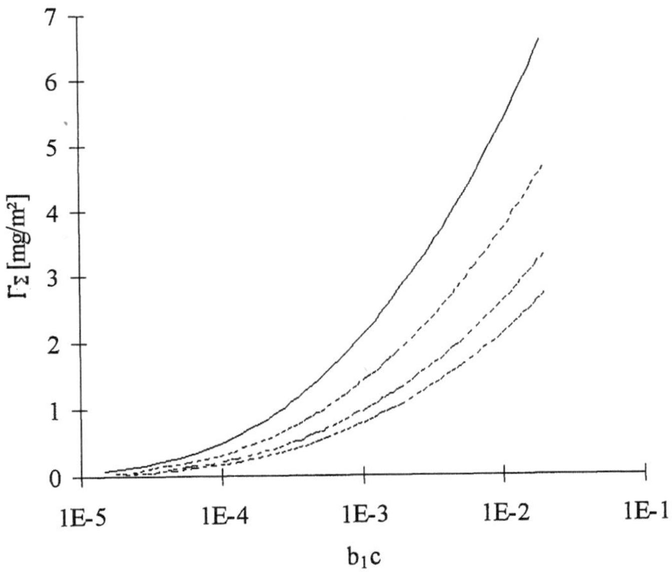

Fig. 5. Dependence of surface pressure on b_1c value for the protein solution ($M = 24000$, $\omega_1 = 2$ nm^2, $\omega_{max} = 60$ nm^2 and $\alpha = 0$) for a = 100 (curve 1), 200 (curve 2), 400 (curve 3) and 600 (curve 4).

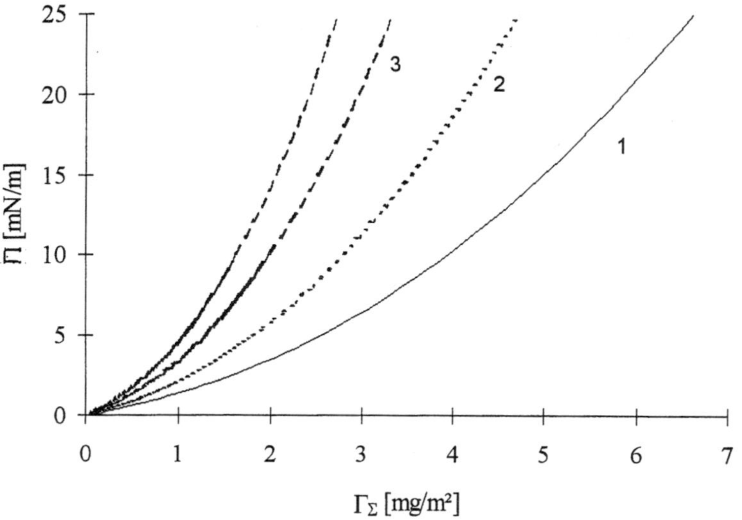

Fig. 6. Dependence of surface pressure on adsorption. The values of the parameters and the notation are the same as in Fig. 5.

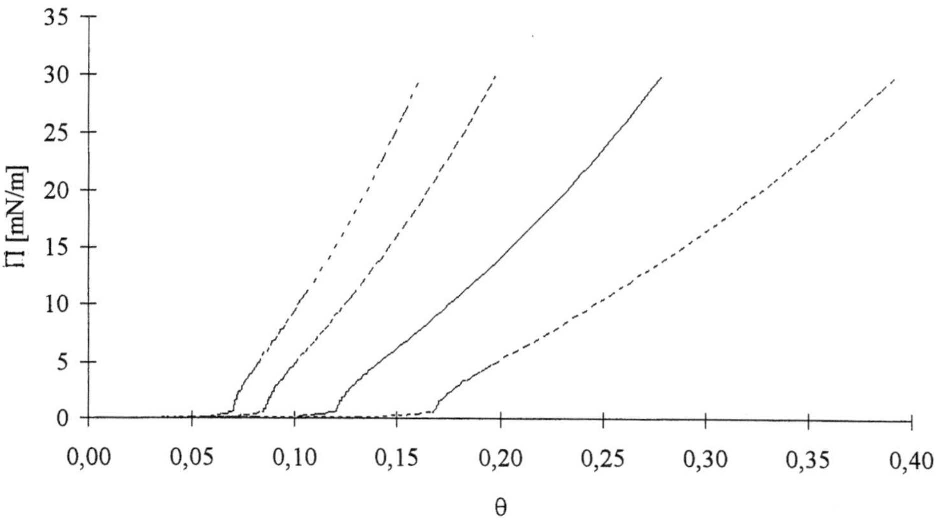

Fig. 7. Dependence of surface pressure on the adsorption layer coverage for the protein solution (M = 24000, $\omega_1 = 2$ nm^2, $\omega_{max} = 60$ nm^2 and $\alpha = 2$) for a = 600 (curve 1), 400 (curve 2), 200 (curve 3) and 100 (curve 4).

The values of the parameters ω_1 and ω_{max} also affect significantly the shape of the dependencies of $\Gamma_\Sigma(b_1c)$ and $\Pi(\Gamma_\Sigma)$. With increasing ω_1, the dependencies of Π and Γ_Σ on b_1c become less steep; a similar effect is observed also for ω_{max}. For a fixed Π the adsorption increases with increasing ω_1. Note that the values of ω_1 and ω_{max} are directly related to the molecular protein mass, its physico-chemical characteristics, and to the properties of the solvent. It can be argued that $\sqrt{\omega_1}$ cannot be smaller than the dimension of the electrostatic blob or the correlation length ξ. The value of ω_{max} is defined as the maximum area which the denatured protein molecule can occupy in the surface layer.

3.8. EVOLUTION OF STATES OF ADSORBED PROTEIN MOLECULES

The main feature of the proposed theoretical model for the adsorption of proteins is the self-regulation of both the state of adsorbed molecules and the adsorption layer thickness in response to the surface pressure. A theory based on this concept was first formulated by Joos and Serrien [9] and differs essentially from the common thermodynamic, statistical and scaling models. The self-regulation mechanism is already included in the Butler's equation [44], from which all main formulas are derived here. It is therefore not necessary to introduce any dependence of the adsorption energy on adsorption, which is common in modern statistical theories [4] to explain a self-regulation effect. Of course, the surface pressure cannot be regarded as the only self-regulating factor, but for the solution/fluid interface this factor is possibly the main one. From Eq. (33) for the model of discrete adsorption states, and from the corresponding equation for the continuous states model, one can calculate the portion of the molecules existing in the respective states ω_i, which is expressed by the ratio Γ_i/Γ_Σ. The distribution function Γ_i or the probability density of Γ_i as a function of the partial molar area ω_i for some given values of Π, is shown in Fig. 8. For very low Π ($\leq$0.1 mN/m) all states exist in the protein adsorption layer; however the part of molecules possessing a maximum area $\omega_i = \omega_{max} = 60$ nm^2 is higher than that of the other states, because a value $\alpha = 2$ was assumed. At $\Pi = 0.5$ mN/m the maximum probability density is achieved for the molecules with $\omega_i \approx 17$ nm^2, while at $\Pi = 1$ mN/m a maximum for the molecules with $\omega_i \approx 10$ nm^2. With further increase in Π the portion of molecules occupying a minimum area increases. For $\Pi \geq 10$ mN/m only a small part of adsorbed molecules occupy an area exceeding the value $\omega_i = \omega_{min} = 2$ nm^2. Therefore, the equilibrium adsorption layer is characterised by an almost complete denaturation at low surface pressures. Note that if $\alpha = 0$, then for $\Pi \approx 0$ all the states of molecules within the surface are equally probable, and the probability density Γ_i is represented by a straight line parallel to the abscissa. With increasing surface pressure unfolded protein molecules refold and segments partially desorb. And finally, for large surface pressures, the equilibrium adsorption layer is comprised of molecules in the state of minimum area only.

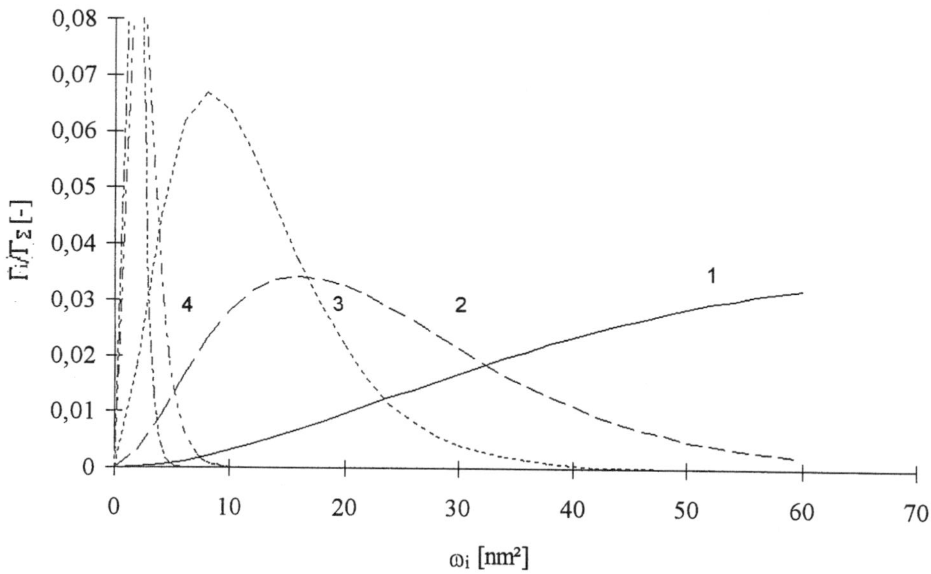

Fig. 8. Dependence of distribution function Γ_i/Γ_Σ for protein solution on ω_i ($M = 24000$, $\omega_1 = 2$ nm^2, $\omega_{max} = 60$ nm^2, $\alpha = 2$ and $a = 600$) for $\Pi = 0.1$ (curve 1), 0.5 (curve 2), 1 (curve 3), 5 (curve 4) and 10 mN/m (curve 5).

3.9. CONCENTRATED SOLUTIONS

The adsorption layer thickness as a function of adsorption is shown in Fig. 9. Here the lower curve is calculated from the Γ_Σ-values with a protein density of $\rho = 1$ g/cm^3, and the upper curve for $\delta = V/\omega_\Sigma$. The molar volume V was taken as 24000 cm^3/mol, which for this example corresponds to $\rho = 1$ g/cm^3. If the adsorption layer coverage θ is taken into account (see Fig. 7), the upper curve in Fig. 9 coincides with the lower one; therefore the upper curve shows the thickness of a loosely packed adsorption layer. It is seen from Fig. 9 that a rapid increase in the thickness of the loosely packed adsorption layer occurs in the adsorption range from 0 to

3 mg/m^2. For higher adsorption values δ increases rather slowly, achieving a limit of $\delta_{max} = V/\omega_1 = 20$ nm. Comparing the calculated results with measured values of a spread β-casein layer [43], we note that the experimental dependence $\delta(\Gamma_\Sigma)$ coincides with the lower curve of Fig. 9. At the same time, the value $\delta \approx 20$ nm^2 was obtained in [43] for $\Gamma_\Sigma = 15$ mg/m^2, which corresponds to the upper curve in Fig. 9 and reflects the actual limiting value of the layer thickness.

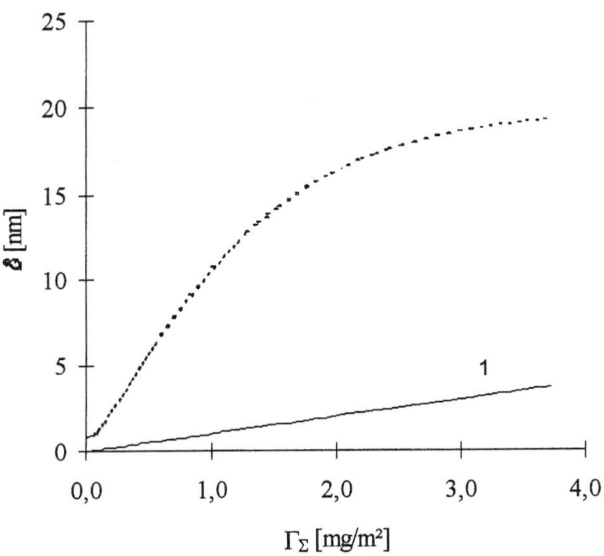

Fig. 9. Thickness of a densely (curve 1) and loosely packed (curve 2) protein adsorption layer. The parameters are the same as in Fig. 8.

The increase in the surface coverage of the adsorbed or spread layer possessing the thickness δ_{max} with increasing protein concentration should lead to the actual dependence of $\delta(\Gamma_\Sigma)$. At the same time, the model described by Eqs. (60) and (35) shows an unrealistically steep surface pressure increase within a very narrow interval b_1c. This contradicts with the experimental data which show that starting from some protein concentration Π remains almost constant, while the adsorption continues to increase, resulting in an increase in the surface coverage up to an almost complete saturation at high protein concentrations. This failure of the theoretical model is possibly due to the fact that $a=a_{el}$ does not remain constant. In fact, the number z of unbounded ions in the protein molecule can decrease when both adsorption and surface coverage increase. Assuming that, starting from some coverage, the value of z is inversely proportional to Γ_Σ, one obtains the approximate dependence $a_{el} \approx 1/(\Gamma_S)^2$. This assumption is an estimation only, however, a relative (per unit polymer concentration) decrease of the osmotic pressure with increasing concentration of the polymer and electrolyte is also predicted by the scaling theory [37].

Let us introduce a critical protein concentration c_c and the corresponding values θ_c and Π_c, above which the value of a decreases. Because the adsorption layer coverage is low at a pressure of $\Pi \approx$ 25-30 mN/m (cf. Fig. 7) only the second term on the right hand side of Eq. (60) can be considered. As for large Π the relation $\omega_\Sigma \approx \omega_1$ holds, one obtains from Eq. (60) the following relation between θ_c and Π_c

$$\theta_c = \left(\frac{\Pi_c \omega_1}{aRT}\right)^{1/2} \tag{74}$$

Assuming that for $\theta = \Gamma_\Sigma \omega_1 > \theta_c$ the value of the constant a is defined by the relation

$$\frac{a^*}{a} = \left(\frac{\theta_{ñ}}{\theta}\right)^2 \tag{75}$$

where a^* is the value of the constant a for $c > c_c$, one obtains from Eqs. (35) and (60) the equation of state for the surface layer and the adsorption isotherm for concentrated protein solutions ($c > c_c$)

$$\Pi = -\frac{RT}{\omega_1}\left[\ln(1-\theta) + a\theta_c^2\right] \tag{76}$$

$$b_1 c = \frac{\theta}{1-\theta} \tag{77}$$

It follows from Eq. (76) that $\Pi \cong \Pi_c$ for $c > c_c$. The calculations performed with Eq. (77) show that, in agreement with experimental data, the adsorption increases significantly for $c > c_c$.

4. COMPARISON WITH EXPERIMENTAL DATA

To demonstrate the validity of the proposed model some examples will be considered here: a solution of a single surfactant, and the two proteins ß-casein and human serum albumin, often used in interfacial studies as model proteins.

4.1. LOW-MOLECULAR SURFACTANTS ADSORBING IN TWO STATES

As a first example we consider the solution of a surfactant which can exist in two states within the adsorption layer: state 1 with $\omega_1 = \omega_{min}$ and state 2 with $\omega_2 = \omega_{max}$. Fig. 10 shows the experimental surface pressure isotherm for an aqueous solution of (N-n-hexadecyl-N,N-dimethylammonio)-acetic acid bromide (BHB16) at pH = 7, reproduced from [66, 67].

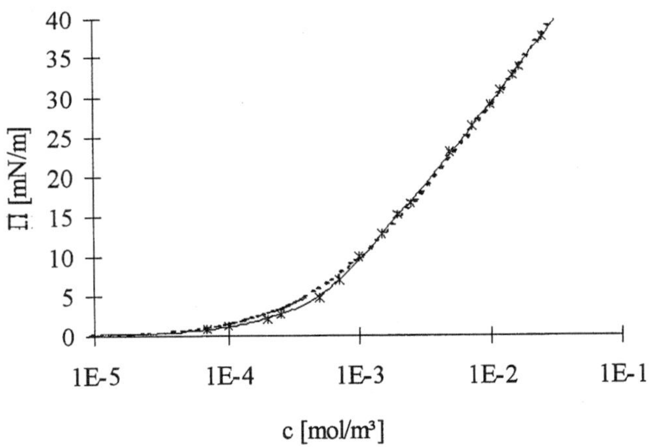

Fig. 10. Surface pressure isotherm for BHB16 at pH = 7. Symbols represent the experimental results [66, 67], solid line is calculated from Eqs. (49) and (50), dashed line is calculated from Eq. (16).

The theoretical curves were calculated from the model proposed here, Eqs. (49) and (50) with $\omega_1 = 2{\cdot}52 \cdot 10^9$ cm^2/mol, $\omega_2 = 1{\cdot}98 \cdot 10^{10}$ cm^2/mol, a = 0, and also from the Szyszkowski-Langmuir equation (16) for $\omega_1 = 2{\cdot}4 \cdot 10^9$ cm^2/mol. The values of these parameters were determined by a best-fit procedure of the theoretical dependencies of $\Pi = \Pi(\lg c)$ to the experimental data. The calculations using Eqs. (49) and (50) yield a deviation between experiment and theory which is

4 times smaller than that obtained with Eq. (16). Thus, considering molecular reorientation within the surface layer and non-ideality of the entropy of mixing due to the difference in the areas ω_1 and ω_2 values agrees well with the experimental results for BHB16 molecules. The values of $\omega_1 = 0.5$ nm^2 and $\omega_2 = 3$ nm^2 per one BHB16 molecule estimated from the surface tension isotherm, coincide with those calculated from a molecular dynamics method [12] for normal and flat orientations of the BHB16 molecule within the surface layer, respectively. In these molecular dynamics calculations the interaction of the BHB16 molecule with water and the thermal motion of the atoms constituting the BHB16 molecule were taken into account. We can conclude that the two-state adsorption model does not only formally better describe the experimental surface tension isotherm, but also leads to realistic values for the molar areas of the two adsorption states.

The evolution of the BHB16 adsorption layer composition is illustrated in Fig. 11. The maximum adsorption for state 2 with a flat orientation occurs at $\Pi \approx 4$ mN/m and 75 % of the surface is occupied by the molecules in state 2. Note that the area occupied by a flatly oriented molecule is 6 times higher than for a vertical orientation.

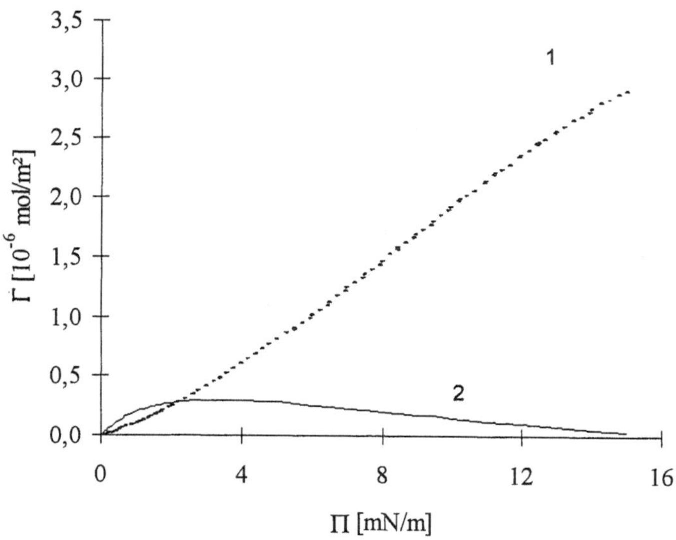

Fig. 11. Dependence of the two adsorption states on surface pressure for BHB16; state 1 (curve 1) state 2 (curve 2)

At $\Pi > 15$ mN/m the state 2 almost completely disappears and the adsorption layer consists only of BHB16 molecules in a vertical orientation.

4.2. β-CASEIN

Graham and Phillips [68] measured the surface tension (Wilhelmy plate method in Langmuir trough) and the adsorption (radioactivity and ellipsometry methods) for aqueous solutions of β-casein. A phosphate buffer with addition of NaCl was used. The isotherms of surface pressure and adsorption were measured independently. The data obtained by Graham and Phillips are in good agreement with the results of other authors [69-71]. Therefore these data have been chosen for a comparison with our theoretical calculations. Both the discrete and continuous state models are employed. In the calculations we assume that $\alpha = 0$ and $\Delta\omega \rightarrow 0$ (or $\Delta\omega = \omega_1$), and the values of $\omega_1 = \omega_{min}$ are varied within the range from 1 to 10 nm^2, ω_{max} in the range from 50 to 100 nm^2 and the value of a within the range from 50 to 800. These three parameters determine the shape of the curves $\Pi = \Pi(c)$ and $\Gamma_\Sigma = \Gamma_\Sigma(c)$, while the adsorption equilibrium constant b_1 defines the position of theoretical curves relative to the abscissa. Note that ω_{max} cannot be varied in a wide range, because this parameter, according to Eq. (39), determines the thickness of the adsorption layer. As the adsorption layer coverage and, consequently, the layer thickness, depend on the intermolecular repulsion constant a, only two parameters of Eqs. (60) and (35), namely ω_1 and a, remain for variation. A special fit program allows to examine various combinations with respect to the agreement between experimental and theoretical data. It was found that the two experimental isotherms $\Pi = \Pi(c)$ and $\Gamma_\Sigma = \Gamma_\Sigma(c)$ correspond satisfactorily to Eqs. (60) and (35) with the following values of the parameters: $\omega_1 = (5 \div 7)$ nm^2, $\omega_{max} = 80$ nm^2 and $a = (80 \div 150)$. In Figs. 12 and 13 the experimental curves obtained by Graham and Phillips [68] are compared with the theoretical results. Within a concentration range up to 0.5 mg/l the agreement is quite good.

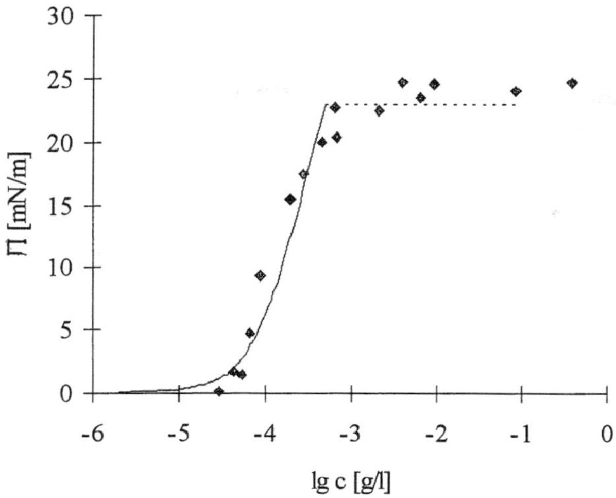

Fig. 12. Dependence of surface pressure on the β-casein concentration; symbols represent the experimental data by Graham and Phillips [68]; lines are calculated for $\omega_1 = 6$ nm^2, $\omega_{max} = 80$ nm^2, a = 120 and $c_c = 0.5$ mg/l.

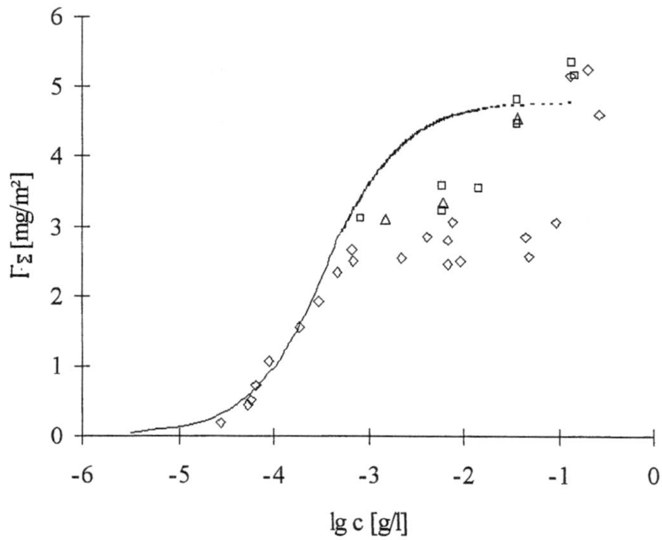

Fig. 13. Dependence of adsorption on β-casein concentration. Experimental data are reproduced from Graham and Phillips [68] (✧ - radioactivity method, □, Δ - ellipsometry), lines are calculated for the same values of the parameters as in Fig. 12.

The value of $\omega_{max} = 80$ nm^2 corresponds to a layer thickness of $\delta_{min} = 0.5$ nm for completely denatured β-casein molecule, or to $\delta \approx 1$ nm if, due to the condition $\alpha = 0$, all states of the molecule within the surface are assumed to be equally probable. These results agree well with a direct measurements of the mass of adsorbed protein described elsewhere [72]. The value of $\omega_1 = 6$ nm^2 for $V = 24000$ cm^3/mol corresponds to the maximum adsorption layer thickness $\delta_{max} = 6.7$ nm. For δ_{max} the expected maximum adsorption should amount to 6.7 mg/m^2. This means that, according to Fig. 13, for a concentration of 0.5 mg/l the adsorption layer coverage below which the agreement between the experimental data and theoretical results takes place, cannot exceed 50 %. This conclusion is indeed supported by the rigorous calculations based on Eqs. (60) and (35), and also follows from the approximation (74). For $c_c = 0.5$ mg/l the surface pressure amounts to $\Pi_c = 23$ mN/m (Fig. 12). As for β-casein it was found $a = 120$ and $\omega_1 = 6$ nm^2 (which corresponds to $3.6 \cdot 10^6$ m^2/mol), then it follows from Eq. (74) that $\theta_c = 0.48$, that is, $\Gamma_{\Sigma c} = 3.2$ mg/m^2. From the plot shown in Fig. 13 for $c_c = 0.5$ mg/l one obtains $\Gamma_{\Sigma c} = 2.9$ mg/m^2. Therefore the maximum coverage ($\theta = 1$) of the 6.7 nm thick adsorption layer should correspond to total adsorption $\Gamma_{\Sigma max}$ of 6.7 mg/m^2. The values of maximum layer thickness and maximum adsorption indicated above agree well with the experimental data [68, 71]. The dependencies of Π and Γ_Σ on c for $c > 0.5$ mg/l calculated from Eqs. (76) and (77) are also shown in Figs. 12 and 13 and satisfactory agreement to the experimental data is obtained. This agreement, however, has not to be overestimated as the model employed to derive Eqs. (76) and (77) is rather crude.

In summary one can say that all three independent experimental sets of data obtained by Graham and Phillips, i.e. $\Gamma_\Sigma = \Gamma_\Sigma(c)$, $\Pi = \Pi(c)$ and $\delta = \delta(c)$, and also the corresponding derived dependencies, e.g., $\Pi = \Pi(\Gamma_\Sigma)$ or $\delta = \delta(\Gamma_\Sigma)$, agree satisfactorily with the multiple molecular state model for protein molecule within the surface for the same set of four main parameters in Eqs. (60) and (35). It was noted above that the parameters ω_1 and a produce opposite effects on the values of δ_{max} and Γ_Σ. Therefore one can choose other values of ω_1 and a which lead to the same satisfactory agreement with the data [68] for $c < c_c$. For example, good agreement is obtained for $\omega_1 = 2$ mn^2 and $a = 400$. For this case however the value of θ_c for $c_c = 0.5$ mg/l decreases to 0.17. At $\theta = 1$ the adsorption layer thickness is 20 nm (which corresponds to

$\omega_1 = 2$ nm^2) this would lead to $\Gamma_{\Sigma max} \approx 20$ mg/m^2, which exceeds significantly that obtained in [68] (see Fig. 13), but agrees satisfactorily with the data [43] for a spread layers of β-casein at maximum compression. The theoretical curves calculated from Eqs. (60) and (35) also agree with the experimental dependencies of $\Pi(\Gamma_\Sigma)$ and $\delta(\Gamma_\Sigma)$ presented in [43]. This difference between the optimum values of the parameters ω_1 and a possibly reflects to some extent the difference in the nature and properties of adsorbed and spread layers of β-casein.

4.3. HUMAN SERUM ALBUMIN

The surface tension isotherms of aqueous Human Serum Albumin (HSA) solutions was studied extensively in [73] using the axisymmetric drop shape analysis (ADSA). The obtained dynamic surface tensions for HSA solutions at various concentrations are shown in Fig. 14.

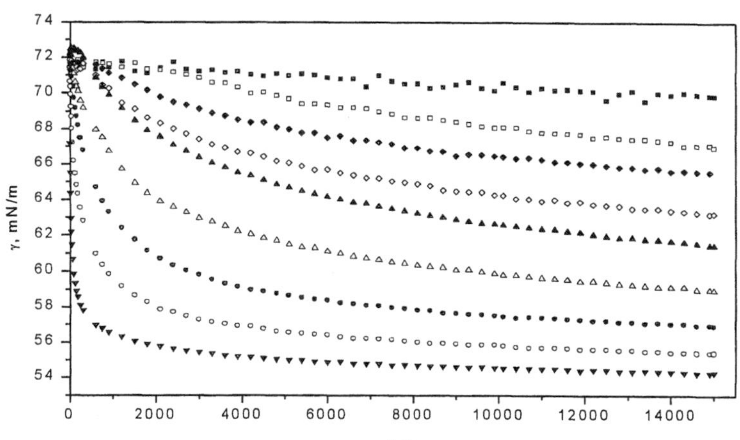

Fig. 14 Dynamic surface tension of HSA solutions at different initial bulk concentrations: $2 \cdot 10^{-8}$ (■), $3 \cdot 10^{-8}$ (□), $5 \cdot 10^{-8}$ (◆), $7 \cdot 10^{-8}$ (◇), 10^{-7} (▲), $2 \cdot 10^{-7}$ (△), $5 \cdot 10^{-7}$ (●), 10^{-6} (O), 10^{-5} mol/l (▼).

In the time range of 4 hours the equilibrium is achieved only for the concentrated HSA solutions ($c > 10^{-6}$ mol/l). This agrees with the data of Gonzalez and MacRitchie [74] obtained for Bovine Serum Albumin (BSA) which has similar structure and properties like HSA. The equilibrium surface tensions for less concentrated HSA solutions were estimated via extrapolation $\gamma(t)|_{t\to\infty}$. For a mixed adsorption mechanism the derivative $d\gamma/dt^{-1/2}$ is defined by the relation [75]

$$\frac{d\gamma}{dt^{-1/2}} = \frac{RT\Gamma_\Sigma^2}{c}\left(\frac{\pi}{4D}\right)^{1/2} + \frac{RT\Gamma_\Sigma^2}{c\beta t^{1/2}} \tag{78}$$

where ß is the adsorption rate constant and D is the diffusion coefficient. Depending on the value of the two terms on the right hand side of Eq. (78), the equilibrium surface tensions can be extrapolated either via $(d\gamma/dt^{-1/2})_{t\to\infty}$ or $(d\gamma/dt^{-1})_{t\to\infty}$. For $ß \to \infty$ which corresponds to the diffusional adsorption mechanism, and for $t\to\infty$ the second term on the right hand side of Eq. (78) vanishes, and the equation transforms into the known relation of Joos and Hansen [75]. On the other hand, if this second term significantly exceeds the first term the extrapolation $d\gamma/dt^{-1}$ is more justified. The equilibrium surface tensions obtained from the two extrapolation procedures yield similar results with differences in most cases lees than 0.5 mN/m [73]. In Fig. 15 the experimental equilibrium surface tension isotherm for HSA at pH=7 is plotted as a function of the initial HSA concentration in the solution. Note that for $c > 10^{-7}$ mol/l the data from [73] agree well with those presented in [68, 74, 75-77] for BSA, while in the region $c < 10^{-7}$ mol/l the values from [73] are lower than those reported by Graham and Phillips [68]. This deviation can be explained by the decrease in protein concentration within the drop due to the adsorption at the drop surface.

Fig. 15 shows the HSA adsorption isotherm fitted to the experimental data by using Eqs. (60) and (35). The parameters of the isotherm (ω_1, ω_{max}, $\Delta\omega$, α and a) were varied such that the maximum HSA adsorption reaches a value of approximately 3 mg/m^2, which corresponds to literature data for BSA [68, 78].

The calculated curve in Fig. 15 refers to the following parameters: $\omega_1 = \omega_{min} = 40$ nm^2, $\omega_{max} = 80$ nm^2, $\Delta\omega = \omega_1$, $a = 80$, $\alpha = 0$ and $b_1 = 2\cdot10^7$ l/mol. These values agree remarkably well with those published in [79-82]. In particular, the minimum area per BSA (or HSA) molecule within the monolayer is between 40 and 50 nm^2. In a spread BSA monolayer the surface pressure starts to increase at an area per protein molecule of 150 x 180 nm^2, which corresponds to a monolayer coverage of ca. 20 % (cf. Fig. 7). Note that for $\Delta\omega = \omega_1$ a variation of ω_{max} within the range from 40 to 200 nm^2 does not effect the theoretical dependencies of Π and Γ_Σ on c. For HSA the adsorption of about 1 mg/m^2 and an adsorption layer thickness of 4 nm the total ion concentration within the surface layer can be estimated as 2 mol/l (assuming a total number of amino-acid residues in a HSA molecule of 580 [82]). Assuming further that the minimum free charge of an albumin molecule $z \cong 20$ [82], the order of magnitude of $a=a_{el}$ in Eq. (57) amounts to few tens which agrees with the results obtained from the fitting. The minimum surface area of a HSA molecule corresponds to the three-domain molecular structure, where each domain comprises of 9 loops connected by sulfide bridges. At pH=7 the size of such a molecule is 14x4x4 nm^3 [82]. This configuration is possibly independent of Π and Γ_Σ and the HSA molecule does not undergo denaturation at the liquid/air interface.

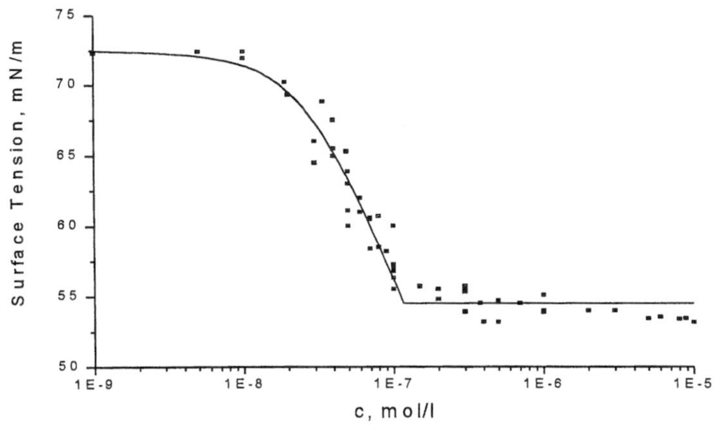

Fig.15. The experimental equilibrium surface tension isotherm of HSA at pH=7 as a function of the initial HSA bulk concentration, and the calculated adsorption isotherm using Eqs. (35), (60) and (76), (77).

From the theoretical dependence of Γ_Σ on c and also from the data given in [68, 78] the concentration decrease in the drop due to HSA adsorption at the drop surface can be estimated. The equilibrium protein concentration within the drop (i.e. the bulk concentration at the adsorption equilibrium state) is related to the initial concentration c_0 via the expression $c = c_0 - \Gamma_\Sigma (A/V)$, where A and V are area and volume of the drop, respectively. Within the concentration range of $10^{-8} < c_0 < 10^{-7}$ mol/l the equilibrium protein concentration in the drop is decreased by 60 to 30 %. Therefore the experimental points in Fig. 15 have to be sifted by 60 to 30 % towards lower concentrations. As this shift is more significant for small Π, the experimental curve becomes less steep. This correction produces minor effect on the optimum parameters of Eqs. (60) and (35), resulting in the decrease of $a=a_{el}$ value from 80 to 60, and a threefold decrease in b_1. At the same time, this change in the shape of the curves leads to an even better agreement of the data in [73] with those obtained by Graham and Phillips [68] at low albumin concentrations, for both dependencies $\Pi(c)$ and $\Gamma_\Sigma(c)$.

5. PROTEIN ADSORPTION KINETICS

The evolution of equilibrium states of adsorbed protein molecules described above can take place if the adsorption process is extremely slow. On the other hand, the reconstruction process of molecular states within the surface has to influence the protein adsorption kinetics. Some consequences of our model with respect to the adsorption kinetics will be considered here.

The state of a protein molecule in the solution bulk depends on the structure of the molecule, the temperature, ionic strength, and the pH value of the solution. It can be assumed generally that a variety of molecular conformations in the bulk exists, which differ from one another in the ω_i values at the moment when the initial contact with the surface takes place. Therefore the total bulk concentration of protein is the sum of the concentrations c_i ($c = \Sigma c_i$), which correspond to the various conformations of the molecules in the bulk. The equilibrium composition of the adsorption layer (Γ_i/Γ_Σ) is defined by Eq. (33) and is controlled by the surface pressure. In general, the composition of the surface layer does not coincide with that of the bulk phase; therefore the ω_i values in the surface layer will differ from that in the bulk. This will lead to a reconformations of states within the adsorption layer according to Eq. (33). Let us consider the data given in Fig. 8 as an example. Assumed that the flux of protein molecules from the solution is comprised mainly of the states possessing $\omega_i = 20$ nm^2. At $\Pi = 0.5$ mN/m the most probable state for the equilibrium composition of surface layer is also the one with $\omega_i = 20$ nm^2. Therefore at $\Pi = 0.5$ mN/m the conformation of the adsorbed molecules within the surface layer will actually remain unchanged. However due to the subsequent increase in adsorption and corresponding increase of surface pressure according to Eq. (60), both the relative and absolute number of equilibrium states with $\omega_i = 20\ nm^2$ will continuously decrease. For example, at $\Pi = 1$ mN/m the most probable state will be the one possessing $\omega_i = 10\ nm^2$. Therefore both the molecules adsorbed earlier, and the molecules with $\omega_i = 20\ nm^2$ just approaching the surface, will undergo reconformations within the surface layer. Part of their segments will be have to desorb. Note that for the initial state of an adsorbed protein molecule a more realistic value would be $\omega_i = (1 \div 2)\cdot\omega_1$. This means that according to our model at small Π all adsorbed molecules will undergo a denaturation within the surface layer.

The reconformation of the states of adsorbed molecules which initially possess the ith state can be represented schematically by

$$\Gamma_{i-1} \underset{k_i^-}{\overset{k_i^+}{\leftrightarrow}} \Gamma_i \underset{k_{i+1}^-}{\overset{k_{i+1}^+}{\leftrightarrow}} \Gamma_{i+1} \tag{79}$$

where the superscripts '+' or '-' at the kinetic constant k denote the forward or backward reaction, respectively. The mass balance equation for the ith state of the adsorbed molecules can be represented in the form

$$\frac{d\Gamma_i}{dt} = -\Gamma_i\left(k_i^- + k_{i+1}^+\right) + \Gamma_{i-1}k_i^+ + \Gamma_{i+1}k_{i+1}^- + I_i \tag{80}$$

where I_i is the diffusion flux of molecules in the ith state from the solution bulk. Therefore the variation rate of the adsorption for the ith state depends on the reconformation rate due to the decrease of ω_i by $\Delta\omega$, the reconformation rate for the closest conformations which differ from the considered one only by $\Delta\omega$, and the diffusion flux of the ith state. For the description of the adsorption kinetics the discrete model of the molecular states within the surface seems to be more suitable. However, the continuous state model can also be used, noting that the number of reaction rate constants in (79) increase almost proportionally to the decrease of the increment $\Delta\omega$. Introducing the relative (with respect to $\Delta\omega = \omega_1$) value of the reaction rate k_0, one obtains

$$k_i = k_{i0}\frac{\omega_1}{\Delta\omega} \tag{81}$$

According to Fig. 8, the process of surface denaturation of protein, that is, the increase of ω_i with respect to the initial value, takes place at very low surface pressures. For low Π the process of protein adsorption seems to be controlled by diffusion [83]. The experimental data presented in [51, 84-88] agree with the diffusion model at least up to $\Pi < 2$ mN/m. It follows from the results obtained in [85, 87, 88] that for protein concentrations in the range $c = 0.001 \div 0.05$ g/l the time t^* at which the surface tension γ starts to decrease are related by the expression $c^2 t^* = \text{const}$, which follows from the simplest diffusion kinetics equation valid for $\Pi \to 0$ [83]

$$\Gamma_{\Sigma(\Pi \to 0)} = 2c\left(\frac{Dt}{\pi}\right)^{1/2} \tag{82}$$

Thus it can be assumed that in low concentrated protein solutions the surface denaturation process have sufficiently long time to be completed, and therefore the composition of the adsorption layer at $\Pi < 2$ mN/m corresponds to the equilibrium composition, Eq. (33). Further reconformation processes of the states within the surface layer depend, according to Fig. 8, on the desorption of segments which were adsorbed beforehand. One can assume as a first approximation that only backward reactions in Eq. (80) affect the $d\Gamma_i/dt$ value

$$\frac{d\Gamma_i}{dt} = \Gamma_{i+1} k_{i+1}^- - \Gamma_i k_i^- + I_i \tag{83}$$

For small deviations from equilibrium (at equilibrium the relations $d\Gamma_i/dt = 0$ and $I_i = 0$ hold), assuming that $\Gamma_i = \Gamma_i^0 + \Delta\Gamma_i$ and $\Gamma_{i+1} = \Gamma_{i+1}^0 + \Delta\Gamma_{i+1}$ one obtains from Eq. (83)

$$\frac{d\Delta\Gamma_i}{dt} = \Delta\Gamma_{i+1} k_{i+1}^- - \Delta\Gamma_i k_i^- + I_i \tag{84}$$

where the superscript '0' refers to the equilibrium states. An important consequence from the theory of equilibrium protein adsorption is that the kinetic constants of the backward reaction for any ith state can be expressed via the kinetic constant for any particular state, say the nth state

$$k_i^- = \frac{i^a}{n} k_n^- \exp\left[-\frac{(n-1)\Pi\omega_1}{RT}\right] \tag{85}$$

The kinetic constants for the forward reactions can be expressed in similar way. As the constants k_i^- and k_i^+ are interrelated via the adsorption equilibrium constant b_i, and all b_i in turn are related to b_1, it follows than to describe the adsorption kinetics in the framework of the proposed model, in addition to the equilibrium adsorption characteristics (ω_1, ω_{max}, a and b_1) one would require only one extra kinetic constant, say, k_n^-, and the protein bulk diffusion coefficient D.

An important practical result follows immediately from Eq. (85). It can be seen from Fig. 8 that for $\Pi > 5$ mN/m the adsorption layer is comprised mainly of the states with $\omega_i < 2\omega_1$. In this case the adsorption rate will be determined by the transition from Γ_2 (with $\omega_2 = 2\omega_1$) to Γ_1, that is, the molecules from the solution can occupy an area at the surface only when molecules in state 2 would transform into state 1, making the required room in the adsorption layer. Thus if the adsorption is controlled by the process $\Gamma_2 \rightarrow \Gamma_1$, then assuming $\Pi \sim \Gamma_\Sigma$ (which is true within a narrow Π range), one obtains from Eqs. (84) and (85)

$$\frac{d\Pi}{dt} = k_0 \exp\left(-\frac{\Pi\omega_1}{RT}\right) \qquad (86)$$

where k_0 is a constant. This equation is just the well-known MacRitchie relation [6, 80, 89], obtained from experiments. The value of ω_1 in Eq. (86) for a number of proteins varies in the range from 0.5 to 2.5 nm^2 [46, 80], which agrees with the estimates of ω_1 as the minimum area occupied by an adsorbed protein molecule, or the increment of the molar area $\Delta\omega$ for the chains possessing limited flexibility.

It is clear that the protein adsorption from more concentrated solutions differs significantly from the process described above. In this case surface denaturation cannot be completed, because the rate of increase in ω_i is limited, and there is no enough room in the surface layer [72, 90]. The evolution of molecular conformations with time and dynamic surface tension for solutions of different concentrations is shown schematically in Fig. 16.

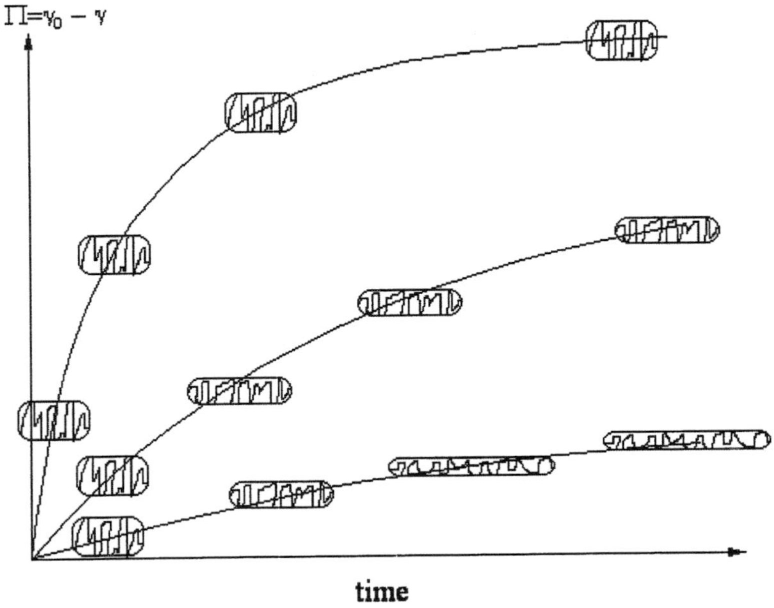

Fig. 16. Sketch of protein structure changes with time and surface pressure, according to [90]

It is seen that, in contrast to dilute solutions where the unfolding of the molecule within the surface is followed by a refolding process, for more concentrated solutions almost no surface denaturation takes place, and the composition of the dynamic adsorption layer is similar to the initial distribution of adsorbed molecules. This explains why the shear elasticity and viscosity for adsorption layers of β-lactoglobulin at low concentrations were found to exceed those measured at large concentrations, while the surface tension of the solutions decreases continuously with the concentration [91]. One can expect that many unusual properties of the dynamic protein adsorption layers, including the adsorption irreversibility mentioned above, and their essentially non-equilibrium nature, can be explained on the basis of the processes of protein reconformation at the surface.

6. REFERENCES

1. H.L. Frisch, R. Simha and F.R. Eirich, J. Chem. Phys., 21(1953)365
2. R. Simha, H.L. Frisch and F.R. Eirich, J. Phys. Chem., 57(1953)584
3. H.L. Frisch and R. Simha, J.Phys.Chem., 58(1954)507.
4. A. Takahashi and M. Kawaguchi, Adv. Polym. Sci., 46(1982)1
5. G.J. Fleer and J.M.H.M. Scheutjens, Adv. Colloid Interface Sci., 16(1982)341
6. F. MacRitchie, Adv. Colloid Interface Sci., 25(1986)341
7. P.G. de Gennes, Adv. Colloid Interface Sci., 27(1987)189
8. M.A. Cohen Stuart, G.J. Fleer, J. Lyklema, W. Norde and J.M.H.M. Scheutjens, Adv. Colloid Interface Sci., 34(1991)477
9. P. Joos and G. Serrien, J. Colloid Interface Sci., 145(1991)291
10. V. B. Fainerman, A. V. Makievski and P. Joos, Colloids & Surfaces A, 90(1994)213.
11. V. B. Fainerman, R. Miller and A. V. Makievski, Langmuir, 11(1995)3054
12. V. B. Fainerman, R. Miller, R. Wüstneck and A. V. Makievski, J. Phys. Chem., 100(1996)7669.
13. V. B. Fainerman, R. Miller and R. Wüstneck, J. Colloid Interface Sci., 183(1996)26.
14. S. J. Singer, J. Chem. Phys., 16(1948)872
15. H. L. Fisch and R. Simha, J. Chem. Phys., 24(1956)652; 27(1957)702.
16. A. Silberberg, J. Phys. Chem., 66(1962)1872; 1884; J. Chem. Phys., 46(1967)1105.
17. R. -J. Roe, J. Chem. Phys., 43(1965)1591; 44(1966)4264.
18. K. Motomura and R. Matuura, J. Chem. Phys., 50(1969)1281.
19. R. J. Rubin, J. Chem. Phys., 43(1965)2392
20. F. L. McCrackin, J. Chem. Phys., 47(1967)1980.

21. M. Lax, Macromolecules, 7(1974)660

22. R. I. Feigin and D. H. Napper, J. Colloid Interface Sci., 71(1979)117

23. C. A. J. Hoeve, J. Chem. Phys., 43(1965)3007; 44(1966)1505

24. C. A. J. Hoeve, J. Polymer Sci. Part C, 30(1970)361; 34(1971)1.

25. A. Silberberg, J. Chem. Phys., 48(1968)2835

26. J. M. H. M. Scheutjens and G. J. Fleer, J. Phys. Chem., 83(1979)1619; 84(1980)178

27. F. Th. Hesselink, J. Colloid Interface Sci., 60(1977)448

28. F.A.M. Leermakers, P.L. Atkinson, E. Dickinson and D.S. Horne, J. Colloid Interface Sci., 178(1996)681.

29. P. G. de Gennes, Scaling Concepts in Polymer Physics, Cornell University Press, Ithaca, New-York, 1979

30. A. R. Khokhlov and K. A. Khachaturian, Polymer, 23(1982)1742

31. S. Alexander, J. Phys. (Paris), 38(1977)983

32. J. Klein and P. Pincus, Macromolecules, 15(1982)1129

33. T. Odijk, J. Polymer Sci. , Polymer Phys. Ed., 15(1977)477

34. J. Skolnick and M. Fixman, Macromolecules, 10(1977)944

35. J. -L. Barrat and J. -F. Joanny, Europhys. Lett., 24(1993)333

36. S. L. Carnie, G. A. Christos and T. P. Creamer, J. Chem. Phys., 89(1988)6484

37. A. V. Dobrynin, R. H. Colby and M. Rubinstein, Macromolecules, 28(1995)1859

38. P. G. de Gennes, Macromolecules, 14(1981)1637; 15(1982)492.

39. M. Daoud and P. G. de Gennes, J. Phys. (Paris), 38(1978)85.

40. D. G. Dalgleish, Colloids & Surfaces, 46(1990)141

41. E. Dickinson, D. S. Horne,J. S. Phipps and R. M. Richardson, Langmuir, 9(1993)242

42. Y. Fang and D. G. Dalgleish, J. Colloid Interface Sci., 156(1993)329

43. R. Douillard, M. Daoud, J. Lefebvre, C. Minier, G. Lecannu and J. Coutret, J. Colloid Interface Sci., 163(1994)277.

44. J. A. V. Butler, Proc. R. Soc. Ser A, 138(1932)348

45. P. Joos, Biochim. Biophis. Acta, 375(1975)1

46. L. Ter-Minassian-Saraga, J. Colloid Interface Sci., 80(1981)393

47. V. B. Fainerman, R. Miller, R. Wüstneck, J. Phys. Chem., 101(1997)6479

48. E. H. Lucassen-Reynders, Colloids & Surfaces A., 91(1994)79

49. P. J. Flory, J. Chem. Phys., 9(1941)660;10(1942)51

50. M. L. Huggins, J. Phys. Chem., 46(1942)151

51. J. Benjamins, J. A. de Feijter, M. T. A. Evans, D. E. Graham and M. C. Phillips, Disc. Faraday Soc.,59(1978)218

52. E. H. Lucassen-Reynders, J. Phys. Chem., 70(1966)1771

53. E. H. Lucassen-Reynders, J. Colloid Interface Sci.,41(1972)156; 85(1982)178.

54. P. van den Bogaert and P. Joos, J. Phys. Chem., 84(1980)190.

55. V. B. Fainerman, Zh. Fiz. Khim., 60(1986)681.

56. P. Joos, Bull. Soc. Chim. Belg., 76(1967)591

57. I. Prigogine, The Molecular Theory of Solutions, North-Holland, Amsterdam, 1968

58. E. A. Guggenheim, Mixtures, Claredon Press, Oxford, 1952

59. R. C. Read, J. M. Prausnitz and T. K. Sherwood, The Properties of Gases and Liquids, 3d. Ed. , McGraw-Hill Inc., New-York,1977.

60. B. B. Damaskin, Izv. AN SSSR, Ser. Khim., (1969)346

61. E. Tronel- Peyroz. , J. Phys. Chem., 88(1984)1491

62. J. Klein and P. Luckham, Nature, 300(1982)429; 308(1984)836

63. J. T Davies, Proc. Roy. Soc., Ser. A, 208(1951)224; 245(1958)417, 419.

64. R. P. Borwankar and D. T. Wasan, Chem. Eng. Sci., 43(1988)1323.

65. V. B. Fainerman, Colloids & Surfaces, 57(1991)249.

66. R. Wüstneck, J. Kriwanek, M. Herbst, G. Wasow and K. Haage, Colloids & Surfaces, 66(1992)1.

67. H. Fiedler, R. Wüstneck, B. Weiland, R. Miller and K. Haage, Langmuir, 10(1994)3959

68. D. E. Graham and M. C. Phillips, J. Colloid Interface Sci., 70(1979)415

69. E. Tornberg and G. Lundh, J. Colloid Interface Sci., 79(1981)76

70. H. Shirahana, J. Lyklema and W. Norde, J. Colloid Interface Sci., 139(1990)177.

71. P.J. Atkinson, E. Dickinson, D. S. Home, and R. M. Richardson, ACS Symp. Ser., 602(1995)311

72. R. Miller, V. B. Fainerman, R. Wüstneck, J. Krägel and D. Trukhin, Colloids & Surfaces A, 1997, in press

73. A.V. Makievski, V.B. Fainerman, M. Bree, R.Wüstneck, J.Krägel and R. Miller, J. Phys. Chem., in press.

74. G. Gonzalez, F. MacRitchie, J. Colloid Interface Sci., 32(1970)55

75. V.B.Fainerman, A,V. Makievski, R. Miller, Colloids Surf. A, 87(1994)

76. A.J.I. Ward, L.H. Regan, J. Colloid Interface Sci., 78(1980)389

77. E. Tornberg, G. Lundh, J. Colloid Interface Sci., 79(1981)76

78. J.A. Feijter, J. Benjamins, F.A. Veer, Biopolymers, 17(1978)1760

79. F. MacRitchie, J. Colloid Interface Sci., 61(1977)223

80. F. MacRitchie, Analytica Chimia Acta, 249(1991)241

81. B.S. Murray, Ph.V. Nelson, Langmuir, 12(1996)5973

82. T. Peters, Adv. Protein Chem., 17(1985)161

83. R. Miller, Trends Polym. Sci., 2(1991)42

84. M. Paulsson and P. Dejmek, J. Colloid Interface Sci.,150(1992)394

85. S. Ghosh and H. B. Bull, Biochemistry, 2(1963)411

86. D. E. Graham and M. C. Phillips, J. Colloid Interface Sci., 70(1979)403

87. K. Kalischevski and K. Schugerl, Colloid Polym. Sci.,257(1979)1099

88. J. A. de Feijter and J. Benjamins, in "Food Emulsios and Foams" (E. Dickinson, Ed.), Special publication no. 58, p. 72. Royal Chem. Soc., London,1987

89. F. MacRitchie, Colloids & Surfaces, 41(1989)25

90. R. Wüstneck, J. Krägel, R. Miller, V. B. Fainerman, P. J. Wilde, D. K. Sarker and D.C. Clark, Food Hydrocolloids, 10(1996)395.

91. J. Krägel, R. Wüstneck, D. Clark, P. Wilde and R. Miller, Colloids & Surfaces A, 98(1995)127.

7. LIST OF SYMBOLS

A	- surface area
a	- intermolecular interaction constant
b	- adsorption equilibrium constant
c	- bulk concentration
D	- diffusion coefficient
d	- dimension of the molecule
F	- free energy
f	- activity coefficient
H	- enthalpy
I	- diffusion flow
i, j	- component or state number
K	- distribution coefficient, equilibrium constant

k	- Boltzmann constant
k^+, k^-	- kinetic constants for forward and backward reaction, respectively
L	- length of monomer chain
M	- molecular mass
n	- total number of components or states
n_i	- ratio of molar surface area values
R	- gas constant
S	- entropy
T	- temperature
t	- time
U	- intermolecular interaction energy
V	- molar volume
x	- volume or molar portion, spatial co-ordinate
z	- co-ordination number
α, β	- constants
Γ	- adsorption
γ	- surface tension
γ_0	- surface tension of the pure solvent
δ	- adsorption layer thickness
$\theta = \omega\Gamma$	- adsorption layer coverage
μ	- chemical potential
ξ	- correlation length
Π	- surface pressure
Π_{os}	- osmotic pressure
ρ	- density
Φ	- number of the macromolecule configurations
ω	- partial molar surface

Proteins at Liquid Interfaces
D. Möbius and R. Miller (Editors)

PROPERTIES OF PROTEIN INTERFACIAL LAYERS AT LIQUID-FLUID INTERFACES

V.N. Izmailova and G.P. Yampolskaya

Department of Colloid Chemistry, Faculty of Chemistry, Moscow State University, Vorob'evy Gory, Moscow, 119899, Russia

Contents

1. INTRODUCTION

Proteins are natural surface active substances. Surface and interfacial phenomena, involving proteins, are wide spread in nature and technology. Detailed study of these phenomena is of great interest for biology, chemical technology, especially for biotechnology, and offers new opportunities in protein and polymer chemistry, separation science, environmental science, for waste minimisation.

Investigations of protein interfacial layers (PIL) are of importance for the understanding of the stability of emulsions and microemulsions when natural high molecular mass surface active substances are applied as stabilisers. Colloid chemistry researchers have yet to elaborate this question.

PIL as considered in this work [1, 2] are films of some thickness formed at an interface between two immiscible liquids (for example, aqueous protein solution-hydrocarbon) due to the surface activity of proteins under equilibrium conditions. The formation of PIL is more complex as compared with the protein adsorption on solids or at surfaces of protein solution with air, and remains poorly investigated. At present adsorption of proteins from solution on solids was studied in details [3, 4], the surface behaviour of proteins was investigated using monolayer techniques [5].

Features of protein interfacial layers and peculiarities of their formation are determined by the thermodynamics of multi-component surfactant-containing systems and depend on specific properties of these biomolecules. Therefore investigating PIL it is useful to take into account the current knowledge of both fields.

The spatial structure of protein molecules governs their essential properties, including surface activity. Native proteins (further, proteins) have a unique structure of molecules, are ideally energetically balanced and created due to principles of structure hierarchy with optimal use of hydrogen bonds, dispersion, electrostatic, as well as hydrophobic interactions to stabilise the compact form of polypeptide chains. All structure elements of a protein molecule, including the secondary structure, domains and globules, as well as protein structures of higher level are specified by amino acid sequences of polypeptide chains. Macromolecular chains of proteins are much different from those of typical synthetic polymers because they are invariably represented by heteropolymers with non-repeating sequences providing an optimal folding of macromolecular chains and compact forms of a protein molecule [6].

Proteins carry functional groups dissociating in water. Properties of proteins are however not equivalent to typical polyelectrolytes because the conformational pH-stability of a native protein with a part of functional groups located inside a globule remains undissociated. For a number of proteins limits of the pH and temperature stability of the native conformation are well investigated.

A number of proteins are to some extent soluble in water (1-20 wt%) resulting in a real molecular solution. However, a protein molecule is a compact particle of a size essentially larger than a water molecule.

For proteins macroscopic concepts can be extended to a molecular level which allows to consider area, volume, protein-solvent interface etc. The average packing density of interior protein atoms is essentially identical to that found in crystals of small organic molecules, and the interior of a protein molecule is not an oil drop but resembles rather a molecular crystal [6]. The surface of proteins has an atomic level with a roughness coefficient of about 1.7 ± 0.2, which is not expected to vary much from one protein to another. Compact molecules of globular proteins have a specific surface topography with asymmetrically located polar and non-polar atomic groups. These surface features determine the amphiphilic character and, consequently, the surface activity of native proteins. About half the surface of an average protein is non-polar yet in contact with water. The surface of membrane proteins (water insoluble) is somewhat more non-polar (about 70-80% of the total surface) than the surface of water soluble proteins. These peculiarities of the protein surface topography, determining the protein surface activity, seem to have functional meaning in biological reactions as well as in the formation of protein structures at higher level (aggregates, dimers and polymers and other condensed forms). The relative stability of the spatial structure and surface topography of proteins allow to consider native proteins as a surface active substance.

The macroscopic approach allows to consider protein solutions as thermodynamically stable dispersions and respectively protein molecules of typical sizes (1 - 10 nm) as colloidal particles.

In this respect protein solutions have generally properties in common with micellar solutions of (low molecular mass) surfactants. Both types of systems are treated often by physicists as "complex fluids" [7]. Of course, unlike micelles of surfactants, in solution proteins behave much more as hard core particles.

The goal of this contribution is a description of the properties of protein interfacial layers (PIL) formed under different conditions, as well as related phenomena proceeding in liquid phases and affecting the formation of PIL and their properties. According to results given in [1, 2, 8-10] the following can be emphasised:

1. Interfacial (surface) phenomena depend on the structure of protein molecules.

2. Two-dimensional phase transitions or structure formations can take place following the protein adsorption at a liquid interface.

3. Solubilisation of hydrocarbons in aqueous protein solution, in contact with a hydrocarbon phase, is one of several processes related to the PIL formation. As the result of solubilisation associates of protein and hydrocarbon molecules are present in the aqueous phase and affect the formation of PIL.

4. Along with the formation of interfacial layers a partitioning of protein between liquid phases is established. It seems, that protein is transferred into organic phase in form of an associate with other components of the system.

Structure-rheological properties of PIL are most sensitive characteristics to any changes in the system. Thus rheological measurements are nearly always performed as a first step to characterise PIL [11]. Beside rheological methods also other techniques are used for the PIL investigation.

To control rheological properties of PIL in the present work different techniques of protein modifications have been applied. To find correlation between PIL properties and association phenomena in aqueous protein system (for example with hydrocarbon molecules) protein complex formation in aqueous solution was forced by different substances (dextran sulphate, lipids, as well as other surfactants) and the effects of complex formation on the properties of PIL were investigated.

To compare the interfacial behaviour of globular proteins (serum albumin, α-chymotrypsin, lysozyme) and water soluble polymers of other structure the rheological parameters of interfacial layers of gelatine (polypeptide chains of collagen) and polyvinyl alcohol (PVA) have been given. As a rule, the properties of PIL described refer to interfaces between aqueous and non-polar phases of approximately equal volumes.

2. ISOTHERMS OF INTERFACE TENSION AND PROTEIN ADSORPTION AT INTERFACES

Water and oil do not mix with each other because of the high energetic cost associated with replacing water-water and oil-oil contacts by water-oil ones. This energetic cost is responsible for the high interfacial tension between bulk oil and water (~30-50 mN/m) and for the sharpness of the interfacial region. The interface in a such system is represented by a coupling of monolayers of water with incorporated oil molecules and of oil with included water molecules [12]. The interfacial tension between immiscible liquids γ_{12} is determined by

$$\gamma_{12} \cong (\sqrt{\gamma_1^d} - \sqrt{\gamma_2^d})^2 + \gamma_{12}^n \qquad (1)$$

where γ_1^d and γ_2^d - are the contributions in interfacial tension from dispersion interactions of the liquid phases 1 and 2, respectively, and γ_{12}^n is the interfacial tension of non-dispersion nature.

Fig. 1 shows equilibrium interface tension isotherms Δγ(log c) for a number of aqueous protein solutions at different interfaces (1′, 2′, 3′ - solution/air, 1, 2 and 3 - solution/benzene, octane, decane, respectively) and with different proteins (1, 1′ α-chymotrypsin, 2′-lysozyme, 2 - cythochrome C, 3, 3′ - bovine serum albumin (BSA)). These isotherms demonstrate that the protein concentration corresponding to the sharp decrease in surface tension or increase in the two-dimensional pressure (Π_s), respectively, is intrinsic for each protein. This value depends on the molecular mass of protein, i. e. its hydrophobicity, and can be considered as a measure of the protein surface activity. The surface activity of a protein can be evaluated as the extrapolated intercept of the isotherm with the abscissa. The comparison of isotherms 1 and 1′ as well as 3 and 3 ′ shows that the interface tension isotherms are shifted towards lower concentrations as compared with the surface tension isotherms, indicating a larger interfacial activity than surface activity.

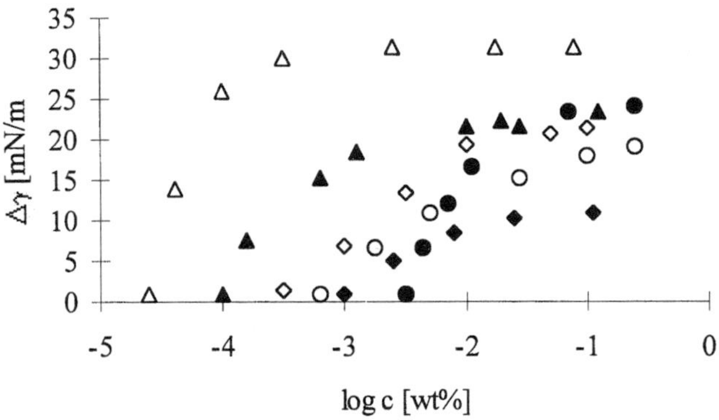

Fig. 1. Surface and interfacial tension isotherms of different globular proteins, 1 - ◆, 1' - ◇, 2 - ●, 2' - ○, 3 - ■, 3' - □, see text for details

In the presence of proteins an increase in the hydrocarbon content in the aqueous phase can be observed. This phenomenon is called solubilisation, and allows to suppose, that hydrocarbons are bound to the protein non-polar sites resulting in a change of the hydrophilic-hydrophobic balance of the protein and a growing ability to act as a surfactant. Therefore, the maximum decrease in the interfacial tension isotherm is larger than in the corresponding surface tension isotherm. This means that under the same conditions the maximum adsorption of protein at liquid/liquid interfaces is higher than at the solution/air surface. It can be anticipated that condensed interfacial layers of some thickness arise with a structure, that is totally different as compared to monolayer formed at the solution/air interface. At equilibrium the protein amount in an interfacial layer depends explicitly on the nature of the non-polar phase, in particular on the size of hydrocarbon molecules (Fig. 2). The decrease in protein adsorption correlates with the loss of hydrocarbon solubilisation in solution, when the length of hydrocarbon molecule is increased (Fig. 2, curves 2 and 3). Hydrocarbon solubilisation reveals the formation of associated particles of protein and hydrocarbon in the aqueous phase with association numbers and, probably, sizes larger for smaller hydrocarbon molecules, and their role in interfacial layer formation.

The longer the hydrocarbon chain, the less is its solubilisation and the amount of protein in the interfacial adsorption layer. In parallel to the solubilisation decrease, adsorption of BSA is reduced approaching a value slightly higher than that at the water-air interface of $2 \cdot 10^{-7}$ mol/m^2 [2].

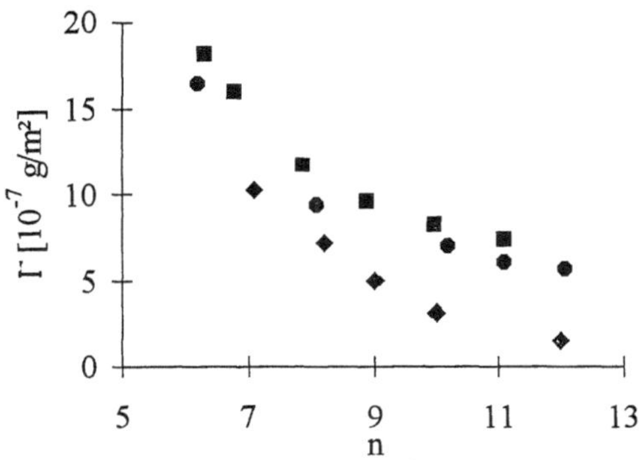

Fig. 2 Dependence of the maximum adsorption (Γ) of protein at aqueous solution-hydrocarbon interfaces on the sizes of aliphatic hydrocarbons; n is the number of carbon atoms in the hydrocarbon molecules; aqueous protein solution: 1-gelatine (■), 2- BSA (●); curve 3 depicts the dependence of solubilisation of aliphatic hydrocarbon by BSA solutions on the length of hydrocarbon molecules (◆).

Thus proteins reduce the excess interfacial energy and adsorb at interface differently, depending on the nature of the organic phase, the protein itself and the ability to solubilise small non-polar molecules.

In the present work the protein adsorption was studied by using different techniques. One of them is the method of radioactive indicators using tritium labelled proteins elaborated in [13]. The method allows to measure the adsorption of proteins in a wide range of concentration down to very low values of 10^{-8} -10^{-5} mol/l. Labelled proteins were obtained by the method of tritium thermal activation. This operation replaces hydrogen atoms in C-H bonds by tritium, which also partially penetrates into labile positions of O-H, N-H and S-H bonds. Labile tritium was eliminated from protein by dialysis. Measurements were performed using MARK-2 (USA) liquid-scintillation counter in a standard dioxane scintillator. Fig. 3 shows results of adsorption measurements for α-chymotrypsin at different interfaces [14]. Theoretically, according to the size of the α-chymotrypsin molecule, a densely packed monolayer, approximately, is consistent with an adsorption of about $5 \cdot 10^{-8}$ mol/m^{2,} however, adsorption is not restricted to a monolayer and further grows with increasing protein concentration. Results given in Fig. 3, in particular the curves 2 and 2′ or 3 and 3′, show the dependence of adsorption on the protein charge. Note that the isoelectric point of α-chymotrypsin is 7.8.

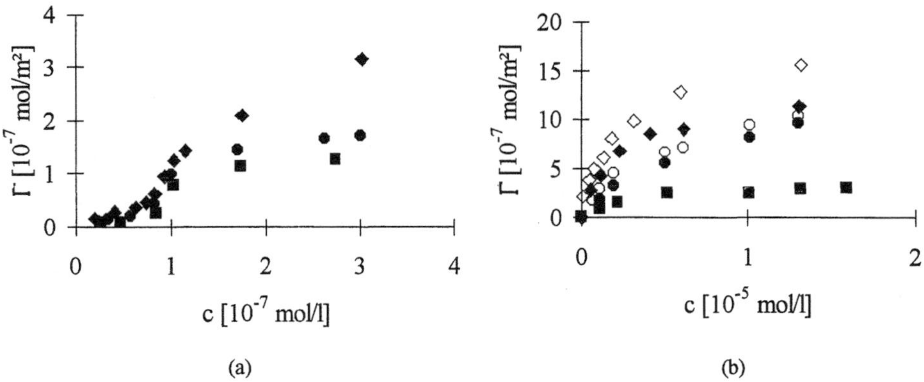

Fig. 3. Adsorption of α-chymotrypsin at liquid-liquid interfaces; low (a) and high protein concentrations (b); curves 2 and 2′ correspond to the water-benzene interface at pH 7.8 and 9.6, respectively, and curves 3 and 3′ to the water-octane interface at pH 7.8 and 11.2, respectively.

In Fig. 3 two regions in the adsorption isotherms can be distinguished, corresponding to α-chymotrypsin concentrations below and above 10^{-7} mol/l. At a concentration of about 10^{-7} mol/l a jump is observed that can be interpreted, due to the homogeneity of liquid interfaces, as a two-dimension phase transition. Moreover, rheology data show, as it will be seen later, that at $C<10^{-7}$ mol/l the adsorption layer behaves newtonian, while at $C\approx10^{-7}$ mol/l it reveals already elastic properties with an elasticity modulus of several mN/m.

At aqueous protein concentrations higher than 10^{-7} M the measured adsorption values indicate the formation of multilayers. The thickness of the PIL is of the order of some ten nm, taking into account only the contents of protein. The phenomenon of multilayer formation at the interface between two liquids cannot be described only in terms of adsorption, but also in terms of phase transitions, and first of all, the formation of coagulation, condensation-crystallisation or other types of structures must be considered (cf. Fig. 4). In this case it is useful to consider the protein molecule as a colloid particle. To emphasise the possibility of an interfacial phase formation, the term "mass accumulation" has been introduced.

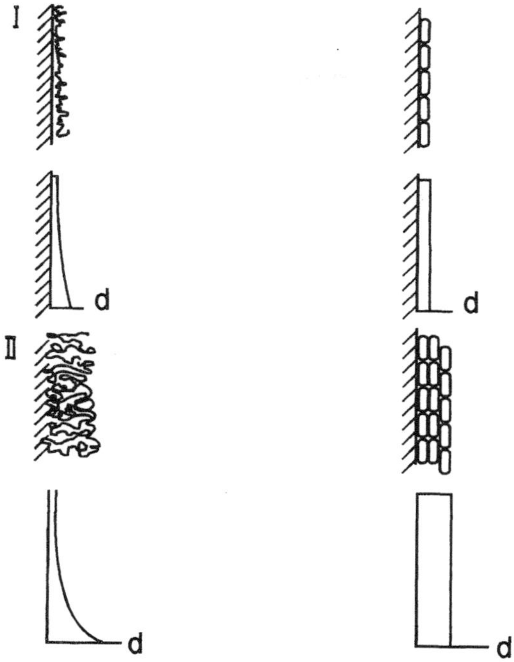

Fig. 4 Scheme of structures of interfacial layers, formed by macromolecules of the different flexibility
A-flexible polypeptide chains (for example, gelatine)
B-rigid particles of globular proteins
I - at low protein concentrations (in cases of globular proteins the dense packed monolayer can be formed)
II - multi-molecular layers of proteins
d - distance from the surface.

3. RHEOLOGICAL PROPERTIES OF PROTEIN INTERFACIAL LAYERS

Rheological behaviour of interfacial protein layers was studied using disk rotating techniques, when the disk is located at an interface of two immiscible liquids and hung on elastic filament making measurements in two regimes [2, 8-11]:

a)measurement of deformation development kinetics, P= const. method

b)studies of kinetics of steady-state flow development, $\varepsilon^{\cdot}$ = const. method

Rheological parameters of the interface layers are measured in surface units. A scheme of the set up is given in Fig. 5.

According to Fig. 6 at very low protein concentrations in the aqueous phase ($4{\cdot}10^{-9}$ - $4{\cdot}10^{-8}$M) and independent of pH and ionic strength the PIL can be regarded as a newtonian liquid with a viscosity from hundreds to several units of surface poises (curves 1-3 and 5). the liquid interface is filled by protein in vary small extent ($\Theta < 1$). Fig. 6a shows dependencies of the shear stress on the rate of deformation; and Fig. 6b shows dependencies of the viscosity on the shear stress.

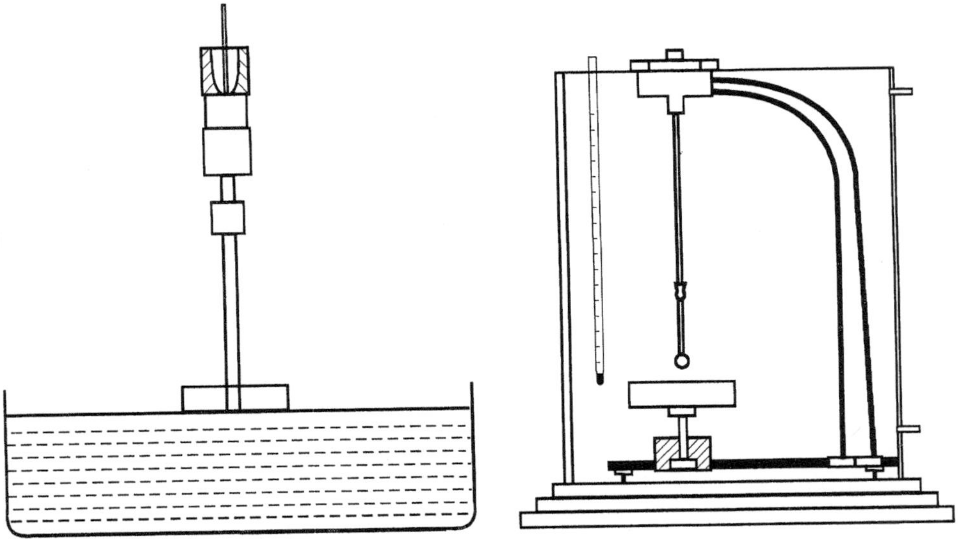

Fig. 5 Scheme of the device used in rotating disk technique: 1- metal plate with a lifting table, 2 - thermostat, 3 - torsion head with limb, 4 - tungsten filament, 5 - the cell with a system under investigation, 6 - optical readout, 7 - the disk

As protein concentration is increased up to 10^{-7}M, the extent of the surface coverage increases too and the interface layers reveal features of a solid-like phase, even for $\Theta < 1$. It is possible to measure flow limits P_{k1} and P_{k2} (curves 4 and 6, Fig. 6a). These data indicate the formation of contacts between particles, which are the structure element, resulting in the development of a two-dimension network. At higher concentrations of about $2 \cdot 10^{-4}$ M the interfacial coverage becomes larger than 1 ($\Theta > 1$) and α-chymotrypsin interfacial layers are characterised by a higher strength: flow limits and elasticity moduli are increased by an order of magnitude. The change of Shvedov's creep viscosities are by 2-3 orders higher than Bingham's viscosities of failure structures. In their turn, the Bingham's viscosities of PIL coincide in a number of experiments with systems, where the PIL was formed under various conditions and are close to the two-dimensional viscosity of newtonian flows for adsorption layers with low coverage. Thus, transitions of PIL from a liquid state (newtonian flow, at $\Theta \ll 1$) to a solid-like network structure ($\Theta \leq 1$) with small number of contacts of low strength, and further to strong multilayer structures ($\Theta > 1$), resembling properties of crystal-like, or liquid-crystalline like structures can be realised for α-chymotrypsin by varying the protein concentration. Rheological parameters of protein interfacial layers are summarised in the Tables 1-7.

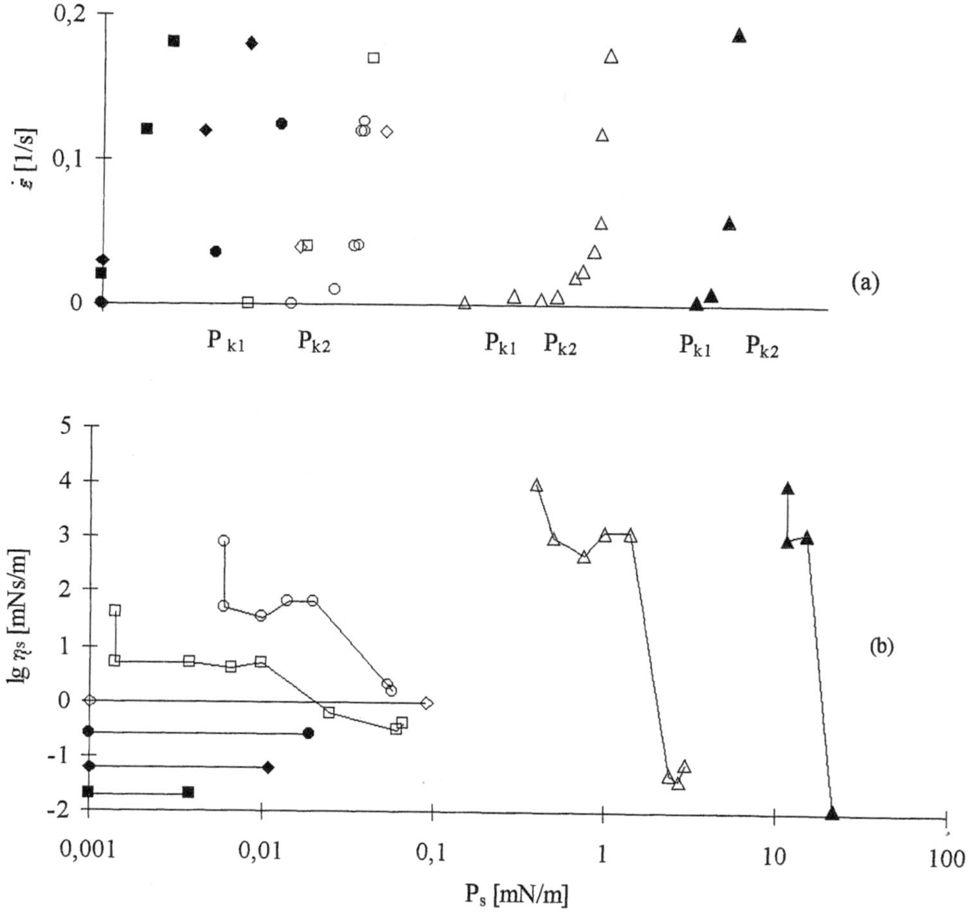

Fig. 6 Rheological behaviour of α-chymotrypsin interfacial layers formed between aqueous solution and octane under various conditions (protein concentration, pH and ionic strength of aqueous phase), as indicated below

Curve	concentration(M)	pH	Ionic strength	Curve	concentration(M)	pH	Ionic strength
1(■)	$4 \cdot 10^{-9}$	6	0.1	2 (◆)	$4 \cdot 10^{-9}$	3	0.5
3 (●)	$4 \cdot 10^{-9}$	3	0.1	4 (□)	$8 \cdot 10^{-8}$	4	0.05
5 (✧)	$8 \cdot 10^{-8}$	6	0.1	6 (○)	$8 \cdot 10^{-8}$	3	0
7 (Δ)	$4 \cdot 10^{-6}$	3	0	8 (▲)	$1.6 \cdot 10^{-4}$	3	0.1

Table 1. Rheological parameters of interface layers of globular proteins, gelatine and polyvinyl alcohol (solution/benzene interface, 20° C).

E_{1s}	E_{2s}	E_s	η_{2s} mN·s/m	θ_2 s	P_{k1}	P_{k2}	η_0*	η_1*
mN/m					mN/m mN· s/m			
Gelatine, C = 0.1 g/100 ml; pH 4.9								
1.7	0.7	0.5	83	109	0.11	0.62	76	0.07
Polyvinyl alcohol, C =1.0								
10.0	2.0	1.7	390	195	1.75	3.6	940	1.6
Serum albumin, C = 0.1; pH 5.1								
3.3	1.0	0.8	360	360	0.3	0.74	56	0.3
Lysozyme, C = 1.0; pH 5.3								
32.3	4.3	3.8	2100	488	1.7	2.4	914	0.4

Table 2. Dependence of rheological parameters on the protein concentration (20°C; water-benzene interface).

c, g/100 ml	E_{1s} mN/m	E_{2s} mN/m	E_s mN/m	λ, %	η_{2s}, mNs/m	θ_2, s	P_{k1} mN/m	P_{k2} mN/m	η_0* mNs/m	η_1* mNs/m
Gelatine, pH 4.9										
0.1	1.7	0.7	0.5	70	83	108	0.11	0.62	76	0.07
0.3	6.0	3.2	2.1	65	360	113	0.25	1.22	398	0.4
0.5	7.6	2.4	1.8	75	620	258	0.30	1.50	302	0.44
Lysozyme, pH 5.2										
0.2	50.1	24.7	16.5	67	5766	233	2.02	4.0	1086	-
1.0	32.3	4.3	3.8	88	2100	488	1.7	2.4	914	0.35

Table 3. Influence of pH on rheological parameters of gelatine interfacial layers at the water-benzene interface, gelatine concentration = 0.3 g/100 ml, 20° C.

pH	E_{1s} mN/m	E_{2s} mN/m	E_s mN/m	λ, %	η_{2s}, mNs/m	θ_2 s	P_{k1} mN/m	P_{k2} mN/m	η_{0s}* mNs/m	η_{1s}* mNs/m
2	3.0	1.7	1.0	64	230	35	0.11	0.76	116	0.3
5	6.0	3.2	2.1	65	360	13	0.25	1.22	398	0.4
8	1.0	0.8	0.5	68	150	88	0.17	0.74	107	0.5

Table 4. Dependence of rheological parameters of gelatine interfacial layers on temperature at the water-benzene interface; gelatine concentration = 0.3 g/100 ml, pH 5.0.

T °C	E_{1s} mN/m	E_{2s} mN/m	E_s mN/m	λ %	η_{2s} mNs/m	θ_2 s	P_{k1} mN/m	P_{k2} mN/m	η_{0s}* mNs/m	η_s* mNs/m
20	6.0	3.2	2.0	65	360	113	0.25	1.22	398	0.4
30	1.8	0.6	0.4	75	128	213	0.11	1.12	134	0.4
40	1.4	0.4	0.3	77	114	285	0.06	0.7	133	0.14

Table 5. Dependence of rheological parameters of PVA interfacial layers on temperature. Concentration of PVA in aqueous solution is 1. 0g/100 ml, equilibrium non-polar phase is benzene.

T °C	E_{1s} mN/m	E_{2s} mN/m	E_s mN/m	λ %	η_{2s}, mNs/m	θ_2 s	P_{k1} mN/m	P_{k2} mN/m	η_{0s}* mNs/m	η_s* mNs/m
20	10.0	2.0	1.7	83	390	195	1.75	3.6	940	1.6
30	7.3	1.7	1.4	81	505	297	1.2	2.8	645	1.1
40	6.0	2.0	1.5	77	650	323	0.85	1.9	457	0.7

Table 6. Influence of temperature on rheological parameters of interfacial layers of globular proteins at the water/benzene interface.

T °C	E_{1s} mN/m	E_{2s} mN /m	E_s mN/m	λ %	η_{2s} mNs/m	θ_2, s	P_{k1} mN/m	P_{k2} mN /m	η_{0s}* mNs/m	η_s* mNs/m
Human serum albumin, pH 5.1; C = 0.1 g/100ml										
20	3.3	1.0	0.78	76	360	360	0.3	0.74	5.6	0.3
40	7.1	1.4	1.2	83	660	471	0.4	1.54	111	0.5
60	13.2	3.5	2.7	77	860	246	0.5	2.4	556	0.9
Lysozyme, pH 5.1; C =1.0 g/100ml										
20	32.3	4.3	3.8	88	2100	488	1.7	2.4	914	0.35
40	73.6	47.9	29.0	60	15700	328	3.0	8.1	1612	4.4
60	76	39	25.7	50	12000	308	5.3	9.3	2849	4.4

Table 7. Effect of the nature of hydrocarbons on the rheological parameters of lysozyme interfacial layers at the water/hydrocarbon interface, lysozyme concentration =1.0 g/100 ml; pH 5.3; 20° C.

Hydro-carbon	E_{1s} mN/m	E_{2s} mN/m	E_s mN/m	λ %	η_{2s}, mNs/m	θ_2 s	P_{k1} mN/m	P_{k2} mN/m	η_{0s}* mNs/m	η_s* mNs/m
Benzene	32.3	4.3	3.8	88	2100	488	1.7	2.4	914	0.4
Penta-decane	12.3	4.1	3.1	76	930	227	0.5	1.6	200	0.16

The symbols used in Tables 1-7 are:

$E_{1s} = P/\varepsilon_o$ is the fast elastic module; where P is the shear stress applied to the system for method with P=const, ε_0 is the fast elastic deformation.

$E_{2s} = P/(\varepsilon_m - \varepsilon_o)$ is the maximal elastic module, where ε_m is the maximum elastic deformation.

$E_s = P/\varepsilon_m$ is the equilibrium elastic module.

$\eta_{2s} = P/ [(d\varepsilon/d\tau)_{max} - (d\varepsilon/d\tau)_{min}]$ is the viscosity of elastic after-effect.

$\theta_2 = \eta_{2s} / E_{2s}$ is the period of elastic after-effect.

P_{k1} and P_{k2} are the limit shear stresses of the Shvedov creep or Bingham flow, respectively.

η_{0s} and η_s are the Shvedov creep or Bingham flow viscosities, respectively.

$\lambda = (\varepsilon_m - \varepsilon_o) \cdot 100\% / \varepsilon_m = E_{1s} \cdot 100\% / (E_{1s} + E_{2s})$ is the degree of elasticity of interfacial layers.

The thickness of protein interfacial layers is about several ten nm [2], and these values should be taken into consideration to convert two-dimensional rheological parameters into bulk parameters. This allows to compare properties of interfacial layers with corresponding bulk structures. From $E_v = E_s\, h^{-1}$ one gets values of rheological parameters of corresponding bulk structures (Nm^{-2}). Here h is the thickness of the protein interfacial layer and E_v and E_s are the elasticity moduli of the bulk and surface respectively. Thus, rheological parameters of interfacial layers of a macromolecular surfactant coincide with those of bulk phases having elasticity moduli of the order of 10^4 -10^6 Nm^{-2}. Note that such values are specific for biological structures. Thus, bilayer lipid membranes (BLM) formed in aqueous phases from cholesterol or egg lecithin solution in decane are characterised by elasticity moduli of the same order of magnitude. Membranes of blood erythrocytes composed of protein have similar elasticity values. This is evidently an indication of the determining role of interfacial rheological properties in the natural processes, involving deformations of membranes, and gives the possibility to simulate mechanic properties of bio-membranes using PIL. In the same time, elasticity values of the order of 10^4-10^6 Nm^{-2} correspond to common elastomers with very strong fine structure.

It is of importance to note that in general independent of the surfactant nature, interfacial structures are stable against dispersion when their elasticity moduli are larger than 10^4-10^5 Nm^{-2} (10^{-12} -10^{-13} erg/molecule) i.e. are much higher than kT.

4. PROTEIN CONFORMATION IN INTERFACIAL LAYERS

Reversibility of adsorption and accumulation of protein at interfaces should be considered taking into account the condensing mass and structure formation. As protein amount at interface increases up to some critical value, inter-particles (protein- protein) interactions grow resulting in a two-dimension structure formation. Conditions of break down of these structures are determined by the competition of energy of the individual contacts between structure elements and the thermal energy of Brownian motion. There is an analogy with the process of peptisation = coagulation of three-dimensional dispersed structures. Two-dimensional structures of PIL are stabilised by hydrophobic interactions, hydrogen and electrostatic bonds between of elements. The advantage of a certain type of contacts is determined by the protein nature, pH and ionic strength of the aqueous solutions, temperature, as well as by the properties of second liquid phase. The optimal value of the free energy controls the number of contacts, the degree of interfacial coverage, structure formation with definite rheological parameters resulting in a hysteresis of the adsorption-desorption process. Desorption rates of proteins from the adsorption layer lower as compared with the adsorption rates are usually connected with the denaturation of proteins at the interfaces. The role of the protein structure in the surface activity on the one hand is evident, on the other one it is not well understood yet [12] and the conformational stability is the most important factor. Depending on the rigidity and surface topography of the protein molecule, the molecules can keep to a large extent their native conformation in the interfacial layer. In many cases schemes of protein interface layers with drastic protein unfolding in the first monolayer at low Θ seem to be strange. Solubilisation of hydrocarbon by protein in aqueous solution does not lead to dramatic loss of catalytic activity for a number of enzymes, such as α-chymotrypsin, trypsin, lysozyme [2]. Because the catalytic activity of proteins is determined by their spatial structure one can conclude, that the proteins preserve a native conformation in associates with hydrocarbon. Now it is of interest to consider data on enzyme activity characterising an influence of the accumulation of protein at the liquid-liquid interface. Enzyme activity was investigated at different coverage of the interface by protein. This type of experiments shows a decrease of catalytic activity of enzymes as compared to the aqueous solution bulk, but a growth of enzyme activity with decreasing interfacial coverage was found. These results were attributed to the hindering of diffusion in the formation of enzyme-substrate complexes [2, 15].

Many results obtained using the dispersion of optical rotation, circular dichroism (CD), and IR attenuated total reflection methods (ATR), show stability of the secondary structure of globular proteins in interfacial layers. Table 8 shows the content (%) of secondary structure of lysozyme in aqueous solution and in adsorption layers [16].

Table 8. Content of secondary structures (%) of lysozyme in aqueous solution and adsorption layers at the water-octane interface.

Protein secondary structure	Aqueous solutions at			Adsorption layers*
	pH 6	pH 2.5	pH 10.5	
α	30±1	28±1	35±1	26±2
β	7±1	8±1	4±1	11±1.5
ρ	63±1	62.5±1.5	61.5±1.5	63±2

*Adsorption layers formed at a "liquid" interface were transferred onto quartz glasses, successively flushed with water and placed in a quartz cell filled with water

Lysozyme interfacial layers formed at interfaces of aqueous protein solution with octane, CCl_4 or hydrophobised quartz were studied at different surface coverage ($\Theta < 1$, $\Theta = 1$ and $\Theta > 1$)

by means of CD. The same method was used for the determination of the conformation of lysozyme in solution in dependence on pH. It follows from Table 8 that the secondary structure of lysozyme remains practically the same, whether the protein is adsorbed at the water-octane interface or is in aqueous solution at a certain pH. According to [16], within the accuracy of measurements the secondary structure of lysozyme adsorbed from solution at interfaces to octane, CCl_4 and the solid quartz surface at different surface coverage remains the same and does not differed from that characteristic to the native protein in water. Results obtained (content of α -helices, β-sheets and ρ- random coiled structures) reveal the lack of dramatic conformation changes in lysozyme interfacial layers.

Nevertheless, one can conclude that small local changes of the protein conformation are possible when the protein interacts with hydrocarbon forming associates, is oriented in the field of surface forces, and forms interface structures. These changes are analogous to liable conformational changes when native proteins "work" - perform usual biological functions, as it is considered in the frame of the theory of dynamic stability of protein molecules.

The adsorption of proteins from aqueous solution at solid surface is the result of the interplay of several processes. One has to discriminate between "hard" proteins which the molecules retain most of their conformation upon adsorption, and "soft" proteins that undergo several structural rearrangements. Adsorption of hard proteins can be interpreted in terms of electrostatic interactions and partial dehydration of the outer sorbent surfaces and the protein. The internal structural changes occurring in the adsorption of soft proteins involve an increase in the conformational entropy of the protein molecule and constitute, therefore, an additional driving force for spontaneous adsorption [3, 12].

It is known that the structure and properties of protein adsorption layers at the interface between two immiscible liquids also depend on the conformational stability of these biological macromolecules. Details on the formation of interfacial layers are still not clear in spite of a large number of studies. To solve this problem it is necessary to apply modern methods used at present for phase behaviour investigations of low molecular mass surfactants. In the works [17-18] NMR spectroscopy was used to elucidate details of interfacial adsorption layer formation, especially the role of interactions between molecules of two different phases and conformational changes of flexible macromolecules. This investigation was performed using gelatine as a model protein with flexible polypeptide chains. For NMR studies oil-in water emulsions were prepared by sonification of gelatine solutions in D_2O in the presence of benzene as non-polar phase under the conditions of a random coiled gelatine in solution. Intensities of proton signals from the gelatine decreased and NMR spectral lines broadened following the interfacial adsorption layers formation, when emulsions are kept at temperatures higher than 313 K. This result is similar to that observed for aqueous gelatine solutions under cooling: random coiled polypeptide chains of gelatine undergo transitions with collageneous helices or gel formation [2]. Data show, that the fraction of mobile gelatine segments in interfacial layers decreases virtually to zero. This may be considered as the formation of collageneous helices resulting probably in a two-dimensional gelation, when gelatine is accumulated at the interface. The intensive increase in the gelatine concentration in thin interfacial layers leads, according to the phase diagram of gelatine, to the formation of thin (about 300 Å) layers of spatial structure stabilised due to a large number of intermolecular bonds. This process is quite analogous to the gelation of gelatine solutions at increasing gelatine concentration up to a critical value, specified for a given temperature. It is of interest to note, that conformational states of gelatine chains are quite different at interfaces between two immiscible liquids and in aqueous solutions under the same temperature. Transfer of large amount of gelatine from the continuous phase in the vicinity of the interface results in the formation of a dense film. Such films are really two-dimensional quasi-solids with a polymer concentration of about 30-40%. Under constant mechanic load these layers in a wide range of shear stresses behave as a typical Kelvin body. Rheological properties of interfacial layers of gelatine provide a strong emulsion stabilisation due to Rehbinder's structural-mechanical barrier [19].

Therefore the appearance of two different signals of benzene in the NMR spectrum suggests that benzene molecules are included into the interface layer, evidently being transported into the interface layer in the composition of associates of gelatine and benzene as the result of a solubilisation.

5. DYNAMICS OF THE FORMATION OF PROTEIN INTERFACIAL LAYERS.

The controlled use of proteins for the stabilisation of dispersed systems requires extensive studies on protein modifications, including their complexes formation with other substances as well as reliable knowledge of dynamics of interfacial layer formation. There is a number of investigations [20-25], where dynamics of proteins adsorption at interfaces of different nature was described in detail, for example with solids, air and hydrocarbons. It was shown that the achievement of adsorption equilibrium at solid surfaces takes few ten seconds while it takes hours at liquid interfaces to air or oils.

The dynamics of interface tension of aqueous BSA solutions and complexes of BSA with dextran sulphate (DS) in contact with n-decane was studied in [23]. Interfacial tension changes with time were measured using a Wilhelmy plate tensiometer (Krüss, Germany) with a continuous record of readings. A roughened surface platinum plate was used for the measurements. Complexes of BSA-DS were obtained by mixing solutions of BSA and DS at pH 8.5 followed by dialysis against acetate buffer (pH 5.6, ionic strength 0.01) over 21 hours. Under these conditions an equilibrium complex is formed and the solution is free of individual molecules. Therefore the complex composition was supposed to be equal to the composition of mixtures and was characterised by the BSA concentration (wt%) as well as relative weight fraction of DS: $W=C_{DS}/(C_{BSA}+C_{DS})$, where C_{DS} and C_{BSA} are the weight concentrations of DS and BSA. Changes of the interfacial pressure $\Pi = \gamma_0 - \gamma(t)$ with time for BSA and BSA-DS complexes are shown in Fig. 7.

Two regions can be distinguished in interfacial pressure curves. The first is determined by the diffusion rate of macromolecules from the solution towards the interface. In the second stage the growth of the two-dimensional pressure is limited by the rate of structure rearrangements in the coagulation structures of the PIL. The initial region of Π changes is described by the following equation

$$\Pi = 2\, C_s\, kT\, (Dt/\pi)^{1/2} \qquad (2)$$

where C_s is the surfactant bulk concentration; k is the Boltzmann constant; D is the diffusion coefficient of the surfactant (protein); T is the temperature (K) and t is the time of interfacial layer formation. The stage of structural rearrangements is described by the equation

$$\ln (\gamma_t-\gamma_\infty)/(\gamma_0-\gamma_\infty) = t/\tau \qquad (3)$$

where τ is the relaxation time of structural rearrangements; t is time; and γ_0, γ_t, and γ_∞ are the interfacial tensions at the initial time t=0, at the time t and at equilibrium, respectively.

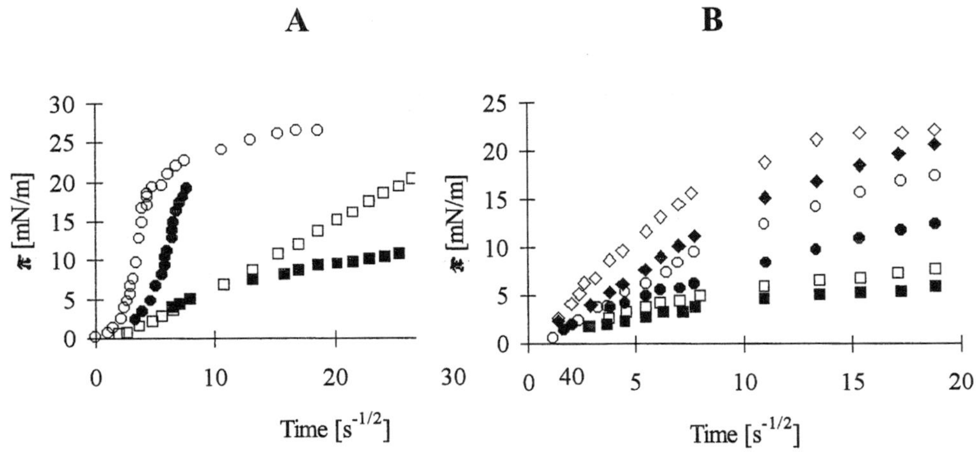

Fig. 7. Dependence of interfacial pressure (at the water-decane interface) on time: A-BSA, B-BSA-DS complexes: 1, 2

The regions of continuous interface pressure growth (over 20 minutes for BSA and 40 minutes for complexes BSA-DS) are controlled by diffusion and occur at concentrations $3 \cdot 10^{-4}$ and $1.5 \cdot 10^{-3}$ wt % for solutions of BSA and complexes BSA-DS, respectively. In these protein concentration regions a linear dependence $d\Pi/d\sqrt{t} = f(C_s)$ is observed (Fig. 8).

The rate of growth of Π at this stage is higher for BSA than for the complex BSA-DS which is in agreement with the respective diffusion coefficients. Fig. 9 shows the dependence of relaxation time τ on protein concentration of BSA and complexes BSA-DS. For the complexes this dependence passes through maximum.

Fig. 10 represents the two-dimension interfacial pressure isotherms $\Pi(C)$ of BSA and BSA-DS complexes. A constant interfacial pressure is achieved for BSA concentrations above $1 \cdot 10^{-4}$ g/100ml and for the complexes above $1.8 \cdot 10^{-3}$ g/100 ml. These results let us conclude that the interfacial tension can be decreased to a large extent.

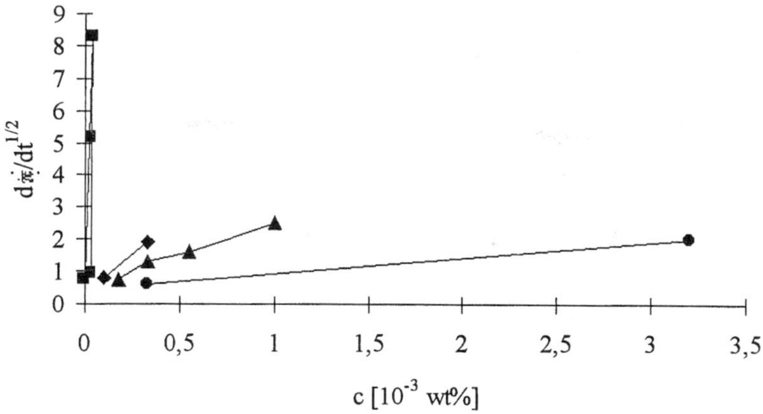

Fig. 8. Dependence of the slope dΠ/d√t as a function of protein concentration at the solution-n-decane interface; BSA (■), BSA-DS complexes at different composition W=0.2 (◆), 0.4 (▲) and 0.6 (●)

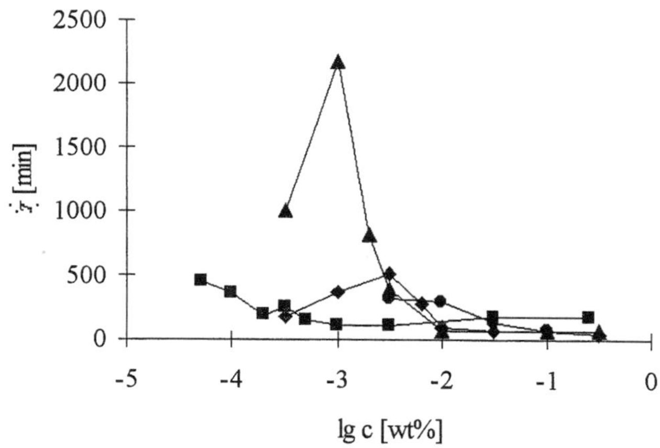

Fig. 9. Concentration dependencies of the relaxation time of structure rearrangements in adsorbed layers at water/n-decane interface; BSA (■), BSA-DS complexes at different composition W=0.2 (◆), 0.4 (▲) and 0.6 (●)

A comparison of the results given in Figs. 7-10 shows that at low protein concentrations regions of continuous interfacial pressure growth are observed, which is controlled by diffusion.

Characteristics of the dynamics of surface layer formation can be obtained by measuring changes of the rheological properties as a function of time. Fig. 11 shows dependencies of the limiting shear stress (P_{rs}) as a function of time for BSA at the water- toluene interface (curve 5). The effect of added salt (alkali metal chlorides) on the dynamics of interface layer formation can be seen easily. The dependencies for LiCl, NaCl, KCl or CsCl (curves 1-4, salt concentration 0.5 M) are analogous to curve 5 (no added salt) and pass through a maximum.

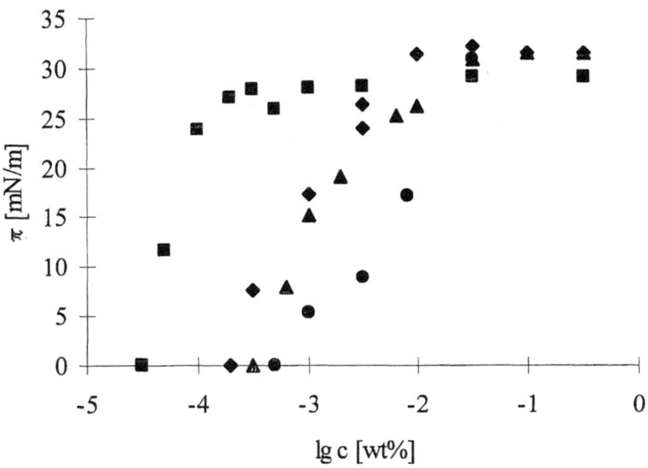

Fig. 10. Two-dimensional pressure isotherms Π(c) of interfacial layers of BSA (■), BSA-DS complexes at different composition W=0.2 (◆), 0.4 (▲) and 0.6 (●)

The type of the dependencies in Fig. 11 reveals a mutual competition of several (at least of two) processes during the interfacial layer formation. It seems that high protein concentrations intensify the mass transfer towards the interface resulting in the formation of a number of random contacts and, because of this, a relatively rigid strained structure is created. This structure is thermodynamically unstable and disappears with time following a mechanical break down with a transition to equilibrium structures. P_{rs} changes are observed during the initial period of 30 minutes in absence of added electrolyte and during a significant longer time of 40 to 60 minutes in the presence of electrolyte (LiCl or KCl and NaCl or CsCl). The addition of salt leads to a decrease in the initial rates of interface layer formation as well as affects the rates of transition to equilibrium interfacial structures. Thus the dynamics of interfacial layer formation characterised in terms of rheological parameters allows to consider the formation of non-equilibrium coagulation structures as an intermediate process leading to the creation of a third phase at the interface in equilibrium with two liquid phases.

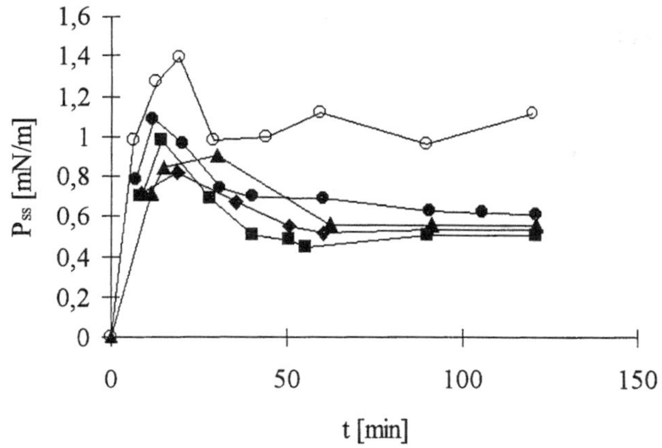

Fig. 11. The dependence of limiting shear stress (P_{rs}) on time for BSA at the water-toluene interface, 0.5 M salt concentration of (■)- NaCl, (◆) -KCl, (▲) -CsCl, (●) -LiCl, (O) -no added salt. $C_{BSA} = 10^{-5}$ M, T=22°C. Values of P_{rs} were measured by the method of constant $\dot{\varepsilon} = 0.042$ rad·s^{-1}

At this stage we can conclude that the dynamics of the formation of protein interfacial layers between two liquids (for example, aqueous protein solution- hydrocarbon) consists of the following kinetic stages:

1. solubilisation of non-polar molecules by proteins resulting in the formation of associates in the aqueous phase,
2. diffusion of solutes from the solution bulk to the subsurface,
3. adsorption of molecules at the interface,
4. relaxation processes connected with the orientation of macromolecules due to the surface topography,
5. formation of coagulation structures at the interface due to interparticle contacts,
6. break-down of coagulation structures and transition to an equilibrium middle phase,
7. protein equilibrium partitioning among the two liquid phases and the interface layer, including mass transfer into the non-polar phase.

6. DISTRIBUTION OF PROTEIN BETWEEN THE LIQUID PHASES

Let us now consider the protein transfer into the hydrocarbon phase and the protein equilibrium partitioning between water, the interface layer and the hydrocarbon [26], using the method of radioactive indicators with a scintillating phase and tritium labelled BSA. The method was developed to study the component distribution and has been described in detail elsewhere [13]. This approach allows processes of protein accumulation at the interface as well as transfer into the hydrocarbon phase to be observed without disturbing the system. BSA at the aqueous solution-toluene and α-chymotrypsin at the aqueous solution-toluene interfaces were investigated in detail in [14, 26]. As the result of these investigations a scheme was proposed accounting for the formation of different kinetic units including protein in both liquid phases and in the PIL.

Systems consisting of aqueous protein solution and hydrocarbon are in general characterised by numerous competing processes leading to the equilibrium, each of them have its own kinetic parameters. Transition processes can be described by the scheme given in Fig. 12.

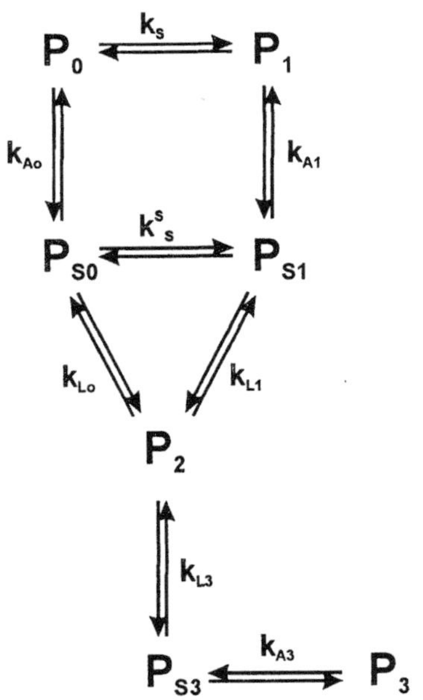

Fig. 12 Scheme of transient processes in the system aqueous protein solution- hydrocarbon: k_S - equilibrium constant of hydrocarbon solubilisation in the aqueous phase, k_S^S - constant of equilibrium solubilisation of hydrocarbon by protein at the interface; k_{A0}, k_{A1}, and k_{A3} are constants of the equilibrium mass transfer to the interface for particles P_0, P_1 and P_3, respectively (details in text); k_{LO}, k_{L1} and k_{L3} - constants of the equilibrium formation of PIL from structure elements P_0^S, P_1^S and P_3^S, respectively. P_2 - PIL of definite composition and structure.

In [27] the average hydrodynamic radius (R) of associates of BSA with non-polar molecules in the adjacent phases of aqueous protein solution and hydrocarbon were determined by the quasi-elastic light scattering method. Results are given in Table 9. Note that the size of scattering particles in the aqueous phase in equilibrium with the hydrocarbon are essentially larger than the size of BSA molecules in the aqueous solution. The size of BSA molecule in aqueous solutions of different concentration were measured using SANS [28]. In low concentrated solutions the diameter of hydrated BSA is 6.1 nm and increases slowly with concentration. The results obtained allow to state that the saturation of the aqueous phase with hydrocarbon leads to the appearance of associates of protein and hydrocarbon molecules due to the solubilisation of the hydrocarbon. The size of these associates reduces with increasing buffer molarity.

Table 9. The influence of the nature and the concentration of lipid and the molarity of an acetic buffer on the diffusion coefficient (D) and the mean hydrodynamic radius (R) of scattering particles in aqueous and benzene phases: pH 6.7, T=298 K, CBSA=1.2.10-5M

Buffer molarity M	The initial composition of the non-polar phase	Water phase			Benzene phase		
		$D\times10^{-7}$ cm^2/s	R nm	error %	$D\times10^{-7}$ cm^2/s	R nm	error %
0.01	Benzene	3.27	7.9	1.5	0.34	107	0.5
	Lecithin in benzene						
	1.3×10^{-4}M	1.52	17.0	0.5	0.28	127	0.7
	1.3×10^{-3}M	1.43	18.0	0.6	5.43	6.6	1.9
	Cholesterol in benzene						
	2.6×10^{-4}M	1.61	16.0	1.0	0.40	90.0	1.3
	2.6×10^{-3}M	1.53	17.0	0.7	0.47	76.0	1.3
0.1	Benzene	4.09	6.3	1.3	0.58	61.0	1.7
	Lecithin in benzene						
	1.3×10^{-4}M	1.84	14.0	1.4	1.29	28.5	1.2
	1.3×10^{-3}M	2.13	12.1	0.9	5.52	6.5	0.7
	Cholesterol in benzene						
	2.6×10^{-4}M	1.94	13.3	0.8	0.72	50.6	1.1
	2.6×10^{-3}M	2.15	12.0	1.7	2.07	17.3	0.9
1	Benzene	4.58	5.9	0.8	0.74	48.3	1.2
	Lecithin in benzene						
	1.3×10^{-4}M	1.84	14.0	0.7	1.89	18.9	1.3
	1.3×10^{-3}M	2.48	10.4	1.5	5.87	6.1	0.6
	Cholesterol in benzene						
	2.6×10^{-4}M	2.21	11.7	1.2	0.81	44.1	0.8
	2.6×10^{-3}M	2.30	11.2	1.0	2.32	15.4	1.2

Therefore large size scattering particles are recorded in benzene. Transfer of protein from the solution into the interface layer as well as into the organic phase was shown by means of the radioactive indicator method [26]. The results show that an equilibrium distribution of protein between both liquid phases and the interfacial layer is achieved. Probably the particles contain protein together with other components of the system. However, it is of interest to understand the reason for the large differences in size of the associates in water and in non-polar phases measured in our experiments. When the number of components in the system is increased, for example by addition of lipids, associates of another sizes appear (Table 9).

Associates of BSA with lipids were recorded after mutual saturation of the adjacent aqueous and benzene phases. The saturation process is accompanied by a transfer of protein and water from the aqueous solution into the benzene phase and simultaneous transfer of lipid and benzene into the aqueous phase. While considering two flows with opposite directions, it is necessary to take into account the process of mass condensation of protein and lipid at the interface. In the moment of contact of water and the oil phase, processes of mass transfer as well as solubilisation and association begin. These processes involve all components and, probably, result in an equilibrium distribution of components between the macrophases and the interface layer. At any rate the achievement of such equilibrium was proven for α-chymotrypsin and BSA. The equilibrium is achieved in about 24 hours, after which aliquots of aqueous and benzene phases were separated. In such multi-component systems the radii of detected associated particles change essentially. The size of particles in the aqueous phase due to the formation of new associates growths approximately by a factor of 2 with the same effect of buffer molarity. In the organic phase a drastic effect on the particle size is observed with increasing lipid concentration. When lecithin is added to the system in concentration $1.3 \cdot 10^{-3}$ M the radii of equilibrium particles become independent of the buffer molarity. The effect of cholesterol is approximately the same. For multi-component systems containing protein, water, hydrocarbon and lipid an essential lyophilisation takes place, and as it was considered above, a drastic decrease in the interfacial tension is recorded (Fig. 1 and 10). When the lipid concentration is increased in the initially two-phase liquid system, but the concentration of protein is kept constant, essential lyophilisation of the system is observed. The amount of protein in the interface layer increases, the interface tension is significantly lower, and the rheological parameters of the PIL are simultaneously reduced, revealing a phase transition of the PIL. This is accompanied by a decrease in the difference of the particles size in the aqueous and organic phases. Such particles must be treated as associates or microemulsion droplets containing at least four components: protein, lipid, benzene and water. The composition of the associates is determined by properties of the phases, in which they are found and by the composition of the contacting liquid phases. But it is evident, that the structure of a particle must correspond to the nature of continuous liquid phases, i.e. reveal attributes of a direct or reverse systems.

Protein in aqueous solution may exist in two different states: as the part of a kinetic unit P_0 (before the contact with a hydrocarbon phase) and as a part of kinetic unit P_1 (after achievement of equilibrium of contacting phases). P_0 is the typical water soluble form of a protein (monomers or dimmers). Associated forms of proteins are produced by globules. P_1 is a larger associate of hydrophobic ("oiled") globules of protein with solubilised hydrocarbon.

The protein conformation in the particle P_1 in contrast to that in the particle P_0 changes insignificantly, as it was discussed above. When protein denaturation takes place due to any reason, this would shift the equilibrium of the system as a whole. A number of examples are known now which show a variation of the phase behaviour of reverse microemulsions formed by native and denaturated proteins [29]. We can suppose that protein at an interface may exist in several main states: as particles P_0 or P_1, oriented and "fixed" at the interface, resulting in the appearance of particles P_0^S or P_1^S involved in the interphase film P_2 of a specific rheological behaviour. According to data of water permeability measured by the radioactive indicator method, the state of P_2 corresponds to various types of phase structures, network, solid-like or liquid-crystalline-like or microemulsion, and depends on the initial protein concentration in the aqueous phase. The main property of P_2, however, is a long range interaction order, responsible for the strength of the interfacial film. Therefore, P_2 can be considered as a continuous spatial associate of structure elements, with non-deformed globules of protein, packed into a structure of a definite type.

Proteins in the organic phase can be included in particles P_3 distinguished from other ones (P_0, P_1, P_0^S, P_1^S or P_2) not only by the amount of interconnected elements of structure, but also by the nature of contacts and structure. Protein globules remain nevertheless the main structural elements.

During the direct transition from the aqueous into the organic phase (in the first region of the kinetic curve) adsorption of P_0 or P_1 takes place resulting in the appearance of P_0^S and P_1^S. The formation of the interfacial layer (interfacial structure) by adsorbed P_0^S and P_1^S is completed in the second stage: $P_0^S \Rightarrow P_2$ as well as $P_1^S \Rightarrow P_2$. As a result of such transition an excess of protein with solubilised hydrocarbon can be transferred either into water as a particle P_1 or into the organic phase as a particle P_3. The processes described indicate the way of involving components into the PIL formation, resulting in an approach of the system to a steady state.

Special experiments to prove the achievement of the equilibrium state of such systems were performed such that an aliquot of saturated organic phase was picked up and spread onto the surface of pure water. The formation of a PIL was recorded for this new system using the radioactive indicator method: $P_3 \Rightarrow P_3^S$. This process is completed with a stable PIL formation: $P_3^S \Rightarrow P_2$. Then a slow stage of P_1 transfer from P_2 into the aqueous phase is observed: $P_2 \Rightarrow P_1$. We can conclude that all the processes may be divided into three groups:

- adsorption and fixation of protein or its associates with other components at the liquid-liquid interface. Evidently, preferential wetting of particles plays a decisive role in this stage [30],
- phase transition from an intermediate PIL into an equilibrium PIL: $P_0^S \Rightarrow P_2$ as well as $P_1^S \Rightarrow P_2$, or $P_3^S \Rightarrow P_2$,
- protein equilibrium distribution between the interfacial layer and the liquid macrophases characterised by the partitioning of the types $P_0 \Leftrightarrow P_1 \Leftrightarrow P_2 \Leftrightarrow P_3$; the sum $P_0 + P_1$ characterises the amount of protein in the aqueous phase, P_2 and P_3 correspond to the amounts of protein in the PIL and the organic phase, respectively.

The establishment of the equilibrium state in aqueous protein solution - hydrocarbon systems as a whole is accompanied by mass transfer processes. Equilibrium, in its turn, is attained as a result of a number of competing processes. In this work, such processes are considered which mainly take into account the distribution of protein. As for hydrocarbons, its solubilisation in the aqueous phase and also NMR data reveal a specific state of benzene in a gelatine interfacial layer. This allows to suppose an equilibrium distribution of hydrocarbon in such systems. It is evident that all components of the system distribute between water, the interface and the oil phase. However, at present the distribution of water in the system is not sufficiently investigated.

7. INFLUENCE OF ADDED SALT ON THE FORMATION AND PROPERTIES OF PROTEIN INTERFACIAL LAYERS AND RELATED PHENOMENA

The thermodynamic equilibrium of a system of two immiscible liquids (water-hydrocarbon) with an added effective surfactant is an important object of investigation in modern colloid chemistry. The phase behaviour determines the stability of emulsions and microemulsions, and phase inversions. One of the main problems is the study of the mechanism of surfactant concentration as a middle phase between liquid phases during the formation of the stabilising surfactant layer during emulsification or development of a middle phase in microemulsions. The method of radioactive indicators allows to study the distribution of components, the PIL formation and the achievement of the equilibrium state in systems of aqueous protein solution-hydrocarbon.

In [31] the influence of added electrolytes was investigated, including the PIL formation and the distribution of BSA between water and hydrocarbon. The necessity of such investigations is caused by the fact that addition of electrolyte affects the phase behaviour of water-hydrocarbon-surfactant systems and is often used to reach ultra low interface tensions. In the same time this problem is not investigated systematically for systems containing proteins as surfactant and alkali metal salts as additives. As example, Figs. 13 and 14 show the dependence of protein mass accumulation at the interface (in units of adsorption) Γ and protein concentration in the organic phase as function of equilibrium protein concentration in the aqueous phase C_e (in bi-logarithmic co-ordinates), when NaCl is added to the system.

Three regions corresponding to various interfacial coverage are separated by vertical and horizontal lines. The curves lg Γ (lg C_e) and lg C_{SPH} (lg C_e) have two linear regions and an intermediate region: the first one corresponds to low interfacial protein packing, i.e. at $\Theta < 1$; the second one corresponds to the model of densely packed monolayer, and the third region corresponds to the formation of the third phase, visually observed. This phase is a thin layer of a middle phase.

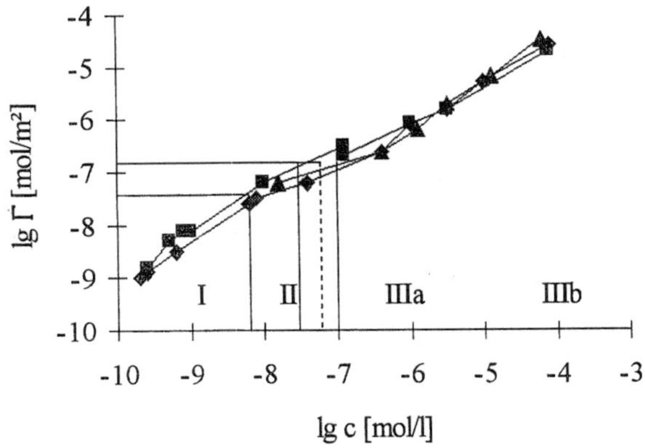

Fig. 13. BSA accumulation at the interface as a function of equilibrium protein concentration in the aqueous phase: (■) - no added salts, (◆▲) - added NaCl in concentrations 0.2 and 1.0 M, respectively; vertical and horizontal lines show limits of monolayers stability accounting for different BSA orientations.

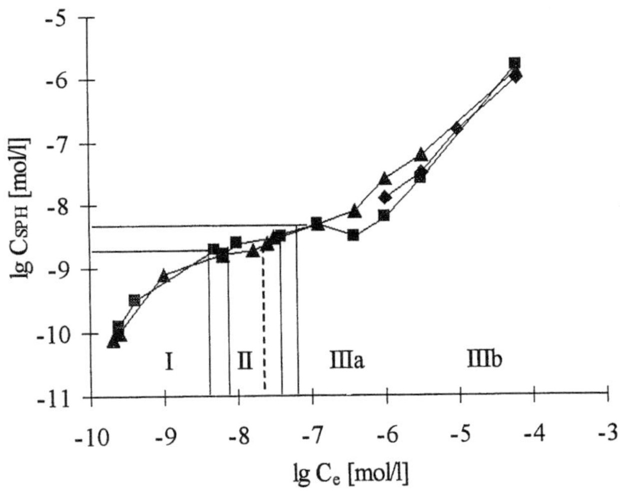

Fig. 14. BSA concentration in toluene as a function of equilibrium protein concentration in the aqueous phase: (▲) - no salts added, (■◆) in the presence of NaCl, 1.0 and 0.2 M, respectively. Vertical and horizontal lines as in Fig. 13.

It should be pointed out that the amount of protein, measured in the hydrocarbon phase is large enough, and is promoted by the contact of the two liquid phases, i.e. by the presence of water and solubilisation of hydrocarbon by protein in aqueous solution. After addition of salts the PIL in the region of a densely packed protein monolayer becomes stable in a wider range of equilibrium protein concentrations, and a transition to multilayer structures is observed at larger protein concentrations. On the other hand the addition of salt leads to essential changes in the rheological properties of the PIL corresponding to a protein multilayer adsorption with a drastic reduction in the rheological parameters (Fig. 15).

A typical curve of the flow of BSA interfacial layer is represented by the dependence (✧) in Fig. 15. The addition of salt decreases the flow limits (LiCl< KCl<CsCl<NaCl), the PIL loses the solid-like properties and becomes a non-newtonian (anomalous) liquids. The dependencies of effective viscosities on the shear stress and differences between viscosities of the PIL with non-destroyed and destroyed structure show a real phase transition of the PIL under salt addition (Fig. 16).

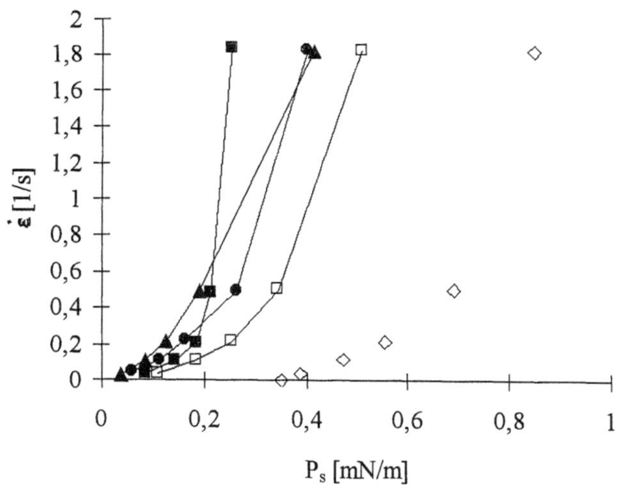

Fig. 15. Influence of added salts on the flow curves of BSA interface layers, formed at the water-toluene interface: ■-NaCl, ●-CsCl, ▲-KCl, □-LiCl, ✧- no added salt; C_{BSA} is $1 \cdot 10^{-5}$ M, age of the PIL is 60 min. Salts are added in concentrations of 0.5 M.

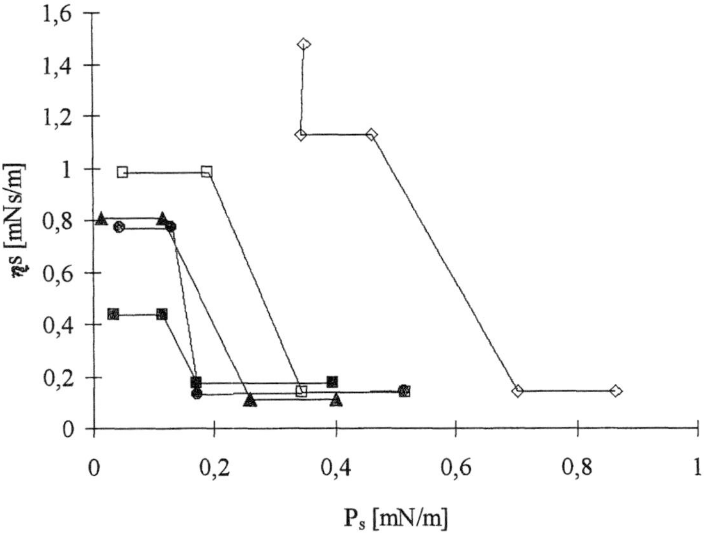

Fig. 16. Effective viscosity of BSA interfacial layers as a function of shear stress. Conditions and symbols are the same, as in Fig. 15.

To some extent the effect of salt is analogous to that observed in similar systems with low molecular mass surfactants: surface active substances are concentrated at the interface resulting in the formation of a middle phase. The characteristic feature of protein containing systems consists in the separation of a more or less thick milk-white film of the third phase which can be observed visually. The formation of this film as a macroscopic transparent liquid phase (as it is specific for three-phase-microemulsions of some low-molecular mass surfactants) was not observed under experimental conditions for proteins. This seems to be obviously the consequence of the intensive transfer of protein containing particles into the organic phase.

8. ROLE OF PIL IN THE STABILISATION OF THIN EMULSION FILMS

The properties of surface and interfacial protein adsorption layers influence the formation of thin (in the limit black) free (foam) and emulsion films. Such thin films are regarded as models to study the stability of foams and emulsions. It was shown elsewhere [32, 33] that black foam films are formed from aqueous protein solutions, when two necessary conditions are fulfilled: the pH of the solution is equal to the isoelectic point of the protein, and the protein concentration is close to the value at which the surface tension has reached a minimum value.

Thin films simulating o/w emulsions are poorly investigated in contrast to BLM. Water soluble globular proteins stabilise o/w thin emulsion films at various conditions. In many cases stable thick films (thicker than black films) are formed. Investigations of thin emulsion films, as discussed in [2], deal with the determination of conditions of the formation of black emulsion films and the physico-chemical factors affecting the film stability or its rupture. All experiments were performed with the Scheludko-Exerowa device and α-chymotrypsin and BSA were used as stabilisers. The first condition for the film stability of emulsion is the formation of protein adsorption layers on the meniscus surfaces during more than 30 minutes before the generation of the film by approaching of menisci. The following parameters have been varied in order to determine a diagram characterising the thickness of stable films: protein concentration, pH and temperature of the aqueous solution. Thinnest (black) films of α-chymotrypsin are formed in a narrow range of protein concentration ($6.6 \cdot 10^{-6}$ -$1.6 \cdot 10^{-5}$ M), pH 6.0-7.2 and temperatures 20-30°C. The formation of thick emulsion films and dimples with large radii were observed in many cases [35-36] and is connected with the polymolecular character of adsorption and structure-rheological properties of the PIL. These thick films affect also the emulsion stability, but the most stabilising effect is connected with the formation of black films. In the case of emulsion black films are formed by the following processes: a repeatedly jump-like film formation of smaller thickness of an initially stable thick film; the PIL flow in the process of film thinning and, at last, dispersion of particles from the PIL into the non-polar phase.

According to the changes in the PIL properties observed when lipid is added to the protein solution the properties and stability of emulsion films change as well. When lecithin is added to the organic phase, grey instead of black films are stable. With further increase in lecithin concentration up to $1.3 \cdot 10^{-4}$ and $1.0 \cdot 10^{-6}$ M for BSA and α-chymotrypsin, respectively, emulsion films become unstable with respect to the loss of elastic properties of PIL and lyophilisation of the system as a whole. Thermodynamic properties of the system affect the film instability.

9. MODIFICATION OF THE INTERFACIAL BEHAVIOUR OF GELATINE USING LOW MOLECULAR MASS SURFACTANTS

A variation of interfacial properties of natural macromolecular substances is possible through an addition of low molecular surfactants resulting in the formation of new surfactants [37].

Characteristics of interfacial adsorption layers (aqueous binary mixtures of gelatine and SDS at the water-decane interface) on various SDS concentrations (sodium dodecyl sulphate) are given in Fig. 17 [38]. Part A shows the interfacial tension isotherm. The interface tension decreases from 29.2 mN/m (at $C_{SDS} = 0$) to 6.1 mN/m at a SDS concentration of $2 \cdot 10^{-4}$ M. According to the γ_∞-values the interface is saturated with gelatine - SDS complexes at SDS concentrations below the CMC. Similar results have been obtained earlier at the water-air interface (cf. Chapter 11). The relaxation time of gelatine-SDS complexes of different composition at the water-decane interface determined by the amount of surfactant bound to gelatine increases from 11 to 30 minutes passing through a minimum (Part B).

The thickness of gelatine interfacial layers was measured in a SDS concentration region where complex formation possess a minimal relaxation time and emulsion films are stable. The results of these measurements show that the presence of SDS decreases the interfacial adsorption layer thickness at the gelatine isoelectric point by about 20 times, from 83 to 4 nm.

Both stable and unstable emulsion film formation depends on the composition of the solution containing gelatine and a surfactant of various content (Fig. 17) [39]. Stable films are formed when stabilised by hydrophobic gelatine-SDS complexes at a ratio $7.7 \cdot 10^{-5}$ to $9.5 \cdot 10^{-3}$ g SDS per g gelatine otherwise unstable films appear. This is in accordance with our understanding of the role of structural-mechanical barriers (Izmailova and Rehbinder [8]). Let us consider now the rheological parameters of interfacial adsorption layers in the range of stable and unstable emulsion films (Fig. 17, Parts C and D). For gelatine-SDS complexes one can observe a non-linear dependence of the rheological parameters on the complex composition. The shear stress P_{ss} and the viscosity η_s at a stationary layer flow reach a maximum in the region of emulsion film stability corresponding to a complex composition of $1.9 \cdot 10^{-4}$ g SDS per g gelatine. Further increase of SDS in the complex results in a decrease of the PIL rheological parameters and finally they appear compatible with PIL parameters in the absence of surfactant, emulsion films loose their stability. In the region of stable emulsion films stabilised by gelatine-SDS complexes P_{ss} values are approximately $2 \cdot 10^{-2}$- $4 \cdot 10^{-2}$ mN/m.

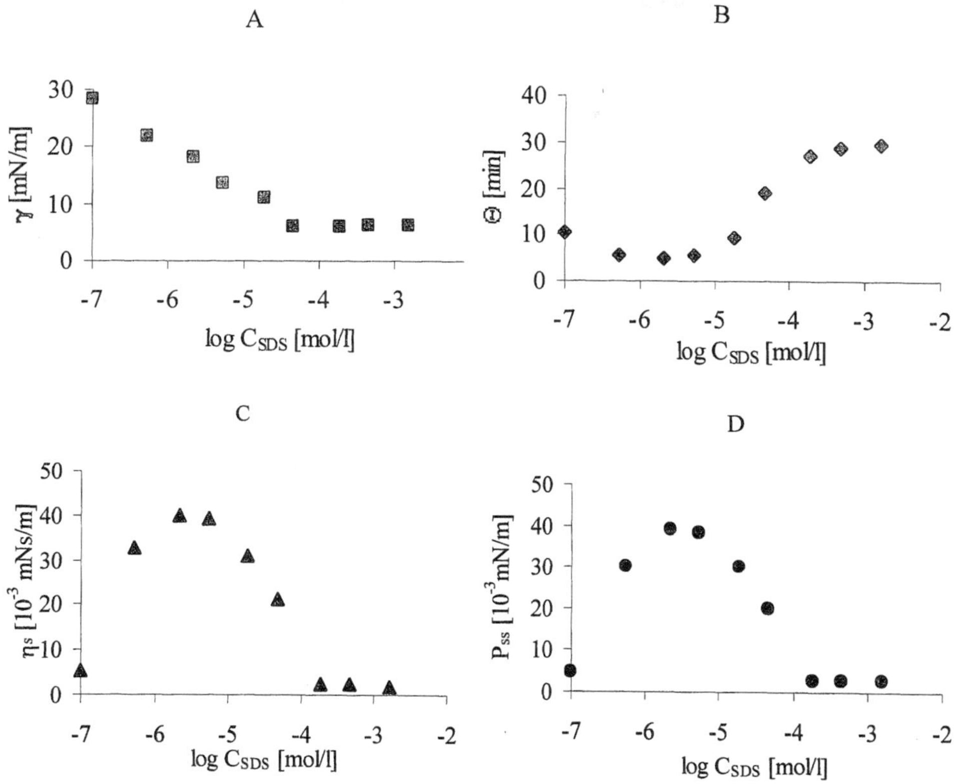

Fig. 17. Rheological parameters of interfacial adsorbed layers (A - interfacial tension γ, B - relaxation time Θ, C - viscosity η_s, D - shear stress P_{ss}) of gelatine - SDS complexes forming stable and unstable emulsion films at the water-decane interface, Cg = 0.3 %, pH 6.0, T = 293 K.

When calculating the rigidity per cm^2 of the gelatine interfacial layer becomes equal to $P_v = 5.6 \cdot 10^4$ mN/m^2, while for PIL formed by gelatine-SDS complexes at a SDS concentration of $2 \cdot 10^{-5}$ mole/l we have $P_v = 7.9 \cdot 10^6$ mN/m^2.

Thus, gelatine-SDS complexes are capable of forming interfacial structures of less thickness which posses increasing strength compared to those formed by gelatine macromolecules alone. Such complexes are good stabilisers of emulsion films. In Fig. 18 the ranges of emulsion stability are given (a) as well as an idea the construction of the mixed SDS/gelatine layers in the emulsion film (b).

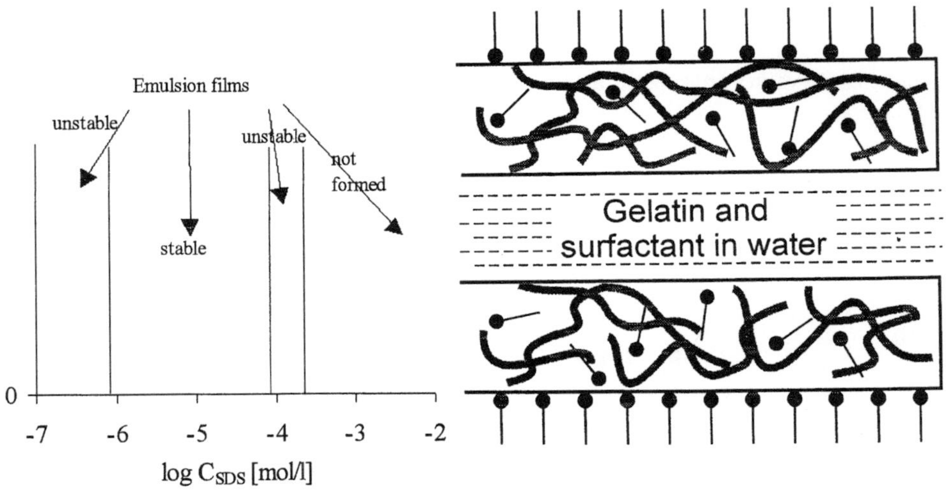

(a) (b)

Fig. 18 Schematic of stability range as a function of the SDS concentration (a) and the structure of an emulsion film stabilised by a SDS/gelatine complexes at the water-decane interface, Cg = 0.3 %, pH 6.0, T = 293 K

10. DISCUSSION

Mixture of surfactant, water and hydrocarbon form a variety of phases. Proteins are surface active substances and accumulate at water-oil interfaces. In dependence on the protein concentration the system initially consisting of equal volumes of aqueous protein solution and hydrocarbon remains two-phasic with a low interfacial packing at protein concentration of about 10^{-7} M, approaching a closely packed monolayer and revealing a phase transition with the formation of a third interfacial phase at protein concentrations larger than 10^{-7} M. In the first case the interface shows properties of a two-dimensional newtonian liquid. In the second case the PIL shows specific rheological properties characteristic for a thin layer of a third phase which appears after the phase transition from a two-dimension liquid to a solid-like state and in turn a transition from a network to another solid-like phase. In certain cases, for example for BSA, a further increase in protein concentration a new two-dimensional phase transition leads to a more liquid state with a loss in elasticity, which is more extensive with addition of salts, lipids etc.

The network formation is clear in terms of a phenomenological physical concept of "complex fluids" [7]. Hard core particles give a meso-phase following a freezing transition, revealing a coexistence of fluid and solid states at volume fractions of 0.49 and 0.55, respectively. Evidently, this analogous transition proceeds when protein is accumulated at the interface in a sufficient quantity.

There is another way to interpret structure formation at low interfacial packing - the consideration of equilibrium processes of coagulation = peptisation [40]. Associated particles, located at the interface, tend to interact and, when their average energy of interaction is larger than 10-15 kT stable coagulation contacts arise stabilising two-dimensional networks.

Freeze-fraction electron microscopy is one [7] among available methods [41-43] for the investigation of phase transitions in the system water-oil-surfactant. This method was used to obtain images of PIL. Fig. 19a shows the network structure of α-chymotrypsin interfacial layer, formed at a protein concentration of about $4.4 \cdot 10^{-8}$ M. Even at such low concentration the formed PIL reveal specific structures of a two-dimension network, in good agreement with rheological measurements of PIL formed under the same conditions. The main element of the network structure are spherical particles, composed by spheres of smaller sizes. In many experiments the smallest particles observed in micrographs correspond to the size of α-chymotrypsin molecules. This indicates a fractal mechanism of the PIL formation.

It is well known that colloidal particles with strong long-range interaction tend to order at a definite concentration resulting in periodic structures [44]. Such type of structure was observed at the liquid-liquid interface for lysozyme at an initial concentration of about 10^{-7} M in the aqueous phase and $\theta < 1$ (Fig. 19b).

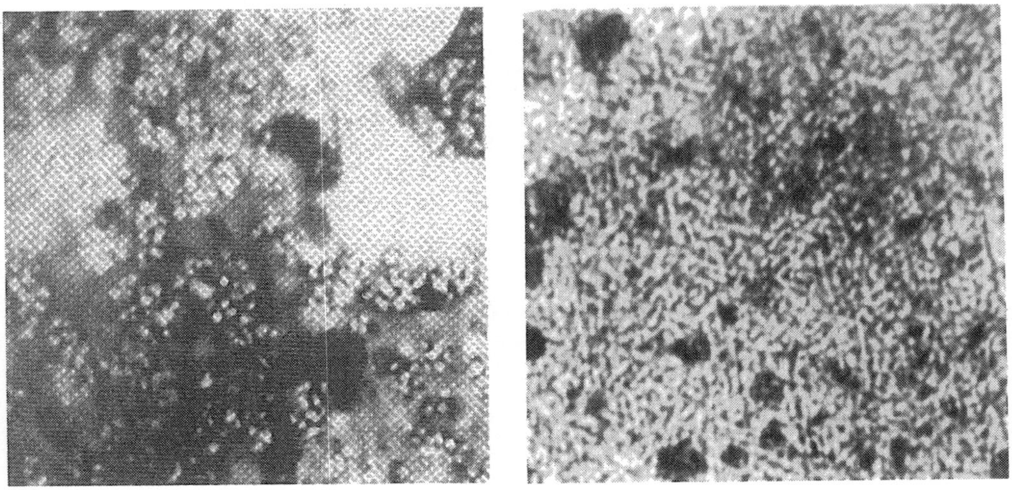

Fig. 19. Elecron micrograph of PIL at the water-octane interface: a-α-chymotrypsin, b-lysozyme.

According to modern concepts, the phase behaviour of the systems under consideration is determined by the tendency of surfactant to accumulate at an interface until saturation, which is accompanied by a depression of the interfacial tension. For such systems saturated surfactant monolayers are the main construction blocks, and their properties determine the type of the equilibrium phase formed between the two liquid phases. Monolayers can acquire various topologies: spherical, cylindrical, planar, inverse spherical and inverse cylindrical, sponge-like, etc. Each topology corresponds to a specific phase, which can be isotropic or anisotropic. During the last decades it became evident, that the topology acquired by monolayers strongly depend on its bending elasticity and spontaneous curvature of surfactant [45]. Not only stretching but also monolayer bending with some radii of curvature costs energy. Deviation of the monolayer curvature from a specific value, intrinsic for the monolayer of a given surfactant, increases the free energy (dG) according to Hooke's equation, that was first taken into account in [45]:

$$dG = [2k_1 (H-H_0)^2 + k_2 J]dS \qquad (4)$$

Here $H=c_1+c_2$ and $J=c_1 \cdot c_2$ are the mean and Gaussian curvatures, respectively, with c_1 and c_2 are the two principle curvatures, H_0 is the spontaneous curvature, dS is the surface area of the monolayer patch, k_1 is the bending modulus, and k_2 is the saddle splay modulus.

The physical meaning of the spontaneous curvature is related to the shape of the surface active molecule. When polar and non-polar parts of the surfactant molecule have approximately the same volume, H_0 tends to zero. Under this condition with increasing surfactant concentration the monolayer forms planar or saddle-like structures, i.e. lamellar liquid-crystalline, cubic-crystalline and sponge-like phases, controlled by the ratio of the size of non-polar and polar protein parts. Note, that the bending modulus k_1 is always positive and characterises the flexibility of the surfactant monolayer. When k_1 is larger than the thermal energy kT, the monolayer of any surfactant is rigid resulting in the formation of more ordered liquid-crystalline phases. If k_1 is comparable with kT new highly disordered phases appear. k_2 can be positive or negative. In the first case saddle-like surfaces are favourable with respect to flat surfaces. At $k_2<0$, the formation of separate disperse particles is preferred.

It seems that the competition of elastic properties of protein monolayers at liquid-liquid interfaces determines the evolution of rheological parameters of PIL with increasing protein concentration and addition of salts or lipids.

At high protein concentration globular particles reveal the tendency to form linear structure elements having the ability to long-range ordering, as this was found in studies of α-chymotrypsin interfacial layers formed on drops in concentrated emulsions of benzene in water (Fig. 20). Analogous processes have been recognised in the work [46] for closely packed protein monolayers at the water-air surface. This can be regarded as a proof of liquid crystalline states of thin films of both types. Onsager [47] has shown that above a certain volume fraction the free energy of the system of colloidal hard rods is minimised by orientational distribution which is no larger uniform but involves a preferred direction for particles axes. This case is differed from "ordinary" colloidal systems of long, rigid macromolecules (e.g. Tobacco mosaic virus and some polyamino acids and their derivatives) formed nematic liquid-crystalline phases. In case of PIL simultaneously with the onset of long-range orientational ordering an increase in the average size of the rods takes place. This is because each colloidal particle here is itself an aggregate comprised of a large number of individual molecules which are in exchange equilibrium with those in all other particles. Accordingly, upon alignment, the system can organise itself into a smaller number of large particles, in order to decrease the loss of orientational entropy (as well as to further decrease their excluded volume).

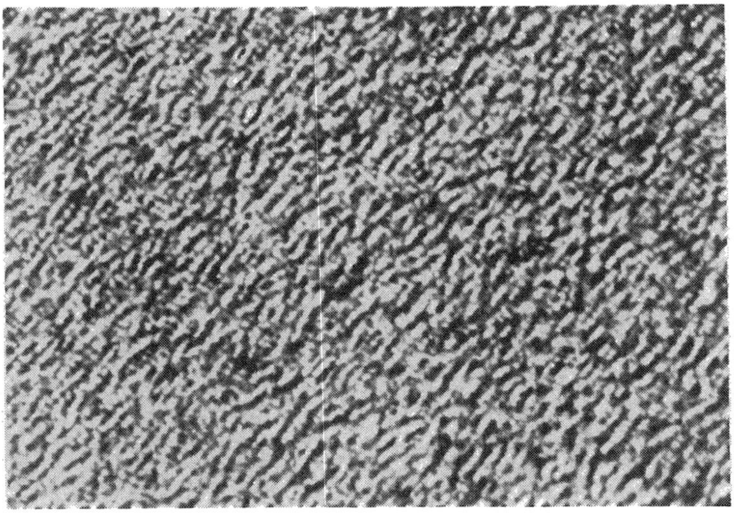

Fig. 20. Electron micrograph of α-chymotrypsin interfacial layers on drops at stabilising an o/w emulsion.

It is accepted, that the phase behaviour of a system can be evaluated by using an average elasticity modulus K calculated for one surfactant molecule [7, 48]. In general $K = 10$ kT for a pure surfactant monolayer. In this case the interface is stiff. Neither kind of surfactants is suitable for the formation of microemulsions and, respectively, to provide very low interfacial tensions. Very strong amphiphiles are preferentially located at the oil-water interface instead of dissolving in one or both of the bulk phases. In this case, once the original water-oil interface is saturated, all additional surfactant participates in the creating of new interfaces, according to the elastic properties of monolayers. Elastic moduli depend sensitively on the chemical composition of the interfacial layer and surfactant nature and size. By mixing the surfactant with a comparable amount of a co-surfactant (for example lipid or protein) K is largely reduced, typically to $K \approx kT$ at room temperature. Such reduction of the modulus provides strong wrinkling of an interface.

Accurate measurements, modelling and calculation of bending moduli are performed recently only for simple surfactants [49, 50]. All characteristics involved in Eq. (4) are unknown for natural surfactants such as proteins or lipids. But for these substances rheological parameters of interfacial layers, monolayers, bilayers and other thin films are well determined. Shear elasticity moduli of saturated PIL are about 1 mN/m. Taking into account the area of an average globular protein in a saturated monolayer, it is possible to determine the corresponding modulus in erg/molecule. When the area of a protein molecule is approximately 30 nm^2, the elasticity modulus is 3 10^{-13} erg/molecule or even several times larger. One can suppose, that such material must reveal a large bending modulus of the same order of magnitude. These evaluations indicate that at high protein concentrations and $\Theta > 1$ the PIL as a rule, must be represented by lamellar phases. Addition of salts and lipids makes these phases unstable.

As it was mentioned above biomembranes are highly elastic films. The composition of the lipid mixture and the protein content in cell membranes are highly specific to the function of the cell. Most cells do not function if the delicate balance of lipids and proteins is perturbed. Such drastic transitions have been shown above when the formation of PIL occurs in the presence of lipids. It is known that interfacial layers of lecithin are characterised by an elasticity modulus of 10 kT, are very rigid and do not form microemulsion in absence of additional components and, thus phospholipids form bilayers and lamellar phases [7]. Addition of a small amount of short-chain alcohols makes the mixed monolayer much more flexible and one can observe a typical microemulsion phase behaviour [51]. When lecithin is mixed with other lipids, surfactants or protein they show typical microemulsion phase behaviour. This is evidently true and for PIL with addition of lipids, and the whole variety of modifications of PIL properties can be interpreted as an reduction of the elasticity modulus resulting in a strong lyophilisation of the interface.

A first description of lyophilic, thermodynamically stable systems were given by Rehbinder and Shchukin [52], connecting the interfacial tension (γ_{12}) and the particle size (δ) of the dispersed phase at the constant temperature:

$$\gamma_{12} \leq \beta \, kT / \delta^2 \qquad (5)$$

Eq. (5) allows to evaluate the entropy gain of dispersion and as a first approximation can explain the dispersion of particles from PIL in the organic phase.

According to the Gibbs' phase rule, the number of thermodynamic degrees of freedom of a macroscopic system increases by one whenever a new molecular species is added. This larger number of composition degrees of freedom in multi-component system is generally accompanied by a richer and more complex phase behaviour. This trend is even more drastically manifested in complex fluid mixtures of self-assembled amphiphiles. Structural transformations and phase transitions in these systems can appear in different forms and on different length scales. Some phase transitions take place within one aggregate, others involve a transition from one aggregation geometry to another but within the same macroscopic phase and certain transitions are accompanied by macroscopic phase separation. Quite often a phase transition takes place simultaneously on all length scales. Proteins are generally more rigid resulting in a considerable reduction of the conformational freedom of chains of the surrounding lipid molecules. This perturbation results in an elastic deformation, and of the free energy increase of the local lipid environment which is proportional to the surface area of the protein facilitating the dispersion formation from the interfacial layer.

11. REFERENCES

1. Izmailova V.N., Yampolskaya G.P. Colloid Chemistry of Proteins, J. of D. I. Mendeleev Chemical Society. 24(1989)81.
2. Izmailova V. N., Yampolskaya G. P., Summ B. D. Surface Phenomena in Protein Systems. Khimia, Moscow, 1988, 240p.
3. Norde W. Cells and Materials, 5(1995)95
4. Claesson, P. M., Blomberg, E., Fromberg, J. C., Nylander, T. and Arnebrant, T., Adv. Colloid Interface Sci., 57(1995)161.
5. Yampolskaya G. P., Nuss P. V., Rasnikova G. Z., Izmailova V. N., Uspehi kolloidnoi khimii Leningrad, Khimia. 1991. P. 292-305.
6. Richards, F.M., Ann. Rev. Biophys. Bioeng. 6(1977)151
7. W.M. Gelbart and Ben-Shaul, A., J. Phys. Chem. 100(1996)13169.
8. Izmailova V.N., Rehbinder P.A. Structure formation in Protein Systems. Moscow, Nauka, 1974. 268 p.
9. Levachev S.M. and Izmailova V.N. Kolloidnyi Zhurnal, 56(1994)146.
10. Alentiev, A.Yu. and Izmailova V.N., Yampolskaya G.P., Kolloidnyi Zhurnal, 53(1991)609.
11. Izmailova, V.N., Progress in Surface and Membrane Sci., 13(1979)143.

12. Magdassi S. and Kamyshny A., Surface Activity of Proteins, S. Magdassi (Ed.), Marcel Dekker, Inc., New York, Basel, Hong Kong. 1996.

13. Alentiev A.Yu., Filatov E.S., Radiohimiya. 33(1991)80.

14. Yampolskaya, G.P., Bogacheva, E.N. and Izmailova, V.N., Kolloidnyi Zhurnal, 44(1982)1151.

15. Izmailova, V.N., Yampolskaya, G.P. and Tulovskaya, Z.D., Biotehnologiya, Itogi nauki i tekhniki, VINITI AN SSSR, Moscow, 4(1987)199.

16. Izmailova, V.N., Yampolskaya, G.P., Lapina, G.P. and Sorokin, M.M., Kolloidnyi Zhurnal, 44(1982)217.

17. Rodin, V.V. and Izmailova, V.N., Kolloidnyi Zhurnal, 56(1994)91.

18. Rodin, V.V. and Izmailova, V.N., Colloid and Surface. A, 106(1996)95.

19. Izmailova, V. N. and Yampolskaya, G. P., Uspekhi kolloidnoi khimii i phisico-khimicheskoi mechaniki, Moscow, Nauka, 1992, pp. 103.

20. Gurov A.N., Nuss P.V., Nahrung, 30(1986)349.

21. Graham D.E., Phillips M.C., J. Colloid Interface Sci., 70(1979)403.

22. Ward A., Regan L., J. Colloid Interface Sci., 78(1980)389.

23. Nuss P.V., Gurov A.N., Kolloidnyi Zhurnal., 54(1992)123.

24. Blomberg E., Claesson P. M., Tilton R. D., J. Colloid Interface Sci., 166(1994)427.

25. Tripp B.C., Magda J.J., Andrade J.D., J. Colloid and Interface Sci., 173(1995)16.

26. Alentiev A.Yu., Izmailova V.N., Yampolskaya G.P., Kolloidnyi Zhurnal, 54(1992)14.

27. Levachev S.M., Izmailova V.N., Kolloidnyi Zhurnal, 56(1994)193.

28. Nossal R., Glinka C.J., Chen, S.-H., Biopolymers, 25(1986)1157

29. Pileni M.P., Larsson K., Trends Org. Chem., 4(1993)793.

30. Widom B., J. Phys. Chem., 100(1996)13190.

31. Pelekh V.V., Alentiev A.Yu., Yampolskaya G.P., Izmailova V.N., Kolloidnyi Zhurnal, 56(1994)73

32. Yampolskaya G.P., Rangelova N.I., Bobrova L.E., Platikanov D.N., Izmailova V.N., Biofisika, 22(1977)939

33. Yampolskaya G.P., Platikanov D.N., Rangelova N.I., Angarska Zh.K., Bobrova L.E., Izmailova V.N., Kolloidnyi Zhurnal, 43(1981)161.

34. Rangelova N.I., Izmailova V.N., Platikanov D.N., Yampolskaya G.P., Tulovskaya Z.D., Kolloidnyi Zhurnal, 52(1990)515.

35. Yampolskaya G.P., Izmailova V.N, Borisova T.K., Kolloidnyi Zhurnal, 46(1984)544.

36. Ivashchuk Yu.A., Izmailova V.N., Yampolskaya G.P., Kolloidnyi Zhurnal, 48(1986)556

37. Izmailova V.N., Derkach S.R., Zotova K.V., Danilova R.G., Kolloidnyi Zhurnal, 55(1993)54.

38. Izmailova V.N., Derkach S.R., Levachev S.M., Tarasevich B.N., Zotova K.V., Poddubnaya O., Kolloidnyi Zhurnal, 59(1997)178.

39. Derkach S.R., Izmailova V.N., Tarasevich B.N., Zotova K.V., Levachev S.M., Zhurnal nauchnoi i prikladnoi fotografii, 42(1997)54.

40. Yamiskii V.V., Pchelin V.A., Amelina E.A., Shchukin E.D., Coagulation Contacts of Dispersions, Moscow, "Khimia", 1982, pp. 185.

41. Glatter O., Strey R., Schubert K.-V., Kaler, E.W., Ber. Bunsenges. Phys. Chem., 100(1996)323.

42. Langevin D., Ber. Bunsenges. Phys. Chem., 100(1996)336.

43. Lindman B., Olsson U., Ber. Bunsenges. Phys. Chem., 100(1996)344.

44. Efremov I.F., Periodic Colloid Structures. L. Khimia, Leningradskoe otdelenie, 1971, pp. 191.

45 W.Z. Helfrich, Naturforsch., 28(1973)693.

46. Yampolskaya G.P., Izmailova V.N., Nuss P.V., Raznikova G.Z., Izvestia Akademii Nauk, ser. fizicheskaya, 59(1995)109.

47. Onsager L., Ann. N. Y. Acad. Sci., 51(1949)627.

48. De Gennes P.G., Taupin C., J. Phys. Chem., 86(1982)2294.

49. Cantor R., J. Chem. Phys., 103(1995)4765.

50. Blokhuis E.M., Ber. Bunsenges. Phys. Chem., 100(1996)313

51. Kabalnov A., Lindman B., Olsson U., Piculell L., Thuresson K., Wennerström H., Colloid Polym. Sci., 274(1996)297

52. Shchukin E.D., Rehbinder P.A., Kolloidnyi Zhurnal., 20(1958)645.

12. LIST OF SYMBOLS AND ABBREVIATIONS

δ	particle size
Γ	adsorption
γ	interfacial tension
γ_0	surface tension of the pure system
γ_{12}	interfacial tension between immiscible liquids
γ^d	disperse contribution to γ
γ^n	non-disperse contribution to γ
η	surface viscosity
λ	degree of elasticity of interfacial layers
Π	two-dimensional pressure
θ	surface coverage
ATR	attenuated total reflection
BLM	bilayer lipid membranes
BSA	bovine serum albumin
C	concentration
D	diffusion coefficient
DS	dextran sulphate
E	surface elasticity
G	free energy
k	Boltzmann constant
K	elasticity modulus of a surfactant molecules
P_i	kinetic units
PIL	protein interfacial layers
P_k	flow limit
P_{rs}	limiting shear stress
R	hydrodynamic radius
S	surface area
SANS	small angle neutron scattering
SDS	sodium dodecyl sulphate
SPH	scintillating phase
T	absolute temperature
W	weight ratio

Proteins at Liquid Interfaces
D. Möbius and R. Miller (Editors)

REVERSIBILITY OF PROTEIN ADSORPTION

F. MacRitchie

CSIRO Plant Industry, P.O. Box 7, North Ryde, NSW 2113, Australia

Contents

1 INTRODUCTION

Among the surface active agents of nature, proteins have an important place. In living systems as well as industrial dispersions (foams, emulsions) proteins are present and, because of the high extensions of interfaces in such systems, frequently play a vital role. As a result, we need to acquire a fundamental understanding of the adsorption process for these biological macromolecules. One aspect that has been controversial is the reversibility of the adsorption process and this will be examined in this chapter.

2 EVIDENCE FOR IRREVERSIBILITY

Much of the early work on protein adsorption at mobile interfaces led to the conclusion that the process was essentially irreversible. The following is a summary of the various pieces of evidence which later will be critically examined:

1. Langmuir and Schaefer [1] applied the Gibbs adsorption equation in its simple form to calculate the increase in solubility that should accompany compression of a surface film of protein; i.e.

$$d\Pi / d\ln c_b = c_s kT \qquad (1)$$

where Π is the surface pressure, c_s the number of molecules per unit area of the surface and c_b the protein concentration in solution in equilibrium with the surface film. It was calculated that, for a protein with a molecular weight of 35,000 (ovalbumin), an increase in surface pressure of 15 mN/m should increase the solubility of the film by a factor of 10^{95}. However, since little tendency for proteins to dissolve from their monolayers is normally observed, this was interpreted to mean that protein adsorption was not a reversible process and that the Gibbs adsorption equation was therefore not applicable.

2. The observation that proteins were stable at fluid interfaces led to studies of spread monolayers [2]. It was found that different proteins gave Π-A isotherms which were very similar when the areas were expressed in m^2/mg. This suggested that the interfacial conformation was similar and not related to molecular size as would be expected if molecules were adsorbed in their globular

solution form. Furthermore, the extrapolated area from the steeply rising portions of the isotherms corresponded to a value of approximately 1 m^2/mg. This was consistent with an unfolding of the molecule in which the polypeptide backbone lay in the interface with polar and nonpolar side-chains predominantly orientated towards and away from the aqueous phase respectively. The drastic unfolding was assumed to indicate an irreversible change in the protein, a change often referred to as surface denaturation.

3. As mentioned in point 1 above, proteins are usually very difficult to desorb from their monolayers at air/water interfaces despite being highly soluble in water. Thus, many protein monolayers can be compressed and, when expanded and recompressed, the Π-A isotherm can be retraced, showing no loss of protein due to dissolution into the subphase. This apparent enigma (i.e., loss of solubility on transferring from bulk solution to interface) has led to the conclusion that there must be an irreversible change caused by the interaction with the surface.

4. If adsorption (or spreading) of proteins is a reversible process, then, on recovery of the film, the protein should retain the same properties such as would be measured by solubility, molecular conformation and biological specificity (e.g., enzyme activity). Attempts to establish this have given rise to results which have been equivocal but, in general, loss of biological activity has been shown in some studies, again pointing to irreversibility.

5. A common phenomenon is that when protein solutions are shaken causing formation of large areas of air/aqueous interface, the protein precipitates and forms a coagulum which does not retain its original solubility properties [3]. Similarly, when protein monolayers are compressed, insoluble fibres are formed which may be removed from the surface [4]. These effects obviously result from interaction with the air/aqueous interface and have also contributed to the conclusion that the surface produces irreversible changes to the protein.

3 EXAMINATION OF EVIDENCE

Despite the above results, evidence has also been presented which tends to support the idea that protein adsorption is reversible; e.g., well-defined adsorption isotherms have been reported in which the steady state surface pressure increases steadily with increasing concentration of protein in solution. Against the background of this apparent conflict, the evidence for irreversibility will now be examined.

1. In principle, the arguments of Langmuir and Schaefer [1] are valid but there are two points that need to be considered in their analysis mainly as a result of subsequent experimental work. First, proteins have been shown to desorb from monolayers under certain conditions and this is further elucidated in point 2 below. Second, it is now well established from studies of polymers that long chain linear molecules show behavior reflecting that of segments of the molecules. This arises from the flexibility of these molecules so that the motion of a given segment is to some extent independent of that of other segments although, of course, it is constrained within the same molecule. One of the earliest observations confirming this concept was that the viscosity of long chain hydrocarbons appeared to be independent of molecular size above a certain chain length [5]. Studies of proteins at interfaces have confirmed that their properties (e.g., results from the surface pressure increment of surface viscosity, pressure displacement and adsorption kinetics) appear to depend on segments of some 5-10 amino acid residues [6]. For example, Table 1 summarises results for surface viscosities of several proteins and one polyamino acid with a range of molecular weights.

TABLE 1: Calculated values for A and ΔG from surface viscosity data for proteins

Protein	MW ($\times10^{-3}$)	A (Å^2)	ΔG (kJ/mole)
Polyalanine	1.5	105	69.5
γ-Globulin	160	110	68.7
Pepsin	34	120	67.4
BSA	67	100	66.1
Lysozyme	15	115	65.3

The free energy of activation for flow and the surface area of the flow unit were calculated from the equation proposed by Moore and Eyring [7] for surface viscosity (η_s) based on the theory of Absolute Reaction Rates:

$$\eta_s = \frac{h}{A}\exp\left(\frac{\Delta G + \Pi A}{kT}\right) \tag{2}$$

where h is the Planck constant, A is the area of flow unit, and ΔG is the free energy of activation for flow.

If we consider that the statistical kinetic unit is not a whole molecule but a segment of only 5-10 amino acid residues, then the argument of Langmuir and Schaefer may need to be modified. The pressure increment of solubility from the Gibbs adsorption equation, if we base our calculations on a segment rather than a whole molecule, would then attain only a moderate value. This taken with the result that desorption from monolayers does occur, could therefore leave the question of reversibility open.

2. There has been a tendency to regard a protein in its solution state as being native and, on adsorption at an interface, as being denatured, mainly as a result of the change from a globular type configuration to that of an extended two-dimensional chain. Denaturation is probably not a useful scientific term to apply to this process. It is well known that polymer molecules adopt a particular configuration depending on their environment. For example, in what is termed a poor solvent (i.e., one in which the free energy of interaction of polymer segments with solvent molecules is higher than between segments with each other), the molecule tends to fold up so as to minimise interactions between segments and solvent molecules. For the opposite case (a good solvent), molecules become extended so as to maximise interactions of polymer segments with solvent. When protein molecules transfer from aqueous solution to a polar/nonpolar interface, they go from a poor solvent to a good one. This is because they can lower the free energy of the system by extending their chains in the interface and orientating their nonpolar side chains predominantly towards the nonpolar phase (e.g. air) and their polar side chains towards the polar phase (e.g., water), consistent with steric constraints. From this point of view, there appears to be no basis for

regarding this as an irreversible process. If the molecule is returned to the bulk phase, it should tend to revert to its original globular conformation, that of lowest free energy for that environment. Of course, the unravelling of the molecule could conceivably expose groups which take up ions or, because of the high concentrations at interfaces, it could favour reactions such as the sulfhydryl-disulfide interchange, thus upsetting the reversibility of the folding process.

3. Although protein monolayers are very stable, desorption from them has nevertheless been reported. Measurements of desorption by decrease in monolayer area (at constant surface pressure) are complicated by the superimposition of other processes causing area decrease, mainly the pressure displacement of segments of molecules [8]. Langmuir and Waugh [9] were the first to report the effects of compression on the stability of protein monolayers. They distinguished between the two processes of pressure displacement and pressure solubility. Pressure displacement refers to the reversible expulsion of segments whereas pressure solubility is the irreversible desorption of whole molecules. Since then, several studies of protein desorption, mainly from air/water interfaces, have been reported [10, 11, 12]. Because the process of desorption is highly relevant to the question of reversibility, this topic will be discussed in greater detail later in the chapter.

4. Although fifty years old, the review by Rothen on the biological activity of protein films [13] summarises a very active period for these studies. Three distinct approaches to investigating the question of reversibility were delineated:

(i) tests made at the water interface where films were spread

(ii) assays using animals after collection of films from interfaces and subsequently suspended or dissolved

(iii) studies *in vitro* of film material after transference to metal or glass slides

The reader is referred to the review for a summary of the many studies.

In regard to the question of reversibility of adsorption, conclusions about recovery of activity will depend on the particular approach. In the first approach (tests made *in situ*), studies have been made either with the enzyme at the surface and the substrate in solution as dealt with by Rothen or, alternatively, with the substrate at the interface and the enzyme in solution as described in more recent studies by Verger and co-workers [14]. In studies where the enzyme (protein) is spread at the interface, complete loss of activity has been observed in certain instances. This is not surprising as the catalytic site of the enzyme often involves a three-dimensional configuration (active site) which locks the substrate into position and this might be expected to disappear when the molecule unfolds. On the other hand, when protein films are compressed, it seems feasible that these three-dimensional structures might be recovered as a result of pressure displacement of protein segments (portions of whole molecules) into the aqueous phase.

The second approach (collection, dispersion and testing activity in animals) is subject to some experimental difficulties. For example, compression and removal of film material can cause precipitation of the protein (see 5 below) which would then lead to an apparent loss of biological activity. The third approach (studies of film material transferred on to plates) has been used extensively not only for checking the preservation of biological activity but for investigating the structure of the film [15]. The same considerations apply as in (i); i.e., a protein in its unfolded conformation may not show biological activity but this does not prove that the process of adsorption is irreversible.

4 INTERFACIAL COAGULATION AND DESORPTION

The previous sections have briefly considered the arguments for and against the reversibility of protein adsorption. Before examining the problem in more detail, it appears useful to clarify aspects of the two processes, interfacial coagulation and desorption, since an understanding of both impinges on the question of reversibility.

4.1 INTERFACIAL COAGULATION

The insoluble precipitate that forms during shaking of protein solutions or after build up of protein films at quiescent interfaces has been studied by a number of workers. The process appears to be a 2D-3D phase change analogous to that which occurs for monomeric compounds when their monolayers are compressed to interfacial pressures above their equilibrium spreading pressures (E.S.P.s). A hypothesis for interpreting the interfacial pressures at which protein coagulation occurs was proposed by MacRitchie and Owens [16]. Briefly, it is based on considering the protein monolayer to be a special case of a duplex film. A substance spreads at an interface providing it can lower the free energy of the interface (γ_a) and is stable until $\gamma_a = \gamma_e$ where γ_e is the interfacial free energy in the presence of the monolayer. Such a film is considered to present two interfaces each of which can be characterised by interfacial energies: γ_b, the interfacial free energy between the nonpolar surface of the monolayer and the non aqueous phase and γ_{ab}, the interfacial free energy between the polar surface of the monolayer and the aqueous phase. Then, for equilibrium,

$$\gamma_a = \gamma_b + \gamma_{ab} + \Pi_e \quad (3)$$

where Π_e is the equilibrium spreading pressure or coagulation pressure.

When Π_e is exceeded, the monolayer precipitates (2D-3D transition) so as to restore the pressure to the value of Π_e. At a quiescent interface, a steady state may be reached where a 3-D film builds up to constant thickness. This occurs when the equilibrium adsorption pressure is greater than the coagulation pressure. Under these conditions, a continuous process of diffusion to the interface followed by adsorption and coagulation proceeds. The rate of build up decreases as the thickness of the film increases and approaches zero when the thick film hinders the diffusion and eventually eliminates the air/water interface. On the other hand, shaking displaces the coagulated protein and creates new interface for the adsorption to occur. Protein therefore continues to come out of solution until the sequence of processes slows down owing to the depletion of soluble protein.

Some explanation is needed for the insolubility of surface-coagulated protein. In solution, the type of proteins we are considering here exist as globular molecules in which nonpolar residues tend to be concentrated in the interior and polar (particularly electrically charged) residues predominate at the surface of the molecule, consistent with steric constraints. As described earlier, molecules adopt an extended configuration at an interface so that surface-coagulated protein contains molecules in this unfolded form. The question arises whether the apparent insolubility occurs because

(i) surface-coagulated protein is a lower free energy state than protein in solution, or

(ii) a high activation energy prevents return from the coagulated state to the solution state at a significantly measurable rate.

This question has not been adequately answered. It seems probable that there would be a high activation energy for the step. In the coagulum, many of the nonpolar groups would be expected to interact with nonpolar groups of other molecules forming a highly entangled network. It would, however, be a more highly orientated network than in heat-coagulated protein because molecules enter from a 2-D conformation. In the transition from the surface-coagulated protein to the folded solution state, nonpolar groups of molecules would need to break links with nonpolar groups of the same or other molecules and transiently interact with the aqueous phase before resuming intramolecular nonpolar links as the molecule folds. This would be a step of high activation energy.

The effects of different variables on precipitation of proteins by shaking and their subsequent dissolution has been studied in detail by Reese and co-workers [17, 18]. Partial dissolution (up to 40%) was found for ß-lactoglobulin by placing the suspension in a bath at 50°C for several hours and the rate was higher at 50°C than at 30°C. Results suggest that there is a gradation of solubility in the surface coagulated protein. It may be that molecules close to the surface of the coagulum are less entangled than those nearer the interior. Another possibility is that disulfide interchange reactions may occur over time, providing cysteine residues are present, leading to a cross-linked structure as has been found for heat coagulated proteins [19]. This would introduce irreversibility into the process. It was found by Reese and Robbins [18] that surface coagulated protein re-dissolved much more easily than heat coagulated protein. It was found that practically all the surface coagulated ß-lactoglobulin could be re-dissolved at pH 2.0 or pH 7.0 and, on readjusting the pH to 5.0, the protein remained in solution. Re-dissolved protein could be coagulated by shaking at the same rate as the original protein. It could also be hydrolysed by pepsin at rates similar to the original protein whereas the surface-coagulated (unfolded) protein was hydrolysed at a much higher rate. The evidence therefore favours a high activation energy to explain the difficulty of re-dissolution rather than a lower free energy state for the surface coagulated protein.

However, it is not clear cut as some of the results suggest that there may be at least partial irreversibility. More well designed experiments are required to fully resolve this issue.

4.2 DESORPTION KINETICS

Although protein films are very stable after spreading or adsorption from solution, some of the earlier studies showed that desorption can be measured at sufficiently high surface pressures [9]. The first quantitative measurements of desorption kinetics for protein monolayers were reported by Gonzalez and MacRitchie [10] for bovine serum albumin (BSA). Measurements of desorption are not so simple for polymeric compounds as for simple surfactants. When protein monolayers are compressed, a relaxation process occurs which manifests itself by a decrease in pressure if the monolayer is compressed to a fixed area or by a decrease in area if the monolayer is held at constant pressure. This behaviour reflects the expulsion of segments of molecules from the surface [20]. The process is reversible since, on expansion, the segments re-enter the surface as seen by increases in pressure (constant area) or area (constant pressure).

In order to separate this reversible process from any irreversible loss of monolayer due to desorption, a procedure was adopted whereby the area of the monolayer at a low pressure (e.g., 5 mN/m) was used as a benchmark to monitor the irreversible loss [10]. At such a low pressure, no irreversible loss of monolayer is observed. The monolayer was maintained at a high pressure for a measured time before expansion and the area adjusted until it reached a constant value at 5 mN/m.

From the decrease in area at 5 mN/m, it was then possible to calculate the irreversible loss. By repeating the experiment for different times at the high pressure, it was possible to obtain curves of the rate of desorption as a function of time. Some curves for BSA are shown in Fig. 1. Two checks were made to deduce whether the permanent losses were due to desorption. One was to measure the rates in the presence of dissolved protein and the other was to measure the adsorption rate when the subphase was stirred. Before interpreting the results, it will be helpful to first consider the mechanism of desorption from an interface.

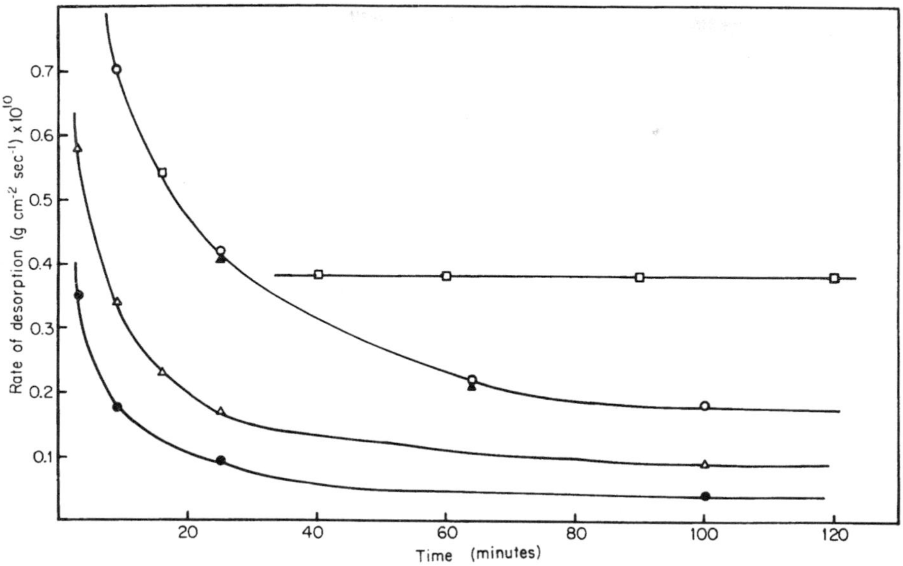

Fig. 1. Rate of desorption-time curves for a BSA monolayer at 25.6 mN/m. (○) no protein in subphase, no stirring; (□), stirring; (Δ) 0.05% BSA in subphase, no stirring; (●) 0.1% BSA in subphase, no stirring, from [10]

It is assumed that, when desorption occurs from a monolayer into a bulk solution which is initially at zero concentration, an equilibrium is rapidly established between the monolayer and an adjacent thin subsurface layer of bulk solution of several molecular diameters in thickness. Also near an interface there is a stationary or stagnant layer of solution in which transport of molecules across it occurs by diffusion. For a liquid subsurface, the thickness of the stationary layer may be of the order of 1 mm in the absence of stirring but this is reduced as the intensity of stirring is increased. The rate determining step in desorption is thus the rate of diffusion from the subsurface layer (of concentration = c_0) through the stationary layer (concentration = 0 at the edge). Under these conditions, assuming no back diffusion, the number of molecules (n) that desorb in time t is given by diffusion theory as:

$$n = 2c_o\sqrt{\frac{Dt}{\pi}} \tag{4}$$

where D is the diffusion coefficient and $\pi = 3.14$.

The rate of desorption is found by differentiation:

$$\frac{dn}{dt} = c_0\sqrt{\frac{D}{\pi t}} \qquad (5)$$

Returning to the interpretation of Fig. 1, it is seen that, as the bulk concentration of BSA is increased, the rate of desorption at 25.6 mN/m shows a corresponding decrease. Similarly, when stirring of the solution was introduced, the rate initially followed the same curve as with stirring absent (diffusion regime) but deviated at an early stage, thereafter becoming constant, consistent with a stationary layer reduced to a smaller thickness by the stirring. Furthermore, when the rate was plotted against $t^{-1/2}$ for several different surface pressures, a linear relationship was found between the rate and $t^{-1/2}$. Results for desorption of BSA at several different surface pressures are shown in Fig. 2. One feature of Fig. 2 is that, although the lines are drawn to pass through the origin, the lines of best fit do not in all cases extrapolate to zero at infinite time as would be predicted for a purely diffusion controlled process. This suggests that a surface barrier to desorption is acting.

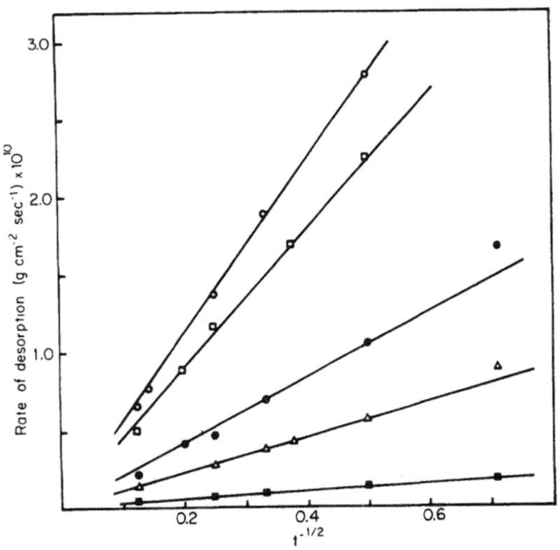

Fig. 2. Rate of desorption vs $t^{-1/2}$ for BSA monolayers at different surface pressures. (○) 28.8 mN/m ,(□) 27.2mN/m; (●) 25.6 mN/m; (Δ) 24.0 mN/m; (■) 22.4 mN/m, from [10]

Further confirmation that these permanent area losses could be ascribed to desorption was obtained by MacRitchie and Ter-Minassian-Saraga [11] using radiolabelled protein. In this work, monolayers of I^{125} -labelled BSA were held for 20 min periods at several pressures and the permanent area losses monitored after expansion to 10 mN/m at which pressure, no measureable desorption was observed. In addition, samples of the monolayer were removed on glass slides at a pressure of 10 mN/m and radioactivity measured in a scintillation counter. An example of the compression cycles is illustrated in Fig. 3 with values of the specific radioactivity recorded (in counts min^{-1} cm^{-2}) at several stages in the experiment.

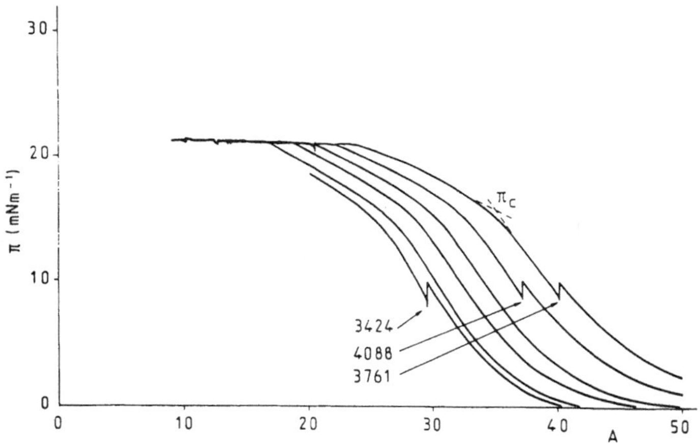

Fig. 3. Compression-expansion (not shown) cycles for a ^{125}I-labelled BSA monolayer held for 20 min periods at 21 mN/m. A is in arbitrary units. Surface area may be calculated from the formula: area = 8.33A + 10.7 cm^2. The numbers correspond to the measured counts min-1 for the films transferred on to glass slides, from [11]

Both radioactivity and film compressibility remained constant within experimental error at all pressures following successive compression cycles. For example, after compression at 27 mN/m, the specific activity was 3720 $^{+}$ 400 counts min^{-1} cm^{-2} compared to an average value of 3620 counts min^{-1} cm^{-2} at 10 mN/m even though 68.6% of the area had been lost. This demonstrates that the monolayer properties are unchanged, consistent with an area loss caused by desorption and not some other process. The radioactivity in the subsolution following the compression cycles was also analysed and found to be of the same order as that expected from the total loss of monolayer calculated from the permanent decrease of area.

4.3 *STUDIES ON DIFFERENT PROTEINS*

Kinetics of desorption have been measured for a number of proteins. The procedure for desorption measurements is illustrated in Fig. 4 [21]. In the left hand diagram, A is the area at 10 mN/m during the first compression of the monolayer.

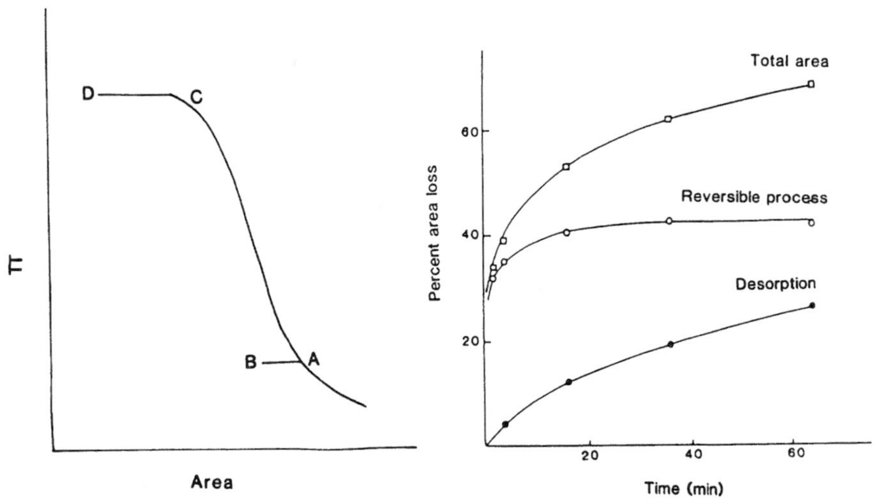

Fig. 4. Schematic illustration of the procedure for separation of reversible and irreversible (desorption) processes in a monolayer of ß-lactoglobulin held at a surface pressure of 30 mN/m. In the diagram on the left, CD is the total area loss at the high pressure. AB is the irreversible loss measured at the benchmark pressure, from [21]

The monolayer was compressed to a pressure of 30 mN/m at point C and maintained at this pressure by decreasing the area. After a measured time, the area was expanded and the constant area at 10 mN/m measured, the point B. By repeating the experiment for different times at 30 mN/m, it was possible to construct curves for the total area loss and the area losses for the reversible (expulsion of segments) and irreversible (desorption) processes. Using the same method, the kinetics of desorption have been measured for a series of proteins of varying molecular weight

(MW) as a function of surface pressure. The results are summarised in Table 2. It is seen that the rate of desorption at a given pressure is related to the molecular weight. Thus, the rate of desorption of insulin (MW=6,000) can be measured at surface pressures as low as 15 mN/m whereas γ-globulin (MW= 160,000) is scarcely detectable below 30 mN/m. In a mixed monolayer of insulin and γ-globulin, the rate of desorption is the same at a given pressure as a pure insulin monolayer, at least until the surface concentration of insulin becomes quite low. This suggests that the steady state concentration of insulin beneath the monolayer depends only on the surface pressure and is independent of the monolayer composition, at least above a certain surface concentration of insulin.

TABLE 2: Rates of desorption of proteins at different surface pressures

Protein	Mol. wt (x 10^{-3})	Rates of desorption (min^{-1} x 10^4)						
		15[a]	20[a]	25[a]	30[a]	35[a]	40[a]	45[a]
Insulin	6	56	530					
ß-Lactoglobulin	17.5		20	50	90			
Myoglobin	17			34	67	144		
γ-Globulin	160				9	20	40	
Catalase						30	70	110

[a]Surface pressure Π (mN/m)

4.4 THEORY OF DESORPTION

A theory of protein desorption needs to account for the following:

a. the rate of desorption is negligible at moderate surface pressures (< 15 mN/m) and is slow at high surface pressures despite the high solubility of proteins in bulk solution.

b. as the molecular weight increases for a series of proteins, the surface pressure at which desorption becomes measurable, correspondingly increases.

The desorption of BSA has been interpreted on the basis of the Π-A isotherm shown in Fig. 5 [22]. If a protein monolayer is compressed sufficiently rapidly so that there is no time for relaxation processes to occur, the dashed line is obtained. However, because segments of molecules are squeezed out of the interface as the monolayer is compressed, the equilibrium Π-A curve follows the full line of Fig. 5. By expelling segments into the bulk phase as loops and tails without desorption of whole molecules, protein (or any polymer) molecules in the monolayer are able to lower free energy in a way not possible for monomer molecules. The effect is to produce an equilibrium Π-A curve having high compressibility at the higher pressures as seen in Fig. 5.

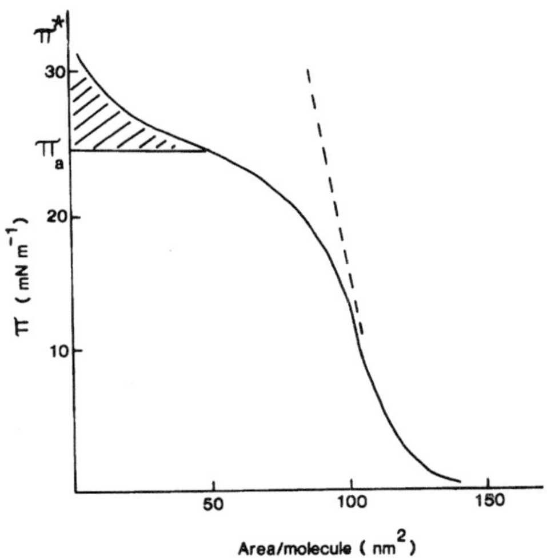

Fig. 5. Equilibrium (full line) and instantaneous (dashed line) Π-A curve for BSA. Hatched area is the free energy of desorption at a surface pressure of Π_a and Π^* is the transition-state pressure for desorption, from [22]

A whole molecule will only desorb when it reaches an activated state where it is no longer stable and then spontaneously leaves the interface. In order for a protein molecule to adsorb, it has been calculated that an area of some 1 - 1.5 nm^2 (6-10 amino acid residues) has to be cleared at the surface [6]. Once this critical area has been cleared, the adsorption process proceeds spontaneously. It seems logical to assume that this is also the critical activation area required to be reached by a molecule in order to desorb. Thus, referring to Fig. 5, the free energy for desorption

(ΔG_{des}) is given by the pressure-area work required to be done on the molecule to bring it from a given pressure

(Π_a) to the pressure corresponding to the critical activation area (Π^*). This can be achieved either by compression of the film such as in a film balance or through fluctuations in configuration of film molecules. Thus,

$$\Delta G_{des} = \int A d\Pi \tag{6}$$

The magnitude of ΔG_{des} may be evaluated from the area under the Π-A curve between the relevant limits of surface pressure. Values of ΔG_{des} were calculated for the BSA monolayer at a series of surface pressures and are recorded in Table 3 together with values of $\Delta G_{des}/kT$ [10].

TABLE 3: Calculated free energies of activation for desorption (ΔG_{des}) from BSA monolayers at different surface pressures

Π (mN/m)	ΔG_{des} (J/molecule)	ΔG/kT
20	$4.3 \cdot 10^{-13}$	106
22.4	$2.6 \cdot 10^{-13}$	64
24.0	$1.7 \cdot 10^{-13}$	42
25.6	$1.0 \cdot 10^{-13}$	24
27.2	$5.8 \cdot 10^{-14}$	14
28.8	$3.6 \cdot 10^{-14}$	9

It is seen that, as the pressure is increased above 20 mN/m, the free energy barrier decreases dramatically. This arises because of the rapidly changing equilibrium configuration of molecules in the monolayer. At pressures below 20 mN/m, ΔG_{des} becomes very large compared to kT, thus explaining the difficulty in detecting desorption at these pressures. Fig. 6 depicts a schematic illustration of how the changing molecular configuration of a protein molecule with increasing pressure facilitates desorption [21]. At low surface pressures when most of the segments are at the interface (in the form of trains), the fluctuation required for a molecule to reach the transition state has an extremely low probability.

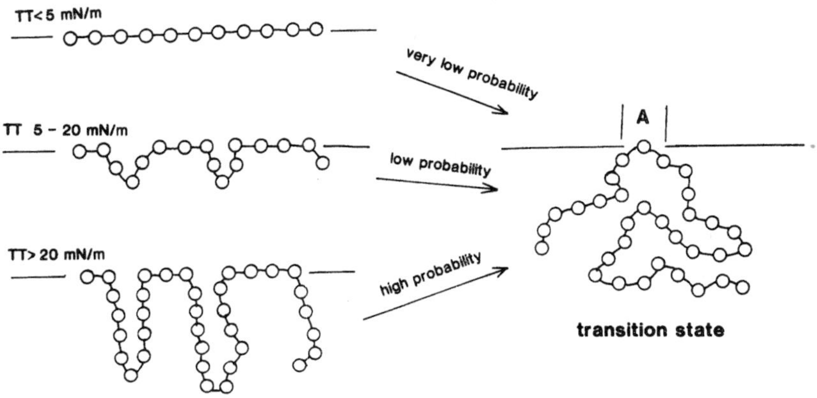

Fig. 6. Schematic illustration of how the probability of a fluctuation attaining the transition state configuration increases with increasing surface pressure, from [21]

5 SOLID/LIQUID INTERFACES

The previous discussion has been based entirely on considerations of adsorption at fluid interfaces. This has been because it was felt that the subject of reversibility could be more adequately discussed on the basis of the large amount of information available from studies at these interfaces, particularly the air/water interface. However, many studies have been made of protein adsorption at solid/liquid interfaces. The question of reversibility has been discussed by Norde [23] in a review in which the following relevant observations were summarised:

(a) either none or only a fraction of adsorbed protein molecules desorb on diluting the system with solvent

(b) additional adsorption can occur by altering conditions such as pH, ionic strength and temperature

(c). exchange of protein molecules between surface and bulk solution has been confirmed

(d). adsorbed protein molecules may be displaced from the solid/liquid interface by molecules of other species, including other proteins

(e). protein molecules that have undergone structural changes on adsorption may not revert to their original conformation after removal from the surface

Observations b, c and d can be reconciled with a case for reversibility of adsorption. The first observation (a), however, appears at first examination to be inconsistent with reversibility. It might be expected that removing protein from solution to create a bulk phase of infinite dilution would lead to complete desorption of the adsorbed species, providing it has a finite solubility in the subphase.

In order to explore this problem more deeply, let us consider the adsorption of a protein at the air/water interface. For any component in adsorption equilibrium between bulk solution and surface, we can apply the condition that chemical potentials of the component must be equal in the two phases, i.e.,

$$\mu_0^b + kT \ln c_b = \mu_0^i + A d\Pi \qquad (7)$$

where μ_0^b can be taken as the standard chemical potential at infinite dilution in bulk solution and μ_0^i is the standard chemical potential at infinite dilution at the interface.

It is possible that there is a value of Π for which the term on the right hand side of the equation will approach the value of μ_0^b. This effectively means that, if one molecule is placed in a system having a bulk phase and an interface, the free energy of the system can be lower if the molecule is at the interface than in solution. What is observed with protein monolayers is that desorption becomes undetectable at low surface pressures (e.g., below about 15 mN/m for moderately sized proteins). Therefore, if a solution of protein is in equilibrium with a monolayer at this pressure and the bulk phase is depleted of protein, there would be no tendency for molecules in the monolayer to desorb. If the surface pressure were higher, it is possible that there would be some desorption initially but this would stop when the surface pressure had been reduced to a certain value. The amount of protein that would desorb would therefore be small relative to the adsorbed amount. Essentially this is what is observed at a solid/liquid interface. When the concentration of protein in solution is reduced to zero, the adsorbed amount is initially decreased but a threshold surface concentration is eventually reached at which no further removal of protein appears to occur.

6 ADSORPTION ISOTHERMS

One of the criteria for reversibility of adsorption is that the component should exhibit well-defined adsorption isotherms. From that point of view, proteins appear to satisfy this requirement. Even though long times are often needed for equilibrium to be attained, adsorption isotherms clearly showing the increase of surface concentration with increase of bulk concentration have been reported at both air/water and solid/water interfaces.

6.1 AIR/WATER INTERFACES

Adsorption isotherms for BSA [10] and ß-lactoglobulin [21] at the air/water interface are shown in Figs 7 and 8 respectively. The surface pressure is plotted as a function of the logarithm of the bulk concentration in each case. At low concentrations for both proteins (below 0.01%), an approximately linear relationship is found. We can apply the Gibbs adsorption equation in its simple form (see Eq. 1 above) to the linear regions of the adsorption isotherms of Figs 7 and 8 , i.e.,

$$d\Pi / d\ln c_b = kT/A \qquad (8)$$

where c_b is the bulk concentration and A should be the area per molecule (i.e. the inverse of the surface concentration). From the slopes in the regions where the Π-log c relationship is linear, we calculate values of $A = 2.5\ nm^2$ for BSA and $2.1\ nm^2$ for ß-lactoglobulin. Values of the same order have been calculated for proteins by other workers using a similar approach [25, 26, 27]. These results tend to confirm the point that was made earlier: i.e., that interfacial properties of proteins are governed by segments of molecules of roughly constant size rather than by whole molecules.

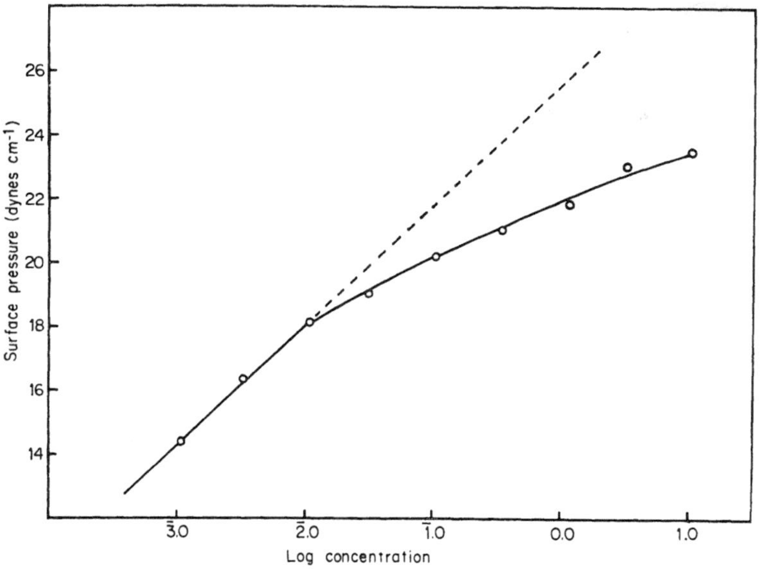

Fig. 7. Adsorption isotherm for BSA on pure water, from [10]

It is useful to compare the bulk concentrations for given surface pressures with those calculated from desorption rates based on the model of Eq. 5. Although the data is limited, some comparisons have been reported for BSA at different surface pressures (Table 4).

It is seen that bulk concentrations calculated from desorption rates are several orders less than those found from the adsorption isotherm. This may be explained by the presence of an activation energy barrier for desorption at the surface. Although a steady state subsurface concentration is attained, this is much lower than the equilibrium concentration. What this means is that diffusion from the sublayer carries away protein at a faster rate than the rate required for the desorption step to build up to the equilibrium concentration. An estimate of the magnitude of the desorption barrier can be made by considering that the diffusional resistance (R_1) and the interfacial resistance (R_2) act in parallel.

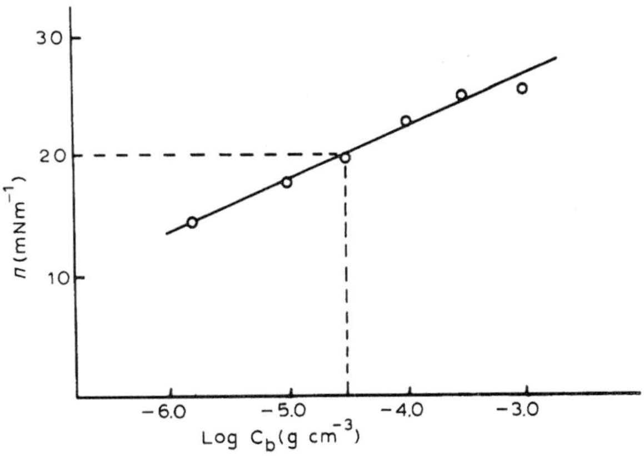

Fig. 8. Adsorption isotherm for ß-lactoglobulin on pure water, from [22]

TABLE 4: Comparison of subsurface concentrations of BSA at different surface pressures obtained from adsorption isotherm and from desorption rates

Π (mN/m)	c_0 (from desorption rate) (g/cm³)	c_0 (from adsorption isotherm) (g/cm³)
28.8	$1.28 \cdot 10^{-6}$	$5.5 \cdot 10^{-2}$
27.2	$1.04 \cdot 10^{-6}$	$2.2 \cdot 10^{-2}$
25.6	$4.86 \cdot 10^{-7}$	$9.1 \cdot 10^{-3}$
24.0	$2.62 \cdot 10^{-7}$	$3.8 \cdot 10^{-3}$
22.4	$5.7 \cdot 10^{-8}$	$1.6 \cdot 10^{-3}$

The overall rate can then be described by the equation:

$$dn/dt = c_0 / (R_1 + R_2) \qquad (9)$$

where c_0 is the equilibrium value of the BSA concentration. R_1 is equal to δ/D (where is the thickness of the diffusion layer) and D is the diffusion coefficient. Taking a surface pressure of 25.6 mN/m from Table 4 and an estimated value for δ of $9 \cdot 10^{-3}$ cm, R_1 is calculated to be $1.3 \cdot 10^4$ s/cm. Substituting values for dn/dt and c_0 in Eq. 9 gives a value of $2.4 \cdot 10^8$ s/cm for R_2. This shows that, although the kinetics of desorption follow a diffusion controlled process, the absolute rate is determined by the very much larger interfacial resistance.

6.2 Solid/Water Interfaces

More information on protein adsorption isotherms at various solid/water interfaces has been obtained than at the air/water interface. Some typical isotherms for adsorption of human serum albumin (HSA) on polystyrene latices are shown in Fig. 9 [24]. This work shows clearly that genuine equilibrium isotherms are obtained and that they are sensitive to alteration of conditions such as pH or temperature. The isotherms are characterised by well-defined plateau adsorptions, a characteristic of protein adsorption at solid/liquid interfaces. As mentioned previously, desorption is almost always found to be small on dilution and a convincing explanation of this phenomenon has not been put forward. The general behaviour illustrated in Fig. 9 has been observed in other studies of protein adsorption on a wide range of solid materials [23].

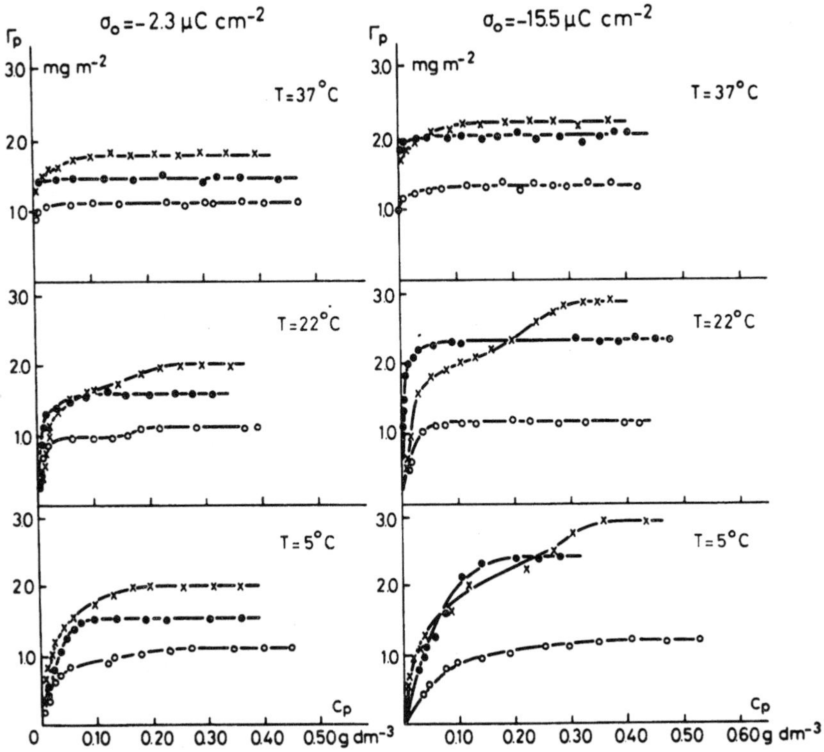

Fig. 9. Adsorption isotherms for human serum albumin on polystyrene latices: (O) pH 4.0; (x) pH 4.7; (●) pH 7.0; KNO_3 concentration = 0.05M, from [24]

6.3 *ELUTABILITY AND EXCHANGE AT SOLID/LIQUID INTERFACES*

The nature of the solid/liquid interface is found to have a significant effect on the amount of protein adsorbed and the ease of desorption. In general, hydrophobic surfaces adsorb more protein and the amount of desorption on dilution is less than at hydrophilic surfaces. For example, an elegant demonstration of this has been reported by Elwing et al. [28]. Using a surface with a hydrophobic gradient, it was shown that the adsorption of fibrinogen decreased with decreasing water contact angle (i.e., decreasing hydrophobicity). The effect of the nature and energy of the solid surface, the reversibility on dilution, the displacement of protein by surfactants and other compounds (displacers) and the exchange of protein molecules between surface and solution are all pertinent to the question of reversibility of adsorption. Some thorough studies of these factors/processes have been carried out by Wahlgren and Arnebrant [29, 30, 31, 32]. An example from their work is shown in Fig. 10 where a comparison is made of the adsorption and desorption on dilution for ß-lactoglobulin at hydrophobic and hydrophilic surfaces measured by ellipsometry.

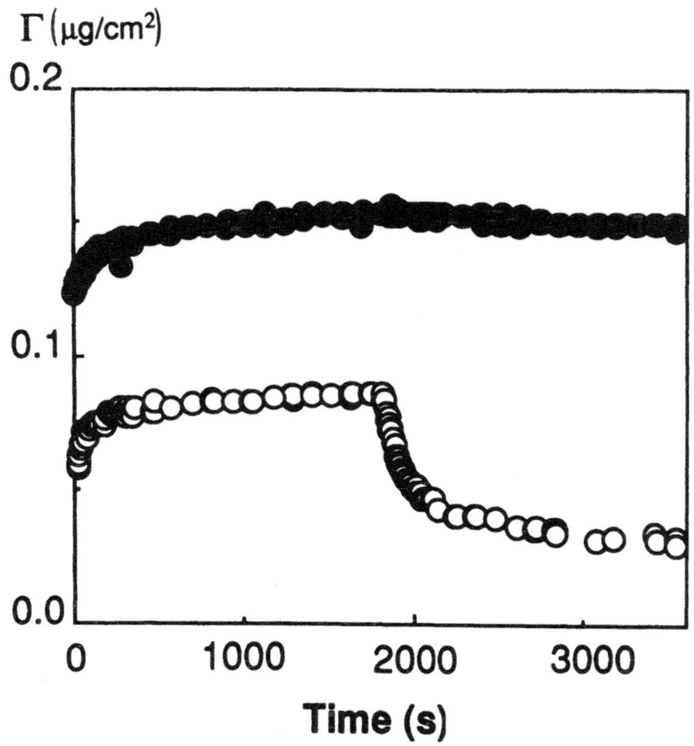

Fig. 10. Adsorption of ß-lactoglobulin (0.1% solution) onto hydrophobic (●) and hydrophilic (○) silica surfaces measured by ellipsometry. Rinsing (dilution) is performed after 30 min, from [29]

The amount adsorbed is less and the amount desorbed on dilution of the bulk phase is greater for the hydrophilic surface, indicating a lower free energy of adsorption. The ease of removal of adsorbed proteins by surfactants, termed the elutability by Rapoza and Horbett [33, 34, 35], has been used as an indicator of the binding or adsorption energy. The amounts of protein removed are influenced by the type of surfactant and by the nature of the solid surface [33]. Although these results point to the adsorption process being reversible, a strong correlation has been found between the elutability of fibrinogen and the residence time for the protein at the surface [33, 34, 35]. Such apparently irreversible effects superimposed on what otherwise appears to be reversible adsorption, presents some difficulties in interpretation. Another process that does not appear to lead to a clear-cut interpretation is the exchange of proteins between surface and bulk. This has been cleverly studied by Brash and co-workers using differently radio-labelled proteins [36, 37]. Usually, it is found that a larger fraction of protein is exchanged than the fraction that is desorbed on dilution. However, it still remains that only a fraction of the protein exchanges whereas, for a completely reversible adsorption, it would be expected that all the protein should be exchangeable.

6.4 Experimental Considerations

Some of the difficulties in interpreting the reversibility of protein adsorption may arise from experimental factors. For example, measurements of surface tensions of protein solutions has often employed the Wilhelmy plate technique. Although this is a simple and effective method, it can be the source of artefacts if drying out of the plate with a consequent change from a zero contact angle occurs during the long times often needed for monitoring adsorption. A successful approach has been to use filter paper as the plate material, thus ensuring that the plate always remains wetted during the measurements. Of course, the presence of impurities is always a danger in interpretation of data from interface studies because of the potentially large effects of minute amounts of surface active components. These may enter the system because of impurities in the protein sample or in the water and its solutes. The critical importance of surface chemical purity has been well demonstrated by the work of Lunkenheimer and co-workers [38, 39]. The problem is compounded at solid interfaces. At least at an air/water interface there is the possibility of monitoring impurities by checking the surface tension and its changes with time/compression. On the other hand, for solid/water interface measurements, there is no simple way of ascertaining whether there has been appreciable adsorption of other surface active impurities before the protein has been introduced into the system. This is particularly true where dispersions are used as the adsorbate.

7 CONCLUSIONS

It is likely that the question of the reversibility of protein adsorption at interfaces will remain open and continue to be discussed in the future. An attempt has been made in this chapter to critically appraise some of the premises which have appeared to point to the irreversibility of the process. Although arguments have been put forward that some of these premises are invalid, there are still experimental aspects of the adsorption/desorption process that are not fully understood. For example, why is it that, although well-defined adsorption isotherms are obtained when the adsorption is approached from the solution side, the isotherms are not followed when the process is reversed; i.e., when the solution concentration is progressively reduced? Many results indicate that, in principle, the adsorption process is reversible. It may be that there are often irreversible processes superimposed. As an example, adsorption potentially opens the way for interchain reactions to proceed. This results from the very high concentrations in the interfacial phase and the unfolded conformation of the protein that undoubtedly occurs at many interfaces. Another question that has not been definitively resolved is whether a protein can recover its original solution conformation and properties after desorption. Many protein solutions can be shaken without any obvious change such as that observed when there is surface coagulation. Shaking of the solutions means that proteins are continuously being adsorbed and desorbed, thus providing the conditions to clarify this question. Some well-designed experiments using techniques that are sensitive to changes in protein conformation are needed to monitor the solutions during the shaking regime.

There is an extensive literature on proteins at interfaces and stimulating papers on the subject continue to be published [e.g., refs 40,41]. The present article may be criticised, probably with some justification, for being too restrictive and for emphasising the author's contributions. In defense, it was thought best not to attempt a complete review on the subject but to focus on work that was specifically aimed at investigating the question of reversibility of adsorption. Criticisms and opposing viewpoints are welcome as it is only by well-designed experiments and rational debate that a better understanding of the topic will be achieved.

8 REFERENCES

1. I. Langmuir and V.J. Schaefer, Chem. Rev., 24(1930)181.
2. H.B. Bull, Adv. Protein Chem. 3(1947)95.
3. H.B. Bull and H. Neurath, J. Biol. Chem., 118(1937)163.
4. J.G. Kaplan and M.H. Frazer, Nature, 171(1953)550.
5. A.L. Van Geet and A.W. Adamson, J. Phys. Chem. 68(1964)238.
6. F. MacRitchie, J. Colloid Interface Sci., 57(1976)393.
7. W.J. Moore and H. Eyring, J. Chem. Phys., 6(1938)391.
8. F. MacRitchie, J. Colloid Sci., 18(1963)555..
9. I. Langmuir and D.F. Waugh, J. Amer. Chem. Soc., 62(1940)2771.
10. G. Gonzalez and F. MacRitchie, J. Colloid Interface Sci., 32(1970)55.
11. F. MacRitchie and L. Ter-Minassian-Saraga, Progr. Colloid & Polymer Sci., 68(1983)14.
12. D.E. Graham and M.C. Phillips, J. Colloid Interface Sci., 70(1979)415.
13. A. Rothen, Adv. Protein Chem., 3(1947)123.
14. R. Verger, M.C.E. Mieras and G.H. de Haas, J. Biol. Chem. 248(1973)4023.
15. B.R. Malcolm, Nature, Proc. R. Soc. London, A305(1968)363.
16. F. MacRitchie and N.F. Owens, J. Colloid Interface Sci., 29(1969)66.
17. E.T. Reese, J. Applied Biochem., 2(1980)36.
18. E.T. Reese and F.M. Robbins, J. Colloid Interface Sci., 83(1981)393.
19. F. MacRitchie, J. Polym. Sci., Symposium No. 55(1976)139.
20. F. MacRitchie, J. Colloid Interface Sci. 79(1981)461.
21. F. MacRitchie, Colloids Surf., 76(1993)159.

22. F. MacRitchie, Analyt. Chim. Acta, 249(1991)241.

23. W. Norde, Adv. Colloid Int. Sci., 25(1986)267.

24. W. Norde and J. Lyklema, J. Colloid Interface Sci. 66(1978)257.

25. N. Benhamou and J. Guastalla, J. Chim. Phys. 57(1960)745.

26. P. Joos, Proc., 5th Int. Congr. Surf. Act., 2(1969)513.

27. M.C. Phillips, M.T.A. Evans, D.E. Graham and D. Oldani, Colloid Polym. Sci., 253(1975)424.

28. H. Elwing, S. Welin, A. Askendal, U. Nilsson and I. Lundström, J. Colloid Interface Sci. 119(1987)119, 203.

29. M. Wahlgren and T. Arnebrant, TIBTECH, 9(1991)201.

30. M. Wahlgren and T. Arnebrant, J. Colloid Interface Sci., 136(1989)259.

31. M. Wahlgren and T. Arnebrant, J. Colloid Interface Sci., 142(1991)503..

32. M. Wahlgren and T. Arnebrant, J. Colloid Interface Sci. 148(1992)201.

33. R.J. Rapoza and T.A. Horbett, J. Colloid Interface Sci., 136(1990)480

34. R.J. Rapoza and T.A. Horbett, J. Biomed. Mat. Research, 24(1990)1263.

35. R.J. Rapoza and T.A. Horbett, J. Biomater. Sci. Polymer Edn., 1(1989)99.

36. J.L. Brash and Q.M. Samak, J. Colloid Interface Sci., 65(1978)495.

37. B.M.C. Chan and J.L. Brash, J. Colloid Interface Sci., 82(1981)217.

38. K. Lunkenheimer and R. Miller, J. Colloid Interface Sci., 120((1987)176.

39. K. Lunkenheimer, H.J. Pergande and H. Krüger, Rev. Sci. Instrum., 58(1987)2313.

40. J.D. Andrade, V. Hlady, L. Feng and K. Tingey, Bioprocess Technol., 23(1996)19

41. J. Lyklema and W. Norde, Progr. Colloid Polym. Sci., 101(1996)9.

Proteins at Liquid Interfaces
D. Möbius and R. Miller (Editors)

INTERFACIAL RHEOLOGY OF MIXED FOOD PROTEIN AND SURFACTANT ADSORPTION LAYERS WITH RESPECT TO EMULSION AND FOAM STABILITY

Brent S. Murray

Food Colloids Group, Procter Department of Food Science, University of Leeds,
Leeds, LS2 9JT, UK, Tel. 44 (0)113 2332962, Fax. 44 (0)113 2332982,
E-mail b.s.murray@cbl.leeds.ac.uk

Contents

1 INTRODUCTION

Interfacial rheology is the study of the relationship between interfacial stress and the resultant deformation of the interface. The subject has a history almost as long as the history of colloid science itself. Ascherson (1840) reported the presence of a visible skin around large emulsion droplets; no doubt this had been observed much earlier but not recorded. However the idea of the idea of a "protective colloid" was established, where here the word colloid was used in the old sense of referring to any high molecular weight substance which did not, at that time, appear follow the same rules as typical low molecular weight species in solution. The protective colloid on the surface of the droplets, etc., protected them from flocculating and thence coalescing into a single bulk phase. It was soon recognised that proteins and proteoglycans in particular formed mechanically very "strong" skins and proteins such as gelatin have long been rated (Freundlich, 1930) as one of the most effective steric stabilisers of colloidal particles. Shotten and Wibberly (1961) published photographs some time ago which graphically illustrate the strong structural coherence of such films[1] .

The study of interfacial rheology has advanced considerably in the last 20 years or so and even a cursory glance at the current colloid literature reveals that this is still (perhaps an increasing) area of interest to colloid scientists seeking to understand the mechanisms of formation and stability of small particles. Areas where interfacial rheology is thought to be important are very diverse, including for example, drug delivery (Niven *et al.*, 1996), lung function (Panaiotov *et al.*, 1996) and anaethesia (Makino *et al.*, 1996), as well as in typical food (and other) emulsions, foams and solid particle dispersions. As a consequence the literature on this subject is rather extensive. (The reader is referred to the other chapters of this book). In the context of food systems the literature over the years mainly 1985 - 1995 has recently been reviewed by Murray

---

[1] The reader may be interested in a simple, illuminating demonstration. Layer a hydrocarbon oil such as n-tetradecane over a solution of 0.1 wt. % lysozyme solution (at pH 7, ionic strength 0.1 mol dm^{-3}) in a large beaker and leave for at least 4 h . If the beaker is carefully placed on an overhead projector and the interface observed, agitation of the interface with a glass rod to create large oil globules will produce transient wrinkles as the interface is compressed and expanded.

and Dickinson (1996). To some extent this review forms the basis for this chapter but with an update on recent new developments and with increased emphasis on the relationship between interfacial rheology and the formation and stability of colloids in food systems.

Without wishing to pre-empt the conclusions of this chapter it is worth saying at the outset that scientists are still far from being able to calculate exactly the effects of the rheological properties of adsorbed layers on the formation and stability of colloidal systems. In part, particularly for food systems, this is because the make-up of the interfacial material is exceedingly complex. It may consist of protein, polysaccharide, lipid or mixtures of all three. Part of the usefulness of interfacial rheology, however, is that it can provide information about the structure and composition of these layers. This in turn may lead to a better understanding of the colloid stabilising effects of such layers at a fundamental, molecular level.

The principal food colloids of commercial interest are emulsions and foams: the air-water (A-W) and oil-water (O-W) interfaces will be discussed in the main in this chapter. (Dispersions of solids, for example fat particles, are also of importance, but have received relatively little attention from the interfacial rheological point of view). Of the ingredients of such emulsions and foams, proteins and low-molecular-weight surfactants are the dominant species at the interface. Proteins are co-polymers of amino acids with hydrophobic and hydrophilic residues; this results in a strong tendency for them to adsorb at both (in fact most) types of interface. Low-molecular-weight surfactants in foods may be natural components, such as phospholipids, glycerides, fatty acids, etc., or synthetic molecules such as the Spans and Tweens. Low-molecular-weight surfactants may adsorb strongly or weakly depending on the chemical structure of the surfactant, i.e., the relative size of the hydrophilic and hydrophobic portions of the molecule (or HLB). Because of the higher molecular weight of proteins and the co-operative nature of their adsorption, protein adsorption and desorption tends to be considerably slower than adsorption and desorption of low-molecular-weight surfactants. Protein adsorption and desorption may involve considerable changes in the three dimensional structure of the molecule, involving alteration of the distribution of intermolecular and intramolecular bonds, unlike with low-molecular-weight surfactants, where such structural changes are minimal. These differences can lead to very different interfacial rheological properties for proteins and low-molecular-weight surfactants and consequently there are large variations in behaviour possible when the two are present in admixture, since both will compete for adsorption at the interface. Most food systems contain mixtures of both proteins and low-molecular-weight surfactants so this competitive adsorption, and any factors which influence it, are of great significance in resolving the relationship between interfacial rheology and the stability of emulsions and foams.

2 SOME DEFINITIONS AND TERMINOLOGY.

The mathematical framework of interfacial[2] rheology has been dealt with in depth by previous authors, most notably by Lucassen (1981) and more recently and comprehensively by Edwards *et al.* (1991) An very readable and elegant introduction to the subject in the food context has recently been provided by Lucassen-Reynders (1993). More general recent reviews have been provided by Miller *et al.* (1996), Sagis and Bedeaux (1996), the authors of other chapters of this book and for films on solid surfaces by Behroozi (1996).

When an emulsion or foam is formed the interface between the two bulk fluid phases is increased by a highly turbulent hydrodynamic flow field. One can divide the types of disturbances which impinge upon the interface into two sorts - those which involve a change in area and those where there is no area change but where the different interfacial elements slip past one another. The former type of disturbance may be likened to pulling at the edges of an infinitesimally thin piece of material with equal stresses on all sides, so that its size changes but the shape of the material remains the same. This is referred to as an interfacial dilatational (or sometimes a dilational) disturbance. If the material is solid-like, there is a tensile resistance to the disturbance, and one can define an interfacial dilatational elasticity or rigidity (see below). If the material is liquid-like, there is flow of material but with a certain resistance to flow, and one can define an interfacial dilatational viscosity. Similarly, distortion of the shape of the infinitesimally thin piece of material, with different stresses on its edges (i.e., a shear deformation) can be described in terms of an interfacial shear viscosity and elasticity. These

---

[2] Throughout we will use the general terms 'interface', 'interfacial', etc., to describe the dividing region in either air-liquid or liquid-liquid two-phase systems, as opposed to the term 'surface', which is reserved specifically for reference to air-liquid systems.

interfacial rheological parameters are the analogues of the three-dimensional equivalents of compressional and shear elasticity and viscosity, though there are complications for interfaces where material can be exchanged between the interface and the bulk phases during the measurement.

All the moduli mentioned above affect the ease with which a deformable interface is distorted to produce droplets or gas bubbles and particularly determine the stability of the newly formed interface once it has formed (Lucassen-Reynders 1993, Lucassen-Reynders and Kuijpers, 1992). The word stability always requires some qualification — here it is meant stability against coalescence of the droplets or bubbles to form a single bulk phase. Coalescence is instigated by the propagation and growth of minute disturbances on the two interfaces and the thin film of fluid separating two droplets or bubbles; this propagation and growth is greatly affected by the interfacial rheology. The film of liquid between interfaces must thin below a certain thickness, however, before film rupture can occur: the rate of film thinning also depends on the interfacial rheology because the bulk flow of fluid is coupled to the flow of elements in the interface.

For deformations in the *x-y* plane we can define the interfacial shear elasticity, G, and the interfacial shear viscosity, η, by

$$p_{xy} = Ge_{xy} \tag{1}$$

$$p_{xy} = \eta \frac{de_{xy}}{dt} \tag{2}$$

where p_{xy} is the interfacial shear stress and e_{xy} is the interfacial shear strain. Attractive interactions between the surface elements (e.g., segments of adsorbed protein molecules or surfactant molecules) leads to an increase in G and η because energy must be expended to overcome these interactions in order to make the surface elements flow past one another.

The interfacial tension, γ, tends to oppose any further increase in the area of the interface and is thus the appropriate surface stress in relation to dilatational rheology. Thus the interfacial dilatational elasticity, ε, and dilatational viscosity, κ, are defined by

$$\varepsilon = \frac{d\gamma}{\frac{dA}{A}} = \frac{d\gamma}{d\ln A} \tag{3}$$

$$\kappa = \frac{d\gamma}{\frac{dlnA}{dt}} \tag{4}$$

If the interface is subjected to a disturbance which tends to try and increase the area of the film at some point and the film expands, carrying the adsorbed surfactant along with it, then a region is created with a lower surface concentration, Γ and therefore with a higher γ. This region of higher γ will then tend to pull back the region of lower γ, restoring the uniform distribution of surfactant in the interface. Thus a force arises which tends to oppose the expansion of the interface. The more the value of γ rises with respect to an increase in A (i.e. the higher the value of ε), then the greater is the restoring force. This is the origin of the so-called Gibbs-Marangoni effect which is an important mechanism in the stabilisation of emulsions and foams. Note that this mechanism does not depend on any specific interactions between the surfactant molecules in the surface and depends on the relationship between Γ and γ, i.e., the surface equation of state.

In any real system the surfactant layer at the interface is always in contact with two bulk phases (i.e., oil and water or water and air), the surfactant having adsorbed from one of the condensed phases. If part of the area of the interface is increased so that Γ changes, then at this point the equilibrium between the bulk and the interface is disturbed, according to the adsorption isotherm. All those factors which affect the rate and extent of surfactant adsorption from the bulk will then also affect the rate and extent to which γ relaxes back to its previous equilibrium value. Thus the interfacial and bulk effects on γ are coupled and this is an extra complicating factor of dilatational rheology compared with shear rheology. With regard to the Gibbs-Marangoni effect, the flow of molecules in the bulk tangential to the interface is coupled to the flow of surfactant molecules in the interface. So the thinning of the thin film of intervening fluid between two interfaces is opposed by the flow of liquid back into the thinning region due to the surfactant flowing back into the expanded interfacial region according to the above mechanism. This helps to retard film drainage (e.g., see Sonin *et al.*, 1994).

The moment one begins to take account of surfactants or other molecules at the interface the notion of an infinitesimally thin interface is lost, since all these molecular elements have a certain physical thickness and strictly we just have a very thin, three dimensional region[3] . Thus one might expect a formal relationship between the mechanical[4] contribution to the dilational moduli and the shear moduli (Trapeznikov *et al.*, 1968). In fact, by analogy with three dimensional moduli (Reiner, 1960), for an isotropic film we have

$$\varepsilon = fG \tag{5}$$

and

$$\kappa = f\eta \tag{6}$$

where $f = 3$ for and incompressible material and $f = 2$ for a completely compressible material. Consequently, for real complex adsorbed films such as occur in food colloids, there are likely to be some direct relationships between the dilatational and shear moduli, so that neither should be considered completely independent of the other.

The difficulties bulk transport effects cause are partly responsible for the division of interfacial rheological studies into two distinct areas (a) studies of dilatational interfacial rheology of mainly simple solutions of low-molecular-weight surfactants, or spread films of insoluble low-molecular-weight surfactants and polymers, and (b) studies of shear interfacial rheology of mainly proteins and other adsorbing polymers. Insoluble films are not usually present in food systems, but studies of spread protein films can still be useful because the effects of exchange with bulk phases is eliminated (Murray, 1997a,b; Murray et al., 1997). Since proteins are generally adsorbed so strongly, a spread film, though still soluble, often effectively behaves as an insoluble (or non-desorbable) film over the time-scale of the rheological experiment. Studies on spread films of insoluble low-molecular-weight surfactants may sometimes provide useful information about the two-dimensional phase behaviour of the corresponding soluble analogues.

---

[3] In fact, two other moduii have been proposed which may be of significance: the tangential elasticity and viscosity. These may be envisaged as retarding the upward and downward motion of the molecules relative to one another at right angles to the plane of the interface (Earnshaw, 1981; Earnshaw and McCoo, 1994).

[4] By *mechanical* contribution it is meant the 'in-film' contribution as opposed to any contribution from bulk diffusion effects.

A survey of the literature also shows that much more is known about interfacial rheology at the A-W interface than at the O-W interface, particularly with regard to dilatational rheology. This is primarily because of the experimental difficulties involved in containing and compressing/expanding a film of molecules at the O-W interface as compared to one at the A-W interface. This is unfortunate from point of view of food colloids since many of the important types of naturally occurring (or added) emulsifying agents, such as the phospholipids or the mono- and di-glycerides may be soluble to some extent in both phases. In addition, the cohesion between hydrophobic parts of adsorbed species, be they protein or simple surfactants, might be expected to be quite different at the A-W interface than at, say, a hydrocarbon-water interface, where the hydrophobic regions of the adsorbate may be interpenetrated by the oil molecules. This effect is in addition to any specific molecular interactions which may occur between the oil solvent molecules and the adsorbing species.

In the following, shear and dilatational rheological studies on proteins, low-molecular-weight surfactants and proteins+low-molecular-weight surfactants mixtures are considered in separate sections, before the relationship to colloid stability is discussed, at the end.

3 INTERFACIAL SHEAR RHEOLOGY

3.1 Measurement considerations

Many methods have been developed over the years and some of these have been reviewed recently by Warburton (1996) and Miller *et al.* (1996). Many methods rely on measuring the rotational motion of a knife edged bob, disc or ring in the interface - effectively the two dimensional equivalent of a typical Couette viscometer. Constant shear rate (Edwards *et al.*, 1991), constant shear stress or creep (Vernon-Carter and Sherman, 1981) and oscillatory (Feng *et al.*, 1991; Lee *et al.*, 1991; Benjamins and van Voorst Vader, 1992) modes of operation have been developed. For studying very delicate films, we have used successfully at Leeds for several years a constant low strain-rate instrument (Dickinson *et al.*, 1985). One point to note about the Couette type viscometers is that they also rely on the condition of no slip at the inner and outer walls, and care must be taken to ensure that this is the case, though this is very rarely tested. Also, in practice, the film rheology may be highly non-linear - particularly for the case of proteins (see section 3.3). This can make results from the damped oscillation methods difficult to interpret. One advantage of the Couette type apparatus is that both the O-W and A-W interfaces (and both soluble or insoluble films) can usually be studied equally easily (subject to the limitations stated above). This is not the case with most of the other experimental methods for investigating interfacial rheology, either dilatational or shear.

The other main type of apparatus for studying shear rheology is the deep channel surface viscometer due to Mannheimer and Schechter (1970). This canal viscometer was developed for measuring values of η for surfactants at the A-W interface, which are generally in a much lower a range (e.g., below 10^{-2} mN s m^{-1}) than η values for proteins (see below). The method is based on the principle of monitoring the motion of tracer particles at an interface contained by a channel, formed by two concentric rings, when subjected to a well-defined flow field. It does not seem to have been widely used, probably because of the associated experimental difficulties, and it is almost exclusively restricted to the A-W interface, although Mohan and Wasan (1976), Nagarajan and Wasan (1994) have described modifications to the technique to enable it to be used at the O-W interface.

3.2 Interfacial shear rheology of low-molecular-weight surfactants.

In general, the magnitude of the values of G and η at the A-W interface for non-ionic and ionic surfactants are several orders of magnitude lower than values of ε and κ under the same conditions (Djabbarah and Wasan, 1982). However, it has been demonstrated that the shear rheology, when measured, is sensitive to the different phases which exists for such surfactant films at different interfacial concentrations (Peng *et al.*, 1993; Sacchetti *et al.*, 1993). Relatively few measurement have been performed at the O-W interface and the dependence of the shear moduli on deformation and rate of deformation, i.e., the non-Newtonian character, has not been greatly explored.

At high salt concentrations the shear rheology of ionic surfactants may be considerably enhanced due to surfactant aggregation phenomena. The effects may be strongly dependent on the salt type, e.g., see Chattopadhyay *et al.*(1992). Such effects may be significant in food concentrates/low moisture content foods.

3.3 Interfacial shear rheology of food proteins.

In contrast to the values for films of low molecular weight surfactants, the shear moduli of adsorbed (or spread) films of protein molecules are generally in a much higher range, e.g., η values from 0.1 to 1000 mN s m^{-1} and G values from 0.1 to 100 mN m^{-1}. The values are also similar or greater in magnitude to the corresponding dilatational moduli values as indicated in the pioneering of Graham and Phillips (1980a,b). The high values of η and G are not very easily measured via canal viscometers. The magnitude of their values reflects the interactions which can occur between adsorbed polypeptide chains: through hydrogen bonding, hydrophobic bonding, covalent bonding (as in the case of disulphide bridges) and salt bridges, etc. Because of this, the development of the viscoelasticity of an adsorbed protein film is intimately connected with the conformational changes of a protein on adsorption. These changes may take place quite slowly, depending upon the conditions, sometimes apparently taking several days (or more) before a steady state is reached. In this respect the resulting film may be better likened to a very thin film of a bulk protein gel, and indeed it behaves in many ways a such. The rheology may be very sensitive to the history of the film, and any variation in parameters which may be expected to affect the conformation of the adsorbed molecule, such as pH, temperature, ionic strength, ion binding, etc., will also tend to affect the values of η and G.

The extreme sensitivity of the interfacial shear viscosity to food protein structure has been demonstrated several times in investigations carried out in this laboratory (Dickinson, 1991; Castle *et al.*, 1987), and it has been highlighted again recently by Benjamins and van Voorst Vader (1992). These latter workers commented on the substantial shear thinning nature of the protein films and the difficulty of obtaining reproducible measurements even from the same protein solution. The high sensitivity to structure, packing and interactions between adsorbed protein molecules may be likened to the sensitivity to shear forces of bulk concentrated dispersions of systems of weakly interacting particles. Since food proteins themselves vary greatly in structure this means that the shear rheology of different individual proteins varies considerably. In general, the greater the internal cohesion and structuring of a protein molecule, then the greater are the values of the shear moduli, as illustrated in Table 1. (Results taken from Murray, 1987; Castle *et al.*, 1987; Dickinson, 1991). Thus the disordered milk protein β–casein, for example, possesses very little internal structuring in bulk solution, compared to the globular milk protein β-lactoglobulin, and it has very much lower interfacial shear viscosity.

Table 1. Comparison of the apparent interfacial shear viscosity, η_{app}, and the apparent interfacial shear elasticity, G_{app}, of selected proteins at the n-tetradecane-water interface after an adsorption time of 24 hours. Solution conditions were 10^{-3} wt. % protein, pH 7.0 phosphate buffer, ionic strength 0.005 mol dm^{-3} and 25 °C, unless indicated otherwise. Measurements were made at shear rate of 1.27 x 10^{-3} rad s^{-1} and a shear strain of 0.03 rad, via the method of Dickinson et al. (1985).

protein	η_{app} [mN s m^{-1}]	G_{app} [mN m^{-1}]
β-casein	0.5	0.1
α_{s1}-casein	4.0	0.3
sodium caseinate	7.4	0.6
gelatin	120	0.6
α-lactalbumin[1]	170	-
κ-casein	180	5.0
lysozyme	630	23.0[4]
β-lactoglobulin[2]	1200	-
myosin[3]	2400	-

[1] ionic strength 0.05 mol dm^{-3}. [2] ionic strength 0.02 mol dm^{-3} imidazole buffer.
[3] 0.5 mol dm^{-3} KCl + 0.02 mol dm^{-3} phosphate buffer, pH 6.5. [4] ionic strength 0.1 mol dm^{-3}

Fig. 1 illustrates the strong time-dependence of the apparent interfacial shear viscosity, η_{app} of a range of food proteins (results taken from Murray, 1987; Castle *et al.* 1987; Dickinson, 1991) due to the adsorbed molecules continuing to unfold and form new intra- and intermolecular cross-links with time.

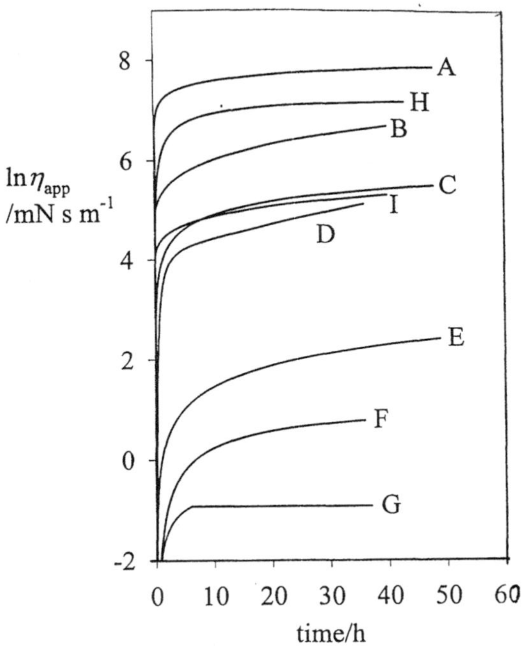

Fig. 1. Dependence of the apparent interfacial shear viscosity η_{app} on the adsorption time for adsorbed films of various proteins at the *n*-tetradecane-water interface. A = myosin, B = lysozyme, C = κ-casein, D = gelatin, E = sodium caseinate, F = α_{s1}-casein, G = β-casein, H = β-lactoglobulin, I = α-lactalbumin.

Solution conditions as in Table 1.

Burgess and Sahin (1994) have reported an increase in the interfacial shear elasticity of β-casein due to ageing of the bulk solutions prior to adsorption at the A-W interface. The protein concentrations used were 1 wt. % or higher, so that the effect may be connected with bulk protein aggregation. Certainly the film elasticities they report are much higher than any others in the literature for this protein. The long time-dependence of η is indicative of the strong forces exerted on protein molecules at interfaces. Consistent with this statement is the observation that disulphide bond interchange is induced on adsorption and ageing of β-lactoglobulin at the O-W interface in emulsions (Dickinson and Matsumura, 1991).

The above comparison of η between proteins is complicated, however, by the fact that the shear rheology may be highly non-Newtonian (extremely shear-thinning). It will be noted that the values in Table 1 and Fig. 1 are quoted as *apparent* interfacial shear elasticities (G_{app}) and viscosities (η_{app}) measured at particular strains and rates of strain. Fig. 2 shows the variation of η_{app} for various films across a range of shear rates and it is seen that the relative values of η_{app} for food proteins may be highly dependent on the shear rate. It should be noted that none of the shear rates employed here (Murray, 1987) were very high compared with those used for the preparation of emulsions or foams, and that those proteins with higher values of η_{app} at low shear rates seemed to be the most shear thinning.

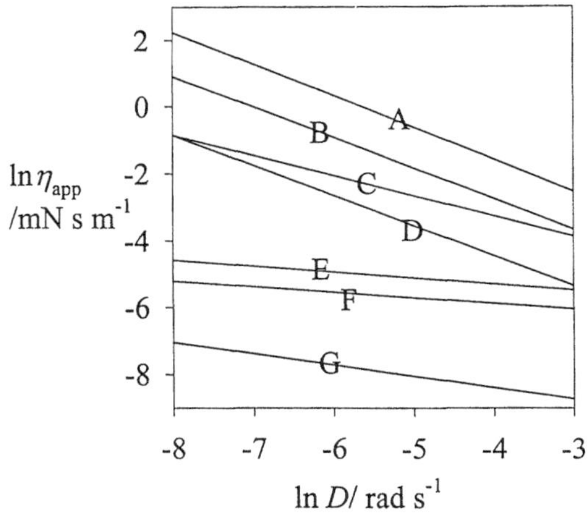

Fig. 2. Dependence of the apparent interfacial shear viscosity η_{app} on the shear rate, D, for adsorbed films of various proteins at the n-tetradecane-water interface. A = myosin, B = lysozyme, C = κ-casein, D = gelatin, E = sodium caseinate, F = α_{s1}-casein, G = β-casein. Solution conditions as in Table 1.

On the other hand, for the results illustrated in Fig. 1, Fig. 2 and Table 1, it was always observed that moduli were found to recover rapidly after each measurement, i.e. there was apparently no irreversible mechanical damage to films despite prolonged shearing. This probably means that the time-dependence of η_{app} genuinely reflects slow molecular rearrangements, whilst the measured viscoelasticity itself is largely dependent on weak molecular interactions (probably hydrogen bonding), which are relatively easily disrupted by shearing. Further evidence for this comes from the fact that, usually during adsorption from low bulk protein concentrations, a lag time may be observed before any measurable η_{app} appears, even though other measurements (e.g., of γ) clearly indicate that appreciable protein adsorption has occurred (Murray, 1987; Murray and Dickinson, 1996). Presumably it is necessary for the interfacial concentration of protein to reach a certain level before the viscoelastic 'gel' network can begin to develop. The fact that shear moduli of protein films generally decrease quite sharply with increasing temperature, in the range approximately 0 to 60 °C, is also an indication that hydrogen bonding between different segments of the adsorbed molecules plays a major role determining the rheological behaviour. At higher temperatures the values of η and G may increase irreversibly with increasing temperature due to formation of new covalent cross-links between the molecules (Avramidis and Jiang, 1991; Dickinson and Hong, 1994) At temperatures where proteins become completely denatured and insoluble in water the molecules will no longer be surface active and the interfacial viscoelasticity will fall.

The distinct differences between η (and G) for different protein films has been exploited in monitoring the competitive adsorption of food proteins (Dickinson, 1991). If a more surface active protein is injected below an adsorbed film of a second protein, then η and G change until they more closely correspond to the interfacial rheology of the more surface active protein. The speed and extent of the change depend on the relative surface activities of the proteins, the residence times of the proteins at the interface and the concentrations of the proteins in the system. In general the predictions of displacement based on such rheological changes are in qualitative agreement with independent measures of the relative concentrations of the different proteins at the interface. (Castle *et al.* 1986, 1987; Dickinson *et al.*, 1988b, 1989; Dickinson, 1991; Hunter *et al.*, 1991; Nylander and Wahlgren, 1994). With the interfacial rheology, however, there are some important additional features. (A) In particular, it is generally observed that the interfacial shear rheology continues to change long after when the measured surface concentrations have *apparently* stabilised. This reflects the slow approach to equilibrium adsorbed configurations for proteins, as noted earlier, but also the fact that the measured Γ of proteins are sometimes in doubt. In a number of cases (Dickinson *et al.*, 1985, 1987; Galazka and Dickinson, 1995) interfacial rheological measurements of the planar interface point to the slow accumulation of secondary adsorbed layers of protein which are not necessarily detected in the measurements of the Γ on the corresponding emulsion droplets. This is because the measurement of Γ on emulsion droplets relies on separating the droplets from the aqueous phase, usually via centrifugation, and measuring the remaining concentration of protein in the aqueous phase or the droplet phase. In this procedure weakly adsorbed secondary layers may be desorbed. Various optical techniques for measuring adsorbed protein concentrations at planar interfaces, e.g., via ellipsometry, may also not be sensitive enough to detect such secondary adsorbed layers. (B) It is also observed that the ease of displacement of one protein by another apparently depends on the age of the interface, the age being reflected in higher η and G (Dickinson, 1991). Thus the most surface active food protein of a mixture does not necessarily dominate the interface immediately; the order and time of exposure proteins to an interface determines the properties of the interface. This has implications for the short and long term stability of emulsions. Exceptions are mixtures of the two disordered milk proteins α_{s1}-casein and β-casein, which approach steady state rather rapidly (Dickinson *et al.*, 1988b) and whose adsorbed layer protein composition is amenable to theoretical interpretation via equilibrium statistical thermodynamics (Dickinson, 1992b).

In some cases the addition of one protein to another may result in an enhancement of η and G, over and above the values of either component on its own. When this occurs it is due to some sort of complex formation between the components at the interface which often reflects a tendency for the components to associate in the bulk as well, e.g., through electrostatic interactions. Examples of systems with this type of behaviour are bovine serum albumin (BSA) + lysozyme (Poole *et al.*, 1984), lysozyme + gelatin (Castle *et al.*, 1987) and gelatin + β-lactoglobulin (Chen and Dickinson, 1995). In the same way the addition of calcium ions to the caseins (Hunt *et al.*, 1993; Dickinson, 1994) which tends to promote their aggregation in the bulk, also enhances the corresponding η of these proteins. Surface complexation can also be detected from changes in η in systems containing mixtures of protein + polysaccharide (Dickinson and Galazka, 1992). In some cases this may be the result of thermodynamic incompatibility of macromolecules in the bulk phase(s).

By deliberately increasing the number and strength of bonds between adsorbed proteins one can increase η above that for the native protein. For example, cross-linking of caseins with the enzyme transglutaminase (Færgemand *et al.*, 1997) produces an increase in η of several orders of magnitude. Færgemand *et al.* (1997) also found, however, that there appears to be an optimum level of cross-linking, otherwise η decreases again, the film appearing to become more susceptible to fracture.

3.4 Interfacial shear rheology of mixtures of food proteins + low-molecular-weight surfactants.

Even low interfacial concentrations of surfactant cause marked disruption of the interfacial protein - protein interactions. As the surfactant : protein molar ratio, R, is increased protein begins to be displaced from the interface, so that the values of η and G usually plummet to the much lower values characteristic of low-molecular-weight surfactants alone. The fall in η as a function of R is generally quite sharp (e.g., see Dickinson and Hong, 1994) - η may be greatly reduced though little protein desorption has occurred (Dickinson, 1991).

Recently Eijt *et al.* (1994) provided evidence that, for a system consisting of a compact globular protein (a fungal lipase) + *p*-nonylbenzene nonaethylene oxide, a composite structure can occur at intermediate R, with surfactant directly at the A-W interface and protein still adsorbed as a secondary layer underneath.

The ease of protein displacement by surfactant apparently depends upon the strength of the protein - protein interactions: globular protein films of higher interfacial shear viscosity, e.g., due to heating (Dickinson and Hong, 1994) or ageing (Chen and Dickinson, 1993) have been shown to be more difficult to displace from the interface. Similar results have been observed with water-in-crude oil emulsions stabilised by polymeric asphaltenes to which low-molecular-weight demulsifiers are added (Mohammed *et al.*, 1994). Euston *et al.* (1995) and Fang and Dalgleish (1996a) have recently made some interesting observations on the relative ease of displacement of different food proteins from the oil-water interface. It appears that for caseinate-stabilised emulsions (at high caseinate concentrations) the β-casein component of caseinate is preferentially desorbed to some extent, this preferential desorption increasing with ageing of the emulsions. This seems surprising since β-casein is the slightly more surface active of the caseins (Castle *et al.*, 1987; Galazka and Dickinson, 1995). Fang and Dalgleish (1996a) speculate that the effect may be connected with different specific interactions between the surfactants and the proteins, but it is interesting to note that β-casein is also the protein with the lowest interfacial shear viscosity (see Table 1). Thus β-casein has the least capacity for self-association and network formation at the interface, which may make it easier to desorb.

Relatively few interfacial rheology experiments have been performed with oil-soluble surfactants, despite their importance in foods. Many phospholipid and glyceride components, for instance, are almost exclusively soluble in the oil phase. However, in one recent study (Chen and Dickinson, 1995) the addition of $C_{12}E_2$ to the oil phase was found at low R to lead to an increase in η, though at long times η fell to values much lower than before the addition. Undoubtedly the initial increase in η was due to some enhanced attractive interaction between the film components. Recently Euston *et al.* (1995) have similarly observed that oil-soluble surfactants did not seem able to desorb caseins from the interface as easily as water soluble surfactants.

The exact value of *R* at which the interfacial shear rheology changes will also depend on the nature of any interactions between the surfactant and the protein in the bulk solution (see Dickinson and Woskett, 1989; Dickinson, 1993; Clark *et al.*, 1993a,b). Any direct binding of surfactant to protein will tend to change the conformation of the protein and consequently alter the interfacial shear rheology of the resultant complex if this is present at the interface. Non-ionic surfactants may bind strongly to proteins, but if so this is generally a highly specific effect involving a small number of hydrophobic binding sites per protein molecule (Coke *et al.*, 1990; Wilde and Clark, 1993; Wilson *et al.*, 1993) For example, in the case of β-lactoglobulin and alcohol ethoxylates the mole ratio of protein to surfactant in any complex is close to 1. It is not known whether such protein-surfactant complexes are more or less surface active than the native protein molecules.

With ionic surfactants the binding to protein may be less specific but with larger numbers of surfactant molecules binding per protein molecule. This means that there may be appreciably less native molecules available in the bulk for adsorption at the interface and that the configuration of the protein-surfactant complex will probably be very different from the native protein; the protein is likely to be in a much more unfolded state. Wüstneck *et al.* (1984, 1987) observed that, at intermediate *R* values, cetyl trimethyl ammonium bromide (CTAB) + gelatin films exhibited enhanced values of η and G due to electrostatic interactions between the oppositely charged protein and surfactant. At higher values of *R*, however, the surfactant alone came to dominate the interface and the interfacial shear rheology reflected the typically low values of η and G for such films. The same picture seems to be true for β-casein and β-lactoglobulin films (Wüstneck *et al.*, 1996a).

The effects on pure protein films of higher-molecular-weight surfactants, particularly those possessing several functional groups, may be more like the effects found for mixed protein films. For example, diacetyl tartaric acid esters of monoglycerides (DATEM) when added to β-lactoglobulin films showed net enhancement of η (Dickinson and Hong, 1994), presumably due to some sort of complex formation at the interface. Similarly, enhancement of η was observed with lysozyme + tristearin at the O–W interface (Ogden and Rosenthal, 1994). Enhancement of interfacial viscosity and rigidity with mixtures of proteins and glycerides has also been observed (Doxastakis and Sherman, 1986; Martinez-Mendoza and Sherman, 1991; Velev *et al.*, 1993). Phospholipids do not seem to be as effective as other surfactants in displacing proteins from interfaces; there is the tendency for associative interactions between protein and phospholipid at the O–W interface (Dickinson and Iveson, 1993; Fang and Dalgleish, 1993; Fang and Dalgleish, 1996a,b).

4 INTERFACIAL DILATATIONAL RHEOLOGY

4.1 Measurement considerations

There are at least as many methods for measuring dilatational rheology as are there are methods for measuring interfacial tension, γ. Franses *et al.* (1996), Miller *et al.* (1996), Kretzschmar and Miller (1991) and Prins (1995) have recently provided some particularly useful reviews and the reader is also referred to the other chapters of this book. All of the dilatational methods are based on the principle of measuring of γ, whilst the interface is subjected to some change in its area. The interface may be macroscopic and bounded by solid walls, as in various 'trough' methods, or the interface may be formed by single small drop or bubble. Trough methods can be applied to both insoluble and soluble films, drop (bubble) methods only to the latter. The area change applied to the film may be oscillatory, a step change or a continuous expansion/compression. In all cases where the strain applied to the film is large and/or is applied rapidly the difficulty arises of non-uniform deformation of the interface, giving rise to gradients in γ throughout the interface which complicate the data analysis, particularly for oscillating drop (bubble) methods and the like (Franses *et al.*, 1996). The problem is exacerbated with films possessing appreciable viscous moduli, which means most food proteins and polymers (Malcolm, 1985; Byattsmith and Malcolm, 1994; Bois and Panaiotov, 1995; Peng and Barnes, 1995).

In this respect methods involving the measurement of the propagation and damping of naturally occurring capillary waves (or 'ripplons') at fluid interfaces (Earnshaw, 1983; Thominet *et al.*, 1988;) have a distinct advantage in being non-perturbative. They can also access a higher range of frequencies than is possible using other methods and capillary waves are primarily responsible for the spontaneous perturbations of the interface which may grow and lead to film rupture. A disadvantage is that the measurements can be difficult for films which are highly damping (e.g., protein films). In addition, the O-W interface presents further theoretical and technical difficulties compared to the A-W interface. When a step change in area is applied to a fluid interface the resultant interfacial tension versus time decay curve can be analysed via Fourier transform methods (Loglio *et al.*, 1979, 1984; Cárdenas-Valera and Bailey, 1993; Kitching *et al.*, 1996). This approach has the advantage that moduli as a function of a range of frequencies can be obtained simultaneously in one experiment.

Techniques where the interface is subjected to continuous expansion are worthy of special mention because they represent a closer approach to the conditions of actual emulsion and foam formation. The method may be applied to drops, as in the dynamic drop volume method (Joos and Van Uffelen, 1995; Horozov and Joos, 1995) and the maximum bubble pressure method (Fainerman *et al.*, 1994), or to films in troughs (Fang *et al.*, 1995; van Aken, 1995; van Aken and Merks, 1996). Joos and co-workers have recently developed the technique for troughs ("peaktensiometry", van Uffelen and Joos, 1994) where the effects of convective mass transport to the interface can also be included (e.g., see Petrov and Joos, 1996a). A novel apparatus, which may be used at both the A-W and O-W interfaces, has recently been developed by Prins and co-workers (Prins, 1995; Prins and Bergink-Martens, 1993; Bergink-Martens *et al.*, 1990, 1994), based on the concept of an overflowing cylinder. Some such experiments have recently been criticised by Lucassen-Reynders and Lucassen (1994), who show that an elastic deformation is always present in the measurements, whereas it is generally assumed that the deformation is purely viscous.

4.2 Dilatational rheology of low-molecular-weight surfactants

A great deal of effort has been expended in making dilatational measurements on surfactant solutions and in physico-chemical interpretation of the results in terms of surfactant transport both within the interface and to/from the interface. A cursory overview will suffice, since the object is to highlight the importance of the interactions of such surfactants with food proteins. Experiments on insoluble surfactant films (i.e., spread at the A-W interface) reveal relationships between the dilatational rheology and the phase behaviour of the monolayers (Kajiyama *et al.*, 1989; Noskov and Zubkova, 1994; Kizling *et al.*, 1995; Smaby *et al.*, 1996). These results may be useful in interpreting the behaviour of the corresponding soluble films but on the whole insoluble surfactant films are of minor relevance to foods. Systems consisting of surfactants which are soluble in one or both of the bulk phases may usefully be considered under two possible conditions (a) where the bulk concentration of low-molecular-weight surfactants is above the critical micelle concentration, CMC, and (b) where the bulk concentration is below the CMC.

4.2.1 Soluble surfactants above the CMC.

As was indicated earlier (section 2) the complicating factor of dilatational rheology is that the interfacial stress (γ) is dependent on both the transport of surfactant within the interface and also to the interface from the adjacent bulk phases. If the interface expands and Γ decreases locally the concentration of surfactant solution immediately adjacent to the expanded area is no longer in equilibrium with the surface. More surfactant molecules will then start to adsorb to this interface according to the adsorption isotherm; there may or may not be an activation barrier to adsorption. The non-convective layer of solution just below the interface will thus become depleted in surfactant compared to the bulk. This will cause diffusion of surfactant molecules towards the interface. If micelles are present in the sub-surface region then the local equilibrium between micelles and free surfactant monomer will be disturbed also. Micelles will tend to dissociate if the local concentration of surfactant monomer falls below the CMC. There will then also be set up a concentration gradient of micelles, with micelles diffusing towards the interface. Thus the observed dependence of γ on time (t) and A will depend on the balance between the rates of movement of surfactant in the interface (where surfactant aggregation/disaggregation phenomena may be important -see Lin *et al.*, 1990), the rates of diffusion of monomers and micelles to the interface, the nature of the surfactant adsorption isotherm, the rates of micelle - monomer exchange and the presence of any activation barriers involved in these processes, e.g., due to electrostatic interactions. Phenomenological values of ε and κ can always be assigned through equations (3) and (4) which characterise the $\gamma(A, t)$ relationship, but attempts have been made to develop models which account for the behaviour in molecular terms. For the most part this has often consisted of trying to find out whether or not the adsorption is diffusion-controlled, or whether or not some energy barrier to adsorption exists which is the rate-controlling factor (e.g. Fainerman *et al.* 1994; Fainerman and Miller 1995). The presence of adsorption barriers, if these existed, would be very significant because adsorption kinetics would then become independent (to some extent) of mass transport to the interface, possibly even during the turbulent conditions of emulsion and foam formation. However, many of the necessary rate constants required to test a particular molecular interpretation of the $\gamma(A, t)$, for example for monomer - micelle exchange, are not available and approximations or assumptions have to be made. With many adjustable parameters in such models, it is perhaps not surprising that agreement has often been lacking between different workers (Jayalakshmi *et al.*, 1995).. Wasan *et al.* (1992) and Chu *et al.* (1994) recently developed a model which suggests that surfactant structuring within a thinning film may be very important - surfactant micelles can apparently become tapped in layered structures within thinning films - giving rise to step-wise thinning of the film. This introduces further complications (the mechanism may also be possible with non-adsorbed food protein molecules).

4.2.2 Soluble surfactants below the CMC

The kinetic situation is considerably simplified below the CMC, and for this reason most work has concentrated on this lower concentration régime, including work on low-molecular-weight surfactants of significance to foods. Even so, the situation can still be quite complicated. For example, recent work has shown that for ionic surfactants the electrostatic interactions between adsorbed and adsorbing molecules can have a dominant effect (Bonfillon *et al.*, 1994; Bonfillon and Langevin, 1993, 1994; Sharpe and Eastoe, 1995, 1996; Earnshaw and Sharpe, 1996; Eastoe *et al.*, 1996). For a recent review of the field see Chang and Franses (1995).

Another reason for most studies concentrating on this lower concentration régime is that ε generally exhibits a maximum below the CMC (e.g., see Kizling *et al.*, 1995). This is because as the surfactant concentration in the bulk increases, any increase in γ due to a decrease in Γ is more rapidly compensated for by adsorption from the bulk. This 'short circuiting' mechanism has the effect of reducing the value of ε and makes the measurement of ε more difficult because the changes in γ for a given strain (change in A) become smaller and relax more rapidly back to the equilibrium value.

In many food colloid systems the concentrations of low-molecular-weight surfactants are likely to exceed the CMC values, particularly for non-ionic surfactants, which tend to have very low CMC values. Thus micelles and other surfactant self-assembled structures, e.g., lipid vesicles, etc., are likely to be present. Consequently, the above types of study at low surfactant concentrations may not be very relevant to practical emulsion and foam stability. Fang *et al.* (1996) have recently modelled the behaviour of fatty acids over a range of concentrations. Another complication is that added food surfactants are often quite broad mixtures containing components which may also have some oil solubility. The case of surface tension relaxation in mixed micellar systems has recently been considered from the theoretical point of view (Miller and Kretzschmar, 1991), but much of the data required for comparison of models with experiments is lacking. As stated earlier, relatively few interfacial dilatational studies have examined the O-W interface and the case of mixtures of oil-soluble and water-soluble surfactants, though these are important in practical situations. The presence of a second bulk phase, offering additional mechanisms for surfactant transport, adds an extra dimension of complexity to the problem. Again, however, the situation has recently been considered from a theoretical point of view (Miller *et al.*, 1991; Joos, 1995). Mixtures of very water-insoluble + water-soluble surfactants at the A-W interface have been considered by Petrov and Joos (1996b).

4.3 Dilatational interfacial rheology of food proteins.

Analysis of the data for protein systems may offer some simplifications compared to that for systems containing low-molecular-weight surfactants, mainly depending upon the time régime, i.e., the rates of strain or frequencies at which measurements are made. Over short time-scales proteins may behave simply as larger surfactant molecules with lower diffusion coefficients. Since protein molecules often appear to unfold slowly at the interface on adsorption, unfolding processes may contribute little to the dilatational moduli over short time-scales. In this respect the interfacial dilational moduli of proteins often appear to be relatively independent of film ageing (Benjamins and van Voorst Vader, 1992) compared to the corresponding shear measurements. This could also be a reflection of the inherent lack of sensitivity of dilatational rheology to film structure compared with shear measurements. On the other hand, recent measurements in this laboratory (Murray, 1997a) have shown that the relaxation of spread monolayers of β-lactoglobulin is unexpectedly fast compared with the behaviour of the corresponding adsorbed films. Some of these results are indicated in Fig. 3.

Thus the dominant slow process in the dilatational rheology of adsorbed films may be the adsorption of new molecules to the interface rather than the rearrangements of existing molecules within the film. Measurements on insoluble synthetic polymer films indicate that surface equilibration is apparently very slow (Nahringbauer, 1995). For long time-scales/low frequency measurements these slow processes must be taken into account (see below).

Micelle formation for proteins obviously need not be considered - though many proteins do aggregate to some extent depending on the conditions such as pH, ionic strength and metal ion concentration. Notable examples of aggregating proteins which are technologically important in foods include milk proteins (caseins and whey proteins) and many plant proteins. In addition, for short time-scales and for small deformations, proteins are generally so strongly adsorbed that they may be considered as irreversibly adsorbed (though it should be stressed that this irreversibility is only notional, see MacRitchie, 1986). Fruhner and Wantke (1996) recently showed that gelatin effectively behaves as an insoluble monolayer at frequencies greater than 50 Hz. Serrien *et al.* (1992) have recently proposed a model for bovine serum albumin (BSA) and milk proteins whereby only the native protein molecules are exchangeable (desorbable) from the interface. Other authors, e.g., Douillard and Lefebvre (1990), have proposed a two-layer model for the adsorption of globular proteins, with a tightly bound first layer and a more loosely bound secondary layer.

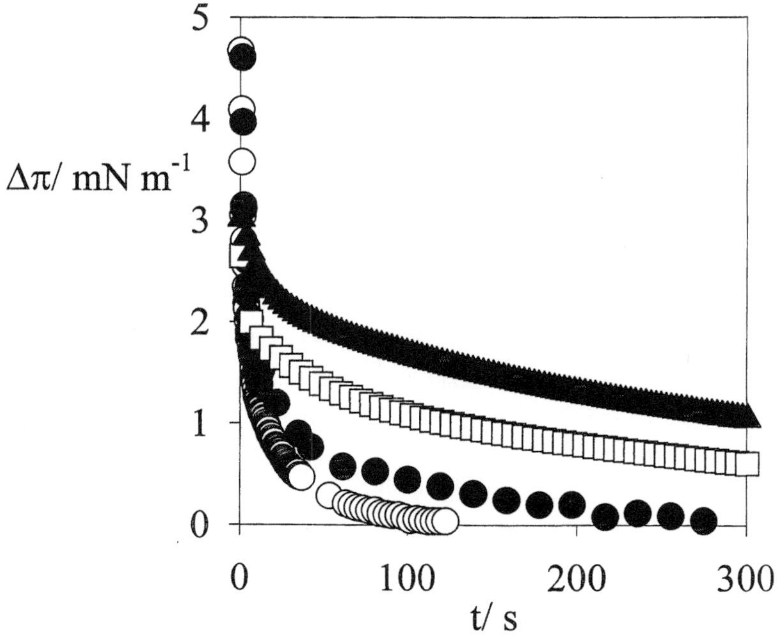

Fig. 3. Change in interfacial pressure, $\Delta\pi$, versus time, t, for spread β-lactoglobulin films expanded by 10 %: (O) A-W film of initial interfacial pressure = 26.0 mN m^{-1}, (●) O-W film of initial interfacial pressure = 23.0 mN m^{-1}. Change in interfacial pressure, $\Delta\pi$, versus time, t, for corresponding adsorbed β-lactoglobulin films expanded by 10 %: (□) A-W interface, (■) O-W interface. (Oil phase = *n*-tetradecane).

It is seen that interpretation of the dilatational rheology of interfaces involving proteins cannot be dissociated from the issues surrounding the measurement and modelling of protein adsorption. The reader is referred to Dickinson and Euston (1992), Fleer *et al.* (1993), Roth and Lenhoff (1993), Dickinson and Matsumura (1994), Haynes and Norde (1994), Horbett and Brash (1995) and Leermakers *et al.* (1996), for detailed discussions of the issues involved. Particularly important issues are the matter of energy barriers to adsorption (e.g., see Miller *et al.*, 1993) and the effects of strong lateral interactions between adsorbed molecules at the interface. For an individual protein molecule to become adsorbed a certain, minimum length of the polypeptide chain must become associated with the surface. When the protein must change its conformation for this to occur, e.g., with a globular protein, then one might expect an energy barrier to adsorption. Alternatively this barrier may be of purely entropic (orientational) origin, where the protein molecule must rotate until the appropriate part of the molecule surface comes into contact with the interface. The more an interface is already covered with protein then the greater will be conformational change required and/or the greater will be the orientational restrictions imposed for further adsorption to occur (Mitchell, 1986). Recently, however, the presence of an electrochemical barrier has been proposed (Xu and Damodaran, 1993, 1994) to explain the pronounced barrier to adsorption of some globular proteins, such as lysozyme, under conditions where the protein is highly charged - though some of these results have recently been called into question (Murray, 1997b). In some cases the adsorption of proteins can be modelled according to a simple adsorption isotherm, though usually agreement is only achieved by assuming unrealistic values of the bulk protein diffusion coefficient (Wüstneck *et al.*, 1996b). Such *effective* diffusion coefficients have no physical significance; they merely indicate that the assumed model does not hold. Unfortunately, there is considerable difficulty in establishing protein adsorption kinetics unequivocally by measuring Γ directly, because there is often considerable disagreement between the results of different methods (Murray 1997b). The kinetics of the relaxation of γ, for step change in area for protein (Ahluwalia, 1996) and polymer (Cárdenas-Valera and Bailey, 1993; Nahringbauer, 1995) films can often be fitted to a sum of exponential decays to obtain a series of relaxation times. However, there may be no sound physical basis for the interpretation of such relaxation times.

At low frequencies (long times) and low shear rates, the viscous part of the dilatational rheology is usually predominant. If this is due to the slow conformational changes of either the protein molecules already adsorbed or those beginning to adsorb, then the same inter- and intramolecular bonding that affects the interfacial shear viscosity can therefore influence the dilatational viscosity. In connection with this, it is often observed that, with a food protein which possesses very little internal structure, such as β-casein, there is a marked change in the dilatational rheology above a certain surface concentration (Graham and Phillips 1980a; Gau *et al.*, 1994; Benjamins *et al.*, 1976). This is at the point where there is no free space available in the interface for molecules to adsorb and unfold without interacting with those already adsorbed. Sarker *et al.* (1996) showed an enhancement in κ (and ε) when proteins were cross-linked by cations. Similarly, large effects of ions have been found on the dilatational rheology of protein films by Chudinova *et al.*, 1995. The dilatational rheology of network forming bread (wheat) proteins has recently been investigated; higher moduli were found with increasing "hardness" of the wheat (Oliver and Sahi, 1995). Increased hardness correlates with a higher content of the cross-linking protein gluten.

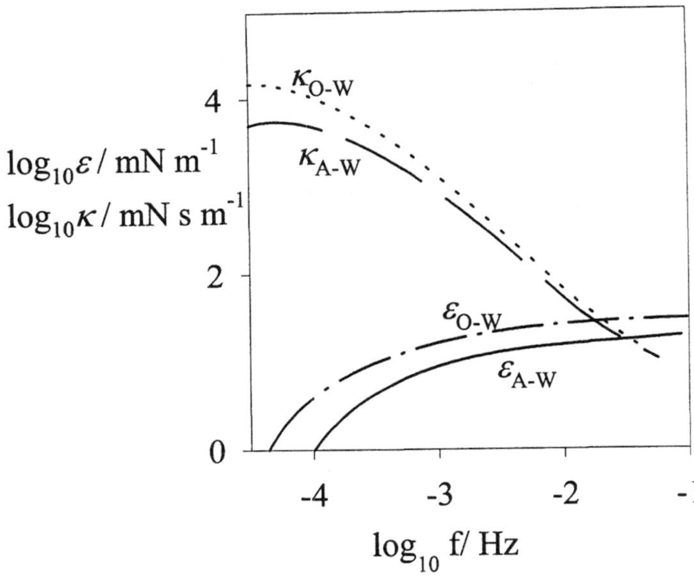

Fig. 4. Comparison of the interfacial dilatational elasticity and viscosity at the A-W interface, $\varepsilon_{A\text{-}W}$ and $\kappa_{A\text{-}W}$, respectively, and at the O-W interface, $\varepsilon_{O\text{-}W}$ and $\kappa_{O\text{-}W}$, respectively, for an adsorbed film of β-lactoglobulin. Solution conditions were 10^{-3} wt. % protein, 20 mM imidazole buffer at pH 7.0, 21 °C, oil phase = *n*-tetradecane, adsorption time 12-15 hours.

For food proteins that form a highly close-packed, cross-linked network at the interface (such as the globular proteins β-lactoglobulin, α-lactalbumin, ovalbumin, soy bean proteins, etc.), a dilatational deformation of the interface might not be expected to result in a uniform change in the surface coverage of molecules. Instead, a much more patchy, localised depletion of adsorbed protein molecules might be expected. In effect, in mechanical terms, the thin protein gel layer might be expected to behave in a 'brittle' fashion with holes and tears appearing. The issue of relaxation within inhomogeneous films seems to have been addressed only by Noskov (1996), though this must have important implications for the interpretation of the measured recovery in γ and the ultimate stability of such films.

Up until very recently there has been very little data available on the dilatational rheology of food proteins at the O-W interface compared with the A-W interface. This is unfortunate because one might expect the dynamics of unfolding to be somewhat different at air and oil interfaces (see section 1 above). Recently we have made some preliminary measurements (Murray *et al.*, 1997) on adsorbed β-lactoglobulin films at the *n*-tetradecane-water interface and the air-water interface using a novel Langmuir trough type apparatus (Murray and Nelson, 1996). Some results are shown in Fig. 4. The results indicate that films are considerably more elastic at the O-W interface compared with the A-W interface. This difference qualitatively agrees with a more detailed study at a paraffin oil-water interface by Williams and Prins (1996), though in a different range of bulk concentrations (using the ring trough due to Kokelaar *et al.*, 1991). Williams and Prins found that the difference between the moduli at the O-W and A-W interfaces decreases markedly as the concentration of protein is increased. Jiang and Chiew (1995) have also recently studied lysozyme at the corn oil -water and A-W interfaces and Benjamins *et al.* (1996) have also recently compared dilatational measurements on sodium caseinate, ovalbumin and bovine serum albumin films adsorbed at O-W and A-W interfaces. Benjamins *et al.* (1996) found that ε and κ were lower at the O-W interface, though the oil used here was chemically quite different - a triacylglycerol oil obtained by purification of sunflower oil. The choice of oil type for such model studies may therefore be quite important in trying to determine the behaviour of real food oil systems. In our own study (Murray *et al.*, 1997) it was found also that the dilatational elasticity, ε, is still non-linear down to fairly low values of the strain (ΔA *ca.* 5%). This may be one explanation of the discrepancy of the results with those of others. Although dilatational measurements are often claimed to be in linear regime this is rarely backed up by published data. If dilational rheology is dominated by the formation and breakage of inter- and intra-molecular bonds (due to conformational changes on adsorption and/or due to in-film rearrangements) it would be surprising to find that the dilatational rheology is perfectly linear whilst the shear rheology is highly non-linear (see section 3.3 earlier). Certainly the dilatational viscosity, κ, is highly non-linear (e.g., see Fig. 4).

Overall, compared to the wide-ranging interfacial shear rheology of proteins, the dilatational film properties of the food proteins examined so far are rather similar to each other. MacRitchie (1986) gave an explanation for this behaviour some time ago in noting the similarity of many surface pressure - area isotherms for spread proteins at the A-W interface. Proteins at the surface can be considered to consist of strings of unit cells, each apparently equivalent to 6 - 8 amino acid residues. Once unfolded, all proteins simply behave as longer or shorter strings of these units. However, proteins which can adsorb quickly and/or rearrange quickly at interfaces, whether due to lower molecular weight (higher diffusion coefficient), or greater flexibility, are expected to give rise to lower dilatational moduli due to the more rapid recovery in γ possible at short times (high frequency). Thus the value of ε for β–casein at the A-W interface is lower than that for BSA, following the general trend that ε increases with decreasing protein flexibility (Benjamins and van Voorst Vader, 1992; Graham and Phillips, 1980a).

4.4 Dilatational interfacial rheology of proteins + low-molecular-weight surfactants

With mixtures of proteins and low-molecular-weight surfactants there are all the possible complications discussed above as well as some new features to consider. Surfactant-protein complexation in the bulk leads to a new set of surface active species (which may be more or less surface active than the individual components). In addition, there is possibility that the equilibrium between proteins, low-molecular-weight surfactants and the protein - surfactant complexes is affected by the changes in the local concentration of surface active species in the vicinity of the interface. Another possibility arising from complexation is that the bulk concentration of surfactant is lowered below the CMC, so that micelles may no longer have an influence on the dynamics of the interfacial relaxation.

Not many measurements have been made on such mixed systems, even though such mixtures are more common than pure surfactant systems in real food colloids. Where measurements have been made they have been restricted almost exclusively to the A-W interface. Recent work by Clark and co-workers (Clark *et al.* 1993b; Clark *et al.*,1994; Sarker *et al.*, 1995a,b) has highlighted the marked heterogeneity of such mixed films, identified the role of a small-molecule cross-linking agent (catechin) in enhancing protein surface viscoelasticity, and has established important differences between behaviour at O-W and A-W interfaces.

Although changes in ε and κ are expected with increasing surfactant-protein molar ratio, R, due to conformational changes and desorption of interfacial protein, the resultant changes in ε and κ are not expected to be as dramatic as with the interfacial shear rheology, because the difference between the moduli for different proteins or between proteins and low-molecular-weight surfactants are not so large in the first place (see earlier). For example, Clark *et al.* (1993b) examined mixtures of Tween 20 + β-lactoglobulin at a constant protein concentration of 0.02 wt. % and R values from 0 to 10. This led to a reduction in the value of ε from 32 to 15 mN m^{-1} whilst κ increased from 1.6 to 4.4 mN s m^{-1}. This contrasts with a relative change in η_{app} of *ca.* 500 to < 0.1 mN m s^{-1} when Tween 20 was added to a β-lactoglobulin film under similar conditions (Chen and Dickinson, 1995; Clark *et al.*, 1995).

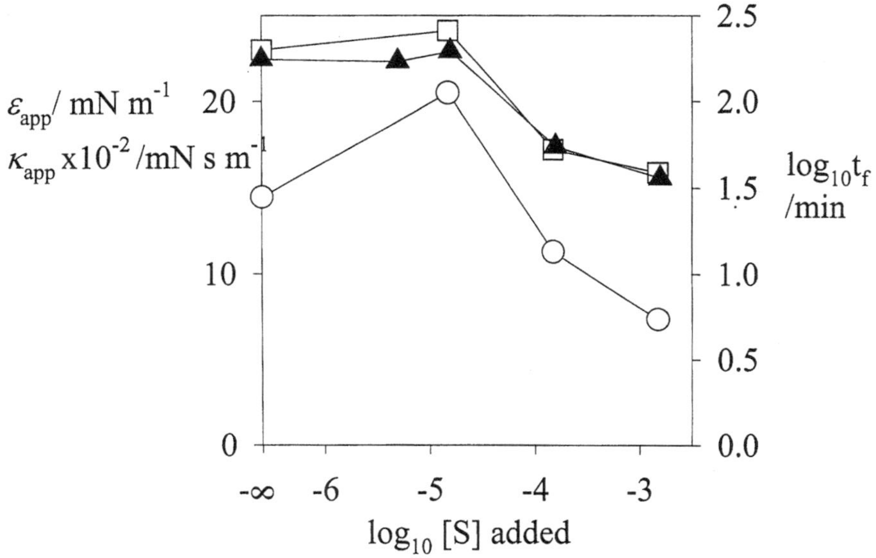

Fig. 5. Relationship between the apparent interfacial dilatational elasticity, ε_{app} (O), apparent interfacial dilatational viscosity, κ_{app} (□), and accelerated foam life-time, t_f(▲) for β-lactoglobulin + $C_{12}E_6$ films at the air-water interface. [S] = concentration of $C_{12}E_6$ added to 10^{-3} wt. % protein solution in 20 mM imidazole buffer at pH 7.0 and 30 °C.

In the combination of Tween 20 + β-lactoglobulin described above the addition of the low-molecular-weight, water-soluble surfactant was observed to decrease ε. We have recently made similar measurements on the β-lactoglobulin + $C_{12}E_6$ system where the addition of surfactant causes a slight enhancement in ε and κ at low R (Murray and Dickinson, 1996; Murray *et al.*, 1997). Some of these results are shown in Fig. 5. The dilatational moduli are quoted as apparent moduli, ε_{app} and κ_{app}, because in these experiments (made with the funnel method of van Hunsel and Joos, 1987), the deformation was considered to be in a non-linear regime ($\Delta A/A$ = 0.3). At higher R, ε_{app} and κ_{app} eventually came to represent the values for the pure surfactant. This emphasises that the effects of surfactants on interfacial rheology can be subtle. At low concentrations of low-molecular-weight surfactants, small effects on protein conformation may shift the different interfacial rheological moduli in different directions.

It has been observed that many other small molecules, which may not themselves be particularly surface-active, can have marked effects on the dilatational interfacial rheology of proteins. Thus, for example Dussaud *et al.* (1994), Dussaud and Vignes-Adler (1994a, b) demonstrated the marked effects of tartaric acid and ethanol on BSA and β–casein films. BSA + ethanol films have also been studied by Chen et al. (1996). Ahmed and Dickinson (1990) and Brierley *et al.* (1996) have studied the effects on a range of pure proteins and proteins from beer, respectively. Ethanol apparently has two modes of action - through its action as a weakly surface active molecule and, like other truly non-surface active components, through changes which it produces in the solvent quality, which alter the conformation of the protein in solution and therefore its adsorption characteristics. Other molecules may have specific bonding interactions with proteins which also affect the conformation of the protein molecules. Thus, for example, diacetyl tartaric acid esters of monoglycerides (DATEM) have been shown to enhance surface dilatational rheology in bread dough systems (Kokelaar *et al.*, 1995).

The above discussions and those in section 3.3 indicate the sensitivity of interfacial rheology (both dilatational and shear) of food protein films to the presence of surfactants. It is worth remarking that this sensitivity is itself a source of difficulty in comparing the results of different workers on different protein samples, since even 'pure' proteins (Clark *et al.*, 1995) and surfactants (Fainerman *et al.*, 1994; Fainerman and Miller, 1995) are often subject to contamination by low levels of other surfactants which can dramatically modify the behaviour of the so-called 'pure' systems.

5 RELATIONSHIP BETWEEN INTERFACIAL RHEOLOGY AND THE STABILITY OF EMULSIONS AND FOAMS.

The question remains whether or not there is a direct, reliable, quantitative link any of the shear or dilatational moduli and the measured emulsion or foam stability.

5.1 Theoretical considerations

The full hydrodynamic theory of emulsion droplet or foam bubble coalescence as set out by Zarpryanov *et al.* (1983), Edwards *et al.* (1991) and others, predicts only a weak dependence on the shear rheological parameters, with the effect of the dilatational elasticity, ε, predominating in most cases, but κ also having a significant effect. This is perhaps not surprising given that the mechanism of film rupture is viewed as the opening up of hole in one or both interfaces, which is treated as purely areal deformation. Such theories have been developed for the cases of systems stabilised by simple, low molecular weight surfactants and assume linear rheological behaviour, i.e., deformations so small or so slow that the moduli are independent of the deformation. All measurements of interfacial rheology (both dilatational and shear) are made on macroscopic interfaces and the range of stresses and strains/rates of strain applied certainly do not reflect the turbulent, non-equilibrium conditions of practical foam formation, emulsification or the processing of the food colloids via mixing, pumping, etc. Measurement at these low strains/low strain rates, or equilibrium conditions, may be of significance to emulsion and foam stability in the quiescent state, such as on shelf storage, and also the partitioning of oil-soluble molecules between emulsion droplets (Dickinson, 1992a; Dickinson *et al.*, 1994). However, measurements at even these relatively low rates of strain indicate considerable non-linear behaviour in the shear moduli for proteins - usually shear thinning (see section 3.3). Shear moduli could possibly increase at very high deformation rates due to a finite limit on the mobility within the interface due to attractive interactions between molecules. Certainly it seems possible that in certain régimes the shear moduli for proteins and low-molecular-weight surfactants could be of similar magnitude. Similar considerations apply for the dilatational moduli, though at high bulk concentrations the lower diffusion coefficients of proteins probably dominate in maintaining high values of ε, since adsorption is too slow for 'short-circuiting' mechanism to operate. Recently a number of methods have begun to probe interfacial rheology (mainly dilatational) under non-equilibrium conditions. Prins (1995) and Prins and Bergink-Martens (1993) have provided some useful reviews of the subject. New techniques include the overflowing cylinder (Prins and Bergink-Martens, 1993; Bergink-Martens *et al.*,1993), which may be used at A-W and O-W interfaces and trough methods for A-W interfaces (van Uffelen and Joos, 1994; Petrov and Joos, 1996a,b; and van Aken, 1995; van Aken and Merks, 1996). Investigation into the significance of micelle structuring (Chu *et al.*, 1995), particle structuring (Chu *et al.*, 1995) or protein structuring (Kozco *et al.*, 1996) within the thinning films is still in its early stages, but clearly this may change the future theoretical description of film thinning and rupture considerably.

During real thin film drainage and/or rupture, shear flow must certainly occur and film molecules must slip past one another. In the case of adsorbed protein films, any dilatational deformation of the cross-linked network at the interface must be coupled to shear deformation of the film. (This is formally expressed in equations (5) and (6) above). Thus if a *correlation* is found between one of the interfacial rheological parameters and stability, this parameter is not necessarily the controlling one. One problem is that quantifying emulsion/foam coalescence under controlled conditions is not simple and many techniques have been developed which are difficult to compare because the stresses and strains imposed on the interface are very different. Most work in the literature refers to macroscopic drops or bubbles well above colloidal size. Another aspect is that coalescence stability is linked with the bulk rheology of the continuous phase. The bulk rheology affects the frequency of particle collisions and thin film drainage. There may be correlations between the interfacial shear rheology and the bulk shear rheology in high particle volume fraction systems (e.g., Castle *et al.*, 1986) which rightly or wrongly gives the impression that the interfacial shear rheology is important in determining the stability of the system.

5.2 Experimental evidence

It will be anticipated that this is somewhat contradictory and there is evidence apparently both supporting and refuting the theoretical viewpoints. As regards systems containing proteins and low-molecular-weight surfactants together, it is often stated in the technical literature that a mixture of surfactants and proteins, or a mixture of low-molecular-weight surfactants, often results in maximum stability characteristics for emulsions and foams. This may be explained (Dickinson, 1992a) as being due to a 'lubricating' effect on the adsorbed protein (or other polymeric stabiliser). In effect, the rather strongly, or rigidly, adsorbed proteins are made to be a little more flexible and mobile at the interface, either due to their partial displacement and/or complexation with the surfactant at the interface. The combination of protein and surfactant may be able to respond more readily to a wide range of rates and extents of deformation, more quickly recovering a surface film capable of preventing film rupture, with positive implications for foam (or emulsion stability).

5.3 Interfacial shear rheology.

Some of the earliest studies on protein films were made by Biswas and Haydon (1960, 1962). These workers and others following on from this (e.g., Izmailova *et al.*, 1970; Davis and Hansrani 1985; Law *et al.* 1986) have found a correlation between the coalescence stability of macroscopic droplets at a planar oil-water interface and the shear moduli of the adsorbed films. Such results and the various criticisms which may be levelled at them have been discussed many times previously (e.g., see Halling, 1981; Dickinson and Stainsby, 1982; Fisher *et al.*, 1987; Dickinson *et al.*, 1988a). One of the most significant criticisms is that the effects of the much smaller droplet sizes in real emulsions systems (Davis and Smith, 1976) is not taken into account in such studies. Dickinson *et al.* (1988a) made some similar measurements on emulsion-sized droplets (a few microns in diameter) and confirmed a general correlation between increased droplet stability and increasing η of the adsorbed protein films. Other more recent investigations demonstrate an apparent link between interfacial shear viscosity and emulsion stability of mixed protein surfactant systems under controlled orthokinetic conditions (Chen *et al.*, 1993; Dickinson *et al.*, 1993) as well as under perikinetic conditions also (Cárdenas-Valera and Bailey, 1993).

5.4 Interfacial dilatational rheology.

Whilst it is probably an oversimplification that the strains and rates of strain which are applied in real systems are such that the interfacial rheology is purely Hookean or Newtonian, it is certainly the case that stable emulsions and foams can be formed with low-molecular-weight surfactants which exhibit only low interfacial shear moduli compared with the corresponding dilatational moduli measured under the usual conditions (Djabbarah and Wasan, 1982). The same would appear to be true for foams. For example, in the study referred to earlier by Clark *et al.* (1993b), mixtures of Tween 20 + β-lactoglobulin at a constant protein concentration of 0.02 wt.% showed a reduction in ε from 32 to 15 mN m^{-1} and an increase in κ of 1.6 to 4.4 mN s m^{-1} as R was increased from 0 to 10, whilst a slight increase in the corresponding foam stability was also observed (Clark *et al.*, 1995). In similar experiments (Murray and Dickinson, 1996; Murray *et al.*, 1997), illustrated in Fig. 5, it is seen that the addition of $C_{12}E_6$ to β-lactoglobulin films at both O-W and A-W interfaces caused a slight enhancement in ε and κ at low R which correlated with an enhanced life-time, t_f, of the corresponding foams.

Several other studies have found a correlation between the dilatational moduli of protein films and stability, such as Sarker et al. (1996) and Kim and Wasan (1996). Jiang and Chiew (1995) have demonstrated this for lysozyme at both the A-W and O-W water interface. Keen and Blake (1996) stress the importance of high κ values in controlling bubble formation, growth and detachment. Miyamoto *et al.* (1996) studied the anti-foaming behaviour of sucrose esters in casein systems - in relation to certain drinks dispensers. Brierley *et al.* (1996) in studying beer foams again emphasised the importance of particle (bubble) size in drawing conclusions about interfacial rheology and stability. Kokelaar *et al.*(1995) and Kokelaar and Prins (1995) have examined the surface dilatational rheology of surfactants (DATEM) + wheat proteins and found an increase in dilatational moduli correlates with retardation of foam collapse, with apparent consequences bread improving properties.

6 CONCLUSIONS

It is seen that both the dilatational rheology and the shear rheology of food protein + surfactant films are sensitive to changes in the structure and composition of such films and any interactions between the components. In food colloid systems the response of such films to interfacial stresses and strains is likely to be exceedingly complex. It is probably a mistake to expect any one particular rheological parameter to describe coalescence stability behaviour over the wide range of conditions that such films are likely to encounter. Nevertheless, measurement of the interfacial rheology gives important clues about the conformation and dynamics of proteins + surfactants. Such structural changes are likely to be reflected in all aspects of the stability of the corresponding emulsions and foams, i.e., not just coalescence but, flocculation, phase separation, etc. To progress the subject further much more experimental work needs to be carried out at the high bulk concentrations typical of real systems, and particularly at the O-W interface and for oil-soluble surfactants. Allied to this, more reliable data is required on the dynamics of protein monolayers and protein adsorption kinetics in order to understand better the behaviour of adsorbed films. And finally, more effort needs to be made to develop and use techniques which are relevant to the high interfacial stresses and strains which occur in real food emulsions and foams, which ultimately determine the fate of the system.

7 REFERENCES

Ahluwalia, A., Stussi, E. and Domenici, C. (1996) *Langmuir*, **12**, 416.

Ahmed, M. and Dickinson, E. (1990) *Colloids Surf.*, **47**, 353.

Ascherson, F. M. (1840) *Arch. Anat. Physiol.*, 44.

Avramidis, K. S. and Jiang, T. S. (1991) *J. Colloid Interface Sci.*, **147**, 262.

Behroozi, F. (1996) *Langmuir*, **12**, 2289.

Benjamins, J. and van Voorst Vader, F. (1992) *Colloids Surf.*, **65**, 161.

Benjamins, J., Cagna, A. and Lucassen-Reynders, E.H. (1996) *Colloids Surf. A.*, **114**, 245.

Benjamins, J., de Feijter, J. A., Evans, M. T. A., Graham, D. E. and Phillips, M. C. (1976) *Faraday Disc.*, **59**, 218.

Bergink-Martens, D. J. M., Bos, H. J. and Prins, A. (1994) *J. Colloid Interface Sci.*, **165**, 221.

Bergink-Martens, D. J. M., Bos, H. J., van Kalsbeek, H. K. A. I. and Prins, A. (1993) in *Food Colloids and Polymers: Stability and Mechanical Properties*, Dickinson, E. and Walstra, P. (Eds) Royal Society of Chemistry, Cambridge, p. 376.

Bergink-Martens, D. J. M., Bos, H. J., Prins, A. and Schulte, B. C.. (1990) *J. Colloid Interface Sci.*, **138**, 1.

Biswas, B. and Haydon, D. A. (1960) *Proc. 3rd Int. Congr. Surfactants*, **Act. II**, 580.

Biswas, B. and Haydon, D. A. (1962) *Kolloid-Z.*, **185**, 31.

Bois, A. G. and Panaiotov, I. (1995) *J. Colloid Interface Sci.*, **170**, 25.

Bonfillon, A. and Langevin, D. (1993) *Langmuir*, **9**, 2172.

Bonfillon, A. and Langevin, D. (1994) *Langmuir*, **10**, 2965.

Bonfillon, A., Sicoli, F. and Langevin, D. (1994) *J. Colloid Interface Sci.*, **168**, 497.

Brierley, E. R., Wilde, P. J., Onishi, A., Hughes, P. S., Simpson, W. J. and Clark, D. C. (1996) *J. Sci. Food Agric.*, **70**, 531.

Burgess, D. J. and Sahin, N. O. (1994) *ACS Symp. Ser.*, **578**, 380.

Byattsmith, J. G. and Malcolm, B. R. (1994) *J. Chem. Soc. Faraday Trans.*, **90**, 493.

Cárdenas-Valera, A. E. and Bailey, A. I. (1993) *Colloids Surf. A*, **79**, 115.

Castle, J., Dickinson, E., Murray, A., Murray, B. S. and Stainsby, G. (1986) in *Gums and Stabilizers for the Food Industry*, Phillips, G. O., Wedlock, D. J., Williams, P. A., Elsevier Applied Science, London, vol. 3, p. 409.

Castle, J., Dickinson, E., Murray, B. S. and Stainsby, G. (1987) *ACS Symp. Ser.*, **343**, 118.

Chang, C. H. and Franses, E. I. (1995) *Colloids Surf. A*, **100**, 1.

Chattopadhyay, A. K., Ghaicha, L., Oh, S. G. and Shah, D. O. (1992) *J. Phys. Chem.*, **96**, 6509.

Chen, J. , Dickinson, E. and Iveson, G. (1993) *Food Structure*, **12**, 135.

Chen, J. and Dickinson, E. (1993) *J. Sci. Food. Agric.*, **62**, 283.

Chen, J. and Dickinson, E. (1995) *Food Hydrocolloids*, **9**, 35.

Chen, P. , Policova, Z., Susnar, S. S., Paceasciak, C. R. and Demin, P. M.and Neumann, A. W. (1996) *Colloids Surf. A*, **114**, 99.

Chu, X. L., Nikolov, A. D. and Wasan, D. T.(1995) *J. Chem. Phys.*, **103**, 6653.

Chu, X. L., Nikolov, A. D. and Wasan, D. T. (1994) *Langmuir*, **10**, 4403.

Chudinova, G. K., Pokrovskaya, O. N. and Savitskii, A. P. (1995) *Russ. Chem. Bull.*, **44**, 1958.

Clark, D. C., Husband, F., Wilde, P. J., Cornec, M., Miller, R., Krägel, J. and Wüstneck, R. (1995) *J. Chem. Soc. Faraday Trans.*, **91**, 199.

Clark, D. C., Mackie, A. R., Wilde, P. J. and Wilson, D. R. (1993a) in *Food Proteins Structure and Functionality*, Schwenke, K. D. and Mothes, R. (Eds) VCH, Weinheim, p.263.

Clark, D. C., Mackie, A. R., Wilde, P. J. and Wilson, D. R. (1994) *Faraday Discuss.*, **98**, 253.

Clark, D. C., Wilde, P. J., Bergink-Martens, D. J. M., Kokelaar, A. J. J. and Prins, A. (1993b) in *Food Colloids and Polymers: Stability and Mechanical Properties*, Dickinson, E. and Walstra, P. (Eds) Royal Society of Chemistry, Cambridge, p. 354.

Coke, M., Wilde, P. J., Russell, E. J. and Clark, D. C. (1990) *J. Colloid Interface Sci.*, **138**, 489.

Courthaudon, J.-L., Dickinson, E., Matsumura, Y. and Clark, D. C. (1991) *Colloids Surf.*, **56**, 293.

Davis, S. S, . and Hansrani, P. (1985) *J. Colloid Interface Sci.*, **108**, 285.

Davis, S. S. and Smith, A. L. (1976) *Colloid Polym. Sci.*, **254**, 82.

Dickinson, E. (1992b) *J. Chem. Soc. Faraday Trans.*, **88**, 3561

Dickinson, E. (1992a) *Pure & Appl. Chem.*, **64**, 1721.

Dickinson, E. (1994) *J. Food Eng.*, **22**, 59.

Dickinson, E. (1991) *ACS Symp. Ser.*, **448**, 114.

Dickinson, E. and Euston, S. R. (1992) *Adv. Colloid Interface Sci.*, **42**, 89.

Dickinson, E. and Galazka, V. B.(1992) in *Gums and Stabilizers for the Food Industry*, Phillips, G. O., Wedlock, D. J., Williams, P. A., IRL Press, Oxford, vol. 6, p. 351.

Dickinson, E. and Hong, S-T. (1994) *J. Agric. Food Chem.*, **42**, 1602.

Dickinson, E. and Iveson, G. (1993) *Food Hydrocolloids*, **6**, 533.

Dickinson, E. and Matsumura, Y. (1994) *Colloids Surf. B*, **3**, 1.

Dickinson, E. and Matsumura, Y. (1991) *Int. J. Biol. Macromol.*, **13**, 26.

Dickinson, E. and Woskett, C. M. (1989) in *Food Colloids*, Bee, R. D., Richmond. P. and Mingins, J. (Eds) Royal Society of Chemistry, Cambridge, p.74.

Dickinson, E., Evison, J., Gramshaw, J. W. and Schwope, D. (1994) *Food Hydrocolloids*, **8**, 63.

Dickinson, E., Murray, B. S. and Stainsby, G. (1985) *J. Colloid Interface Sci.*, **106**, 259.

Dickinson, E., Murray, B. S. and Stainsby, G. (1988a) *J. Chem Soc. Faraday Trans. I*, **84**, 871.

Dickinson, E., Murray, B. S. and Stainsby, G. (1988b) in *Advances in Food Emulsions and Foams*, Dickinson, E. and Stainsby, G. (Eds) Elsevier, London, p. 123.

Dickinson, E., Owusu, R. K. and Williams, A. (1993) *J. Chem. Soc. Faraday Trans.*, **89**, 865.

Dickinson, E., Rolfe, S. E. and Dalgleish, D. G. (1988b) *Food Hydrocolloids*, **2**, 397.

Dickinson, E., Rolfe, S. E. and Dalgleish, D. G. (1989) *Food Hydrocolloids*, **3**, 193.

Djabbarah, N. F. and Wasan, D. T. (1982) *Chem. Eng. Sci.*, **37**, 175.

Douillard, R. and Lefebvre, J. (1990) *J. Colloid Interface Sci.*, **139**, 488.

Doxastakis, G. and Sherman, P. (1986) *Colloid Polym. Sci.*, **264**, 254.

Dussaud, A and Vignes-Adler, M. (1994a) *J. Colloid Interface Sci.*, **167**, 256.

Dussaud, A and Vignes-Adler, M. (1994b) *J. Colloid Interface Sci.*, **167**, 266.

Dussaud, A., Han, G. B., Ter Minassian-Saraga, L. and Vignes-Adler, M. (1994) *J. Colloid Interface Sci.*, **167**, 247.

Earnshaw, J. C. (1983) in *The Application of Laser Light Scattering to the Study of Biological Motion*, Earnshaw, J.C.and Steer, M. W. (Eds) , Plenum, New York, p.275.

Earnshaw, J. C. (1981) *Nature (London)*, **292**, 138.

Earnshaw, J. C. and McCoo, E. (1995) *Langmuir*, **11**, 1087.

Earnshaw, J. R. and Sharpe, D. J. (1996) *J. Chem. Soc. Faraday Trans.*, **92**, 611.

Eastoe, J., Dalton, J., Rogueda, P., Sharpe, D., Dong, J. F. and Webster, J. R. P. (1996) *Langmuir*, **12**, 2706

Edwards, D. A., Brenner, H. and Wasan, D. T. (1991) *Interfacial Transport Processes and Rheology*, Butterworth-Heinemann, Boston.

Eijt, S. W. H., Wittebrood, M. M., Devillers, A. C. and Rasing, Th. (1994) *Langmuir*, **10**, 4498.

Euston, S. E., Singh, H., Munro, P. A. and Dalgleish, D. G. (1995) *J. Food Sci.*, **60**, 1124.

Færgemand, M., Murray, B. S. and Dickinson, E. (1997) in press.

Fainerman, V. B. and Miller, R. (1995) *Colloids Surf. A*, **97**, 65.

Fainerman, V. B. and Miller, R.` (1995) *J. Colloid Interface Sci.*, **175**, 118.

Fainerman, V. B., Makievski, A. V. and Miller, R. (1994) *Colloids Surf. A*, **87**, 61.

Fang, J. P., Wantke, K. D. and Lunkenheimer, K. (1996) *J. Colloid Interface Sci.*, **182**, 31.

Fang, J., Joos, P., Lunkenheimer, K. and Van Uffelen, M. (1995) *Colloids Surf. A*, **97**, 271.

Fang, Y and Dalgleish, D. G. (1995) *Langmuir*, **11**, 75.

Fang, Y and Dalgleish, D. G. (1993) *Colloids Surf. B*, **1**, 357.

Fang, Y. and Dalgleish (1996a) *J. Agric. Food Chem.*, **44**, 59.

Fang, Y. and Dalgleish (1996b) *J. Am. Oil Chem. Soc.*, **73**, 437.

Feng, S. S., MacDonald, R. C. and Abraham, B. M. (1991) *Langmuir*, **7**, 572.

Fisher, L. R., Mitchell, E. E. and Parker, N. S. (1987) *J. Colloid Interface Sci.*, **128**, 35.

Fleer, G. J., Cohen-Stuart, M. A., Scheutjens, J. M. H. M., Cosgrove, T. and Vincent, B. (1993) *Polymers at Interfaces*, Chapman and Hall, London.

Franses, E. I., Basaran, O. A. and Chang, C. H. (1996) *Current Opinion Colloid Interface Sci.*, **1**, 296.

Freundlich, H. (1930) *Kapillarchemie,* Akademische Verlagsgeselleschaft, Leipzig, vol. II

Fruhner, H. and Wantke, K. D. (1996) *Colloids Surf. A*, **114**, 53.

Galazka, V. B. and Dickinson, E. (1995) *J. Texture Stud.*, **26**, 401.

Gau, C-S., Yu, H. and Zografi, G. (1994) *J. Colloid Interface Sci.*, **162**, 214.

Graham, D. E. and Phillips, M. C. (1980a) *J. Colloid Interface Sci.*, **76**, 227.

Graham, D. E. and Phillips, M. C. (1980b) *J. Colloid Interface Sci.*, **76**, 240.

Halling, P. J. (1981) *CRC Crit. Rev. Food. Sci. Nutr.*, **15**, 155.

Haynes, C. A. and Norde, W. (1994) *Colloids Surf. B*, **2**, 517.

Horbett, T. A. and Brash, J. L. (1995) Eds, *Proteins at Interfaces II Fundamentals and Applications*, ACS, Washington.

Horozov, T. and Joos, P. (1995) *J. Colloid Interface Sci.*, **173**, 334.

Hunt, J. A., Dickinson, E. and Horne, D. S. (1993) *Colloids Surf. A*, **71**, 197.

Hunter, J. R. , Carbonell, R. G. and Kilpatrick, P. K. (1991) *J. Colloid Interface Sci.*, **143**, 37.

Izmailova, V. N., Tulovskaya, Z. D., El'shimi, A. F., Nadel, L. G. and Alekseeva, I. G. (1970) *Dokl. Akad. Nauk SSR*, **191**, 1081.

Jayalakshmi, Y., Ozanne, L. and Langevin, D. (1995) *J. Colloid Interface Sci.*, **170**, 358.

Jiang, Q. and Chew, Y. C. (1995) *ACS Symp. Ser.*, **610**, 183.

Joos, P. (1995) *J. Colloid Interface Sci.*, **171**, 399.

Joos, P. and Van Uffelen, M. (1995) *J. Colloid Interface Sci.*, **171**, 297.

Kajiyama, T., Morotomi, N., Uchida, M. and Oishi, Y. (1989) *Chemistry Letts.*, 1047.

Keen, G. S. and Blake, J. R. (1996) *J. Colloid Interface Sci.*, **180**, 625.

Kim, Y. H. and Wasan, D. T. (1996) *Ind. Eng. Chem. Res.*, **35**, 1141.

Kitching, S., Johnson, G. D. W., Midmore, B. R. and Herrington, T. M.(1996) *J. Colloid Interface Sci.*, **177**, 58.

Kizling, J., Stenius, P., Eriksson, J. C. and Ljunggren, S. (1995) *J. Colloid Interface Sci.*, **171**, 162.

Kokelaar, J. J. and Prins, A. (1995) *J. Cereal Sci.*, **22**, 53.

Kokelaar, J. J., Garritsen, J. A. and Prins, A. (1995) *Colloids Surf. A*, **95**, 69.

Kokelaar, J. J., Prins, A. and De Gee, M. (1991) *J. Colloid Interface Sci.*, **146**, 507.

Koskov, K., Nikolov, A. D., Wasan, D. T., Borwankar, R. P. and Gonsalves, A.(1996) *J. Colloid Interface Sci.*, **178**, 694.

Kretzschmar, G. and Miller, R. (1991) *Adv. Colloid Interface Sci.*, **36**, 65.

Law, T. K., Florence, A. T. and Whateley, T. L. (1986) *Colloid Polym. Sci.*, **264**, 167.

Lee, H., Jiang, T. S. and Avramidis, K. S. (1991) *J. Colloid Interface Sci.*, **146**, 90.

Leermakers, F. A. M., Atkinson, P.J., Dickinson, E. and Horne, D. S. (1996) *J. Colloid Interface Sci.*, **178** , 681.

Lin, S-Y., McKeigue, K. and Maldarelli, C. (1990) *AIChE J.*, **36**, 1785.

Loglio, G., Tesei, U., and Cini, R. (1979) *J. Colloid Interface Sci.*, **71**, 316.

Loglio, G., Tesei, U., and Cini, R. (1984) *J. Colloid Interface Sci.*, **100**, 393.

Lucassen, J. 1981) in *Anionic Surfactants. Physical Chemistry of Surfactant Action*, Lucassen-Reynders, E. H. (Ed) Marcel Dekker, New York, p. 217.

Lucassen-Reynders, E. H. (1993) *Food Structure*, **12**, 1.

Lucassen-Reynders, E. H. and Kuijpers, K. A. (1992) *Colloids Surf.*, **65**, 175.

Lucassen-Reynders, E. H. and Lucassen, J. (1994) *Colloids Surf.*, **85**, 211.

MacRitchie, F. (1986) *Adv. Colloid and Interface Sci.*, **25**, 341.

Makino, M., Kamiya, M., Nakajo, N. and Yoshikawa, K. (1996) *Langmuir*, **12**, 4211.

Malcolm, B. R. (1985) *J. Colloid Interface Sci.*, **104**, 520.

Mannheimer, R. J. and Schechter, R. S. (1970) *J. Colloid Interface Sci.*, **32**, 195.

Martinez-Mendoza, A. and Sherman, P. (1991) *ACS Symp. Ser.*, **448**, 130.

Miller, R. and Kretzschmar, G. (1991) *Adv. Colloid Interface Sci.*, **37**, 97.

Miller, R., Loglio, G., Tesei, U. and Schano, K. H. (1991) *Adv. Colloid Interface Sci.*, **37**, 73.

Miller, R., Policova, Z., Sedev, R. and Neumann, A. W. (1993) *Colloids Surf. A*, **76**, 179.

Miller, R., Wüstneck, R., Krägel, J. and Kretzschmar, G. (1996) *Colloids Surf. A*, **111**, 75.

Mitchell, J. R.(1986) in *Developments in Food Proteins*, Hudson, B. J. F. (Ed) Elsevier Appied Science, London, vol 4, p. 291.

Miyamoto, T., Miyama, S., Nishiyama, S. and Esumi, K. (1996) *J. Jap. Soc. Food Sci. Technol.*, **43**, 806.

Mohammed, R. A., Bailey, A. I., Luckham, P. F. and Taylor, S. E. (1994) *Colloids Surf. A*, **91**, 129.

Mohan, V. and Wasan, D. T. (1976) *Colloid Interface Sci.*, [Proc. Int. Conf.], 50th, 4, 439.

Murray, B. S. (1997a) *Colloids Surf. A* (in press)

Murray, B. S. (1997b) *Langmuir* (in press) .

Murray, B. S. (1987) *Ph. D. Thesis*, University of Leeds.

Murray, B. S. and Dickinson, E. (1996) *Food Sci. Technol. Int.*, **2**, 131.

Murray, B. S. and Nelson, P. V. (1996) *Langmuir*, **12**, 5973.

Murray, B. S., Ventura, A., Lallemant, C. (1997) *Colloids Surf. A* (in press) .

Nagarajan, R. and Wasan, D. T.(1994) *Rev. Sci. Instr.*, **65**, 2675.

Nauhringbauer, I. (1995) *J. Colloid Interface Sci.*, **176**, 318.

Niven, R. W., Prestrelski, S. J., Treuheit, M. J., Ip, A. Y. and Arakawa, T. (1996) *Int. J. Pharmaceutics*, **127**, 191.

Noskov, B. A. and Zubkova, T. U. (1994) *J. Colloid Interface Sci.*, **170**, 1.

Noskov, B. A., (1996) *Colloid J.(USSR)* , **58**, 58.

Nylander, T. and Wahlgren, N. M. (1994) *J. Colloid Interface Sci.*, **162**, 151.

Ogden, L. G. and Rosenthal, A. J. (1994) *J. Colloid Interface Sci.*, **168**, 539.

Oliver, G. and Sahi, S. S. (1995) *J. Sci. Food Agric.*, **67**, 221.

Panaiotov, I., Ivanova, T., Proust, J., Boury, F., Denizot, B., Keough, K. and Teneva, S. (1996) *Colloids Surf. B*, **6**, 243.

Peng, J. B. and Barnes, G. T. (1995) *Colloids Surf. A*, **102**, 75.

Peng, J. B., Barnes, G. T. and Abraham, B. M. (1993) *Langmuir*, **9**, 3574.

Petrov, P. and Joos, P. (1996b) *J. Colloid Interface Sci.*, **182**, 214.

Petrov, P. and Joos, P. (1996a) *J. Colloid Interface Sci.*, **181**, 530.

Poole, S., West, S. I. and Walters, C. L. (1984) *J. Sci. Food Agric.*, **35**, 701.

Prins, A. (1995) in *New Physico-Chemical Techniques for the Characterization of Complex Food Systems*, Dickinson, E. (Ed) Chapman and Hall, Cambridge, p.214.

Prins, A. and Berginke-Martens, D. J. M. (1993) in *Food Colloids and Polymers: Stability and Mechanical Properties*, Dickinson, E. and Walstra, P. (Eds) Royal Society of Chemistry, Cambridge, p. 291.

Reiner, M.. (1960) *Deformation and Flow*, H. K. Lewis, London, p. 178

Roth, C. M. and Lenhoff, A. M. (1993) *Langmuir*, **9**, 962.

Sachetti, M., Yu, H. and Zografi, G. (1993) *Langmuir*, **9**, 2168.

Sagis, L. M. C. and Bedeaux, D. (1996) *Physica A*, **230**, 437.

Sarker, D. K., Wilde, P. J. and Clark, D. C. (1995a) *Colloids Surf. B*, **3**, 349.

Sarker, D. K., Wilde, P. J. and Clark, D. C. (1995b) *J. Agric. Food Chem.*, **43**, 295.

Sarker, D. K., Wilde, P. J. and Clark, D. C. (1996) *Colloids Surf. A*, **114**, 227.

Serrien, G., Geeraerts, G., Ghosh, L. and Joos, P. (1992) *Colloids Surf.*, **68**, 219.

Sharpe, D. and Eastoe, J. (1996) *Langmuir*, **12**, 2303.

Sharpe, D. and Eastoe, J. (1995) *Langmuir*, **11**, 4636.

Shotten, E. and Wibberly, K. (1961) *Boll. Chim. Farm.*, **100**, 802.

Smaby, J. M., Kulkarni, V. S., Momsen, M. and Brown, R. E. (1996) *Biophys. J.*, **70**, 868.

Sonin, A. A., Bonfillon, A. and Langevin, D. (1994) *J. Colloid Interface Sci.*, **162**, 323.

Thominet, V., Stenvot, C. and Langevin, D. (1988) *J. Colloid Interface Sci.*, **126**, 54.

Trapeznikov, A. A., Volkova, N. G. and Shirokova, T. Yu. (1968) *Doklady Akademii Nauk. SSR*, **276**, 903.

van Aken, G. A.(1995) in *Food Macromolecules and Colloids*, Dickinson, E. and Lorient, D. (Eds) Royal Society of Chemistry, Cambridge, p. 43.

van Aken, G. A. and Merks, M. T. E. (1996) *Colloids Surf. A*, **114**, 221.

Van Hunsel, J. and Joos, P. (1987) *Colloids Surf.*, **25**, 251.

van Uffelen, M. and Joos, P. (1994) *Colloids Surf. A*, **85**, 119.

Velev, O. D., Nikolov, A. D., Denkov, N. D., Doxastakis, G., Kiosseoglu, V. and Stalidis, G. (1993) *Food Hydrocolloids*, **7**, 55.

Vernon-Carter, E. J. and Sherman, P. (1981) *J. Dispersion Sci. Technol.*, **2**, 399.

Warburton, B. (1996) *Current Opinion Colloid Interface Sci.*, **1**, 481.

Wasan, D. T., Nikolov, A. D., Lobo, L. A., Koczo, K. and Edwards, D. A. (1992) *Prog. Surf. Sci.*, **39**, 119.

Wilde, P. J. and Clark, D. C. (1993) *J. Colloid Interface Sci.*, **155**, 48.

Williams, A. and Prins, A. (1996) *Colloids Surf. A.*, **114**, 267.

Wilson, D. R., Wilde, P. J. and Clark, D. C. (1993) in *Food Colloids and Polymers: Stability and Mechanical Properties*, Dickinson, E. and Walstra, P. (Eds) Royal Society of Chemistry, Cambridge, p. 415.

Wüstneck, R., Hermel, H. and Kretzschmar, G. (1984) *Colloid Polym. Sci.*, **262**, 827.

Wüstneck, R., Krägel, J., Miller, R., Fainerman, V. B., Wilde, P. J., Sarker, D. K. and Clark, D. C.(1996b) *Food Hydrocolloids*, **10**, 395.

Wüstneck, R., Krägel, J., Miller, R., Wilde, P. J. and Clark, D. C.(1996a) *Colloids Surf. A*, **114**, 255.

Wüstneck, R., Kretzschmar, G. and Zastrow, L. (1987) *Colloid J. USSR*, **49**, 207.

Xu, S. and Damodaran, S. (1994) *Langmuir*, **10**, 472.

Xu, S. and Damodaran, S. (1993) in *Food Proteins Structure and Functionality*, Schwenke, K. D. and Mothes, R. (Eds) VCH, Weinheim, p. 270.

Zapryanov, Z., Malhotra, A. K., Anderangi, N. and Wasan, D. T. (1983) *Int. J. Multiphase Flow*, **9**, 105.

Proteins at Liquid Interfaces
D. Möbius and R. Miller (Editors)

RELATION BETWEEN SURFACE RHEOLOGY AND FOAMING BEHAVIOUR OF AQUEOUS PROTEIN SOLUTIONS

A. Prins, M.A. Bos, F.J.G. Boerboom, H.K.A.I. van Kalsbeek

Dept. of Food Science, Division of Integrated Food Science & Food Physics,

Wageningen Agricultural University, P.O. Box 8129, 6700 EV, Wageningen, The Netherlands.

Contents

1 INTRODUCTION

Protein foams are important for several categories of foods like desserts and bakery products and have therefore been studied over many years. These studies have demonstrated that numerous factors affect the foaming behaviour of proteins, among which are pH, temperature, ionic strength, the presence of other molecules like sugars or lipids and the protein source. Other studies have concentrated on the relationship between the foaming properties and the physical or chemical properties of proteins and surfaces, e.g. surface pressure, surface concentration, surface elasticity, surface viscosity and protein surface hydrophobicity, protein solubility, protein charge, respectively.

Although the above mentioned properties play an important role in foam formation, they are also important in foam stability. Physical processes occurring in foam and determining the foam stability are coalescence, Ostwald ripening (also called disproportionation) and drainage. Coalescence is the fusion of two gas bubbles to one bigger bubble. It occurs when the thin film of continuous phase between two contacting bubbles breaks. Ostwald ripening is the diffusion of dispersed phase material (gas) from smaller to larger gas bubbles. This is due to the difference in Laplace pressure between both bubbles which creates a concentration gradient of soluble gas in the continuous phase. Disproportionation does not occur when the gas is not soluble in the continuous phase. Drainage is the rate at which the liquid flows through the foam to the bottom of the foam, resulting in a higher gas fraction in the foam. Depending on the bubble size distribution, the natural source of protein and the amount of protein, the above mentioned mechanisms are responsible for the physical stability or instability. In general one can conclude that proteins are able to slow down drainage sufficiently so that differences in stability mainly depend on the ability of the adsorbed proteins to slow down disproportionation.

Proteins have in common with low molecular weight surfactants that as a result of their adsorption at an aqueous surface the surface tension of water is lowered. Proteins are built up by a large number of amino acids of various nature. In solution these macromolecules have an often complicated three dimensional structure, which is determined among others by the amino acid sequence of the peptide chain, the presence of sulphur bridges and the way this chain is folded into structural motifs like α-helix and β-sheets.

A protein molecule can interact with other molecules and surfaces in a great number of ways. This is mainly due to the complexity of the protein molecule which comprises positive and negative charges, groups with hydrogen bonding capacity as well as polar regions. Intermolecular and surface forces can thus occur via electrostatic interactions, hydration forces, acid-base interactions, hydrogen bonding and the van der Waals interactions. The hydrophobic and ionic interactions together with the gain in entropy due to conformational changes are often regarded as the driving forces for protein adsorption.

When proteins are adsorbed at an aqueous surface they can change their conformation. The time scale of these conformational changes and the extent in which the conformation is changed, strongly depends on the above mentioned intermolecular and surface forces as well as the intramolecular forces. An example is given in Fig. 1. A native protein molecule (globular) can adsorb in such a way that its structure is maintained (case a) or that it unfolds at the surface (case b). Also the orientation of the protein may affect its interfacial properties. A classification between "hard" and "soft" proteins, referring to high or low internal stability respectively, as used by Arai and Norde (1990), can sometimes be beneficial. However, it is impossible to generalise the behaviour of proteins at interfaces due to their size, flexibility as well as their susceptibility to changes in solution conditions and properties of the interface.

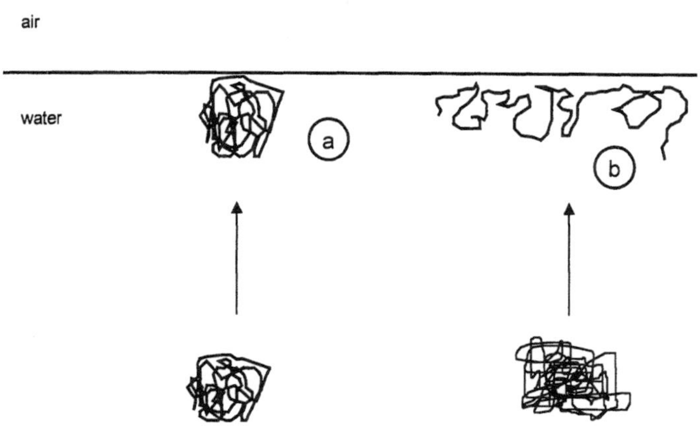

Fig. 1 Possible modes of adsorption of proteins onto an Air/Water interface. Highly schematic.

The consequences of these conformational changes for the lowering of the surface tension are considerable because it depends on the ability of the protein molecule to expose its various chain elements to the surface. The more the protein molecule is able to unfold its structure and the more surface active these unfolded chain elements are, the more the surface tension will be lowered. This lowering of the surface tension is one of the driving forces for the unfolding process.

Another contribution to the unfolding process is that hydrophilic and hydrophobic parts of the molecule can move to an environment which is energetically more favourable; the hydrophilic parts to the aqueous phase and the hydrophobic parts to the oil or gas phase.

Once adsorbed the desorption of a protein molecule from an interface is difficult because in order to do so all the numerous adsorbed parts of the molecule have to desorb from the interface simultaneously and the chance that this occurs is extremely small. In addition to this the concentration gradient which can be the driving force for such a desorption is very low. Desorption of proteins from an interface proceeds faster when they are replaced by small molecular surfactants such as sodium lauryl sulphate.

From all this it follows that in the equilibrium between adsorbed protein molecules and molecules in solution a kind of one way traffic exists in the transport of protein molecules: to the surface is easier than from the surface to the solution.

The unfolding of protein molecules at a water surface opens the possibility that the exposed innerside of the molecule interacts with the exposed innerside of a neighbouring protein molecule. These intermolecular lateral interactions as described above can give rise to a kind of network structure in the surface resulting in an extra contribution to the mechanical properties - rheological properties - of the surface. This network structure especially manifests itself when the surface is subjected to shearing deformations: it gives rise to measurable surface shear elasticity and -viscosity. These properties are in general not present in the surface of soluble low molecular weight (LMW) surfactant solutions. Low molecular surfactants, soaps, detergents or emulsifiers do adsorb at the aqueous surface and lower the surface tension but these molecules itself cannot spread over the surface because they usually have only one active group. The final conclusion is that the relaxation processes taking place in surfaces stabilised with proteins or detergents are different in nature, including the time scale.

The content of this chapter is focussed on the behaviour of proteins at surfaces in expansion or compression, however, comparison with the behaviour of LMW surfactants at these surfaces acts as a leitmotiv in the whole chapter. Aspects of adsorption of molecules at expanding and/or compressed surfaces, which are important for foaming and emulsification behaviour, are the kinetics of adsorption, the adsorbed amount and related to that, the rigidity of the adsorbed layer. Furthermore, the dynamic surface tension and the ability of a protein/surfactant solution to generate surface tension gradients at interfaces are important tools to study interfacial behaviour of proteins and surfactants. Therefore, in section 2 a model will be presented which describes the behaviour of protein molecules at an expanding surface, assuming that unfolding of a protein molecule at an interface is a first order kinetic mechanism. The next section deals with the rigidity of the adsorbed layers at expanding and compressed interfaces. We will show, by using various techniques, that protein solutions are able to form a kind of stagnant layer at the surface ("skin formation"), whereas surfactant solutions are generally unable to do so. How important such a stagnant layer is will be made clear in paragraph 3 were we will present data dealing with emulsion droplet break-up. In section 4 the foaming behaviour of proteins and surfactants will be discussed in terms of coalescence, drainage and disproportionation. Finally, in the last section 5 the most important aspects related with the comparison between the behaviour of proteins and surfactants will be summarised.

2 Modelling mechanical properties of liquid surfaces from unfolding parameters of proteins.

2.1 Introduction

In the general introduction it has been stated that molecular properties affect the dynamic surface behaviour (dynamic surface tension, surface concentration) and eventually also the mechanical properties of the surface (surface tension gradient, relative rate of expansion). Knowledge about the molecular characteristics of the proteins is therefore important. Despite numerous studies on proteins adsorbed at liquid surfaces however, hardly any information is available which can serve as a direct link between molecular properties and surface rheological properties. In addition to this not much information is available on how to translate molecular properties into mechanical properties of the surface.

Here an attempt will be presented, which proposes a model in which semi-empirical parameters describing the molecular properties of protein molecules are translated into mechanical properties of the surface. In this case only dynamic properties of surfaces in continuous steady state dilation on a radially expanding surface have been considered. A more general approach will be presented elsewhere (Boerboom, 1997).

The hypothesis used here to describe the behaviour of adsorbed protein molecules assumes that two processes are important in the surface active behaviour in dynamic situations. The first process is the transport to the surface by means of diffusion and/or convection and the second process is the rearrangement of the proteins adsorbed at the surface. These two sub-processes lead to two time scales. These time scales together determine the resulting surface properties. Here it is assumed also that the behaviour of proteins adsorbed at a surface in the time scale considered can be described as being a transition between two extreme states which will be denoted as native and unfolded.

Measurements of surface concentration (adsorption), relative rate of expansion and surface tension of radially expanding surfaces were performed in an overflowing cylinder by means of respectively ellipsometry, laser Doppler anemometry and the Wilhelmy plate technique. The results of these measurements were fitted to the model by four adjustable parameters. Data obtained by these adjustable parameters were compared to measurements on protein solutions of the same protein in other circumstances in order to verify the validity of the theory and these adjustable parameters. The mechanical properties which could be derived from these predictions and measurements were investigated for a protein and a low molecular surfactant as well. Good agreement was found between these fitted and measured data.

2.2 Theory

In this section we aim to propose an approach which can be used to relate molecular properties of proteins to dynamic surface properties. In order to do this the characteristics of the parameters needed to describe protein molecules have to be defined. This will be treated extensively elsewhere (Boerboom (1997)). When these parameters have been defined, these need to be related to surface tension and eventually to the surface tension gradient. In the subsequent sections the way of reasoning is explained and some characteristic results are shown.

In order to facilitate the creation of a model some simplifying assumptions need to be made in order to enable the calculation of the different properties. Here three considerations underlying the model are summarised.

First of all the unfolding of proteins is simplified considerably. Here the hypothesis of Serrien et al.(1992) will be used. This hypothesis entails that the adsorbed proteins can occur in only two conformations; native and unfolded. The transition between these two states occurs by means of first order kinetics. In reality however the protein molecules may be present at the surface in a large number of conformations. It is assumed that, since the averaging takes place over a big number of molecules, the fraction of unfolded molecules can serve as a "mean conformation".

Secondly, essentially the same consideration also applies to the calculation of surface tension. Since there are only two extreme states in which the molecule can exist, the ratio of the two extremes and the total adsorbed amount determines the surface tension. These extremes with respect to surface tension are: the surface tension due to hard sphere interactions for native molecules and the minimum surface tension reached at total unfolding at the surface concentration considered for unfolded molecules. The surface tension is calculated as a weighted mean of these two surface tensions.

This kinetic model for the adsorption and desorption can be motivated from the extrapolation of a wide range of possible conformations to only two extreme conformations. It is because of this assumption for instance that unfolded molecules cannot desorb. This is a consequence of the fact that when proteins are in a totally unfolded conformation, the desorption is highly improbable. However in the situation that only very few amino acid segments are adsorbed, the chance of desorption is much higher.

Thirdly, if these simplifications are assumed, the surface properties can be calculated by means of a numerical procedure described more completely elsewhere, Boerboom (1997). This is accomplished by assuming a relative rate of expansion of the surface a priori. The surface properties following from these parameters are calculated by means of the model described in the subsequent paragraphs and are checked for agreement with hydrodynamics regarding the flow in the overflowing cylinder set-up.

The modus operandi followed with this model is as follows: First a bulk concentration and a relative rate of expansion are assumed a priori. Then a number of measurements of relative rate of expansion, surface tension and surface concentration are performed, at the bulk concentration studied. Then four adjustable parameters are calculated from these measurements. Then, by using these parameters, various surface properties are calculated under different circumstances. These properties: surface tension, surface concentration, relative rate of expansion, surface and tension gradient are compared to measurements by varying bulk concentration, and relative rate of expansion.

In the following paragraphs the mathematical translation of the physical considerations underlying the model is presented.

2.2.1 UNFOLDING OF ADSORBED PROTEINS

In Fig. 2 a schematic representation of the model of unfolding of proteins at interfaces has been shown.

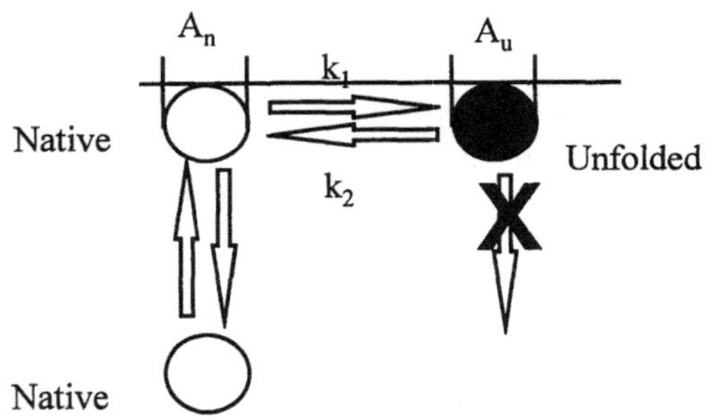

Fig. 2 Schematic representation of the unfolding of protein molecules.

This has been presented in a paper by Serrien et. al. (1992). In the model presented there, it can be seen that first proteins adhere to the surface in a native conformation, after that, these proteins may unfold to an unfolded conformation. The rate of unfolding is determined by a first order rate constant k_1. The transition from an unfolded to a native molecule is also possible. This process is assumed to be determined also by a first order rate constant k_2. The unfolded molecules cannot desorb from the surface in this unfolded conformation. Also the immediate transition of a native molecule in bulk to an unfolded molecule in the surface is assumed to be impossible.

In addition to the model of Serrien et al. (1992), another intricacy has been assumed in the model presented here. This is the area the molecules occupy in the surface: it has been assumed that as a consequence of the unfolding of the molecules the geometrical occupied area, (which is the area not accessible to proteins for geometrical reasons) of the surface, changes.

The rate constants k_1 and k_2 and the time molecules have available for unfolding determined by the time scale considered, now determine the distribution of the molecules over the native and unfolded conformation. The cross sectional areas A_n and A_u respectively determine the areas these molecules occupy in the surface. Essentially, k_1, k_2, A_n, and A_u are the parameters which determine the unfolding behaviour of the proteins at the surface, the aim of the model is to predict these, by matching the constants to data obtained from characteristic measurements of relations between relative rate of expansion, surface concentration and surface tension.

2.2.2 Transport to the surface

In Fig. 3 a schematic representation of the model has been given. In this figure the underlying considerations concerning the transport have been given. Some preliminary calculations and the results obtained by Bergink et. al.(1990, 1994) have indicated that in the case of a radially expanding surface in a steady state, the transport to the surface can be described best by convective diffusion (Van Voorst Vader, 1964). In the solution, the basic continuity equation for the steady state (dc/dt=0) is:

$$D\frac{\delta^2 c}{\delta z^2} = V_z \frac{\delta c}{\delta z} \qquad [1]$$

In this equation D is the diffusion constant, c is the concentration in bulk and z is the co-ordinate perpendicular to the surface. From this the concentration gradient $(dc/dz)_{z=0}$ and the flux towards the surface $D(dc/dz)_{z=0}$ is found to be: $D(c_b - c_s)(2\theta/\pi D)^{\frac{1}{2}}$, where c_s and c_b respectively mean subsurface concentration and bulk concentration, θ means the relative rate of expansion of the surface dlnA/dt. At the expanding surface, the mass balance is given by $\frac{d\Gamma}{dt} = D(\frac{dc}{dz})_{z=0} - \Gamma\theta$, hence in steady state we have:

$$\Phi = D(\frac{\delta c}{\delta z})_{z=0} = \Gamma \cdot \theta \qquad [2]$$

Here, Γ means surface concentration.

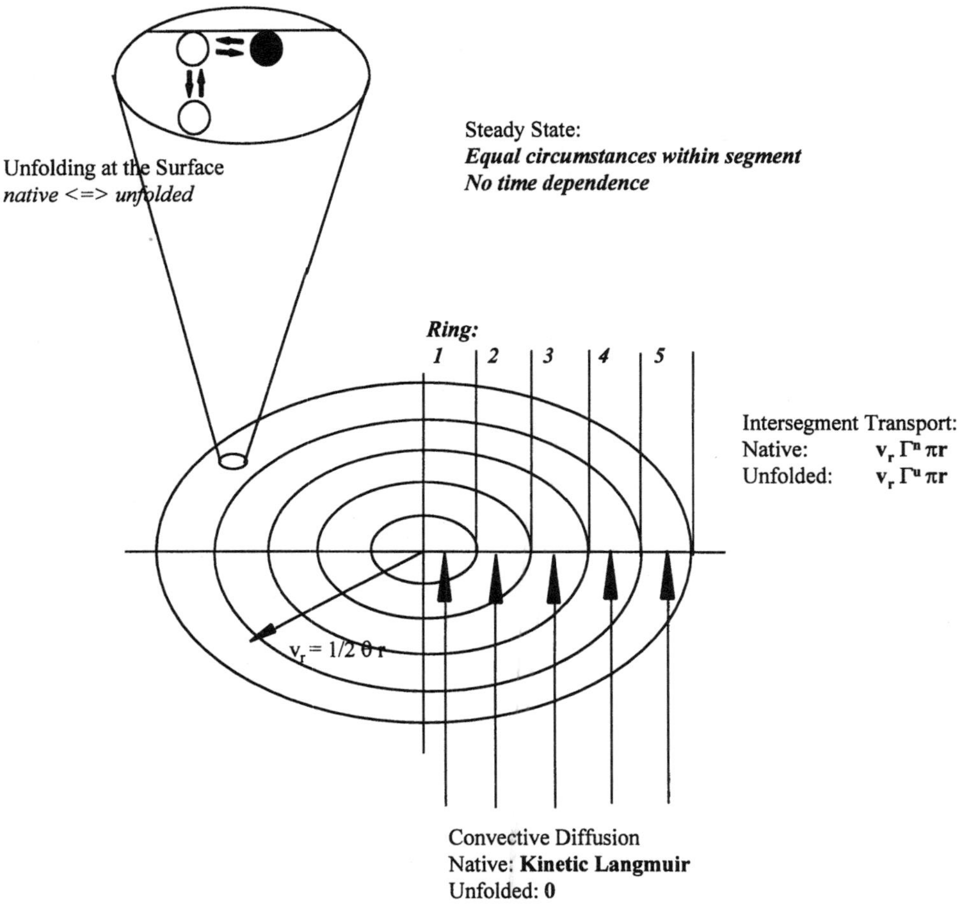

Fig. 3 Schematic representation of the model for the unfolding of proteins.

In order to calculate the steady state value of the surface concentration a relation between the concentration and the flux Φ of proteins in [kg m^{-3} s^{-1}] is needed. This flux has been assumed to be described by the kinetic Langmuir equation.

$$\Phi = \beta\, c_s(1 - A_n\Gamma^n - A_u\Gamma^u) - \alpha\,\Gamma^n \qquad [3]$$

In this equation A_n and A_u respectively mean the surface areas of the native and unfolded conformation, c_s means the subsurface concentration and Γ means surface concentration. The superscripts for Γ, n and u indicate the native respectively unfolded state of the protein α and β are parameters calibrating the flux to the surface. This flux describes the net transport of proteins to a surface not in equilibrium with the bulk solution underneath. In this equation, the first term on the right side, describes the attachment to the surface of native molecules, the second term on the right describes the dissociation of native proteins from the surface. From the combination of Eqs. 2 and 3 the following equation is found for the mass flux.

$$\Phi = \beta\, c_b(1 - A_n\Gamma^n - A_u\Gamma^u) - \beta\Gamma(1 - A_n\Gamma^n - A_u\Gamma^u)\sqrt{\frac{\pi\theta}{2D}} - \alpha\frac{\Gamma^u}{\Gamma_\infty} \qquad [4]$$

This yields a relation between Γ and Γ^n and Γ^u.

2.2.3 *Mass balances*

Now the basic equations for mass transport to the surface and discharge from the surface have been defined in equations 1 to 4 . These equations can now be applied to the mass balances by which the radially expanding surface is described. The schematic drawing in Fig. 4a summarises the different contributions of transport and unfolding in and to the surface.

In this model the circle shaped expanding surface which is considered here is divided in concentric rings of equal width. Over each concentric ring, the surface concentration, the fraction of unfolded molecules and the surface tension is considered to be constant. In addition to this the surface is considered to be in a steady state, which means that none of the properties changes as a function of time. In this way the properties of the surface can be calculated numerically. Such an expanding surface can for instance be found in an experimental set-up of the overflowing cylinder or in a bubble moving with respect to the surrounding liquid.

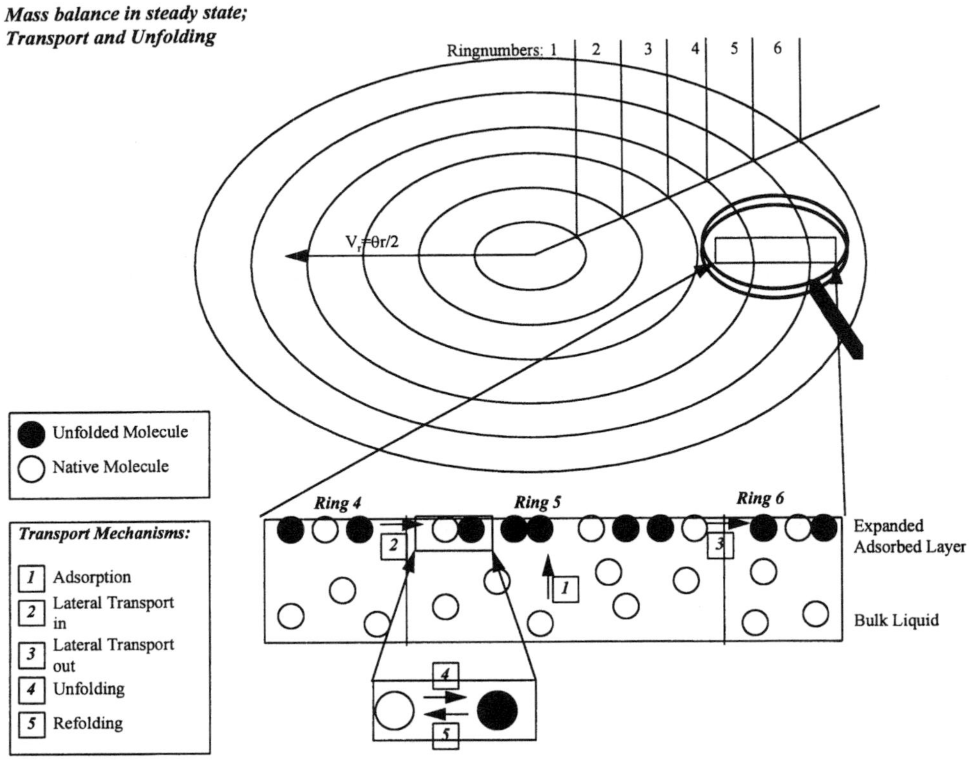

Fig. 4a Schematic representation of contributions to transport and unfolding at the surface

For each concentric ring it is possible to calculate the fraction of unfolded molecules, and the surface concentration by solving two mass balances over each ring.

This mass balance describes the balance between the supply and discharge of the proteins to and from the surface.

In Fig. 4a and 4b the mass balances underlying the model for respectively native and unfolded molecules are explained. In these figures it can be seen that the mass balance of native molecules over each concentric ring can be described by means of six contributions which can be written as:

$$\pi(r_n^{\,2} - r_{n-1}^{\,2})\,\Phi - \pi r_n^{\,2}\theta\Gamma_n^n + \pi r_{n-1}^2\theta\Gamma_{n-1}^n + \pi(r_n^{\,2} - r_{n-1}^{\,2})k_2\Gamma_n^u - \pi(r_n^{\,2} - r_{n-1}^{\,2})k_1\Gamma_n^n = 0 \quad [5]$$

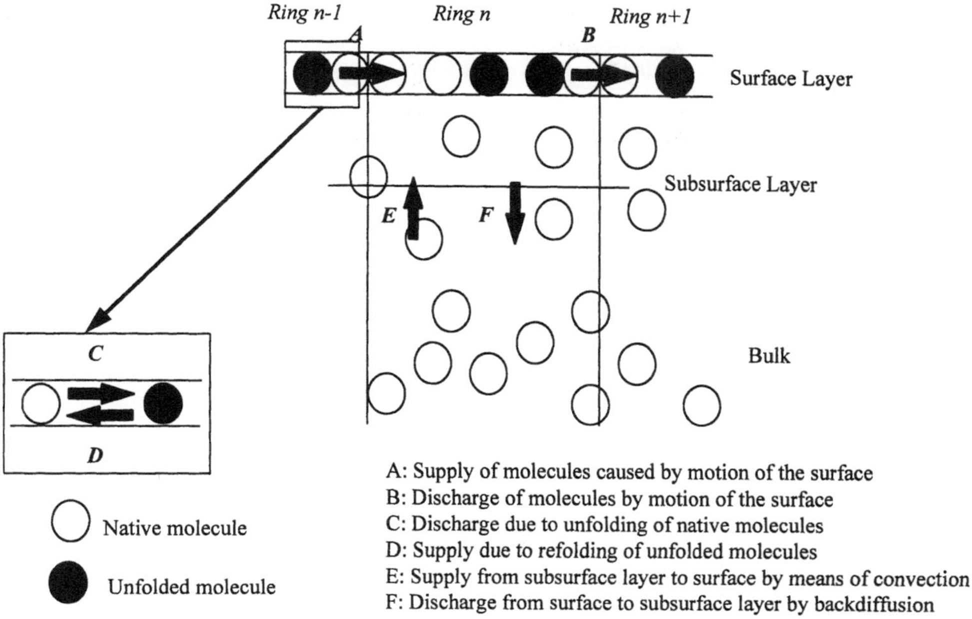

Fig. 4b Schematic representation of mass balance of native molecules.

Here Φ means the net flux to the surface, which is considered to consist only of native molecules. The superscript n for Γ means that this only applies to the proteins in their native conformation. The superscript u means that it applies to proteins in the unfolded conformation. *k1* and *k2* are the rate constants for the conformation transitions Here the subscripts n and n-1 for Γ indicates the ring number, this parameter refers to. So n-1 means the segment previous to the ring which is studied at that moment, closer to the stagnation point, (see Fig. 4a). n Is the ring which is under study at that moment. In this equation, *Φ* means the rate of adherence (flux) at the surface of native protein molecules this is the net transport from the bulk to the surface.

In this figure E and F together represent the net flux of native molecules to the surface, which is the first term $\pi(r_{n-1}^2 - r_n^2)\Phi$ in Eq. 5. The second and third contribution in the equation are represented by respectively A and B in Fig. 4b. The fourth and fifth term are represented by C and D in this figure.

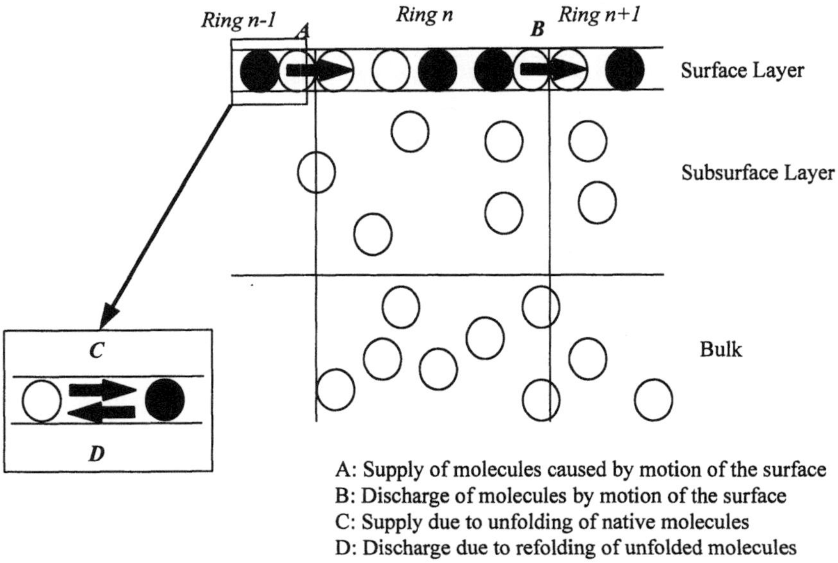

Fig. 4c Schematic representation of mass balance of unfolded molecules.

For the unfolded conformation the mass balance can be written as follows:

$$\pi r_{n-1}^{2}\theta\Gamma_{n-1}^{u} - \pi r_n^{2}\theta\Gamma_n^{u} + \pi(r_n^{2} - r_{n-1}^{2})k_1\Gamma_n^{n} - \pi(r_n^{2} - r_{n-1}^{2})k_2\Gamma_n^{u} = 0 \qquad [5]$$

The notations used in this balance are the same as the notation used in equation 4. In Fig. 4c this mass balance is explained schematically. The first and second contribution are represented by respectively A and B. The third and fourth contribution are represented by respectively C and D.

2.2.4 SURFACE TENSION

In the previous paragraphs the transport to the surface and the unfolding of proteins at the surface have been outlined. From the equations described there, the surface concentration of native and unfolded molecules can be calculated.

In order to calculate surface tension a relation between surface concentration and fraction of unfolded proteins needs to be known. From theoretical considerations it is known that the surface pressure π, (which is the difference between the surface tension of the pure solvent γ_0

and the surface tension which is measured γ), can be considered to consist of two contributions. An ideal gas contribution $\pi = \gamma_0 - \gamma = kT\Gamma$, in which k is the Boltzmann constant, T is the absolute temperature and Γ is the surface concentration in moles per square meter. The second contribution is due to interaction of the molecules in the surface and is considered to be a function of the fraction of the surface occupied by the adsorbed molecules and surface concentration. The ideal gas contribution is negligible in the case of proteins because of their high molecular weight. Therefore it is assumed that the second contribution totally determines the surface tension. When protein molecules interact as hard disks, de Feijter and Benjamins (1982) demonstrated that the surface pressure can be predicted by the Helfland equation:

$$\pi = \frac{kT\Gamma^n}{(1-\Gamma^n A_n)^2} \qquad [7]$$

Here π is surface pressure, kT is the Boltzmann constant multiplied by the absolute temperature Γ is the surface concentration and Γ^n is the fraction of occupied area. Here we will assume that molecules in their native conformation behave like hard particles, which satisfy the Helfland equation. When proteins unfold however the interaction between the particles change from hard disk behaviour to the behaviour of compressible particles. This is provided by the equation of state of proteins in equilibrium as predicted by the model of van Aken (1995).

The surface tension of an adsorbed layer of proteins under dynamic conditions $\pi_{\Gamma\,dyn}$ in a partially unfolded state is assumed to be a linear combination of these surface pressures as given in equation 8.

$$\pi_{\Gamma dyn} = \pi_{\Gamma helf} + \frac{\Gamma^u}{\Gamma}\left(\pi_{\Gamma SS} - \pi_{\Gamma helf}\right) \qquad [8]$$

Here $\pi_{\Gamma\,helf}$ means the surface pressure of hard particles at surface concentration Γ, $\pi_{\Gamma\,SS}$ means the surface tension at surface concentration Γ in equilibrium.

2.2.5 Mechanical Properties

Up to now it has been assumed that both bulk concentration and relative rate of expansion can be chosen at will. The mechanical properties however dictate a relation between the relative rate of expansion, surface tension gradient and the velocity gradient existing at a surface. In this section these relations are described.

As the surface tension is now known at n different places at the surface, the surface tension gradient can be calculated also by using the following simple equation:

$$\frac{d\gamma}{dr} = \frac{\gamma_{n-1} - \gamma_n}{\Delta r} \qquad [9]$$

Here the subscripts n and n-1 indicate the ring number of which the surface tension is considered, Δr means the width of one ring. From hydrodynamic theory it can be deduced that since the surface is a free surface, all stresses must cancel there this implies that the relation between surface tension gradient and velocity gradient in the bulk liquid close to the surface can be written as follows:

$$\frac{d\gamma}{dr} = \eta_b \left(\frac{\delta v_r}{\delta z}\right)_{z=0} \qquad [10]$$

here η_b means the bulk viscosity of the solution.

Up to now θ has been considered to be a parameter which could be varied at will. However in the case of the overflowing cylinder there is a relationship between the force imposed on a surface expressed in a stress in the shape of dv/dz or dγ/dr and the relative rate of expansion which is the response of the system on this stress. In the paper by Bergink et. al. (1994) an equation is derived which describes the relation between this stress and the relative rate of expansion for the overflowing cylinder in the case that the surface is propelled by a surface tension gradient. This results in the relation between the velocity gradient and θ in formula 11

$$\frac{1}{\eta_b}\frac{d\gamma}{dr} = \frac{dv_z}{dz} = 0.415r(\theta - \theta_0)^{\frac{3}{2}}\sqrt{\frac{\rho}{\eta_b}} \qquad [11]$$

Here r is the distance to the centre of the surface, ρ and η are respectively the density and the bulk viscosity and θ_0 is the relative rate of expansion of a pure water surface. Now there is only one combination of surface tension, surface tension gradient, and relative rate of expansion that is possible for a given situation in an overflowing cylinder.

2.3 Comparison of model calculations to experiments in an overflowing cylinder

2.3.1 Overflowing Cylinder Technique

Measurements on expanding surfaces were performed by using an overflowing cylinder. This device has been studied extensively by Bergink et. al. (1990, 1993, 1994). A schematic representation of the set-up is displayed in Fig. 5. The operation of this apparatus can be summarised as follows. Liquid is pumped up in the inner cylinder and flows over the rim into the gap between the inner and outer cylinder. This liquid flows down along the outside of the inner cylinder in a thin film propelled by gravity. Due to this falling motion, the surface at the top of this film is expanded. The resulting increased surface tension causes a surface tension gradient along the horizontal surface at the top of the cylinder. When the surface tension on the top surface deviates from the surface tension of the film, a surface tension gradient establishes itself, which gives rise to a stress in the direction from the stagnation point which causes an increase in the relative expansion rate at the surface. By manipulating the weight of the free falling film by changing the height of the film, the relative expansion rate can be changed. At the top of the free falling film the surface tension is increased with respect to the equilibrium surface tension. This surface tension is called dynamic surface tension.

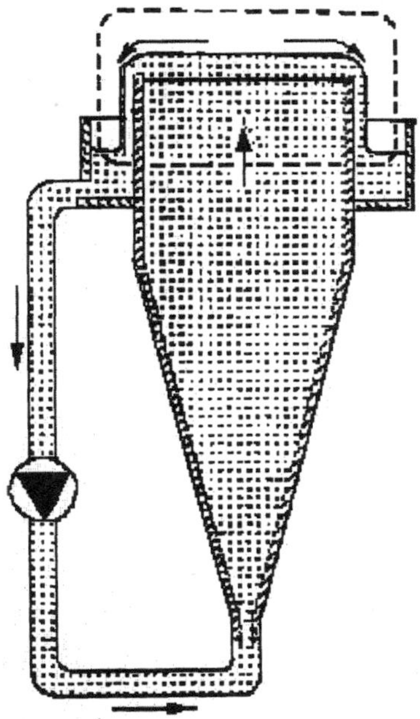

Fig. 5 Schematic representation of the overflowing cylinder.

2.3.2 EXPERIMENTS

In order to check if the model could predict mechanical properties from the molecular properties outlined in previous sections, experiments were performed at expanding surfaces in the overflowing cylinder. Two different experiments were performed using the same solutions of β-casein under the same conditions. In the first series of measurements, the surface concentration and surface tension were determined by means of ellipsometry and the Wilhelmy plate technique. In the second series of experiments the surface tension and the relative rate of expansion of the surface were measured as a function of the falling film height. By means of the surface tension these experiments could be related to each other. So the relation between surface tension, surface concentration, falling film height and relative rate of expansion could be obtained.

2.3.3 COMPARISON TO EXPERIMENTAL DATA

The predictions of the model with respect to the transport to the surface were verified by modelling the low molecular surfactant CTAB. This could be done by applying the same model as for the protein by leaving the out the unfolding and refolding steps. Bergink (1993) and Manning-Benson and Darton (1997) investigated the dynamic properties of this surfactant. Data with respect to the equation of state of this surfactant were obtained from Prins et. al. (1967). Therefore these data could be compared with predicted values. In Fig. 6 equilibrium surface tension, dynamic surface tension and the magnitude of the surface tension gradient are shown originating from experiment and theoretical predictions. In order to make these predictions possible, the relative rate of expansion was varied over the surface in order to match the situation in the overflowing cylinder.

In this graph it is found that the decrease in dynamic surface tension takes place at a different concentration for the theoretical curve compared to the measured curve. Therefore the maximum in the relative rate of expansion, which coincides with the strongest decrease in dynamic surface tension, is also found at a different concentration. Also the magnitude of the relative rate of expansion is not predicted well by the model. It can be seen also that the measured equilibrium surface tension deviates significantly from the predicted equilibrium curve. There are probably two causes for this poor match between theory and experiment. The first is that in the model the presence of surfactant in relative rate of expansion occurs above the CMC. Second there can be serious doubt about the purity of the CTAB used in the experiments. This may explain also why the equilibrium surface tension deviates so strongly between measured and predicted values, presumably there are some minor components present in the CTAB sample.

From the data from the model it can be found that the concentration of a very pure sample of CTAB needs to be increased a factor 3 in order to obtain the same surface tension, due to depletion of the surface by the expansion while in the experiment the concentration needs to be increased by a factor of 40.

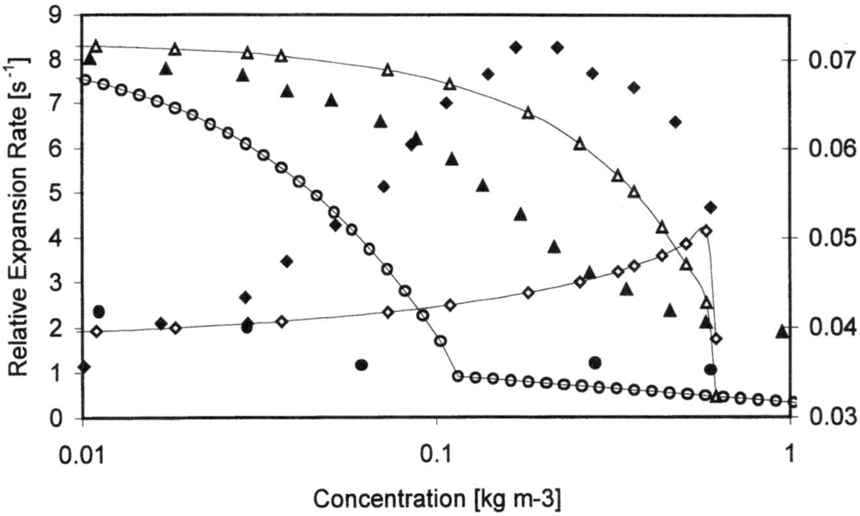

Fig. 6 Equilibrium surface tension (predicted) ○, equilibrium surface tension (measured) ●, dynamic surface tension (predicted) Δ, dynamic surface tension (measured) ▲, relative rate of expansion (predicted) ◊ and relative rate of expansion (measured) ◆ as a function of concentration for CTAB. Measured values from Bergink (1993).

The model has successively been applied to β-casein in order to model the unfolding properties. The fit parameters were obtained by modelling experimental data to the model. In the following table the parameters used to fit these data have been given

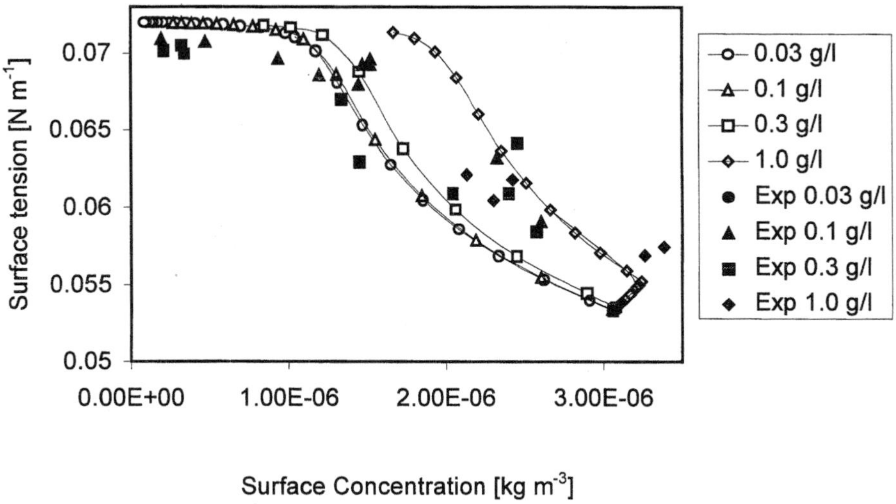

Fig. 7 Relation between surface tension and surface concentration for different bulk concentrations for β-casein 0.075 M imidazole buffer pH 6.7 (measured and calculated).

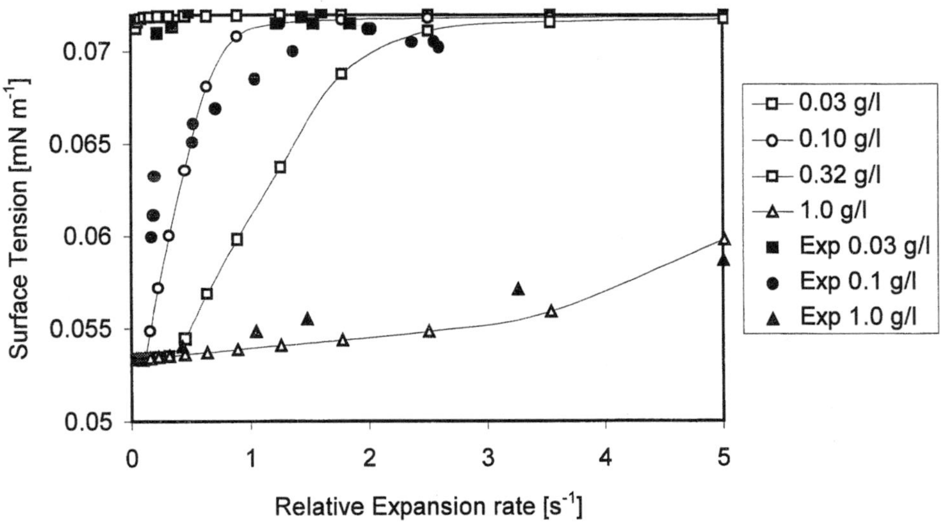

Fig. 8 Relation between relative rate of expansion and surface tension for β-casein 0.075 M imidazole buffer pH 6.7 for different bulk concentrations (predictions and experiment).

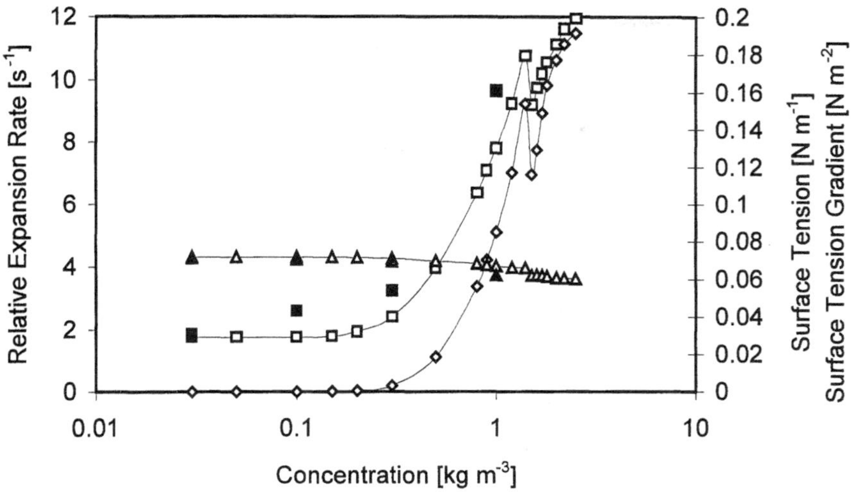

Fig. 9 Relation between the concentration in bulk and the maximum relative rate of expansion, calculated ❑ and measured ■, surface tension gradient calculated ◊, and dynamic surface tension, calculated Δ and measured ▲, β-casein 0.075 M imidazole buffer pH 6.7.

Table 1: Parameters obtained from the model for β-casein.

Parameters	
k_1 (unfolding constant)	8.5 s^{-1}
k_2 (refolding constant)	4 s^{-1}
A_n (dimensionless area native molecule)	0.6
A_u (dimensionless area unfolded molecule)	1

In Fig. 7 and 8 the relations between surface tension and surface concentration for β-casein is given as predicted by the model, along with experimental data and the surface tension is plotted as a function of relative rate of expansion from experiment and from experimental data respectively. In Fig. 9 experimental data for the relative rate of expansion and surface tension gradient are given for as a function of concentration in the bulk liquid.

In the results good agreement has been found between the properties observed in experiment and predicted by the model. Now the model can be placed into perspective with respect to it's practical significance. If it should be possible to model the unfolding properties of proteins with respect to the surface properties and mechanical properties of the surface, the importance of the unfolding for the mechanical properties can be quantified as a function of imposed stress and time scale of a certain surface of a particular proteinaceous solution.

In essence it is now possible to determine which mechanism is limiting under certain circumstances. The ratio of the unfolding rate and reciprocal of the dilation rate expressed in the relative expansion rate determines the fraction of unfolded molecules at the surface. For β-casein, the unfolding can be considered to be limiting above dilation rates around 8.5 s^{-1} since at that point k_1 is equal to θ. This means that regardless of concentration, which determines the transport to the surface, the surface tension can be considered to be the surface tension of the pure solvent at expansion rates higher than 8.5 s^{-1}.

In general this means that in the case of proteins the rates of two processes; transport and unfolding, are decisive for the total behaviour which may be the most fundamental reason why dispersion properties of proteins are different from the dispersion properties of low molecular surfactants where only the transport process is limiting.

Another parameter following from the model is the surface area occupied by the molecule. It is not surprising that the area is smaller for native molecules. It should be mentioned however that the area's used in the model are geometrical parameters accounting for the inhibition of adsorption due to the presence of other molecules. Hence from the decrease of this parameter as a consequence of unfolding it can be concluded that the presence of unfolded molecules inhibits the adsorption of other molecules more than the presence of native molecules. The reasons for this inhibition may be multilayer-formation, distribution of hydrophobic and hydrophilic segments over the adsorbed layer, distribution of charge over the adsorbed layer and many other factors.

If we now generalise these observations to a situations in which a different protein is present, it can be understood why proteins which unfold slowly, cannot contribute to the creation of surface tension gradient, even if the transport to the surface is fast. On an empty surface the time scale of expansion will be dictated by the properties of the pure solvent under these circumstances. Furthermore it can be concluded that if the rate of unfolding is smaller than the rate of expansion the unfolding of the protein is limiting irrespective of the rate of transport to the surface.

2.3.4 CONCLUDING REMARKS

From a simple population model predicting the adsorption and unfolding of proteins at liquid surfaces, data were predicted which show to be in good agreement with experimental values. The results show that the surface tension in expansion is dependent on both the transport to the surfaces as on the unfolding of proteins at the surface. The dependence on the time scale of unfolding is likely to determine the difference in surface behaviour between low molecular surfactants and macromolecules.

3 STAGNANT LAYER FORMATION OF PROTEINS AT LIQUID INTERFACES.

3.1 Introduction

It is well known that liquid surfaces can be motionless as a result of a surface tension gradient, large enough to compensate the shear stress exerted by the moving liquid on that surface. Such motionless surfaces can be observed in every day life at the surface of liquid food stuffs like soup and milk. When relaxation in such surfaces is absent, they behave purely elastic, comparable with an insoluble monolayer. Presence of a "skin" is another example of a stagnant layer. Here under "skin" we think of a relative thick layer (thicker than molecular dimensions) on the surface which has a certain amount of rigidity/cohesion. An example of such skin formation is the formation of bubble ghosts in beer when carbon dioxide bubbles shrink due to the dissolution of gas in beer (Ronteltap, 1990). Macromolecular surfactants such as proteins and polysaccharides are thought to promote skin formation. The coherence of such a skin is derived from the presence of a two-dimensional network, or as postulated by Prins et al. (1996), a kind of gel layer formation in the surface. Therefore, it is to be expected that as well as shearing deformations, dilational deformations give information on the mechanical properties of such stagnant layers.

Surface rheological parameters which are often used and which are physically well described and defined are the surface dilational modulus, the surface dilational viscosity, the surface shear modulus and the surface shear viscosity. All four parameters depend on the time scale of the measurement: the moduli depend on the frequency of the sinusoidal deformation and the surface viscosity depends on the rate of deformation. However, they have in common that in these rheological properties the surface tension acts as a "force", which is assumed to be uniform over the surface of investigation. Now one can ask which surface rheological technique or surface rheological parameter is important for studying the occurrence and properties of such skins? Therefore, in this section of the chapter we will describe and discuss results obtained with various techniques for a protein solution able to create a "skin" and a surfactant solution. The techniques used are the overflowing cylinder to measure the dilational viscosity and the relative expansion rates of the surfaces (Bergink-Martens, 1993), the ring through technique and the channel method for the measurement of the surface dilational modulus (Kokelaar, 1991; Prins et al., 1996). Finally, the consequences for a practical system will be discussed for the droplet break-up in an oil/water emulsion stabilised by two proteins and a surfactant.

3.2 The overflowing cylinder.

Let us now consider the behaviour of surfactants and proteins in the overflowing cylinder (see Fig. 10). As explained in the previous paragraph, low molecular surfactants are able to create a bigger surface tension gradient than proteins. This can be inferred from the higher dlnA/dt value obtained for surfactants (order of 10 s^{-1}) compared with the one for proteins (4 s^{-1}) at a high length of the falling film (L) at the outside of the cylinder. Considering the curve for surfactant one can conclude that already at small L the surfactant is able to cause higher dlnA/dt values in comparison to the pure solvent, water, causes. In contradiction, proteins need a much longer falling film before any relative expansion rate can be measured, which then is rapidly rising to its maximum value with increasing L. Thus, for proteins the dlnA/dt values are at low L are much lower than for pure water. What does this mean? Apparently, proteins adsorbed at the interface are able to resist the force which is pulling on them to move them over the rim of the cylinder, whereas surfactants are not able to. In the case of surfactants, the interaction between the surface and the bulk solution takes place via transport of surface active material to and from the interface. From a hydrodynamic point of view one can say that in both cases there exists a velocity difference between the surface and the liquid. Under these conditions as a result of the surface rheological properties defined above, the surface is able to generate a surface tension gradient $d\gamma/dx$ along the surface.

This surface stress (σ (Pa)) is able to compensate the viscous stress exerted on the surface by means of the liquid that moves parallel to the surface:

$$\sigma \equiv \eta_b \left(\frac{dv_x}{dz} \right)_{z=0} - \frac{d\gamma}{dx} = f\left(\frac{\mathrm{dlnA}}{\mathrm{dt}} \right) \qquad [11]$$

where η_b is the bulk viscosity (Pa·s), v_x is the velocity in the x-direction and z the coordinate perpendicular to the liquid surface. Depending on the signs of the surface tension gradient and the velocity gradient two situations can be distinguished. Both situations are given in Fig.(11). In the first situation, Fig. 11a, corresponds with values for dlnA/dt which are higher than those of water (occurring at high L values), whereas Fig. 11b is applicable for dlnA/dt values lower that those of water (occurring at low L values).

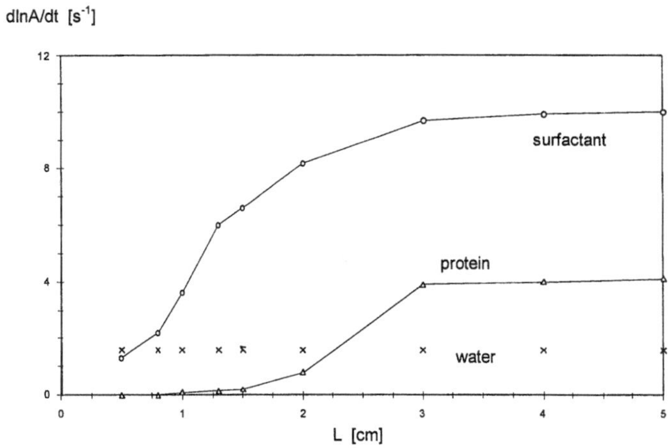

Fig. 10 The relative expansion rate dlnA/dt as function of the length of the falling film of the overflowing cylinder (L) for water, a milk protein solution and a surfactant solution. Note that the figure is schematic.

In the first case the surface tension gradient drives the liquid flow parallel to the surface and in the latter the velocity gradient in the liquid drives the movement of the surface and creates a counteracting surface tension gradient. It is noteworthy that both cases are applicable for both protein and surfactant solutions, however, for surfactants the velocity gradient driven surface movement is very small expressed in L values. For proteins the L region where the velocity gradient drives the surface movement is much larger. This difference cannot be accounted for by surface tension gradients alone. Probably there will be some kind of additional surface force which counteracts the surface velocity and is typical for proteins. It is therefore reasonable to speak of a kind of network formation of the proteins at the surface which can withstand the pulling force of the moving liquid.

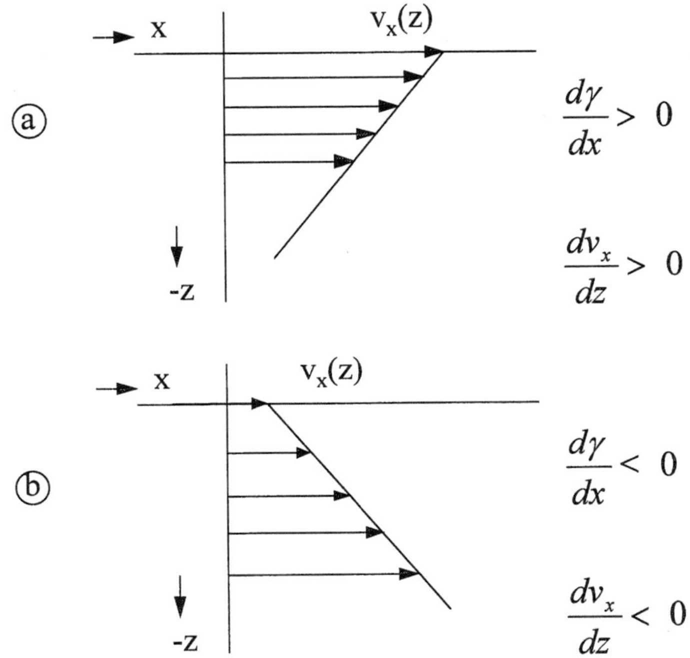

Fig. 11 The two situations occurring in the overflowing cylinder. a). The surface tension gradient drives the liquid flow parallel to the surface; b) the velocity gradient in the liquid drives the surface.

3.3 The canal method

Let us now introduce the next technique (the canal method) in which stagnant layer formation can be studied. This technique does not use the Wilhelmy plate method to measure the change in surface tension. This might be advantageous because when the Wilhelmy plate is used it is uncertain if forces are transferred to the plate in a proper way, for instance due to the rupture of the skin close to the plate. In the canal technique a stress is exerted on the liquid surface by means of a streaming liquid in an open rectangular canal (see Fig. 12). The liquid flow is controlled by means of a pump, tap and flow meter. The resulting deformation is then measured by means of the displacement of marked spots (small piece of paper or talc powder) on the surface.

The stress on the motionless surface is calculated by means of the following relationship valid for a rectangular canal with a square cross-section:

$$\sigma = 3.57\eta Q\frac{b+h}{(b+h)^2} \qquad [12]$$

where η is the bulk viscosity, Q the liquid flux and b and h the width and height of the wetted part of the canal.

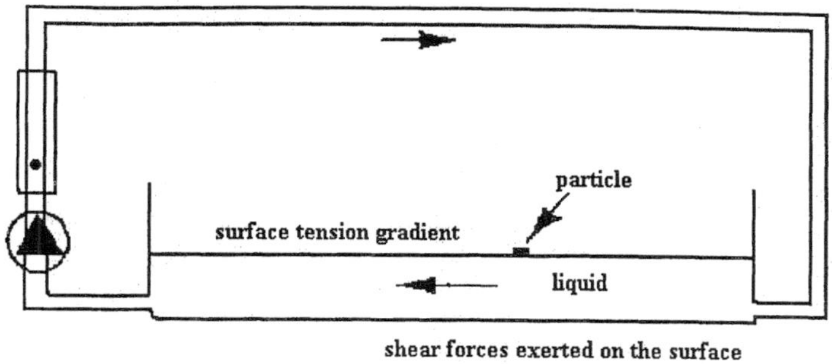

Fig. 12 Schematic representation of the canal method showing a cross-section through the canal. The dimensions of the canal are: length 50 cm, width 5 cm and height of the liquid level 5cm.

This relation can be rewritten to obtain an expression for the surface dilational modulus E [Prins et al., 1996]:

$$E = \frac{ld\sigma}{dlnl} \qquad [13]$$

in which l the length of the surface on the upstream side of the particle. By changing the flux Q by a known amount, the viscous drag on the surface changes by an amount $d\sigma$ according to equation (12). The corresponding displacement of the particle can be measured with a cathetometer whereas l follows from the distance to the inlet of the canal.

What happens in the case of protein solutions? It appears that in spite of the liquid flow through the canal the marker at the surface is completely motionless. This behaviour has been found for various proteins like egg white, β-lactoglobulin and bovine serum albumin. By increasing Q with a certain amount, the marker is displaced and is found to return to its original position when the original flow rate is re-established. This clearly shows that the surface behaves purely elastic and that relaxation does not occur. In Fig. 13 an example is given for the surface dilational modulus determination with the canal method for egg white as a function of the applied stress. As can be seen, the values of E increases with ageing of the surface. However, the values obtained are comparable with those obtained under dynamic conditions in a ring trough (Williams and Prins, 1996).

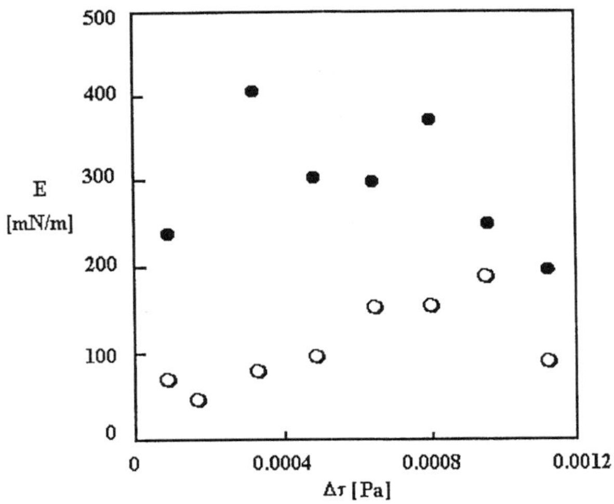

Fig. 13 Surface dilational modulus as measured with the canal technique as a function of the applied change in stress for a 0.48% w/w aqueous egg white solution; ○ before ageing, ● after ageing.

What will be the behaviour of a surfactant solution? For surfactants a stagnant layer was never observed, not even at the lowest Q values. The marker is always moving when a stress is exerted on the surface by liquid motion. Therefore, surface dilational modulus measurements of surfactant solutions can not be performed in the canal.

3.4 The ring trough technique

The ring trough technique is a technique in which close to equilibrium circumstances, the surface dilational properties of protein and surfactant solutions can be determined. This technique was fully described by Kokelaar et al [1991] for the water-air interface and was later adapted for measurements at the oil-water interface by Williams and Prins [1996]. The schematic diagram of the ring trough modified for oil-water measurements, is given in Fig. 14. By moving a fully wetted roughened glass ring sinusoidally up and down in a liquid surface, an expansion of the surface is obtained by the upwards movement of the ring and during downwards movement of the ring compression is established. The Wilhelmy plate can be used to provide real-time measurements of the surface tension. From the variation in the interfacial tension with interfacial area the values for the surface dilational properties: dilational modulus ($|E|$), storage modulus ($E_{d,}$) and loss modulus ($\omega\eta_d$) and loss tangent ($\tan\theta$) can be calculated.

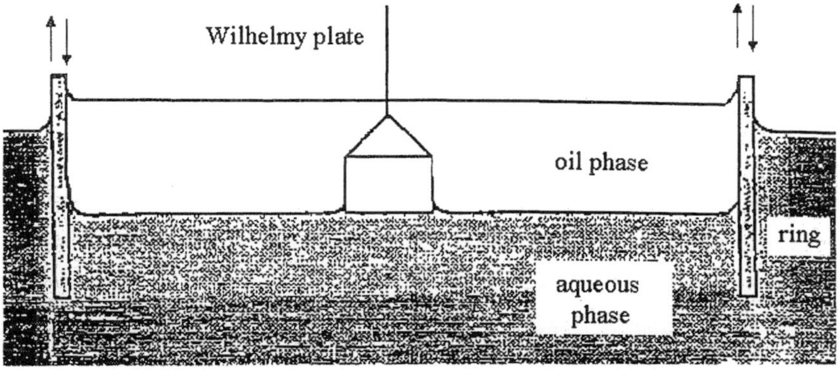

Fig. 14 Schematic diagram of the ring-trough apparatus with oil layer present. Side elevation.

In Figs. 15a and b the surface dilational modulus and $\tan\theta$ for β-casein and β-lactoglobulin are given as function of the concentration for the oil-water interface obtained by Williams and Prins [1996]. Comparison of both proteins shows that at low concentrations the moduli and the loss tangents have approximately the same value. Low concentrations allow more unfolding of the molecules at the interface and so the structure of the two proteins may be fairly similar. The diffusional exchange with the bulk is small for these concentrations and therefore the conformational changes of the molecules are responsible for the relaxation in the surface. At high concentrations the modulus for β-lactoglobulin (60 mN/m) is very much higher than that for β-casein (10 mN/m), whereas the relative viscosity is high for β-casein but remains low for the β-lactoglobulin. This is in qualitative agreement with what has been found by other researchers for globular proteins (β-lactoglobulin) and random coil proteins (β-casein) [Benjamins and van Voorst Vader, 1992]. The globular protein is thought to form a strong cohesive network within the interface, which will limit the amount of diffusional and conformational relaxation. The modulus for β-casein remains constant, even at high concentrations. This suggests that the protein forms a network structure in the interface, although this is likely to be weak compared with that formed by globular proteins.

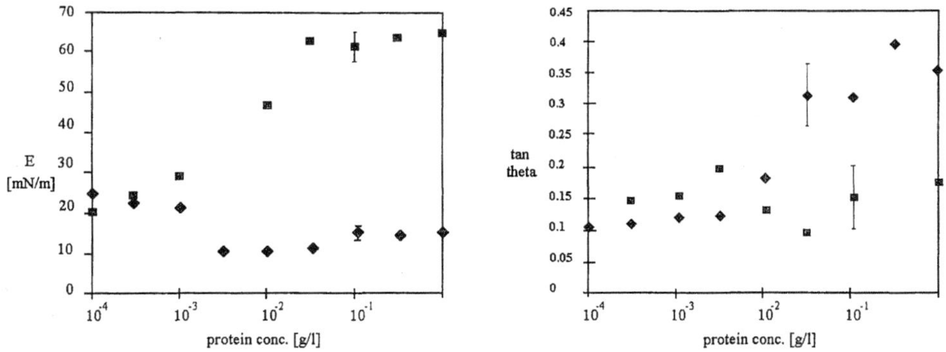

Fig. 15 a) The interfacial modulus and b) the loss tangent as a function of protein concentration. Comparison of β-lactoglobulin (■) and β-casein (◆) at the oil water interface.

Similar results have been found for both proteins at the air water interface, showing that the interfacial structures are rather the same at the air-water and the oil-water interface. It also implies that hardly any protrusion of non-polar amino acid residues in the oil phase occurs, otherwise the surface dilational modulus should be higher for the air-water interface. Therefore, it might be possible to use the dilational data obtained for proteins at the air-water interface and to apply them to the oil-water systems, particularly the frequency range can be extended for measurements toward the values relevant for emulsification.

3.5 Droplet break-up

In this section we will address to the practical relevance of skin (network) formation. Therefore we will consider the break-up of an oil droplet in simple shear flow as can be studied in a couette device. Depending on the shear force G, the viscosity of the disperse and continuous phase η_d and η_c, the radius of the undeformed droplet r and the equilibrium interfacial tension γ_{eq}, two independent variables, the capillary number Ω and the viscosity ratio λ, can be defined

$$\Omega = \frac{\eta_c G r}{\gamma_{eq}} \qquad [14]$$

$$\lambda = \frac{\eta_d}{\eta_c} \qquad [15]$$

The capillary number is the ratio between the viscous stress and the stabilising Laplace pressure. For an emulsifier free system there is a critical capillary number as a function of the viscosity ratio as shown in Fig. 16 by the drawn line, where droplets are just stable to break-up. It should be noted here that the shape of droplet break-up depends upon the viscosity ratio. The results reported here in this section corresponds to break-up via the dumb-bel shape.

In Fig. 16 also the data are given for β-casein (1.0 and 0.1 g/l) and β-lactoglobulin (1.0 and 0.1 g/l). Two remarkable results can be seen. Firstly, by adding β-casein (1.0 and 0.1 g/l) or low concentrations of β-lactoglobulin (< 0.1 g/l) to the continuous phase it becomes more difficult to break up the oil droplets, something one would not expect by adding an emulsifier because it should make break-up more easily. Why? This "difficult" break-up can be explained as postulated by Janssen et al., 1994.

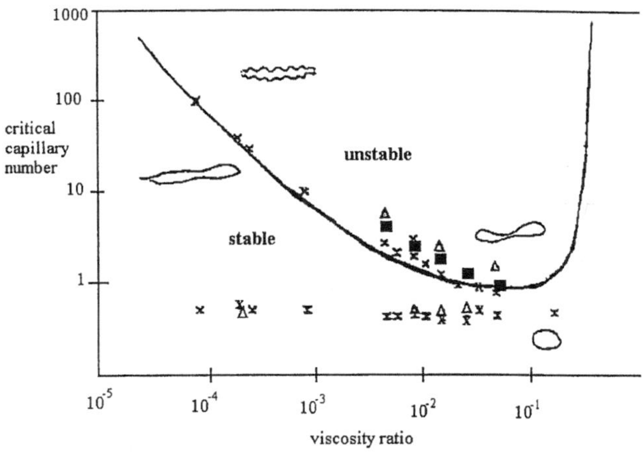

Fig. 16 Critical capillary number as a function of the viscosity ratio for three different systems of investigations: Drawn line represent the theoretical values for an emulsifier free system, * β-lactoglobulin (1 g/l); Δ β-lactoglobulin (0.1 g/l); ■ β-casein (1 g/l); × β-casein (0.1 g/l).

They have shown that the dynamics of droplet break-up has been successfully linked to the elasticity modulus of the adsorbed emulsifiers at a deforming planar interface. The result is that at droplet break-up the effective interfacial tension for the droplet is higher than the equilibrium interfacial tension measured under quiescent conditions and therefore the break-up will be more difficult. In the case of high concentrations (1.0 and 0.1 g/l) of β-lactoglobulin the critical capillary number is independent of the viscosity ratio, showing that the viscosity of the dispersed phase becomes unimportant and that the break-up is governed by properties of the interfacial layer. Furthermore, it can be seen that breaking up of an oil droplet is much more easily to perform. Therefore this is called "easy" break-up. In practise this means that the maximum droplet size which can exist in a certain flow field is around 200 times smaller than expected from the equilibrium interfacial tension (for the lowest viscosity ratio). The explanation for this behaviour has been given by Williams and Prins, 1997. It can not be the fact that droplet deformations and shear forces generate surface tension gradients which can cause such low surface tensions (<0.5 mN/m) to get small droplet formation via tip streaming; which was not observed for β-lactoglobulin. Furthermore, in experiments with large deformation surface tension values lower than 10 mN/m were never observed. They introduced the term "grip" on the interface to explain the "easy" droplet break up. They already found in experiments performed with the ring trough technique that high concentrations of β-lactoglobulin were able to generate a relative high surface modulus, which they correlated with the ability to generate a rigid network layer at the oil/water interface (see section 3.3). They correlate this network formation with ability to break-up droplets more easily to the fact that the shear force has more "grip" on this network. The energy transfer on to the rigid interface is much more effective, in other words, the surface is not able to dissipate the energy in the surface layer or into the bulk of the droplet. In the case of β-casein also a network can be formed, but this network isn't rigid but much more mobile. Therefore energy dissipation can occur via the interfacial layer and the bulk of the droplet. As a result, more effort has to be made to break up the oil droplet.

3.6 Summarizing

The break-up of a liquid droplet or gas bubble when it is subjected to a given shear stress in a flowing liquid is studied. The conclusion is that the more able the surface of such a particle is to remain motionless under these circumstances, the more efficient the liquid action on the particle is to cause the break-up of that particle. One essential part of this way of reasoning is that the order to realise break-up, the surface of the particle has to be broken too. However, the higher the shear stress at which break-up occurs, the more efficient the particle can break-up.

4 FOAMING BEHAVIOUR OF PROTEIN AND SURFACTANT SOLUTIONS

4.1 Introduction

The stability of foams is a dynamic process, which means that the structure of a foam changes continuously. These changes are induced by different processes, which are creaming, drainage, coalescence and Ostwald ripening (Prins, 1988). Creaming is the rise of bubbles in a liquid under influence of a density difference and drainage is the flow of liquid the other way round. Coalescence is the coarsening of bubbles caused by rupture of the films and Ostwald ripening, also called disproportionation, is the coarsening which is caused by gas transfer between bubbles of different size. All these different factors interact with each other in a complicated way, for instance drainage thins the films and increases coalescence and Ostwald ripening. Also coarsening of a foam increases the degree of drainage because the number of motionless (film) surfaces becomes smaller.

Considering foam stability, two rather different types can be distinguished, which are 'transient' and 'stable' foams. 'Transient' foams are short-living foams, like beverages or alcohol-water mixtures, which disappear in a time-scale of seconds. 'Stable' foams have a life-time from minutes to days.

In this part about foam stability some differences between foams stabilised by proteins and by low-molecular surfactants will be discussed. Studies of foam stability of proteins of different conformations is in progress (Van Kalsbeek and Boerboom, to be published). Special attention will be paid to the role of a stagnant surface network in the stabilisation of a foam. Several authors have mentioned the importance of a protein network adsorbed at the air/water interface, often called 'protein film', to the stability of foams (Graham and Philips, 1976). However, it is not often mentioned to what type of stability the network is contributing to, e.g. to stability against drainage or stability against Ostwald ripening or coalescence. Furthermore most studies on network properties are carried out close to equilibrium, for instance studies of shear and dilational moduli in the surface.

In this foam stability study a protein (β-lactoglobulin) and a low-molecular weight surfactant (Sodium lauryl sulphate) are foamed by mechanical agitation with a stirrer (φ 4.6 cm). 100 ml of solution is stirred in a glass vessel (φ 6 cm) for 70 seconds at 2500 rpm. The different types of stability of these foams will be discussed below.

4.2 Creaming and drainage.

Creaming of gas bubbles through a liquid and drainage of liquid from a foam both describe the relative motion of bubbles and continuous phase. In creaming the movement of the bubbles is more important than the liquid flow and in drainage it is the other way round. In the case of aqueous foams with a viscosity of the continuous phase of 10^{-3} Pa·s, creaming usually results in the formation of a creamed bubble-layer in a few seconds. The creaming rate is however also determined by the amount of gas which is incorporated in the foam. When this amount becomes high, e.g. more than 65%, the bubbles can not rise through the liquid undisturbed and form a packed bubble-layer. The creaming rate becomes much smaller and it will be difficult to distinguish creaming from drainage. The bubble velocity during creaming in a liquid depends on the surface dilational viscosity (Prins, 1988).

Drainage of liquid from a foam is dependent on the foam structure. Two extreme situations can be distinguished which are wet foams and polyhedral foams. In a wet foam the bubbles are present like spheres and the liquid fraction in the foam is higher than about 5%. In a polyhedral foam the bubbles are deformed to a honeycomb structure which usually takes place if the fraction of liquid in the foam is < about 5 % and the bubbles are larger than about 1 mm. Between the bubbles almost flat liquid films are formed which are surrounded by the so-called Plateau borders.

Drainage of liquid through a close packed foam layer can be caused by two different processes, which are flow of liquid through the films and Plateau borders and the so-called marginal regeneration. In the latter process thick and thin layers inside a film are exchanged with the Plateau border. There are some limitations in the occurrence of this process because the film parts which are exchanged with one another have a size of about 1 mm^2 . This means that in a foam with bubble sizes smaller than 1 mm, as in the case of the stirred foam used in this study, the process does not occur. Furthermore marginal regeneration does not occur in films which have shear properties like proteins, because the interchange of thin and thick film parts can not take place due to the resistance of a protein covered surface to shear deformation.

Since the fresh formed foam in this study has bubbles sizes smaller than about 200 μm it is still a wet foam. The liquid flows around the spheres of the foam and can be slowed down by the presence of a force which makes the surface of a bubble motionless. This can for instance be accomplished by the presence of a surface tension gradient over the bubble surface (Prins, 1988).

A study to the influence of surface tension gradients on the foamability of alcohol-water solutions, shows a clear correlation between the ability of a system to create a gradient and the life-time of the alcohol stabilised foam. The ability of creating a surface tension gradient is measured by means of the overflowing cylinder technique which is discussed earlier (Tuinier et al., 1996). The amount of foam that can be produced is correlated to the possibility of slowing down the drainage rate, since the aqueous alcohol films will coalesce if they become too thin, because the disjoining pressure of alcohol stabilised surfaces is too small to prevent coalescence. It must be mentioned that this concerns a transient foam which is very unstable. One of the reasons for this instability is that the surface tension gradient is very small. Once an alcohol stabilised film is formed the surface tension gradient will diminish by transport of the volatile alcohol to the gas phase followed by adsorption at parts of the surface with a low coverage of alcohol molecules.

In the more stable foams formed by SLS and β-lactoglobulin such a transport process can not take place. However, relaxation of the surface tension gradient can take place by either transport through the surface or exchange between the surface and the bulk liquid. In the overflowing cylinder two differences between proteins and surfactants can be observed (see section 3.1). The first is that at high value of the falling film ($L > 2$ cm) the ability to create surface tension gradients is usually higher for surfactants than for proteins. The magnitude of this difference is about a factor 2. Another observation is that at low height of the falling film ($L < 2$ cm) proteins are able to slow down the surface expansion rate to very small values ($d\ln A/dt \approx 0.001\ s^{-1}$). It is thought that this is caused by the formation of a stagnant layer in the surface and that this layer is partly formed by a surface tension gradient and partly by the presence of a cohesive network. The lowering of the expansion rate of the surface in the overflowing cylinder by proteins is usually a few orders of magnitude higher than the lowering of the expansion rate by surfactants.

It is thought that the mentioned different factors, which are the formation of a stagnant layer at low values of L and the ability of creating surface tension gradients at high values of L, may contribute to the stability of a bubble surface against drainage of down flowing liquid.

The drainage rate of foams stabilised by β-lactoglobulin and SLS is investigated by measurement of the amount of liquid in the foams in time. For both systems these measurements are given in Fig. 17.

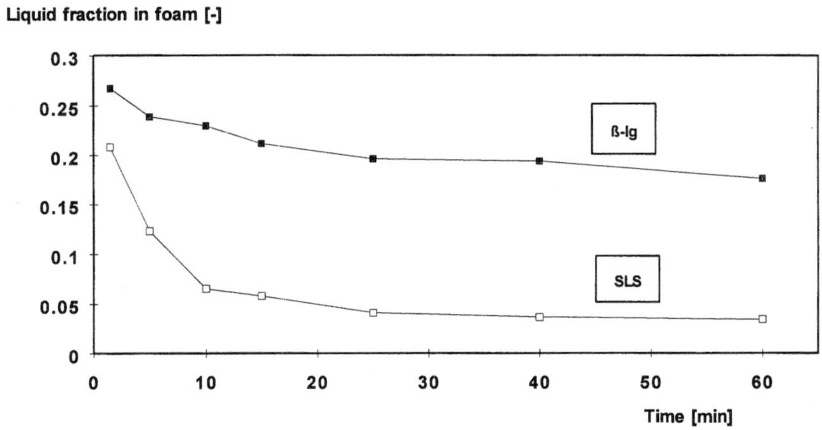

Fig. 17 Fraction of liquid in foams stabilised by SLS (0.25 g/l, I=1 M NaCl) and β-lactoglobulin (0.25 g/l, pH = 6.7, I = 75 mM).

This Fig. 17 clearly shows that the SLS-stabilised foam drains faster than the foam stabilised by β-lg. This would implicate that the formation of a stagnant surface layer at low values of L is more important in slowing down drainage than the ability of creating a surface tension gradient at high values of L.

It must be mentioned that the rate of drainage is dependent on the bubble size distribution, in the way that it is lower for smaller bubble sizes. It appeared that the initial bubble size is more or less equally for both systems, but that the SLS stabilised foam coarsens much faster which will be discussed later. This means that the faster drainage rate of SLS may partly be caused by the faster coarsening. It is not known to which extent this coarsening contributes to the faster drainage rate in the SLS-stabilised foam.

4.3 Coalescence

Coalescence is the coarsening of a foam caused by the breaking of liquid films between bubbles. When a liquid drains through a foam film, the film will become thinner and (parts of) the two surfaces in the film will get in close neighbourhood of each other. If the distance between these two surfaces is very small the adsorbed molecules will interact with one another and will repel each other. Absence of this repulsion will make it possible for the two bubble surfaces to reach critical film thickness at which spontaneous film rupture due to thermal motion will take place. If the repulsion is strong enough to compensate for the drainage in the film (including Plateau border suction) and the Van der Waals attraction between the surfaces, an equilibrium film can be formed. For low molecular weight ionic surfactants like SLS, which derive repulsion by their charged surfaces, equilibrium films are formed in the order of 5-10 nm. High molecular weight surfactants like proteins repel each other sterically as well as by electric charges and usually reach an equilibrium thickness of about 25 nm (Clark et al., 1989, 1991).

There is no general accepted theory on the mechanism which causes coalescence and different aspects of a certain system play a role in the breaking process. It is seen in experiments that drainage plays an important role as an inducer of coalescence. When drainage proceeds fast the surface will contain little surfactant due to expansion of the surface at the top of the bubble. This may lead to coalescence if the liquid film between the bubbles becomes too thin and there is not enough surfactant present to provide repulsion of the surfaces.

Drainage may cause disturbances in a film, for instance by the formation of small waves (Vrij and Overbeek, 1968). These disturbances increase the surface tension locally and result in a surface tension gradient. This surface tension gradient plays an important role in damping the disturbances, but it is also possible that a molecular network, as in the case of proteins, gives resistance to dilation. This agrees with the idea that a network structure is able to slow down drainage of liquid. A question is whether dilation of a protein network results in a change (increase) of the surface tension or not. Further more relaxation of the surface tension of the expanded and compressed surfaces of the film may be important in providing a surface tension gradient which remains its value at longer time scales.

The impression is that the foam stabilised by SLS is more vulnerable to coalescence than the foam stabilised with β-lg. However, it seems that the drainage rate is higher in the SLS-foam and this might be the initiator of the coalescence. So to be more sure that one system is less stable against coalescence than another system, the drainage rates should be the same.

4.4 Ostwald ripening

This is the process whereby gas diffuses from small bubbles into bigger ones. The driving force for this process is the Laplace pressure difference over the curved surface of the bubble:

$$\Delta P = \frac{2\gamma}{r} \qquad [16]$$

The Laplace pressure is higher for smaller bubbles than for bigger ones. As a result of this the solubility of a gas in the liquid close to the bubble is higher for smaller bubbles:

$$\frac{S_r}{S_\infty} = e^{\frac{2\gamma V}{RTr}} \qquad [17]$$

in which S_r is the solubility of the gas of a bubble with radius r in the continuous phase, S_∞ is the solubility when $r = \infty$, γ is the surface tension of the bubble surface, V is the molar volume of the gas, R is the gas constant and T is the temperature.

The diffusion of gas from smaller bubbles to the bigger ones gives rise to coarsening of a foam. The dissolution of gas from a small bubble into the liquid is given by de Vries (1958):

$$r_t^2 = r_0^2 - \frac{4RTDS\gamma}{P_{atm}\delta} t \qquad [18]$$

in which t is time, r_0 is the bubble radius at time t=0, r_t is the bubble radius at time t, D is the diffusion coefficient, S is the gas solubility, P_{atm} is the atmospheric pressure and δ is the film thickness between the bubbles.

In this calculation a constant value is used for γ, but in liquids containing a surface active material this is not the case: the value of γ decreases as a result of compression of the bubble surface, during shrinking of the bubble. For surface active molecules with a low desorption rate, for instance proteins, the surface tension in compression, γ_{comp}, differs from the surface tension of the equilibrium surface, γ_{eg}. The decrease of γ_{comp} at a certain compression rate can be expressed by the surface dilational viscosity η_d^s:

$$\eta_s^d = \frac{\gamma_{eg} - \gamma_{comp}}{d \ln A / dt} \qquad [19]$$

This influence of η_d^s has been examined for different types of beer by Ronteltap and Prins (1990). The surface tension of the beer was studied in compression in a Langmuir trough equipped with a caterpillar belt. For different beers it was found that the beer type with a higher η_d^s slowed down the dissolution of the bubble more than the one with a smaller value of η_d^s.

For β-lg and SLS the surface dilational viscosity is given as a function of the relative expansion rate in Fig. 18. From this figure it follows that the surface dilational viscosity of β-lg is an order of magnitude higher than the one of SLS. The reason for this is probably that the protein molecules are adsorbed more irreversible than low-molecular weight surfactants and that the proteins will not be pressed out of the surface easily.

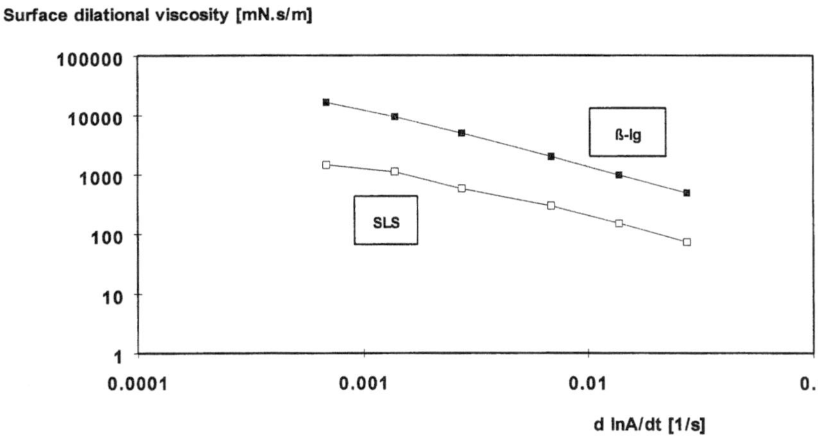

Fig. 18 Surface dilational viscosity of compressed surfaces of SLS (2.5 g/l, I=0.1 M NaCl) and β-lg (0.25 g/l, pH=6.7, I= 75 mM). Compressions are performed in a trough equipped with a caterpillar belt.

In contradiction the SLS molecules may form a more compact layer at the surface in which the molecules are more closely packed than in the case of proteins. When such a surface layer is compressed the SLS molecules will desorb if the concentration of molecules in the surface becomes too high and therefore the surface tension can not decrease much in comparison to the equilibrium value.

The behaviour of an adsorbed layer in compression in a caterpillar belt trough is probably related to the behaviour in the compressed part of the falling film of the overflowing cylinder. In this compressed part a stagnant layer is formed which is slowed down much more for proteins than for low-molecular weight surfactants. (see section 3.1).

The influence of the surface dilational viscosity to stability against Ostwald ripening is studied in a foam as a function of time. The foaming method is the same as described earlier. Photographs of the foams are taken at the glass wall of the foaming column. The bubble size is measured at 1 cm^2 at the upper part of the foam. The bubble sizes of bubbles larger than 0.2 mm are measured in classes having a width of 100 μm. The mean bubble size of each class is given in the legend. The volume of the bubbles in these classes is calculated and is given in Fig. 20, as a function of time. Each histogram in one graph shows the volume distribution at a certain time. This is the volume of bubbles present in each class in the first row of bubbles against the glass wall. This volume gives a three-dimensional picture of the bubble size distribution close to the glass wall and it is as a measure of the coarsening of the foam. One example of a histogram is given in Fig. 19.

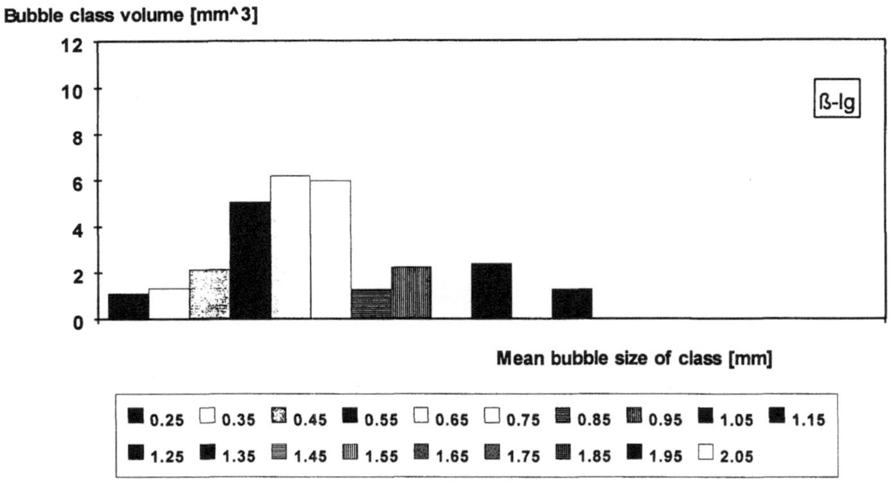

Fig. 19: Volume of bubble size classes in foam stabilised by β-lactoglobulin, 60 minutes after foam making. Bubbles > 0.2 mm are measured. The numbers in the legend are the mean diameters of each class in mm.

In Fig. 20 histograms of different times after foam making are given next to each other in one figure. It can be seen in Fig. 20 that directly after foam making the volume of the bubbles bigger than 0.2 mm is still very small and the volume in these size classes increases in time. The SLS stabilised foam coarsens much faster in time than the β-lg stabilised foam. Although some coalescence took place in the SLS foam, the coarsening is mainly caused by the higher Ostwald ripening rate in the SLS foam. Further more the number of small bubbles in the SLS foam decreases faster than in the β-lg foam, which is also an indication that the Ostwald ripening rate is higher in the SLS stabilised foam. It must be mentioned that the liquid content in the SLS foam is smaller than in the β-lg foam (Fig. 17), which means that the gas has to diffuse over smaller distances from the small to the big bubbles.

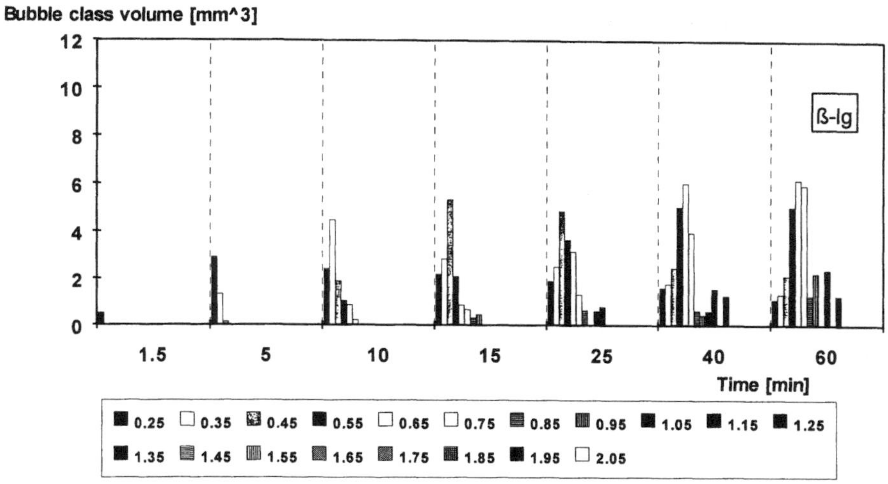

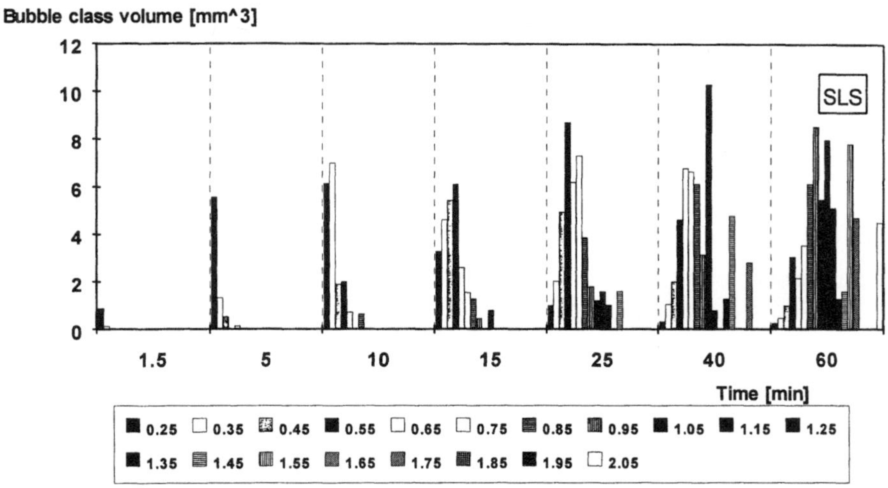

Fig. 20 Volume of bubble size classes in foams stabilised by SLS (0.25 g/l, 1 M NaCl) and β-lg (0.25 g/l, pH=6.7, I=75 mM), as a function of time. The bubbles are measured at 1 cm^2 in the top of the foam for bubbles > 0.2 mm. The numbers in the legend are the mean diameters of each class given in mm.

The physical interpretation of the difference in dissolving behaviour of the bubbles is the same as in the case of a compressed layer. The adsorbed molecules behave different in respect with reversibility of the adsorption. When a SLS stabilised bubble shrinks the molecules will be pressed out of the surface and the surface tension will not be decreased much in comparison to the equilibrium surface tension. Adsorbed β-lg molecules will not desorb as easily as SLS molecules. The surface tension can become much lower than the equilibrium value and the shrinking rate will be slowed down in a higher extent.

5 CONCLUDING REMARKS

One of the most pronounced differences in surface behaviour between proteins and low molecular surfactants is that surfactants can lower the equilibrium surface tension to a much higher extent (around 30 mNm^{-1}) than proteins can (around 45 mNm^{-1}). Although the surface load (in mg m^{-2}) for both components are more or less the same, the number of molecules present at the interface is for surfactants about two orders bigger. Furthermore, desorption of adsorbed surfactant molecules can occur whereas protein molecules hardly desorb. The latter is probably due the number of bonds between the protein molecules and the conformational changes. All these differences play a role in the rheological behaviour of both systems and can counteract or support each other depending on whether the measurements are far from or close to equilibrium, in dilation or in shear, in compression or in expansion. In Table 2 the main differences important for foam formation and foam stability are given.

From a simple population model predicting the adsorption and unfolding of proteins at liquid surfaces, data were predicted which show to be in good agreement with experimental values. The results show that the surface tension in expansion is dependent on both the transport to the surfaces as on the unfolding of proteins at the surface. The dependence on the time scale of unfolding is likely to determine the difference in surface behaviour between low molecular surfactants and macromolecules.

Comparing the ability of adsorbed molecules to make the surface motionless it is observed that adsorbed proteins produce adsorbed layers which are more able to be motionless. The first consequence of this observation is that in a agitated liquid bubbles with a more motionless surface (more rigid surface) can be more easily deformed. This means that smaller bubbles are produced than when the surface is less motionless. The second consequence is that the drainage rate from a low molecular weight surfactant foam is much faster than from a protein foam.

Table 2 Main differences between proteins and surfactants regarding foaming and emulsifying properties.

Parameter	Proteins	Surfactants
γ_{eq} (mN m^{-1})	45	28
Γ_{max} (mol m^{-2})	low	high
Γ_{max} (mg m^{-2})	2-3	2
film thickness in equilibrium	40 nm	4 nm
adsorption	irreversible	reversible
conformational changes	yes	no
skin formation	yes	no
bubble size distribution	more smaller bubbles	more bigger bubbles
drainage	slow	fast
Ostwald ripening	slow	fast

6 LIST OF SYMBOLS

A	Surface area protein (3)	[m^2 kg^{-1}]
b	width of the canal	[m]
c_b	bulk concentration	[kg m^{-3}]
c_s	subsurface concentration	[kg m^{-3}]
D	diffusion coefficient	[m^2 s]
G	shear force	[s^{-1}]
h	height of the canal	[m]
I	ionic strength	[Moles m^{-3}]
k	Boltzmann constant	[J K^{1-}]
k_1	unfolding constant	[s^{-1}]
k_2	refolding constant	[s^{-1}]
l	length of the canal	[m]
L	length of the falling film	[m]
P_{atm}	atmospheric pressure	[N m^{-2}]
ΔP	Laplace pressure difference	[N m^{-2}]
Q	flux of liquid	[m^3 s^{-1}]
R	gas constant	[J mol^{-1} K^{-1}]
r	distance from centre (stagnation point)	[m]

Δr	ring thickness	[m]
S	solubility of the gas	[mol m^{-3}]
SLS	sodium dodecyl sulphate	
t	time	[s]
T	absolute temperature	[K]
v	velocity	[m s^{-1}]
V	molar volume	[m^3 mol^{-1}]
z	vertical co-ordinate	[m]

Greek symbols

α	constant mass flux desorption	[kg s^{-1} m^{-2}]
β	constant mass flux adsorption	[m s^{-1}]
β-lg	β-lactoglobulin	
δ	film thickness	[m]
γ	surface tension	[N m^{-1}]
γ_{eq}	equilibrium surface tension	[N m^{-1}]
γ_{comp}	surface tension in compression	[N m^{-1}]
γ_d	dynamic surface tension	[N m^{-1}]
Γ	surface concentration	[kg m^{-2}]
Γ_∞	maximum adsorption	[kg m^{-2}]
η_b	bulk viscosity	[N m^{-2} s^{-1}]
η^d_s	surface dilational viscosity	[N m^{-2} s]
η_d	viscosity of the disperse phase	[N m^{-2} s^{-1}]
η_c	viscosity of the continuous phase	[N m^{-2} s^{-1}]
λ	viscosity ratio	[-]
π	surface pressure	[N m^{-2}]
σ	surface stress	[N m^{-2}]
θ	relative rate of expansion	[s^{-1}]
ρ	specific density	[kg m^{-3}]
Φ	mass flux	[kg s^{-1} m^{-2}]
Ω	capillary number	[-]

Subscripts

n	ring number (4) (5)	[-]

Superscripts

n	native conformation
u	unfolded conformation

7 REFERENCES

J. Benjamins and F. van Voorst Vader, Colloids Surfaces, 65 (1992) 161

D.C. Clark, A.R. Mackie, L.J. Smith and D.R. Wilson, in Food Colloids, eds. R.D. Bee, P. Richmond and J. Mingins, (1989) 39.

D.C. Clark, P.J. Wilde and D.R. Wilson, Colloids and Surfaces 59, 1991, 209.

D.J.M. Bergink Martens, H.J. Bos, A. Prins, and B.C. Schulte, J. Colloid and Interface Sci., 138 (1990) 1.

D.J.M. Bergink Martens, H.J. Bos and A. Prins, J. Colloid and Interface Sci., 165 (1994) 221.

D.J.M. Bergink Martens. Ph.D. thesis, Agricultural University, Wageningen, (1993), ISBN 90-5485-128-7.

F.J.G. Boerboom, Ph. D. thesis in preparation (expected 1997).

J.A. de Feyter and J. Benjamins, J. Colloid and Interface Sci., 90 (1981) 289.

D.E. Graham and M.C. Philips, in Foams, ed. R.J. Akers, 1976, 227.

J.J.M. Janssen, A. Boon and W.G.M. Agterof, AIChEJ. 40 (1994) 1929

H.K.A.I. van Kalsbeek and J.F.G. Boerboom, to be published

J.J. Kokelaar, A. Prins and M. de Gee, J. Colloid Interface Sci., 146 (1991) 507

S. Manning-Benson and C.D. Bain, J. Colloid and Interface Sci., in press.

A. Prins, C. Arcuri and M. van den Tempel, Journal of Colloid and Interface Science, 24 (1967) 84.

A. Prins, in "Advances in food emulsions and foams", eds. E. Dickinson and G. Stainsby, (1988)91.

A. Prins, A.M.P. Jochems, H.K.A.I. van Kalsbeek, J.F.G. Boerboom, M.E. Wijnen and A. Williams, (1996), Prog. Colloid Polym. Sci, 100, 321-327

A.D. Ronteltap, and A. Prins, Colloids and Surfaces 47(1990) 269.

G. Serrien, G. Geeraerts, L.Ghosh and P. Joos, Colloids and Surfaces 68 (1992) 219.

R. Tuinier, C.G.J. Bisperink, C. van den Berg and A. Prins, J. Colloid and Interface Sci., 179 (1996) 324.

G.A. van Aken, in E. Dickinson and D. Lorient (Eds.), Food Macromolecules and Colloids, The Royal Society of Chemistry, (1995) 43.

F. Van Voorst Vader, Th. F. Erkens and M. van den Tempel, Trans. Faraday Soc., 60(1964)1170.

A. Vrij and J.T.G. Overbeek, J. American Chem. Soc. 12 (1968) 3074.

A. Williams and A. Prins, Colloids Surfaces A, 114 (1996) 267.

A. Williams, J.J.M. Janssen and A. Prins, submitted to Colloids Surfaces A

Proteins at Liquid Interfaces
D. Möbius and R. Miller (Editors)

MOBILITY OF ADSORBED PROTEIN MOLECULES AS STUDIED BY FLUORESCENCE RECOVERY AFTER PHOTOBLEACHING (FRAP)

David C. Clark and Pete J. Wilde

DMV International, NCB-laan 80, P.O. Box 13, 5460 BA Veghel, The Netherlands
Institute of Food Research, Norwich Laboratory, Colney, Norwich, NR4 7UA, UK

Contents

1 INTRODUCTION

It has long been recognised that the structure and properties of the adsorbed layers of surface active molecules that form at the junctions between different phases in dispersions such as foams or emulsions are critical factors in defining the stability of the dispersion. For many years fruitful research has been performed on macroscopic interfaces and a vast amount of data and knowledge has been derived (Graham & Phillips 1979). In recent years methods have been developed which allow the surface rheological properties of interfacial layers to be investigated using either shear or dilational approaches (Castle et al. 1981, Krägel et al. 1993, 1994, Kokelaar et al. 1991). Again considerable insight has been gained into the elastic and viscous properties of macroscopic interfaces but only following mechanical perturbation of the adsorbed layer. In many cases, the data obtained from such methods are contingent upon the attachment of the interfacial layer to mechanical device that is used to generate the perturbation, be it a rotating sharp edged bob in a surface shear rheometer or an oscillating barrier in a Langmuir trough modified for surface dilational rheology studies. A further limitation of such methods is confinement to the single macroscopic interface. When considering foam and emulsion stability to coalescence, the structure of most interest is that which separates the dispersed phase namely, the microscopic foam film or the oil-water-oil film in an oil-in-water emulsion. Only optical techniques offer the possibility of probing the properties of such small and delicate structures without mechanical disturbance.

One such method that allows investigation of interactions occurring at the molecular level, diffusion properties and hence indirect access to rheological properties in microscopic samples is fluorescence recovery after photobleaching (FRAP). This method, also referred to as fluorescence photobleaching recovery (FPR) or fluorescence microphotolysis, was originally developed in the mid seventies and was extensively applied to the investigation of diffusion of lipids and proteins in cell membranes (Axelrod et al. 1976, Peters et al. 1974). The methods relies upon the irreversible photobleaching or fading of fluorescence in a spatially inhomogeneous manner (i.e. the perturbation). The rate of spatial mixing of bleached and unbleached fluorophore can then be monitored photometrically and related to a diffusion coefficient or linear flow speed.

2 THE FRAP METHOD

Several variations of the FRAP technique exist (Axelrod 1985) but the simplest are based on the principles outlined in the schematic diagram shown in Fig. 1. The method requires that the molecular species of interest is fluorescent labelled or alternatively that an independent fluorescent probe molecule is partitioned into the environment of interest. In the case of thin films, the surface diffusion properties of a given protein in the adsorbed layer can be measured by forming a thin film (diameter 100-200mm), from a solution which contains the fluorescent-labelled protein or molecule of interest. An attenuated laser beam is used to illuminate a small spot (approximate diameter 5mm) on the surface of the thin film, eliciting fluorescence from labelled molecules contained within the spot, which is recorded (Fig. 1(a)).

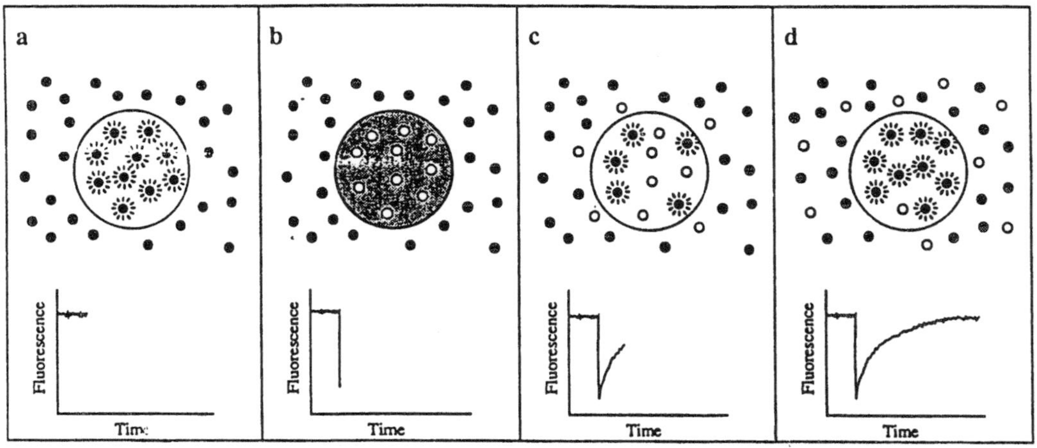

Fig. 1 A schematic representation of the various stages in a typical spot FRAP experiment. See text for explanation (Reproduced from Clark (1995) with the permission of Elsevier)

These fluorescent molecules are then irreversibly photodestroyed (bleached) by increasing the intensity of the laser beam approximately 1000x for a few milliseconds (Fig. 1(b)), before returning the laser intensity to its attenuated level. Fluorescence returns to the bleached spot only if the bleached molecules are free to diffuse laterally away from the spot to be replaced by non-bleached molecules in the surrounding film diffusing into the spot (Fig. 1(c) and(d)). Measurement of the time dependence of this process and knowledge of the dimensions of the bleached spot, allows calculation of the surface diffusion coefficient.

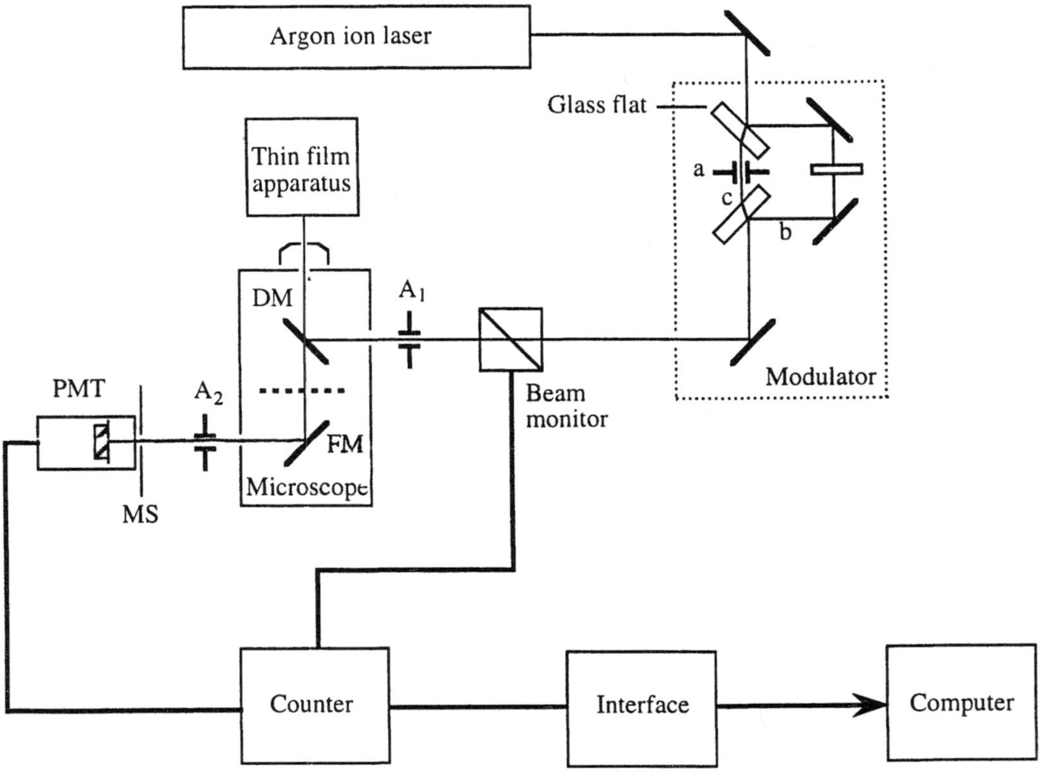

Fig. 2 A schematic diagram of a simple spot FRAP apparatus. The components are described in the text.

Several reviews and books have described the construction and use of various forms of FRAP apparatus and the reader is referred to these works for appropriate details (Wolf 1989; Clark 1995). Briefly, the 488 nm line of a high power (10W) argon ion laser (Coherent Innova 100-10) is attenuated by a device such as the optical modulator shown in Fig. 2. This device is comprised of a fast electronic shutter (a), which when closed allows only the monitoring laser beam (b), which has been attenuated by multiple reflection and a neutral density filter to illuminate the sample. When the shutter is open the intense bleaching beam (c) which is transmitted through two of the glass flats passes through to the sample. The crucial factor with this modulator arrangement is beam alignment to ensure that both attenuated and bleach beams are coincident at the sample.

The beam provided by the modulator passes through a beam monitor (beam splitter and photodiode), the signal from which is used to electronically compensate for minor fluctuations in the laser beam intensity. The beam is then launched through a pinhole aperture (A_1) located at the image plane, at the entrance port of the epi-illumination attachment of the fluorescence microscope. Our apparatus can be used with both upright (Nikon Optiphot) or inverted (Nikon Diaphot TMD) microscopes but the latter is most convenient for foam or emulsion thin film measurements. The filter block in the epi-illumination attachment is selected to match the laser line used for excitation and the emission peak of the fluorescent probe. For example, a 510nm dichroic mirror (DM), mounted at an angle of 45°, is suitable for reflection of the 488nm excitation beam necessary to excite fluorescein based fluorophores. The reflected beam is focussed through the extra long working distance objective (Nikon) lens (magnification x40 or x20) of the microscope onto the sample. The 510 nm dichroic mirror allows transmission of the emitted light from the above fluorescent labels. A 520nm long pass filter removes stray excitation light and prevents it reaching the photon counting photomultiplier tube (PMT; Thorn-EMI 9816B) positioned at the cine camera port of the inverted microscope. The PMT is protected during the bleaching pulse by an electronic gating circuit and a mechanical shutter (MS). Prior to entering the detector, the emitted light beam passes through a second aperture (A_2) again positioned at the image plane. The two apertures have equivalent diameters (eg 200 mm) and serve to make the apparatus confocal. System timing and control, data acquisition and data analysis are performed using a VME microcomputer system (Motorola 68020). The diameter of the focused laser spot on the sample was measured using a beam profile measuring device (BeamScan, Model 2180; Photon Inc.) and its intensity profile was Gaussian. FRAP data curves were analysed by a non-linear least squares fit to the expression (Axelrod 1985):

$$F(t) = \frac{F_0 + F_\infty (t/\beta\tau_D)}{1 + t/\beta\tau_D} \qquad (1)$$

where β is related to the proportion of bleached sample in the illuminated spot, P (ie. the prebleach fluorescence intensity - F_0 divided by the prebleach fluorescence intensity). In practice, the value of β is obtained from a lookup table in the analysis programme.

The lateral diffusion coefficient, D, is given by

$$D = \omega^2/4\tau_D \qquad (2)$$

where ω is the radius of the circular spot and τ_D is the characteristic diffusion time.

3 APPLICATIONS OF FRAP IN COLLOID CHEMISTRY

3.1 Early application of FRAP in colloid studies

The FRAP method was applied along classical surface chemistry lines by Peters' group in the early 1980s in measurements of translational diffusion of phospholipid spread on the surface of a modified Langmuir trough (Peters & Beck 1984, Beck & Peters 1985). These monomolecular films at the air-water interface were used as a model system for single leaflets of biological phospholipid membranes and allowed parameters such as packing density, surface pressure, surface viscosity and surface potential, ionic strength and temperature of the subphase and phospholipid composition to be varied and monitored in a controlled manner. The Langmuir trough was adapted to eliminate surface streaming effects. An important finding from this study was that translational diffusion decreased with decreasing mean molecular area and decreased sharply between surface pressures of 8-10 mN/m corresponding to the transition of the monolayer from a fluid to a crystalline state. In addition, evidence of phase separation was observed upon compression of the phospholipid (DPPC) monolayer doped with the fluorescent NBD-egg phosphatidyl ethanolamine in the range of surface pressures from 2.5 to 15 mN/m. At the low end of the range, homogeneous fluorescence was observed. At intermediate values of surface pressure dark spots appeared which were thought to be lipid in crystalline form. These grew in size with increasing surface pressure and fluorescence was only observed in the surrounding 'continuous' liquid phase. At the highest surface pressures the fluorescence became homogeneous throughout the film but the diffusion coefficient was significantly reduced by almost 100 fold.

Following these experiments there have been no reports of the FRAP technique being applied to investigate interfaces from a colloid stability perspective until the work of Clark's group was initiated in the 1990s (Clark et al 1990a, 1990b, Clark 1995, Wilde & Clark, 1993). In this work, FRAP measurements were applied to isolated foam films and films formed between two emulsion droplets. These structures are good models for well drained foams or creamed emulsions since the interlamellar phase is of similarly limited volume and the orientation of the surface active material is identical to that the real dispersion.

3.2 Preparation of foam films

Foam films are self-organised, ordered liquid films which can exhibit variability in their thicknesses, hydrodynamic, kinetic and other properties dependent upon the surfactant and solution properties (ionic strength, pH etc) from which they formed. In nature they can exist, for example, as the structures that separate oil droplets in creamed emulsions, concentrated particles in suspensions or the gas bubbles in foams. They can be easily prepared by several techniques *in vitro* and used as model systems to study the above phenomena.

We can separate foam films into three types, namely, thick equilibrium, common black and Newton black (Fig. 3) foam films. According to the theory of Derjaguin, Landau, Verwey and Overbeek (DLVO) (Derjaguin & Landau 1941, Verwey & Overbeek, 1948), thick equilibrium foam films are obtained at low ionic strength, where the electrostatic double layer repulsion between the monolayers is high and arrests drainage and film thinning. The distance between the monolayers in thick foam films is approximately 20-40 times larger than the length of the adsorbed molecules. The film thicknesses are controlled by the bulk electrolyte concentration and can be measured by an interferometric technique. Common black films are obtained from the thick foam films by the spontaneous appearance and growth of black spots in samples that contain intermediate electrolyte concentration (Fig. 3). Common black films have a characteristic thickness of less than 20 nm. The next equilibrium state of thin films is the Newton black film, which possess a higher degree of order and comprise a bilayer with almost no free solvent in between the monolayers. Hence such thin films are considerably thinner than common black films often with thicknesses in the range 7-9nm. Newton black films are formed under conditions of high concentrations of electrolyte. The properties of common and Newton black films composed of different amphiphiles including phospholipids have been widely investigated.

Although methods were available to prepare and investigate isolated air-suspended thin liquid films many years ago (Mysels et al. 1959), they have only been developed further comparatively recently. The most extensive studies have been performed on surfactant-stabilised films using molecules such as sodium dodecyl sulphate (Scheludko 1967). The foam film apparatus used in our laboratory was developed from the film holders used in earlier work by this Bulgarian group.

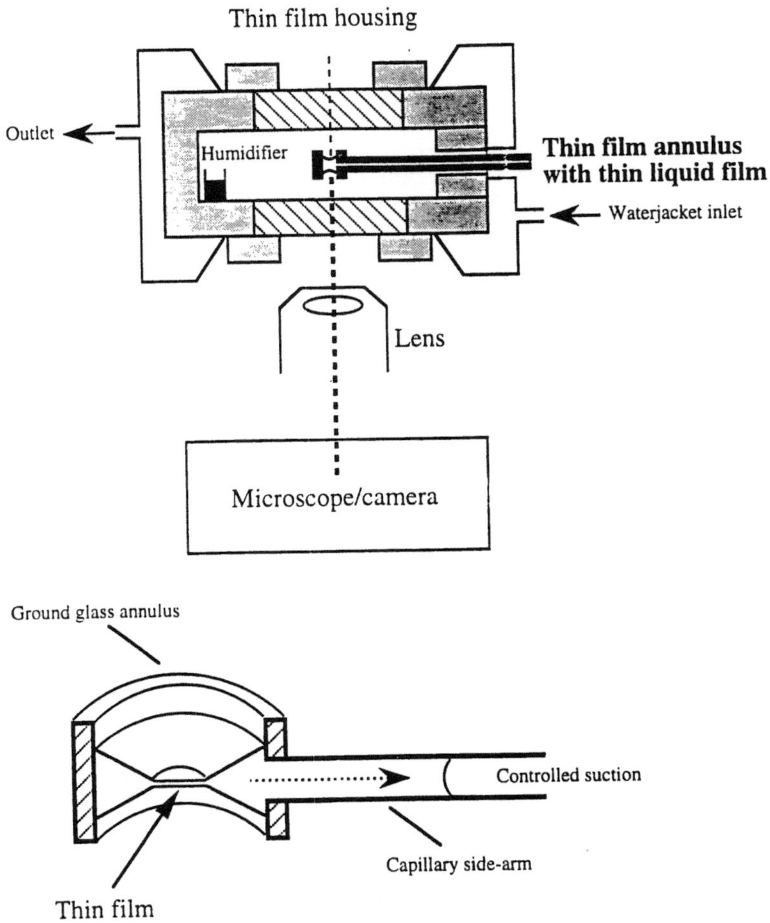

Fig. 3 A schematic cross section of the different foam film types with accompanying photographs of actual foam films.

Microscopic thin films (Clark et al. 1989, 1990a) have generally been formed by introduction of a droplet of solution into a ground glass supporting ring or annulus (Fig. 4). This device is crudely analogous to a miniaturised version of the loop children use to blow bubbles from soap solutions. However, rather than relying on gravity to drain off surplus bulk liquid as in the child's toy, film formation is initiated by withdrawal of liquid by applying controlled suction. This is achieved via a capillary sidearm that is connected to the film supporting ring. Liquid withdrawal is stopped once a thick horizontal planar film of appropriate diameter (eg. 0.3mm) has been formed. Drainage of the thick film proceeds from this point mainly as a result of suction from the adjoining wedge shaped region that surrounds the film. This region is referred to as the Plateau border.

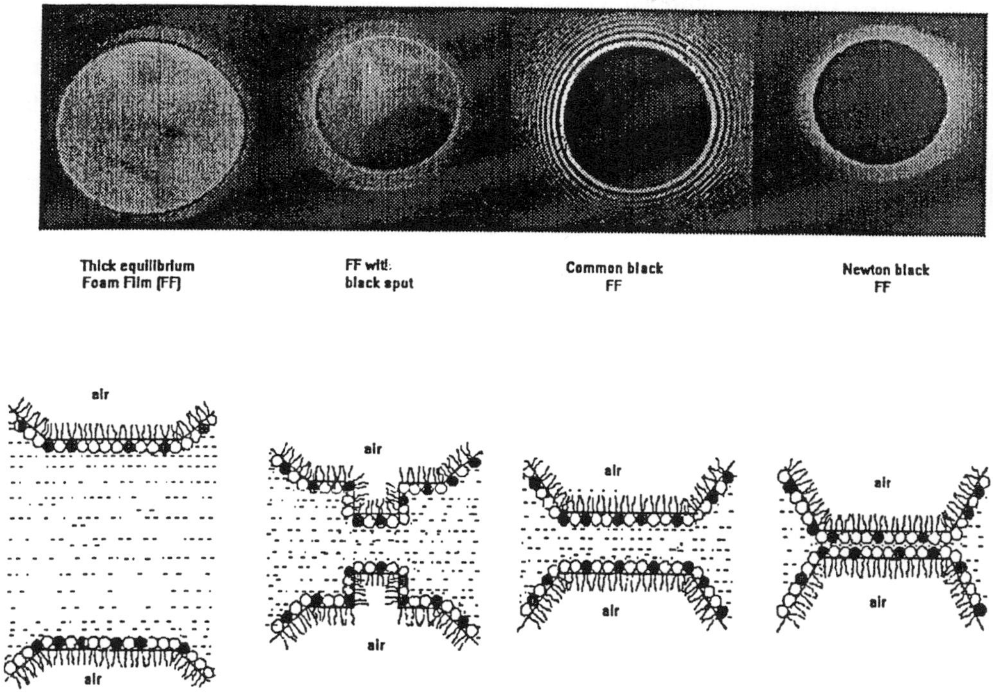

Fig. 4 An apparatus for the formation of foam films. (A) Foam film chamber with optical windows for film observation. (B) Cross section of a foam film ring comprised of a roughened glass annulus.

Oil-water-oil thin films are representative models of close packed oil in water emulsions and can also be prepared by flooding a specially designed vessel equipped with an optical flat as a base plate with the aqueous solution of interest. One oil droplet can be anchored to the base of the vessel by making a small patch on the inner surface of the optical flat hydrophobic by silanation. A second oil droplet can then be suspended from a hypodermic syringe and forced down onto the upper surface of the captive droplet by micromanipulators. Details of such experimental arrangements have been published previously (Clark et al. 1992, Wilde & Clark 1993).

Microscopic thin films are relatively fragile structures and are highly sensitive to changes in temperature, mechanical disturbance and evaporation. We have designed a dedicated chamber that allows the necessary control of the environment surrounding the film whilst not impeding observation of film drainage and measurement of equilibrium thickness or surface diffusion in the adsorbed layer.

3.3 FRAP studies on phospholipid stabilised foam films

Surface diffusion studies on low molecular weight surface active molecules have been carried forward in recent years (Clark et al. 1990a) with particular emphasis on studies of phospholipid surface diffusion (Lalchev et al, 1994a, 1994b, 1995). Lateral diffusion of lipids adsorbed at interfaces is of physiological and technological importance, for example in cell membrane function and food foam and emulsion stability. The recent work has been carried out almost exclusively using phospholipid stabilised black foam films (BFFs). This allows the findings to be contrasted with those performed on phospholipid samples spread on a Langmuir trough (Peters and Beck, 1983; Beck and Peters, 1985). Far more data exists in the literature reporting lateral diffusion coefficients of lipid in bilayer structures suspended in water, such as phospholipid bilayer vesicles (Sackman & Traube, 1972), liquid crystal multilayers (Lee et al., 1973), black lipid membranes (BLMs) (Fahey et al., 1977; Poo & Cone, 1974; Ladha et al., 1997) and the membranes of living cells.

FRAP measurements of surface lateral diffusion in phospholipid (lecithin) BFFs were first reported by Lalchev et al. (1991; 1994a). Measurements of the properties of phospholipid BFFs are of particular interest as they provide an alternative model system for biological membranes along with phospholipid monolayers at the air/water interface, black lipid membranes (BLMs), single unilamellar vesicles (SUVs) and multilamellar vesicles (MLVs). A phospholipid foam film can be obtained as a stable bilayer (lamella), consisting of two phospholipid monolayers at the upper and lower air/water interfaces enclosing a thin liquid core (interlamellar liquid). Alternatively, they can exist in a multilayer form, consisting of several lamellae parallel to the primary adsorbed layer, extending into the liquid core. The latter structures are referred to as stratified foam films. Previous investigations of the properties of BFFs composed of pure phospholipids (Exerowa et al, 1984; Cohen et al., 1992), or of lipid-protein mixtures (Exerowa et al., 1986; Lalchev et al.1992)) showed that they can be used as model systems for the study of liposome-liposome interactions, BLMs, MLVs, inter membrane interactions (Exerowa and Lalchev 1986), cell fusion (Naydenova et al, 1990), lung physiology and coalescence phenomena in food emulsions and foams (Clark et al., 1989; 1991; Wilde and Clark, 1993; Wilde et al., 1993). A unique feature of BFFs is that the molecular orientation in the bilayer structure is precisely the reverse of that in BLMs (Fig. 5) and also cell membranes. However, BFFs exhibit several properties similar to BLMs including the thickness, refractive index, hydrodynamic behaviour, electrical properties, and stability. BFFs are spontaneously formed (in vitro and in vivo), are very stable (usually more so than BLMs) and convenient for systematic investigations.

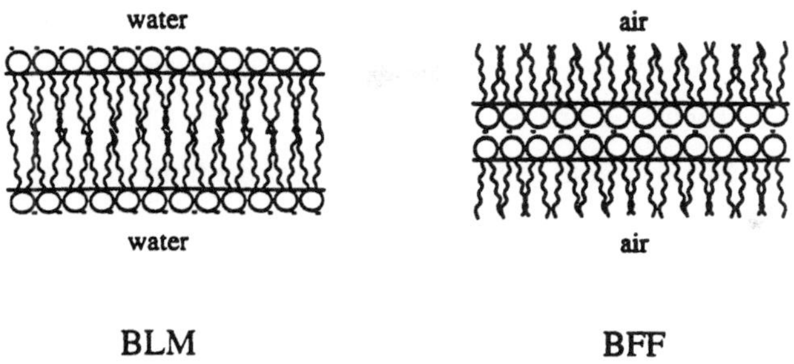

Fig. 5 A schematic representation of the molecular orientation of phospholipid molecules in black lipid membranes (BLMs) and black foam films (BFFs).

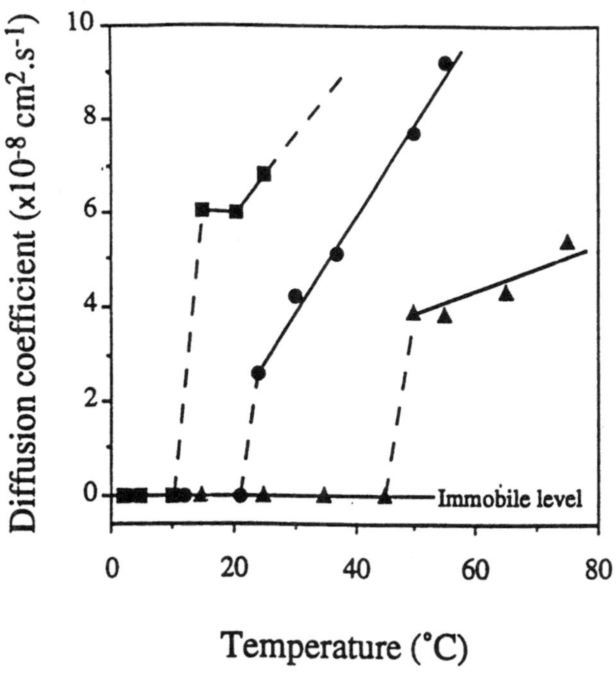

Fig. 6 The temperature dependence of the diffusion coefficient, D of surface adsorbed 5-N-(octadecanoyl) aminofluorescein in BFFs stabilised by DMPG (■), DMPC (●) and DMPE (▲). C_{lipid}=13 mg/ml ;pH=6.7 ; C_{el}=140 mM NaCl. (Reproduced from Lalchev et al. (1994) with the permission of Academic Press)

Thick equilibrium, common black and Newton black foam films can be prepared from lipid dispersions containing 5mM, 140 mM and 1.0 M NaCl respectively. Electrolyte concentrations in these ranges (Kolarov et al. 1968, Huisman & Mysels 1969, Kolarov et al. 1989) conform with the concept of a critical electrolyte concentration for the common-Newton black film transition (C_{el} =0.33 M NaCl) for foam films stabilised by detergent-like surfactants. Typically, a decrease in the film thickness of about 3-4 nm is seen between thin films formed from solutions containing 140 mM NaCl and 1.0 M NaCl, i.e. corresponding to the films of common black (12-15nm thick) and Newton black (7-9nm thick) type respectively (Lalchev et al, 1995). Data have been reported by Lalchev et al. (1994) concerning the effect of the hydrocarbon chain length, fatty acid unsaturation, polar head charge and the temperature dependence of the lateral diffusion coefficient of phospholipid molecules in BFFs. In these studies surface diffusion was monitored by FRAP in BFFs doped with the fluorescent lipid analogue, 5-N-(octadecanoyl) aminofluorescein, as a reporter group. Measurements were performed across the temperature range of 8-85°C, in order to measure the lipid probe mobility below and above the gel-to-liquid crystalline phase transition of the phospholipid.

Three phospholipid types, L-α-phosphatidyl-DL-glycerol dimyristoyl (DMPG), L-α-phosphatidylcholine dimyristoyl DMPC, and L-α-phosphatidyl ethanolamine dimyristoyl DMPE) were investigated to estimate the role of the different polar headgroup. The temperature dependence of the diffusion coefficient is shown in Fig. 6.

Each lipid examined showed very limited mobility (immobile level) below a certain critical temperature. At this level, the molecular diffusion was due to flow rather than to diffusion and may be compared with the convective flow problems experienced by Beck and Peters (1985) in their trough studies. Such flow behaviour could be easily identified from the shape of the FRAP data curve and was signified by the absence of a change in slope of the recovery curve. Under such conditions, the computed fit of equation (1) to the FRAP recovery curves was poor, consistent with pure flow or a mixture of flow and diffusion in the presence of a partially (or fully) immobile fraction. Alternatively, curves did not recover to the prebleach level and often had the appearance of a step function indicating complete immobility over the timescale of the experiment.

The dashed lines in Fig. 6 signify the onset of surface diffusion in the adsorbed layer by connecting the immobile level with the first measured diffusion coefficient that returned an acceptable computed fit to the data. The temperature at which this occurred was dependent on the phospholipid type and followed the order DMPG (15°C), DMPC (24°C) and DMPE (50°C). The first measurable diffusion in the DMPC and DMPE thin films occurred at temperatures near to where the transition to the liquid crystalline $L\alpha$ phase is completed. However, this was not the case for DMPG, where diffusion ($D=6x10^{-8}$ cm^2/s) was detected at temperatures (15 and 20°C) where gel $P_{b'}$ and/or $L_{b'}$ phase may exist. Measurements of the diffusion coefficient of ODAF in DMPG-stabilised films was not possible above 25°C as the fluorophore, ODAF was excluded from the BFFs above this temperature (dashed line above 25°C for DMPG curve in Fig. 6). This behaviour was only observed with this specific phospholipid in this study.

The surface diffusion behaviour of other phospholipids including DOPE (L-α-Phosphatidyl Ethanolamine Dioleoyl), DPPA (DL-α-Phosphatidic Acid Dipalmitoyl), DLPE (L-α-Phosphatidyl Ethanolamine Dilauroyl), DMPE (L-α-Phosphatidyl Ethanolamine Dimyristoyl) and DPPE (DL-α-Phosphatidyl Ethanolamine Dipalmitoyl) and various phospholipid mixtures was also measured in BFFs. The temperature of onset of diffusion of these lipids in BFFs is compared with key phase transition temperatures in bulk aqueous suspensions in Table 1.

Table 1 Comparison of the onset of surface diffusion in BFFs with bulk phase transition temperatures

Phospholipid	Onset of surface diffusion in BFF (°C)	Phase transition temperature in bulk (°C)	Phase transition
DMPG	15	19.1, 25.6	$L'_b(L_{sg}) \rightarrow P'_b \rightarrow L_\alpha$
DOPE	22	8	$L_\alpha \rightarrow H_{II}$
DMPC	24	15.5, 24.2	gel state→L_α
DPPA	37	64	gel state→L_α
DLPE	43	43	gel state→L_α
DMPE	50	56	gel state→L_α
DPPE	50	66.4	gel state→L_α
DMPC/DMPG (1:9)	20		
DLPE/DPPA(8.5:1.5)	35		

Examination of Table 1 shows that there is some correlation between the temperature of onset of surface diffusion of the phospholipid in the BFF and phase transition in aqueous solution to $L\alpha$ liquid crystalline phase of the lipid. Certainly, reasonable to good correlation is seen with DMPC,

DLPE and DMPE. However, poor correlation is seen with DMPG, DPPA, DOPE and DPPE. This may be explained in the following way. It has been suggested that diffusion in DMPG and DPPA at temperatures where the gel state may be expected to be present may relate to the negative charge o the headgroups of these lipids. Repulsion effects of negatively charged head groups may result in a reduction in the surface concentration of lipid in the adsorbed layer. This reduction in surface molecular density per unit area of the BFF monolayers could account for diffusion of these negatively charged lipids in the gel state. In the case of DOPE, this lipid undergoes a direct transition from $L\alpha$ to the inverted hexagonal phase at 8°C, thus the phase state of this monounsaturated phospholipid is rather different to the other lipids studied in this investigation. Finally, DPPE BFFs only formed when the lipid had been fully hydrated by heating to 70°C, followed by cooling to the desired temperature of measurement. It is possible that the DPPE did not have sufficient time to relax back to the gel state prior to measurement.

The influence of the fatty acid chain length on D as a function of temperature was systematically investigated independently from head group with the saturated phospholipids, DLPE, DMPE and DPPE. Surface diffusion was first measurable (i.e. the transition from the immobile level to diffusion) at temperatures >45°C. The initial value of the diffusion coefficient followed the order DLPE>DMPE>DPPE at any given temperature and increased with similar slope with increasing temperature. These phospholipids are expected to be in L_{α} lamellar liquid crystalline phase in the temperature range studied (55-85°) (Axelrod, 1985). Therefore, the results obtained show that in this phase state, the diffusion coefficient increases with decreasing saturated chain length of phospholipids. A mixed effect is observed with the unsaturated phospholipid, DOPE (C18:1). On the one hand, the increased chain length results in a decrease in D to below that observed for DLPE but above that for DMPE and DPPE. Logically, this is due to the unsaturated bond in the C18 chain. On the other hand, the chain unsaturation extends the measurable diffusion down to 22.7°C compared to approximately 45°C for the saturated phospholipids. It is worth noting that above 8°C, DOPE dispersed in water exists in a $H_{\|}$ inverted hexagonal phase. However at 15°C, the shape of the FRAP recovery curve for DOPE was typical for those with immobile fraction and flow with only minor contributions from true diffusion. The conclusion from these experiments is that the main determinant of D is chain unsaturation, which in turn determines the lipid phase state at a given temperature. Fatty acid chain length also influences the magnitude of D but to a lesser extent. Provided the phospholipid phase state does not change across a given temperature interval, the chain length controls the magnitude of D.

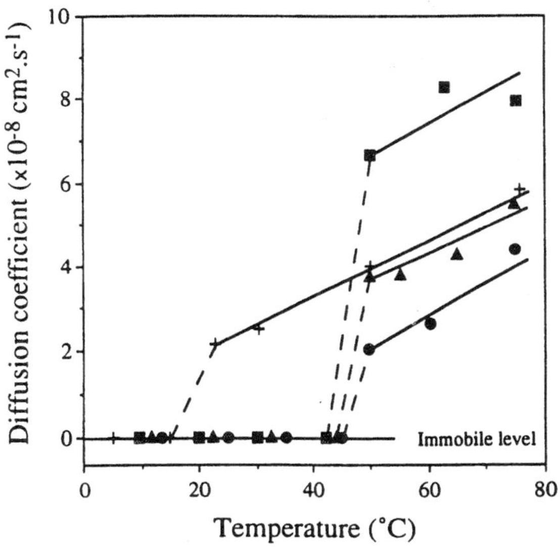

Fig. 7 The temperature and chain length dependence of the diffusion coefficient, D of surface adsorbed 5-N-(octadecanoyl) aminofluorescein in BFFs stabilised by DLPE (■), DMPE (▲), DPPE (●) and DOPE (+). C_{lipid}=13 mg/ml ;pH=6.7 ; C_{el}=140 mM NaCl. (Reproduced from Lalchev et al. (1994) with the permission of Academic Press)

Within the limits of its phase state, the lateral diffusion of a given phospholipid in BFFs depends predominantly on its molecular characteristics (eg. chain length and saturation, polar head charge and size. It has proved possible to rank the relative importance of some molecular parameters on the diffusion coefficients using mixtures of defined phospholipids. For instance, a very strong effect of charge in the mixture DMPC/DMPG (Table 1) in the temperature range where DMPG is in both gel and lamellar liquid crystalline phase states. It is notable that in the temperature windows where phase transformations take place, the scatter in the measured D values is higher, consistent with coexistence of two phases in the adsorbed layer similar to that reported by Beck and Peters (1985). In the case of the mixed DLPE/DPPA system, the chain length and the nature of polar head rather than the charge appeared to be the main determinant of D. It is possible that hydration state of the polar head and potential for hydrogen bonding (especially in case of PE), could also be important factors that may influence D.

In addition, comparisons can also be made with lipid diffusion in black lipid membranes (BLMs), where the molecular orientation is the exact mirror image of that of phospholipid in a

BFF (Fig. 5). It is notable that reported values for D of phospholipids in BLMs are approximately one order of magnitude greater (i.e. faster) than that of the same lipid in a BFF (Fahey and Webb, 1978; Ladha et al., 1997). The most obvious conclusion to be drawn from this observation is that repulsive electrostatic forces between the adsorbed lipid at the interfaces of the BFF have a much more dominant effect on D than the interactions between acyl chains. Thus the orientation of the hydrocarbon chains in a BLM results in the formation of an electrically insulating layer which effectively eliminates cross membrane leaflet, headgroup-headgroup interactions. In contrast, repulsive forces between lipid headgroups associated with opposing interfaces of a BFF can operate through the electrically conductive interlamellar liquid of a BFF and have a major effect on D.

The range over which inter-interface headgroup interactions was investigated in foam films covering a range of thickness controlled by ionic strength. The diffusion coefficient of ODAF in approximately 100nm thick equilibrium films is approximately five-fold greater than in the black films due to the absence of short-range interactions normal to the interfaces in the thick films. The diffusion coefficients of ODAF in common black films are higher than in Newton films composed of the same lipid(s) due to the absence of an interlamellar solvent layer between the monolayers and the specific properties of the latter films.

It is interesting to relate the above studies to the drainage and stability of phospholipid-stabilised foams. This was recently done in experiments in which a microconductivity method (Wilde, 1996) was used to study the foams under conditions where three different foam film types could be formed - thick foam films, common black foam films and Newton black foam films (Lalchev et al., 1997). The foaming properties of DMPG were investigated at 20 and 28°C where this lipid is in the gel and liquid crystalline states, respectively. Higher conductivity signals were observed at the higher temperature where DMPG is in the liquid crystalline state. This is indicative of wetter and/or more stable foams under these conditions. This effect was observed independent of foam film type. However, for a given phase state, the type of foam films formed significantly influence the stability and rate of drainage of the foam. Indeed, the water content of the films is ranked in the order TFF > CBF > NBF. Typical drainage curves for foams formed under CBF and NBF conditions are shown in Fig. 8. Further characterisation of foam film characteristics showed a decrease in film thickness and an increase in film lifetime and surface molecular diffusion coefficient (D) in the adsorbed layer for CBF and NBF at the higher temperature (i.e. in the liquid crystalline state). It is likely that the fluidity of the interfacial layer is the most important factor contributing to foam stabilisation.

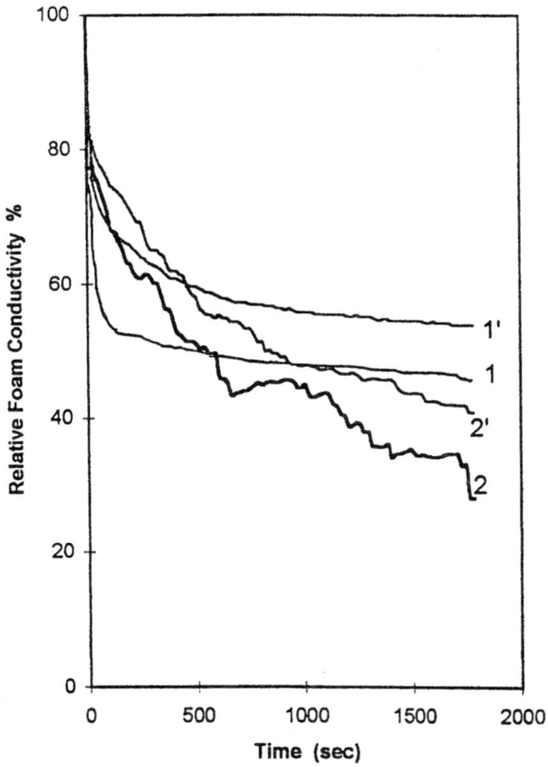

Fig 8 The relative foam conductivity as a function of time for CBF- and NBF-foams stabilised by DMPG in gel and liquid-crystalline phase state. Curve 1- CBF-foam at T=20°C ; curve 1'- CBF at T=28°C ; curve 2 - NBF-foam at T=20°C; curve 2' - NBF at T=28°C.

Finally, significant hysteresis in the diffusion coefficient was observed during temperature cycling and the effect is larger in common black films than in Newton films (Lalchev et al 1995). It is possible that a change in the surface composition occurs during temperature cycling in mixed films which alters the character (and area) of the hysteresis curve and is likely determined by the kinetics of both molecular exchange (adsorption-desorption) and phase transition processes. The phase transition and hydrogen bonding capacity of the lipid has a strong influence on the lateral diffusion of ODAF in the monolayers of the films during temperature cycling.

3.4 FRAP studies on foam films stabilised by protein alone

The surface diffusion properties of low molecular weight surfactants such as SDS or the phospholipids as described above can be contrasted with that of proteins in foam films. Proteins being larger than surfactants can be readily derivatised with fluorescent moieties such as fluorescein through covalent bonding to reactive amino acid side chain groups, such as the primary amine of lysine. Solution diffusion studies by the FRAP method have shown that fluorescein isothiocyanate labelled BSA (FITC-BSA)(Clark et al 1990b) diffused freely with a diffusion coefficient of approximately $3x10^{-7}cm^2/s$. This was in reasonable agreement with previously published values (Barisas & Leuther 1979). FRAP measurements have also been performed on foam films stabilised by FITC-BSA. The films were allowed to drain to equilibrium thickness before measurements were initiated. Thin films covering a range of different thicknesses were studied by careful adjustment of solution conditions. FITC-BSA stabilised films that had thicknesses up to 40nm showed no evidence of surface diffusion as there was no return of fluorescence after the bleach pulse in the recovery part of the FRAP curve (Fig. 9 (c)). In contrast, experiments performed with thin films that were >80 nm thick showed partial recovery (55%) of the prebleach level of fluorescence (Fig. 9 (b)). This suggested the presence of two classes of protein in the film; one fraction in an environment where it was unable to diffuse laterally, as seen with the films of thicknesses <45 nm, and a second fraction that was able to diffuse with a calculated diffusion coefficient of $1x10^{-7}cm^2/s$. This latter diffusion coefficient was 3 times slower than that observed for FITC-BSA in solution.

Care needs to be exercised in the interpretation of these data. Firstly, the slow drainage of the protein films especially once the perimeter of the film reaches black thicknesses suggests that these films contain very little interlamellar liquid. Therefore it is reasonable to assume that the vast majority of the fluorescence signal from the <45 nm thick films originates from protein in the adsorbed layer. The complete immobility of the fluorescent-labelled protein in these structures over the timescale of our experiments suggests that diffusion in the interlamellar liquid region is very restricted or highly compartmentalised. Indeed, it is possible that protein molecules bridge between the two interfaces (Velev et al 1993).

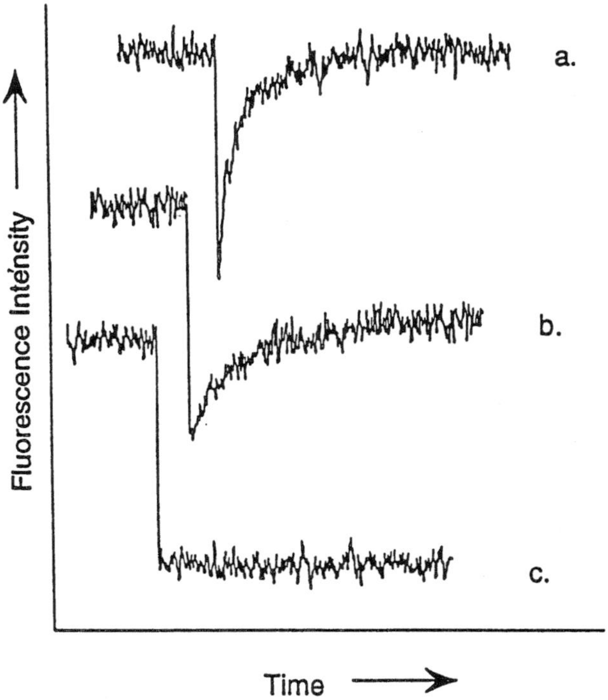

Fig. 9 Typical FRAP data curves obtained with (a) 2mM SDS in 2 mM sodium phosphate buffer, pH 7.0 containing 0.1M NaCl and 14μM ODAF (5—N-(octadecanoyl)-aminofluorescein); (b) FITC-BSA (0.5 mg/ml) in distilled water, pH 8.0 at an equilibrium foam film thickness of 83 nm; (c) FITC-BSA (0.2 mg/ml) in 50 mM Na acetate buffer, pH 5.4 at an equilibrium foam film thickness of 14 nm. (Reproduced from Clark (1995) with the permission of Elsevier)

The partial recovery observed in films >80nm thick (Fig. 9 (b)), is consistent with abolition or a significant reduction in the impediments to diffusion in such films. However, the diffusion coefficient is significantly lower than that observed in aqueous solution. Calculations predict a significant enhancement (several orders of magnitude of concentration) of protein in the adsorbed layer compared to the interlamellar solution. Therefore, it is necessary to define a mechanism that can account for an increase in protein concentration in the interlamellar space to explain the observed 55% recovery, whilst impeding protein diffusion compared to bulk solution. One hypothesis involves a low affinity interaction and exchange of protein adsorbed in the secondary layers with that in the interlamellar space. This would be consistent with a previous FRAP result of mobile and immobile fractions of BSA bound at a macroscopic quartz-water interface

(Burghardt & Axelrod 1981). In this study, partial recovery was attributed to adsorption/desorption processes in the adsorbed multilayers.

3.5 FRAP studies on mixed protein and emulsifier stabilised thin films

As discussed in the previous sections, there are fundamental differences between proteins and surfactants regarding their interfacial mobility (Clark et al 1990a, 1990b). It is these clear differences which allows quantification of the surface mobility by FRAP, in order to determine which molecular species is dominating and stabilising the interface of protein/surfactant mixtures (Coke et al, 1990, Clark et al, 1991a, 1994; Wilde & Clark, 1993; Wilson et al, 1993; Mackie et al, 1993, 1996; Wilde et al 1993; Sarker et al, 1995a; Cornec et al, 1996). Probing the interfacial occupancy by tensiometry and related techniques is often successful with simple non-interacting mixtures of surfactant or lipids. However, with mixtures of proteins and surfactants, competitive adsorption is not a simple case of tensioactive competition. Proteins have many complex levels of interactions, between protein molecules at the interface, and also with surfactants, both at the surface and in the bulk (Coke et al, 1990; Clark et al, 1992; Wilde et al 1993). This leads to major complications in the adsorption process, and often the relative occupancy of the proteins and surfactants is a dynamic process depending on the relative concentrations, surface activity and levels of interaction (Clark et al, 1995; Krägel et al, 1995). Measurement of the interfacial fluidity by FRAP quantifies the dominant species at the interface and the subsequent stability mechanisms. Transitions from one state to another, (i.e. an interface populated with immobile, interacting surface adsorbed protein, to one where the adsorbed species are freely diffusing and capable of responding to surface tension gradients) can occur over a small concentration range, or time period, with only small changes in the surface tension and other interfacial characteristics. The following examples show how measurements of the interfacial molecular dynamics can relate to the stability of dispersions containing protein/surfactant mixtures.

As stated before, there are two main interfacial adsorbed layer structures; in the case of proteins, a viscoelastic, gel-like film (sometimes referred to as a skin) formed by most proteins and a fluid adsorbed layer containing molecules capable of diffusion and able to stabilise interfacial layers by migration against surface tension gradients as described by the Marangoni mechanism. Mixtures of proteins and surfactants will lead not only to competition between the molecular species, but also between the stability mechanisms. This can have disastrous effects (Coke et al, 1990; Clark et al, 1991a; Wilson et al, 1993; Cornec et al, 1996). Fig. 10 shows the stability of a foam stabilised by BSA in the presence of increasing amounts of the surfactant Tween 20 (polyoxyethylene sorbitan monolaurate). In the absence of Tween 20 , BSA formed a very stable, slow draining foam. Increasing the concentration of Tween 20 in the initial solution reduced the foam stability, but at still higher concentrations of Tween 20, once again stable foams were formed. This return of stability is due to the displacement of BSA at the interface by Tween 20. Above 200mM Tween 20, the foam stability was the same both in the presence and absence of BSA. Note that at pH 4.2, low levels of Tween 20 had an immediate negative effect on the stability of the BSA foams.

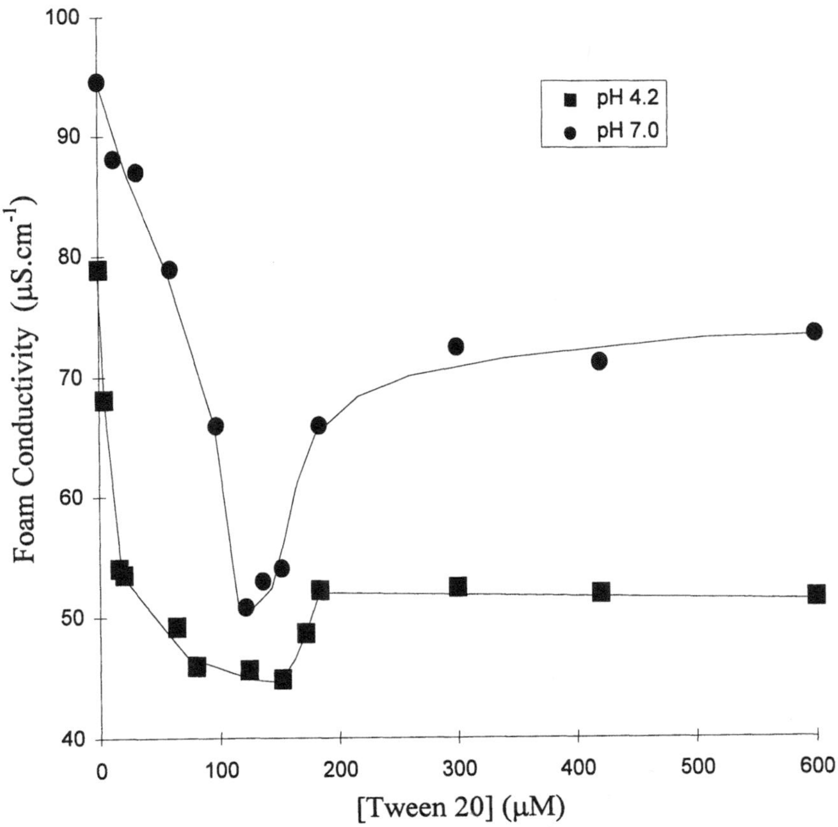

Fig. 10 Foam stability, expressed as foam conductivity after 5 minutes drainage, for foams stabilised by 4.08 mg/ml BSA as a function of Tween 20 concentration. In 10mM citrate buffer pH4.2 (■): and pH 7.0 (●).

The surface lateral diffusion coefficient (D) of the fluorophore, DiO-NBD (4-(N,N-dioctyl)amino-7-notrobenz-2-oxa-1,3-diazole in foam films formed from solutions of BSA and Tween 20 at a range of molar ratios is shown in Fig. 11. At pH 4.2, a transition between the immobile interfacial film (D=0) of adsorbed BSA to a mobile, rapidly diffusing interface containing significant adsorbed Tween 20 occurred around 40 mM Tween 20. In contrast, the transition at pH 7, occurred at higher Tween 20 concentrations. This corresponds well with the observed foaming behaviour (Fig. 10). Also, the values of D are lower at pH7, suggesting that a significant amount of BSA was still present at the surface, which again is reflected in the lower rate of decline of the foam stability at pH 7 (Fig. 10), suggesting that the adsorbed BSA molecules resisted displacement at the higher pH.

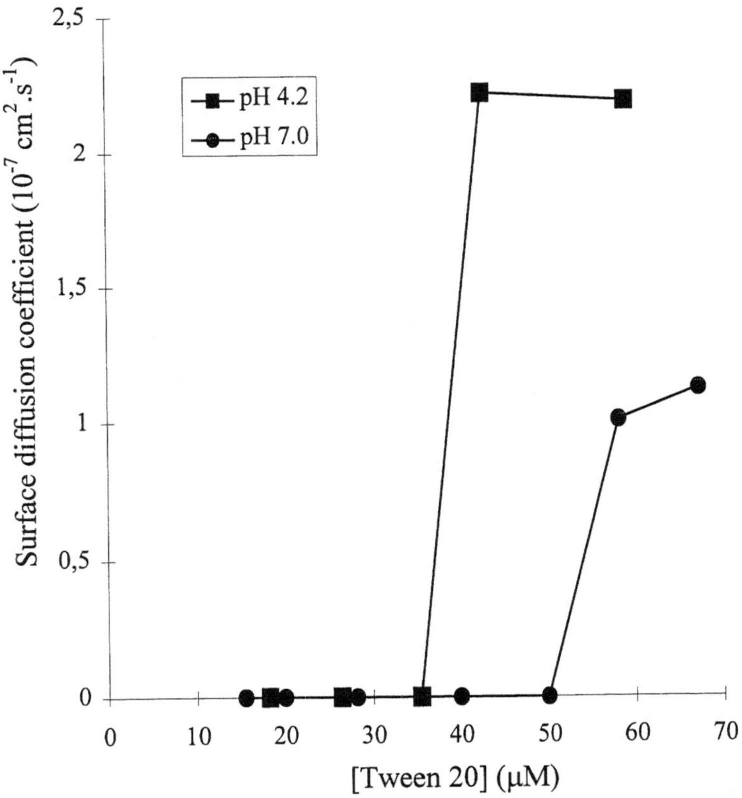

Fig. 11 Surface lateral diffusion coefficient of thin films stabilised by 8.16 mg.ml^{-1} BSA as a function of added Tween 20 concentration. In 10mM citrate buffer pH4.2 (■): and pH 7.0 (●).

The foam drainage properties as a function of time for BSA in the presence and absence of 90mM Tween 20 are shown in Fig. 13. The foam conductivity is a measure of the amount of liquid in the foam. It clearly demonstrates that BSA forms a very stable, slow draining foam, by virtue of its interfacial viscoelastic properties as defined by a surface diffusion coefficient of zero. The non-mobile adsorbed protein layer retards the flow of liquid from the foam, and allows the foam lamellae to be sterically stabilised and prevent coalescence. On the other hand, in the presence of Tween 20, the molecules that comprise the adsorbed surface layer freely diffuse and hence the foam displays extensive drainage.

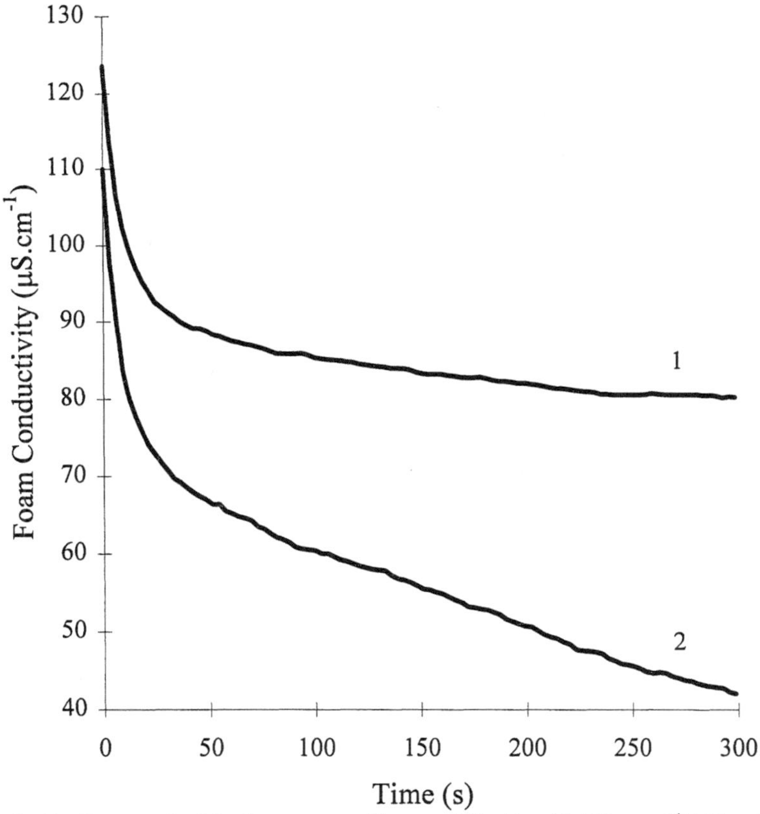

Fig. 12 Foam conductivity decay curves of foams stabilised by: (1) 4.08 mg.ml^{-1} BSA and (2) 4.08 mg.ml^{-1} BSA + 90mM Tween 20 in 10 mM citrate buffer pH 4.2.

The mobility of the molecules in the adsorbed layer promotes foam drainage. Of more critical importance, independent observations have shown that breakdown or inhibition of formation of the interactions between adsorbed protein molecules in the surface layer results in only limited protein displacement from the interface (Clark et al.1994). It is evident that the non-interacting diffusing protein molecules that remain in the adsorbed layer are significantly less efficient at stabilisation of the foam lamellae by the Marangoni mechanism, so coalescence events within the foam are also more likely. This clearly demonstrates the effect of the mobilisation proteins at interfaces by competitive adsorption of low molecular weight surfactant and the effect on foam stability.

Whilst the above findings are representative of the majority of findings with mixed surfactant/protein systems, the link between surface dynamics and foam stability is not always as easy to explain. In another case, a basic lipid-binding protein from wheat - Puroindoline, was studied in the presence of a water-soluble lipid analogue, Lyso-palmitoyl phosphatidylcholine (LPPC) (Wilde et al, 1993). The foam stability of this mixture is shown in Fig. 13. Unusually for protein:surfactant mixtures, the foam stability increased as the lipid was added. The stability eventually decreased back to a level similar to that of the LPPC alone. As the LPPC was added to the puroindoline, the drainage characteristics of the thin films changed, and the surface mobility, also shown in Fig. 13, becomes apparent at an LPPC concentration of around 20mM. At this concentration, the foam stability ceases to rise, and then decreases. The increase in foam stability as small amounts of LPPC were added, was attributed to the break up of puroindoline aggregates observed in the surface of thin films, and the altered charge properties of the protein:lipid complex. However, as seen in the previous example, an indicator of destabilisation of the foam appears to be the onset of surface diffusion of adsorbed protein. Once the protein molecules are mobilised, they cease to become an efficient foam stabiliser, and subsequently, the foam stability is reduced.

Measurements of stability and diffusion phenomenon at the oil water interface are technically more difficult, and so significantly fewer data are available (Mackie et al, 1996; Cornec et al, 1996; Mackie et al, 1993; Wilde et al, 1993). However, very similar effects have been observed. This can be illustrated by reference to a model system which has a great deal of exposure in recent years, that of β-lactoglobulin (BLG) and Tween 20 (Wilde et al, 1993; Cornec et al, 1996). Fig. 14 shows the relative foam stability (conductivity signal) and emulsion stability (droplet coalescence at a planar interface). The stability of the emulsion system appeared to be affected at much lower molar ratios of Tween 20 than the foam system. This is perhaps surprising, as only the hydrophobic phase has changed, and both protein and surfactant are solubilised in the aqueous phase.

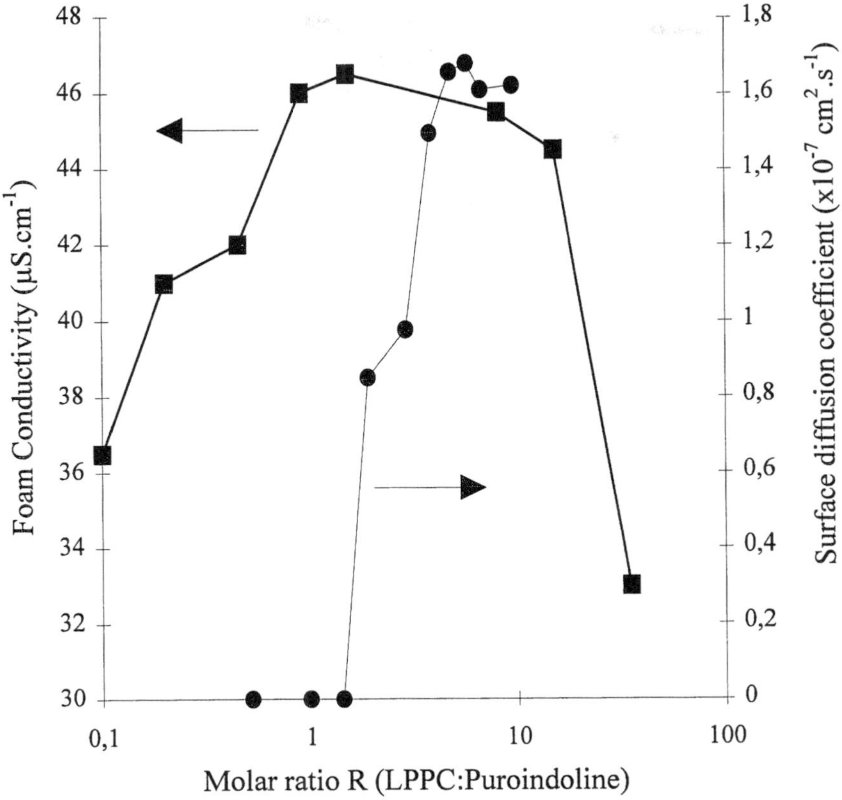

Fig. 13 Foam stability, expressed as foam conductivity after 5 minutes drainage (■) and surface lateral diffusion coefficient (●) of foams/thin films stabilised by 0.1 $mg.ml^{-1}$ puroindoline with increasing concentrations of lyso-palmitoyl phosphatidylcholine (LPPC). In 10 mM phosphate buffer pH 7.0.

However, measurements of the interfacial diffusion coefficient by FRAP (Fig. 15) show that there is clearly a difference between the two systems. Surface diffusion of BLG at the oil-water interface occurred at almost an order of magnitude lower concentration than at the air-water interface. One possibility is that the hydrocarbon chain of the Tween 20 molecule has a much greater affinity for the oil-water interface than the air-water, whereas for the protein, perhaps the opposite is true. Alternatively, the observation could be explained by the difference in penetration of the protein into the air and oil phases. If hydrophobic loops of the adsorbed protein extend further into the air phase than the oil phase, one could speculate that interactions between neighbouring hydrophobic protein loops would be more resistant to solubilisation by the low molecular weight surfactant resulting in higher concentrations of surfactant being required to induce diffusion in protein layers at the air-water interface. Whatever the reason for the observed difference between the levels of Tween 20 required to induce protein diffusion at the two interfaces, the relative affinities of the protein and surfactant for the interface effectively alter when the dispersed phase is changed. The result is that the onset of surface diffusion of BLG and destabilisation of the dispersion occurs at lower concentrations of Tween 20 at the oil-water interface than at the air-water interface.

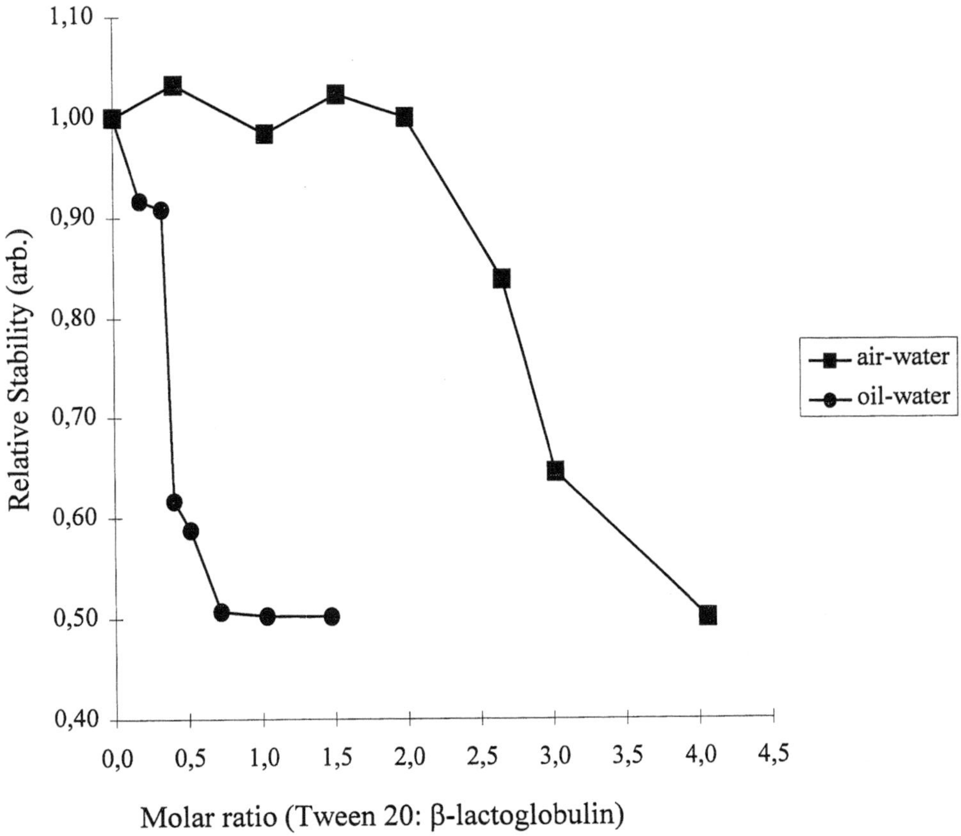

Fig. 14 Relative stability of foams (■), and emulsion droplets (●), stabilised by β-lactoglobulin in the presence of increasing concentrations of Tween 20, expressed as molar ratio of Tween 20 : β-lactoglobulin.

Measurements which can only be performed at the oil-water interface are those which involve the competitive adsorption between proteins and oil soluble emulsifiers. Fig. 16 shows the stability to coalescence at the oil-water interface of solutions containing BLG with increasing concentrations of the oil soluble emulsifier Span 80 (sorbitan monooleate). Reduction in the stability to coalescence of the emulsion began to appear in emulsions containing a molar ratio (R) of surfactant to protein of 2. It is interesting to note that the molar ratio of the water soluble surfactant Tween 20 required to cause the same destabilisation is around 0.5 (Fig. 14). This demonstrates that the destabilisation events provoked by competitive adsorption are more effective when both the surfactant and protein are present in the same phase. Fig. 16 also shows the surface diffusion data for the same system. The onset of surface diffusion also occurs around a molar ratio of 2.

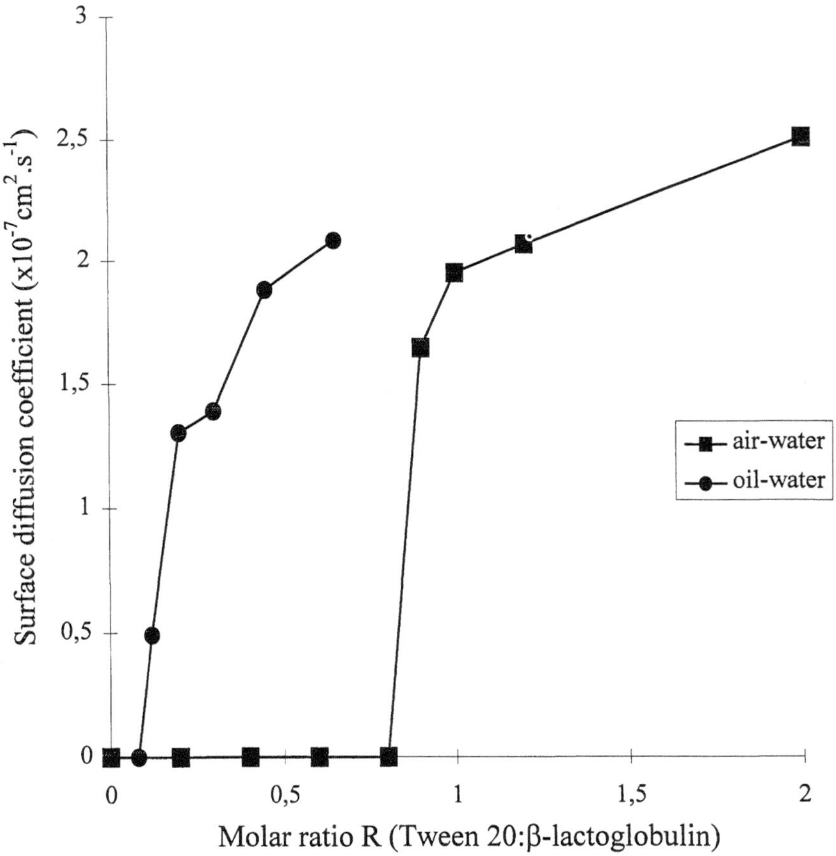

Fig. 15 Surface lateral diffusion in thin films stabilised by 0.2 mg.ml^{-1} β-lactoglobulin as a function of Tween 20 concentration at the air water (■), and oil-water (●) interfaces. In 10 mM phosphate buffer pH 7.0.

The above provides evidence that the mechanisms determining dispersion stability, particularly to coalescence, of both foams and emulsions stabilised by mixed protein - surfactant/emulsifier systems, are linked to the mobility of the protein molecules at the interface. The proteins rely on their ability to self interact and build a strong cohesive network (recently referred to as a 'skin' by the Prins group) at the interface to retard drainage and stabilise against coalescence. Breakdown of this network increases drainage and induces coalescence and is one of the primary causes of dispersion instability caused by the presence of surfactants, lipids or emulsifiers.

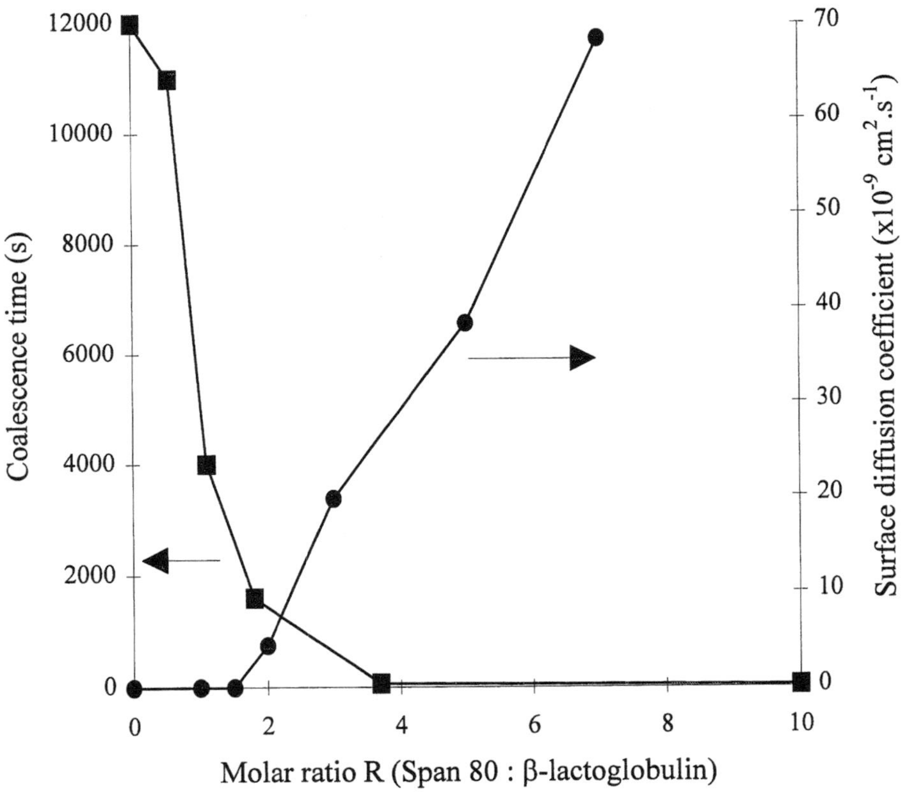

Fig. 16 Coalescence time of emulsion droplets (■), and surface lateral diffusion coefficient of thin films (●), stabilised by 0.2 mg.ml^{-1} β-lactoglobulin with increasing concentrations of Span 80. In 10 mM phosphate buffer pH 7.0.

4 CONTROLLING STABILITY IN MIXED EMULSIFIER/PROTEIN FOAMS WITH CROSSLINKING AGENTS

In the previous section we gave examples of how the onset and increase in the surface lateral mobility of protein stabilised systems resulted in destabilisation. It would be consummate therefore if we could demonstrate that the reverse were also true. The hypothesis is simple, breaking down the protein-protein interactions and subsequent mobilisation of the remaining adsorbed protein molecules results in instability, reforming the broken links by means of a crosslinking agent should restore stability.

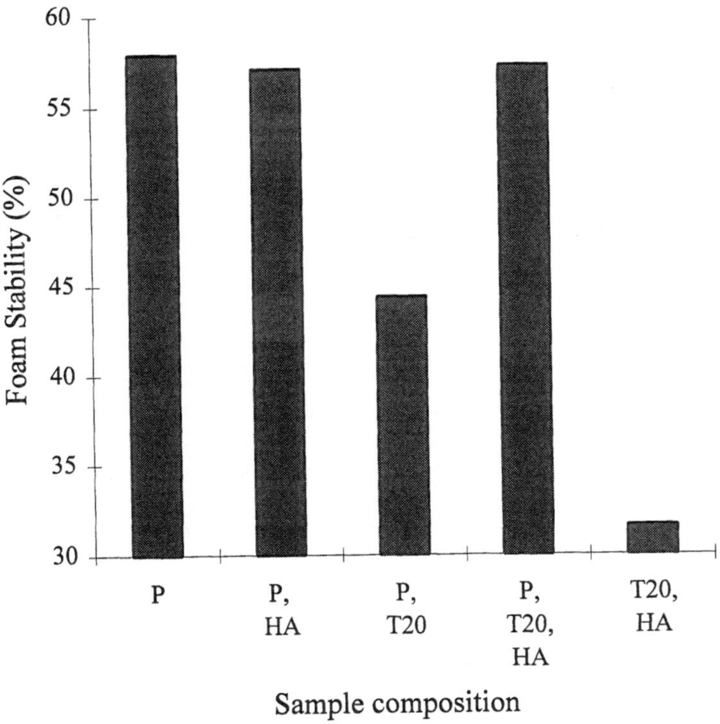

Fig. 17 Foam stability of various mixtures of 0.2mg.ml^{-1} β-lactoglobulin (P), 11mM Tween 20 (T20), and 40 ppm hop acids (HA). In 10 mM phosphate buffer pH 7.0.

This was first addressed in the area of beer foam stability, where it had been demonstrated empirically that the presence of hop acids in beer helped stabilise the foam by the crosslinking of proteins at the interface. This hypothesis was tested more thoroughly by studying the foam stability of a model foam system(Clark et al, 1991c). Fig. 17 shows the foam stability of a series of solutions containing BLG, Tween 20, mixed hop acids and mixtures thereof. The inclusion of Tween 20 in the BLG solutions resulted in unstable foams but this destabilisation could be eliminated by the inclusion of hop acids. The surface diffusion data is shown in Fig. 18. The control system is a solution containing both BLG and Tween 20. Upon the addition of the hop acids, the diffusion coefficient decreased, indicating that the diffusing body (i.e. fluorescently labelled protein) is increased in size by hop acid induced crosslinking. This results in larger and larger surface aggregate formation and reduces surface mobility and liquid drainage from the lamellae, resulting in a more stable foam.

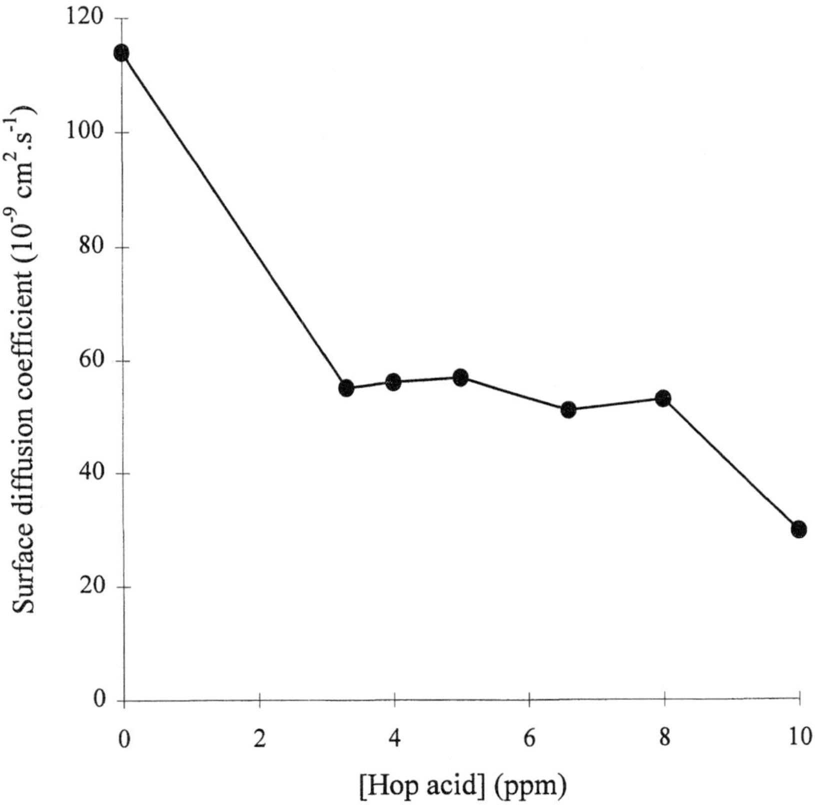

Fig. 18 Surface lateral diffusion coefficient of thin films stabilised by $0.2mg.ml^{-1}$ β-lactoglobulin and 11mM Tween 20, as a function of hop acid concentration.

Further evidence of this effect came from similar experiments involving crosslinking agents such as polyphenols (Sarker et al 1995b), polyvalent ions (Sarker et al 1996) pentosans and PGA. Fig. 19 shows the effect of adding Al^{3+} to a BLG stabilised foam which has been destabilised by Tween 20. An interesting feature of this data is the maximum in foam stability which was obtained around 4mM Al^{3+}. At higher Al^{3+} concentrations the stability dropped slightly. This is thought to be due to saturation of the protein-Al^{3+} sites of interaction, blocking further interactions, or even causing bulk aggregation phenomenon, reducing the amount of protein monomer available for surface adsorption.

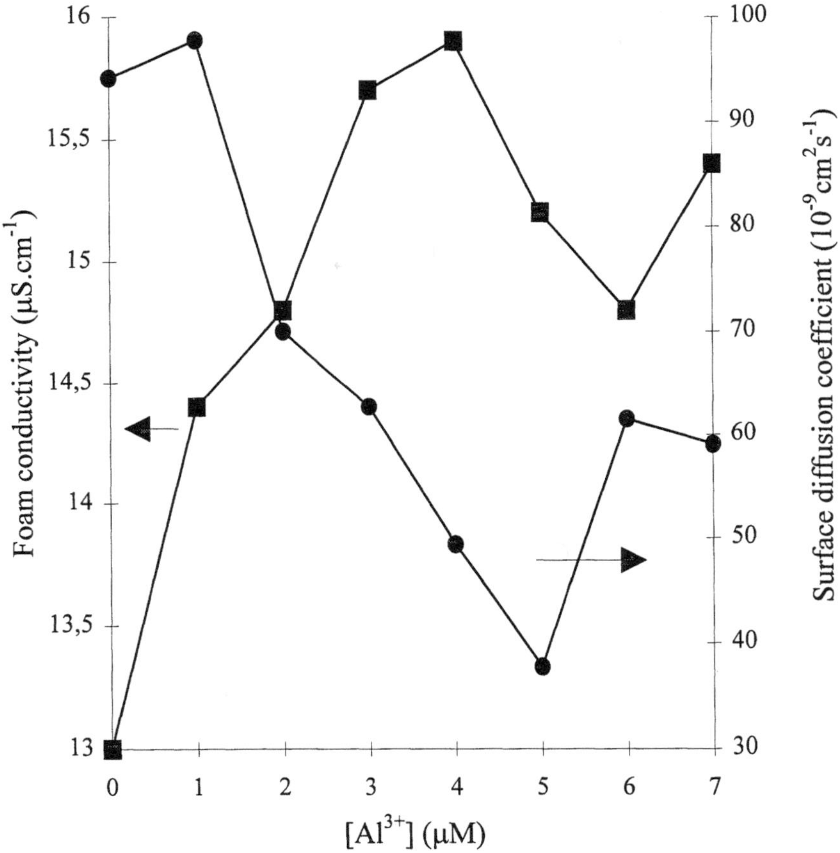

Fig. 19 Foam stability, expressed as foam conductivity after 5 minutes drainage (■) and surface lateral diffusion coefficient (●) of foams/thin films stabilised by 1.0 mg.ml^{-1} β-lactoglobulin and 11mM Tween 20, as a function of increasing concentrations of Al^{3+}. In 10 mM phosphate buffer pH 7.0.

The surface mobility data of the same system is also shown in Fig. 19. It is clear that at the optimal foam stability, D is at a minimum, in fact at 5mM D is about a third of the control value. The evidence for the weaker protein interactions or lower surface protein concentrations, at higher Al^{3+} concentrations is displayed by the return of mobility. This is a common feature of the crosslinker restabilised systems, particularly in model systems where many of the variables are well controlled and there is little excess protein or surfactant to cause secondary effects. In real systems where there is usually an excess of surface active components, particularly of the protein, the destabilising effects of the crosslinking agents are not as apparent.

5 CONCLUSIONS

The FRAP method can be used to investigate molecular dynamics in foam and emulsion films. These delicate, fragile elements are critical structures that determine the stability of foams and emulsions but that are not accessible to investigation by dilational of shear rheology. Systematic studies of phospholipid stabilised foam films have been undertaken and demonstrate the influence of lipid structure, orientation and phase state on measured surface diffusion in foam films. Foam and film drainage properties are clearly influenced by the diffusion properties of the surface adsorbed phospholipid layer. Foam films stabilised by protein alone have highly rigid adsorbed layers where surface diffusion is not detectable. Only when a critical concentration of added low molecular weight surfactant is exceeded is diffusion detected in these adsorbed layers and this onset of diffusion in most cases correlates with a decrease in stability of the dispersion (foam or emulsion). FRAP measurements have shown that addition of protein crosslinking components can re-establish the gel-like nature of the adsorbed layer in foam films formed from mixtures of surfactant and protein. This process correlates well with restoration of foam stability. Further work is required to establish that the same is true with emulsions. Whatever the outcome, the use of FRAP to follow molecular dynamics in interfacial layers has demonstrated the power of this technique to correlate processes at a molecular level with those at the macroscopic scale of foam and emulsion stability.

6 ACKNOWLEDGEMENTS

The authors would like to acknowledge the involvement of Alan Mackie and Dr. Andrew Pinder in the design, construction and continued development of the FRAP apparatus. The experimental results described in this paper were obtained in collaboration with Drs. Mark Coke, Michel Cornec Dipak Sarker, Zdravko Lalchev and Mr David Wilson. Much of this work was funded by the BBSRC and EC Network project.

7 REFERENCES

Axelrod, D., in "Spectroscopy and Dynamics of Molecular Biological Systems" P.M. Bayley and R.E.Dale (Eds.) Academic Press London (1985)163

Axelrod, D., Koppel, D.E., Schlessinger, J., Elson, E. and Webb, W.W., Biophys. J., 16(1976)1055

Barisas, G. and Leuther, M.L., Biophys. Chem., 10 (1979) 221

Beck, K. and Peters, R., "Spectroscopy and the Dynamics of Molecular Biological Systems" P.M.Bayley and R.E. Dale, (Eds.), Academic Press, London, (1985) 177

Burghardt, T.P. and Axelrod, D., Biophys. J., 33(1981)455

Castle, J., Dickinson, E., Murray, B.S. and Stainsby, G., ACS Symp. Ser., 343 (1987)118

Clark, D.C., "Characterization of Food - Emerging Methods" A.G. Gaonkar (Ed.), Elsevier (1995) 23

Clark, D.C., Mackie, A.R., Smith, L.J. and Wilson, D.R., "Food Colloids" R.D.Bee, P.Richmond and J.Mingins (Eds.), Royal Society Special Publication No.75, Cambridge (1989)97

Clark, D.C., Coke, M., Mackie, A.R., Pinder, A.C. and Wilson, D.R., J.Colloid Interf. Sci., 138(1990b)207

Clark, D.C., Dann, R., Mackie, A.R., Mingins, J., Pinder, A.C., Purdy, P.W., Russell, E.J., Smith, L.J. & Wilson, D.R., J. Colloid Interf. Sci., 138(1990a)195

Clark, D.C., Coke, M., Wilde, P.J. and Wilson, D.R., "Food Polymers, Gels and Colloids", E.Dickinson, (Ed.), Royal Society of Chemistry Special Publication No.82, Cambridge, (1991a)272

Clark, D.C., Wilde, P.J. and Wilson, D.R., Coll. Surf., 59(1991b)209

Clark, D.C., Wilde, P.J. and Wilson, D.R., J. Institute of Brewing, 97 (1991c)169

Clark, D.C. and Wilde, P.J., "Gums and Stabilisers for the Food Industry - 6" G.O.Phillips, D.J. Wedlock and P.A. Williams (Eds) Oxford Press, Oxford, (1992)343

Clark, D.C., Wilde, P.J., Wilson, D.R. and Wüstneck, R., Food Hydrocolloids, 6(1992)173

Clark, D.C., Mackie, A.R., Wilde, P.J., Wilson, D.R., Royal Soc. Chem. Faraday Discussion, 98(1994)253

Clark, D.C., Husband, F., Wilde, P.J, Cornec, M., Miller, R., Krägel, J., Wüstneck, R., J. Chem.

Soc. Faraday Trans., 91(1995)1991

Cohen, R., Exerowa, D., Kolarov, T., Yamanaka, T. and Muller, V., Colloids Surf., 65(1992)201

Coke, M., Wilde, P.J., Russell, E.J. and Clark, D.C. J. Colloid Interface Sci., 138(1990)489

Cornec, M., Mackie, A.R., Wilde, P.J and Clark, D.C. Colloids and Surfaces A: Physicochemical and Engineering Aspects,114(1996)237

Derjaguin, B.V. and Landau, L., Acta Physicochim. USSR, 14(1941)633

Exerowa, D. and Lalchev, Z., Langmuir 2(1986)668

Exerowa, D., Lalchev, Z., and Kashchiev, D., Colloids Surf. 10(1984)113

Exerowa, D., Lalchev, Z., Marinov, B., and Ognyanov, K., Langmuir 2(1986)664

Fahey, P., Koppel, D.,Barak, L.,Wolf, D.,Elson, E. and Webb, W., Science ,195(1977)305

Fahey, P.F. and Webb, W.W., Biochemistry, 17(1978)3046

Graham, D.E. and Phillips, M.C., J.Colloid Interf. Sci., 70 (1979) 403

Huisman, F. and Mysels, J., J. Phys. Chem., 73(1969)489

Kokelaar, J.J., Prins, A. and Gee, M. J., Colloid Interf. Sci., 146(1991)507

Krägel J., Siegel, S., Miller, R., Born, M., Ehmke, B. and Schano, K.H., Prog. Colloid Polym. Sci., 93(1993)283

Krägel J., Siegel, S., Miller, R., Born, M. and Schano, K.H., Colloid. Surf., 91(1994)169

Krägel, J., Miller, R., Wüstneck, R., Clark, D.C. and Wilde, P.J., Progress in Colloid and Polymer Science, 98(1995)239.

Kolarov, T., Scheludko, A. and Exerowa, D., Trans. Faraday Soc., 64(1968)2864

Kolarov, T., Exerowa, D. and Cohen, R., Colloids Surf., 42(1989)49

Ladha, S., Mackie, A.R., Harvey, L.J., Clark, D.C., Lea, E.J.A., Brullemans M. and Duclohier, H., Biophys. J., 71(1997)1364

Lalchev, Z., Christova, Y., Todorov, R., Alexandrov, V., Stoichev, P. and Petkov, R., Appl. Cardiopulmon. Pathophysiol., 4(1992)315

Lalchev, Z., Ishida, H. and Nakazawa, H., "Colloid and Molecular Electro-optics" B.R. Jennings and S.P.Stoylov (Eds.) Institute of Physics Publishing, Bristol, UK, (1991)239

Lalchev, Z., Todorov, R., Ishida, H. and Nakazawa, H., Eur. Biophys. J., 23(1994a)145

Lalchev, Z., Wilde, P. and Clark, D. , J. Colloid Interface Sci., 167(1994b)80

Lalchev, Z., Wilde, P. and Clark, D. , J. Colloid Interface Sci., 174(1995)283

Lalchev, Z., Wilde, P. and Clark, D. , J. Colloid Interface Sci., 1997, in press

Lee, A., Birdsall, N. and Metcalfe, J., Biochemistry 12(1973)1650

Mackie, A.R., Wilde, P.J., Wilson, D.R. and Clark, D.C. Royal Chem. Soc. Faraday Trans., 89(1993)2755

Mackie, A.R., Nativel, S., Wilson, D.R., Ladha, S. and Clark, D.C. J. Sci. Food Agric., 70 (1996)413

Mysels, K., Shinoda. K. and Frankel S. "'Soap Films" Pergamon Press, New York, (1959) Naydenova, S., Lalchev, Z., Petrov, A. and Exerowa, D., Eur. Biophys. J., 17(1990)34

Peters, R ., Peters, J., Tews, K.H. and Bahr, W., Biochim. Biophys. Acta, 367(1974)282

Peters,R and Beck, K, Proc. Natl. Acad. Sci. USA, 80(1983)7183

Sackman, E. and Traube, H. J. Am. Chem. Soc., 94(1972)4482

Sarker, D.K., Wilde, P.J. and Clark, D.C. Colloids and Surfaces B: Biointerfaces, 3(1995)349

Sarker, D.K., Wilde, P.J. and Clark, D.C. J.Agric. Food Chem., 43(1995b)295

Sarker, D.K., Wilde, P.J. and Clark, D.C. Colloids and Surfaces A: Physicochemical and Engineering Aspects, 114(1996)227

Scheludko, A. Adv. Colloid Interf. Sci., 1 (1967) 391

Verwey, E.J.W and Overbeek, Th.G. "Theory of stability of lyophobic colloids", Elsevier, Amsterdam (1948)

Wilde, P.J., J. Coll. Interf. Sci., 178(1996)733

Wilde, P.J. and Clark, D.C., J. Coll. Interf. Sci., 155(1993)48

Wilde, P.J., Clark, D.C. and Marion, D., J. Agric. Food Chem., 41(1993)1570

Wilson, D.R., Wilde, P.J. and Clark, D.C. "Food Colloids and Polymers: Structure and Dynamics", RSC Special Publication. Royal Society of Chemistry, London, (1993)415

Wolf, D.E., "Fluorescence Microscopy of Living Cells in Culture. Part B." D.L.Lansing Taylor and Y.Wang (Eds.),Academic Press, San Diego, (1989)271

Proteins at Liquid Interfaces
D. Möbius and R. Miller (Editors)

INTERFACIAL TENSIONS OF PROTEIN SOLUTIONS USING AXISYMMETRIC DROP SHAPE ANALYSIS

P. Chen, R.M. Prokop, S.S. Susnar and A.W. Neumann

Department of Mechanical and Industrial Engineering
University of Toronto, Toronto, Ontario, Canada M5S 3G8

Contents

1 INTRODUCTION

Proteins are biological macromolecules that are often present at interfaces such as cell membranes, blood vessel walls and implant surfaces. Protein adsorption at the interface plays an important role in many biological processes. Examples are wound healing, blood clotting, tissue integration of biomaterials and adhesion of infectious microorganisms. In addition, protein adsorption is relevant to medicine and the chemical industry in areas such as drug delivery, the development of biomedical devices and chromatographic separation processes. Studies of protein adsorption have been carried out for many years, yet the underlying mechanisms remain unclear [1].

A protein consists of a chain of hundreds or thousands of amino acid units along with their side groups; its complicated physical chemistry is related to its unique molecular structure. Adsorption at the interface may involve three distinct steps: (a) diffusion, where protein molecules migrate from the bulk phase to the layer next to the interface (the so-called subsurface); (b) overcoming the energy barrier between the subsurface and the interface; (c) conformational change after adsorbing at the interface. These microprocesses are manifest in the macroscopic properties of the interface. The most sought after property is surface (interfacial) tension which, for present purposes, is equal to surface free energy of the interface [2]. The purpose of this chapter is to elucidate the surface tension of protein solutions, and the methodology used is axisymmetric drop shape analysis (ADSA) [3-5]. With ADSA both static surface tension and dynamic surface tension can be measured; this provides information about the microscopic mechanism of molecular adsorption at the interface.

In a sense, the static surface tension measurement is the determination of the equilibrium surface tension; however, from the perspective of this chapter, this is a kinetic (time-dependent) measurement, and it includes monitoring the surface tension variation over the equilibration time of the surface [6]. Since the equilibration of protein at liquid-fluid interfaces can last as long as several hours or even days [7,8], it is necessary to develop an efficient scheme to determine the equilibrium surface tension. Therefore, criteria for obtaining equilibrium surface tension will be discussed in the static measurement. Subsequently, we will examine the surface tension as a function of temperature and protein concentration in an aqueous solution.

In the dynamic surface tension measurement, the surface area of an interface is changed, and the pattern of the surface tension response is analyzed. This pattern reflects the dynamics of molecular movements and interactions at the interface, and insight into adsorption mechanisms can be gained [6].

Different techniques have been employed to study the surface tension of proteins at the air-water and oil-water interfaces, such as the Wilhelmy plate [9,10], the du Noüy ring tensiometer [11], and those based on the volume [9,12], weight or shape of a pendant drop [13,14]. In the ring method, the force required to pull a ring from the surface of a liquid is determined. This method has the disadvantage of enlarging the surface area during the measurement process, which leads to alteration of the adsorption state of the proteins. Viscoelastic effects in addition to surface tension effects may come into play. The Wilhelmy plate technique requires the establishment of a zero contact angle that is difficult to guarantee with systems involving protein solutions, due to adsorption onto the plate. Moreover, this is even more difficult in liquid-liquid systems which, on the other hand, are relevant to many biological processes. The ring method also suffers further complications in liquid-liquid systems. The calculation of interfacial tension with the du Noüy tensiometer requires a correction factor for the weight of the column of liquid while the ring is removed. In liquid-liquid systems, consideration of the density difference across the interface is required for an accurate correction. The drop volume technique relies on the volume of a liquid drop detaching from a capillary tube to determine the interfacial tension. Although it is applicable to liquid-liquid systems, it requires extremely careful manipulations for the determination of the volume of the detaching drop. Also, to perform time-dependent studies, the detachment of the drop at the desired time must be elicited by the rate in which the drop is grown. This in itself inflicts an added disturbance to the system. Pendant drop methods, on the other hand, rely on the shape of a drop for interfacial tension determinations as dictated by the Laplace equation of capillarity. In its simplest form, the drop shape is defined according to characteristic dimensions such as the height and diameter [15] or a few preselected points, such as the apex and inflection points, along the drop profile [16]. A more versatile technique, axisymmetric drop shape analysis (ADSA) [3-5, 17], utilizes the whole drop profile, with equal importance attached to every profile coordinate [6]. With the advent of image analysis schemes, the drop profile may be obtained with subpixel resolution leading to measurements with a high degree of accuracy. ADSA may be applied to both liquid-air [14,18] and liquid-liquid systems [8,19]. It may also be used to study the pressure [19], temperature [8], and time [8,14] dependence of the interfacial tensions.

In this chapter, axisymmetric drop shape analysis-profile (ADSA-P) [3-6] was employed to study three aspects of protein surface tension behavior: (1) the temperature dependence of the interfacial tension of human serum albumin at the water-decane interface; (2) concentration dependence of the interfacial pressure of human serum albumin at the water-decane interface; (3) dynamic surface tension response to surface area change of mixed solutions of a protein and small or medium-sized organic molecules.

2 Axisymmetric Drop Shape Analysis-Profile (ADSA-P)

A schematic of the ADSA-P experimental setup is given in Fig. 1, the basic components of which are as follows: With the use of a microsyringe (Hamilton Gastight syringe, Chromatographic Specialties Inc., Brockville, ON, Canada), a pendant drop of the protein solution was formed at the tip of a vertical Teflon capillary of circular cross-section (inner diameter, 1.5 mm), thus producing an axisymmetric boundary for the drop. The drop was enclosed in a sealed quartz cuvette (model 330984, 10×10×30 mm^3, Hellma Canada Ltd., Concord, ON, Canada) which contained decane or air. The cuvette was mounted in an environmental chamber (model 100-07, Ramé Hart, Inc., Mountain Lakes, NJ, USA). The chamber was linked to a thermostatted water bath (Lauda K-2/R, Brinkmann Instruments) maintaining the temperature of the set-up to the accuracy of ± 0.1°C. The drops were illuminated with a white light source (model V-WLP 1000, Newport Corp., Irvine, CA, USA) shining through a heavily frosted diffuser. Images of the drop were obtained by a microscope (Leitz Apozoom, Leica, Willowdale, ON, Canada) linked to a monochrome charge-coupled device video camera (Cohu 4810, Infrascan, Inc., Richmond, BC, Canada). The video signal of the drop was transmitted to a digital video processor (Xvideo board, Parallax Graphics Inc., Santa Clara, CA, USA) which performed the frame grabbing and digitization of the image to 640 × 480 pixels with 256 grey levels.

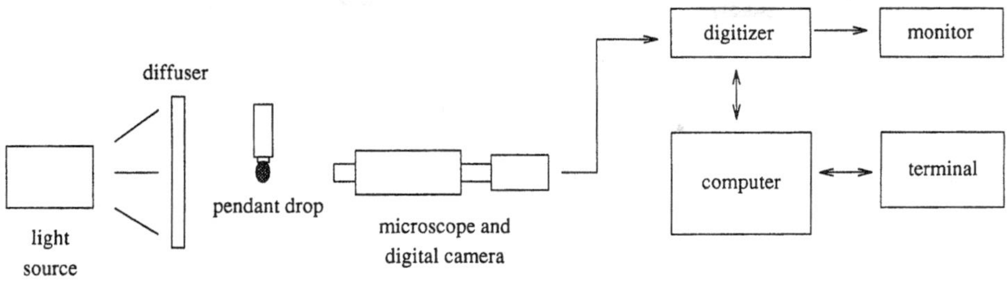

Fig. 1 Schematic of an experimental set-up for ADSA-P.

In static surface tension studies, the experiment was continued until an approximately constant interfacial tension was obtained for a sequence of measurements. For each run, images were captured at 1 s intervals initially and progressively less rapidly (up to 150 s intervals) near the end of the experiment. In dynamic surface tension studies, the experiment was continued until repeated cycles were observed in the surface tension response to the surface area perturbation. For each run, images were captured at a reasonably fast pace (up to 0.5 s intervals) so that the features of dynamic surface tension could to be obtained. To produce a controlled surface area perturbation, the microsyringe was connected to a stepper motor (Model 18515, Oriel Corp., Stratford, Conn, USA) which was computer-controlled. The motion of the syringe plunger changed the volume of the drop and hence changed the surface area [4,6].

A workstation (Sun SPARCstation 10, Sun Microsystems, Mountain View, CA, USA) was used to acquire the images from the digitization board. Image analysis schemes were used to determine the drop profile coordinates with subpixel resolution and to correct for optical distortion [4]. The entire set-up, except for the water bath and the workstation, were placed on a vibration-free table (Technical Manufacturing Corp., Peabody, MA, USA) to isolate the system from external disturbances.

ADSA-P fits a theoretical profile given by Laplace equation of capillarity to the experimental profile of a drop. An objective function is formed which describes the deviation of the experimental profile from the theoretical one. This function is minimized by a non-linear least squares regression procedure yielding the interfacial tension (and the contact angle in the case of a sessile drop [3-6]). The program also provides the volume, surface area, and the radius of curvature at the apex of the drop; for sessile drops the contact radius and the contact angle are also given. The program requires several arbitrary coordinate points along the drop profile, the value of the density difference across the interface, and the magnitude of the local gravitational constant as input. Each single image of a drop is analyzed ten times with twenty different randomly chosen profile coordinate points each time. The average resulting 95% confidence limit for each measurement is better than ± 0.2 mJ/m^2 in this work (although greater accuracy, approximately ± 0.04 mJ/m^2, has been obtained routinely using this procedure for non-protein solutions).

3 TEMPERATURE DEPENDENCE OF THE INTERFACIAL TENSION OF HUMAN SERUM ALBUMIN AT THE WATER-DECANE INTERFACE

Temperature dependent studies of interfacial tension allow the detection of conformational changes of proteins. Recently, conformational changes of bovine serum albumin (BSA) have been reported below 60°C in the bulk of the solution using differential scanning calorimetry [11]. An interfacial tension study is an obvious follow-up [8]. Since the temperature coefficient of surface tension represents the surface entropy, such measurements would contain information about surface molecular structure. Moreover, knowledge of the interfacial tension of an aqueous protein solution-hydrocarbon interface is clearly biologically relevant at temperatures near that of the body.

As mentioned above, after formation of a pendant drop of a protein solution, adsorption and conformational changes occur. The equilibration of such an interface can be very slow, on the order of hours, and in some extreme cases, several days [13]. It is difficult to run a meaningful experiment for such a long period. Therefore, the equilibrium value of interfacial tension is not readily obtained. It is necessary to find an "experimental equilibrium" value which meets certain criteria while allowing for a shorter duration of the experiment. An example of achieving this is to use extrapolation of the measured interfacial tension values. The resulting "experimental equilibrium" values will be used to study the temperature dependence of the interfacial tension, from which the interfacial pressure (the interfacial tension difference between the pure water-decane interface and the human serum albumin (HSA) solution-decane interface) can be derived.

3.1 Materials

The HSA used was from Sigma (Sigma Chemical Corp., St. Louis, MO, USA). The sample contained 15.4% nitrogen, was free from fatty acids and had an average molecular weight of 65,000. The sample was used without further purification. The aqueous solutions were prepared with distilled water, de-aerated by vacuum before use. The decane was supplied by Caledon Laboratories Ltd. (Georgetown, ON, Canada), and had been distilled in glass and certified for gas chromatography (Code 3301-2). Before use, decane was mixed with an equal amount of distilled water and vigorously shaken, in order to saturate with water. The protein concentration in its aqueous solution was 0.02 mg HSA per ml of water.

3.2 Results

3.2.1 *Experimental Isotherm*

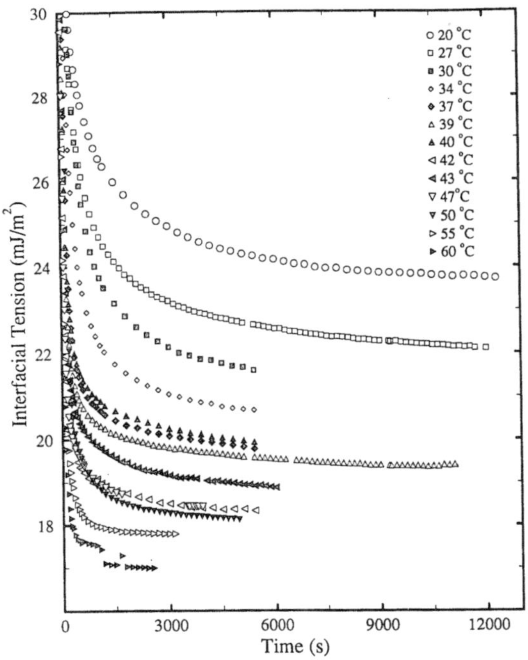

Fig. 2 Interfacial tension between 0.02 mg/ml aqueous solutions of HSA and decane at 20, 27, 30, 34, 37, 39, 40, 42, 43, 47, 50, 55 and 60 °C.

The interfacial tension of the 0.02 mg/ml aqueous HSA solution-decane was measured in the temperature range from 20 to 60°C. The error limits were ± 0.2 mJ/m^2 at the 95% confidence level. Figure 2 shows the interfacial tension as a function of time.

From Fig. 2, we can distinguish two domains in the isotherms: one at the beginning (first few minutes) in which the interfacial tension decreases quickly; the other, thereafter, in which the interfacial tension changes slowly. Two trends are observed as the temperature increases: (a) the second domain appears earlier and (b) the isotherm reaches lower values of the interfacial tension in the second domain.

In order to compare the data obtained at different temperatures, we need to establish experimental equilibrium criteria. Hence, the emphasis of this section is on the second domain of the interfacial tension as a function of time.

3.2.2 *EXTRAPOLATION OF THE PLOT OF γ VERSUS $1/\sqrt{t}$*

It has been suggested [20,21] that an extrapolation to zero in the plot of interfacial tension γ versus $1/\sqrt{t}$ can be useful for estimating the equilibrium value of γ. Based on a transport-controlled mechanism, this linear extrapolation has been shown to give an interfacial tension value close to the equilibrium one [22]. Figure 3 shows the interfacial tension γ versus $1/\sqrt{t}$ for some temperatures in the second domain of the isotherms.

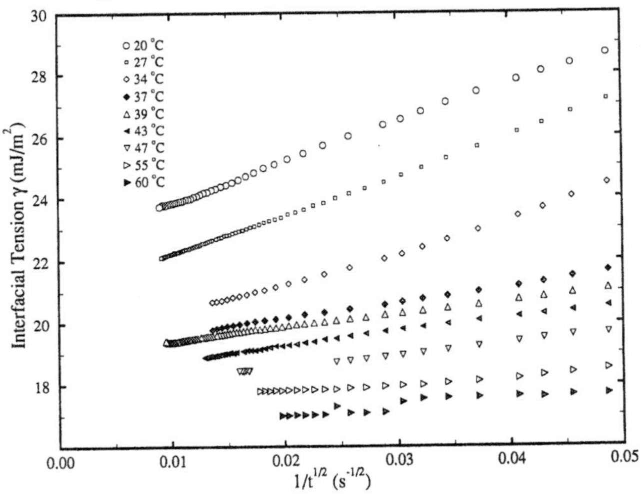

Fig. 3 Interfacial tension versus $1/\sqrt{t}$ for selected temperatures.

The data were fitted to a straight line by linear regression. Extrapolation to zero (i.e., $t \rightarrow \infty$) gives an estimate of the equilibrium interfacial tension γ_∞. The results are presented in column 2 of Table 1.

Table 1 Equilibrium values γ_∞ obtained by two methods: A) extrapolation and B) minimum slope

	A)	$1/\sqrt{t}$	B)	$\lvert d\gamma/dt \rvert$
temperature T (°C)	interfacial tension γ_∞ (mJ/m^2)	correlation coefficient r	interfacial tension γ_∞ (mJ/m^2)	time (s) needed to obtain $\lvert d\gamma/dt \rvert < 10^{-4}$ (mJ/m^2 s)
20	21.50	0.996	23.98	7275
27	21.01	1.000	22.41	7275
30	20.04	0.996	21.77	4500
34	19.47	1.000	20.74	4950
37	19.14	0.995	19.93	4200
39	18.91	0.996	19.72	3800
40	19.91	0.996	20.03	4800
42	17.78	0.997	18.48	3450
43	18.30	0.998	19.03	4100
47	17.80	0.999	18.48	1650
50	17.47	0.995	18.25	1575
55	17.47	0.986	17.85	1650
60	16.74	0.982	17.04	1875

3.2.3 Slope $d\gamma/dt$

In the second domain of the isotherms (Fig. 2), we can see that the slope decreases with time. When $d\gamma/dt = 0$, the isotherm would have reached equilibrium. Experimentally, satisfaction of this condition is not easy to obtain: as mentioned above, the equilibration time may be very long. Further, the experimental measurements were not performed in real time, and it was not practical to determine precisely when to terminate the experiment. Thus, it was decided to identify the smallest value of $d\gamma/dt$ reached at all temperatures, and to consider the corresponding γ values as the equilibrium values. The "equilibrium" interfacial tension so obtained may be expected to be comparable from concentration to concentration. If the cut-off value is made reasonably small, then the systems will not experience a large decrease in interfacial tension after this cut-off point. Hence, the interfacial tension obtained will be a reasonable approximation of the true equilibrium interfacial tension.

A further complication exists in that the isotherms are not totally smooth, but rather show small oscillations, possibly due to small temperature fluctuations. Since the values of γ have fluctuations, $d\gamma/dt$ also has fluctuations of about 5×10^{-5} mJ/m^{2}s when approaching equilibrium (Fig. 4). It has been pointed out [9] that these small slopes could not be distinguished from the observed artifacts. Therefore, the data were smoothed with respect to time. The procedure was as follows: A polynomial was fitted to the experimental interfacial tensions as a function of time, and the derivatives of the smoothed data with respect to time were calculated. The lowest value of $|d\gamma/dt|$ reached is 1×10^{-4} mJ/m^{2}s for 37 and 40°C, and this value is also reached for all the other temperatures before the termination of the experiments. Therefore, 1×10^{-4} mJ/m^{2}s was selected as the "limiting slope" (note: this value is greater than the random slope fluctuations of 5×10^{-5} mJ/m^{2}s). This slope is reached in the isotherms at different times for different temperatures. The results are summarized in Table 1. The general trend is a decrease in the time required for reaching $|d\gamma/dt| = 1 \times 10^{-4}$ mJ/m^{2}s with an increase in temperature.

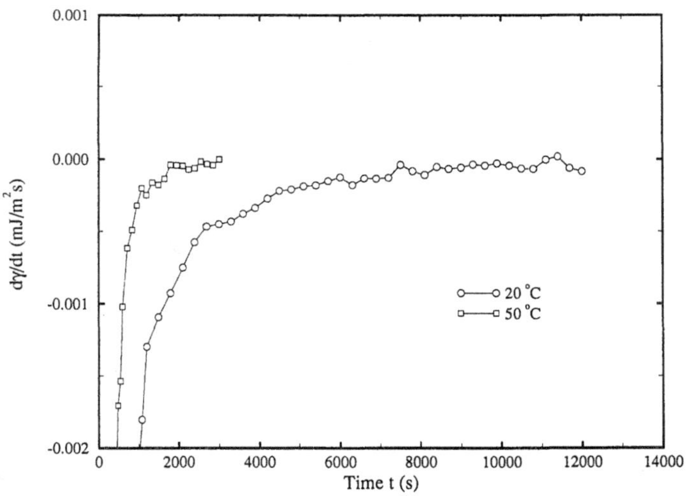

Fig. 4 Slope of interfacial tension with respect to time versus time for selected temperatures.

3.2.4 *Interfacial Tension as a Function of Temperature*

With the data of the interfacial tension γ_∞ obtained from the two procedures above, the results are plotted in Fig. 5 as a function of temperature, T. It can be seen that the minimum slope criterion gives higher values for γ_∞ than does the extrapolation method.

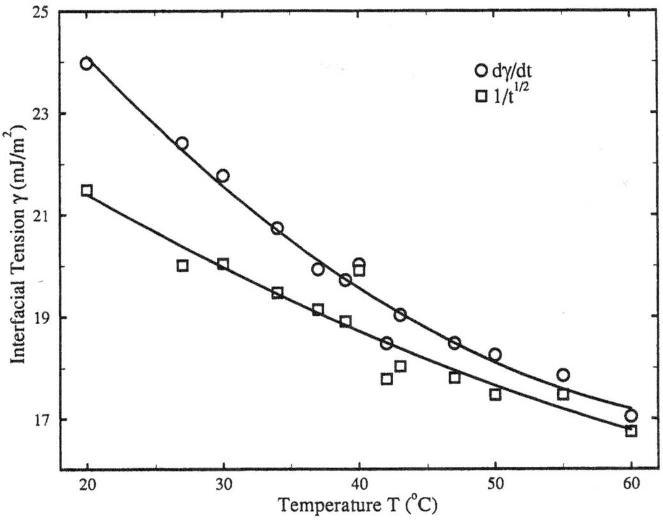

Fig. 5 Equilibrium interfacial tension versus temperature; both criteria for estimating γ_∞ are shown.

3.2.5 *Interfacial Pressure as a Function of Temperature*

The interfacial pressure is defined as $\pi = \gamma_0 - \gamma$, where γ_0 is the interfacial tension of pure water-decane, and γ is the interfacial tension of the same interface in the presence of HSA. The temperature dependence of interfacial pressure of the solution-decane system is a combination of two effects: the variation in the interfacial tension of pure water-decane and the change in the adsorption of HSA from the bulk solution to the interface, i.e., the interfacial tension of solution-decane due to a change in temperature. To establish the interfacial pressure, measurements of the interfacial tension of water-decane have to be performed over the same range of temperatures. The results are shown in Fig. 6. It is noted that the change in the interfacial tension of water-decane is not linear with respect to temperature.

Figure 7 shows π versus T for both equilibrium criteria. While the overall trend is the same for the two curves, the interfacial pressure changes by almost 4 mJ/m^2 from 20 to 45°C for the data obtained from the minimum slope criterion, and by only 2 mJ/m^2 for the data obtained from the extrapolation. At higher temperatures, the difference in π between the two methods becomes small.

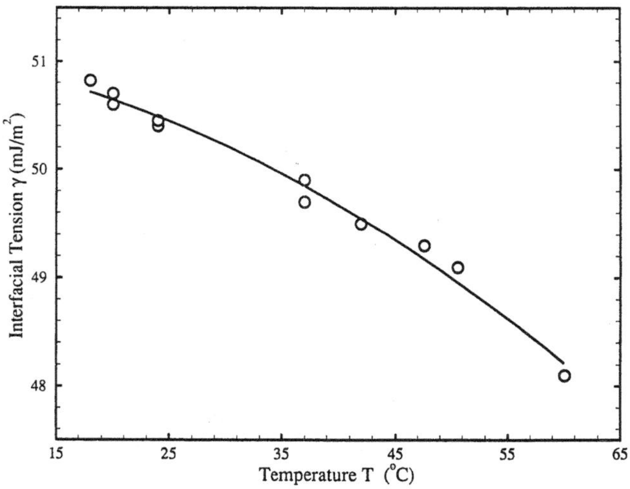

Fig. 6 Interfacial tension of pure water-decane versus temperature.

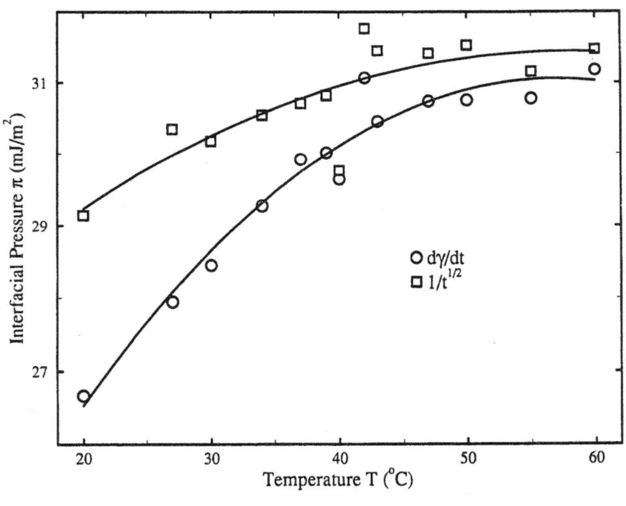

Fig. 7 Equilibrium interfacial pressure for the solution-decane interface versus temperature; both criteria for estimating γ_∞ are shown.

3.3 Discussion

The experimental equilibrium interfacial pressure is found to be 10 mJ/m^2 higher at the HSA solution-decane interface than at the HSA solution-air interface under the same experimental conditions [18]. Similar behavior has been found for BSA at the water-iso-octane interface [13]. At first sight this finding might be surprising, as one might expect the water-air interface to have a larger potential for interfacial tension reduction due to adsorption than the water-hydrocarbon interface with its lower interfacial tension. However, it should be realized that a surfactant reduces the surface tension of water only to, say, 30 mJ/m^2 whereas the interfacial tension between water and hydrocarbon can be readily reduced to near zero, e.g., by an emulsifier. Thus, the driving force for protein adsorption, i.e., the maximum surface or interfacial tension reduction, for the two types of systems may be nearly equal, or somewhat larger for the aqueous/hydrocarbon system. Therefore, it is not surprising that the spreading (interfacial) pressure is actually larger at the aqueous solution-decane interface than at the aqueous solution-air interface [8]. It is an interesting question whether such observations and comparison might be useful to compare the overall hydrophobic character of different proteins, as well as the flexibility of protein molecules and mobility of side chains.

From Fig. 2 and Table 1, we can see that, for higher temperatures, the isotherms not only reach lower values in the interfacial tension but also need less time to approach equilibrium. This behavior may be the consequence of several factors: An increase in the diffusion coefficient with temperature, which controls the process initially, and an increase in the adsorption area of the HSA molecule with temperature [13]. Also, conformational changes could occur faster or the adsorption energy barrier could be overcome more easily.

From the concentration dependence of the interfacial tension [23], we can see that the protein concentration used, 0.02 mg/ml, is close to the minimum concentration required to saturate the water-decane interface with HSA, at room temperature (≈ 25°C). This means that the interface can be saturated by two different mechanisms: (a) an increase in the bulk concentration and consequently an increase in the adsorption density, Γ, (protein interfacial concentration); (b) an increase in the temperature and consequently an increase in the interfacial area occupied by the protein molecules. This second mechanism requires a lower Γ to obtain the same change in the interfacial tension, and consequently less time to approach equilibrium. Thus, it appears that the plateau reached by the interfacial pressure at temperatures above 45°C reflects saturation of the interface with HSA [8].

The present study highlights existing difficulties in the determination of the equilibrium interfacial tension γ_∞. There is a difference between the results obtained by the two methods, of approximately 2 mJ/m^2 at room temperature. It is clear that the choice of a limiting slope of

$|d\gamma/dt| = 1 \times 10^{-4}$ mJ/m^{2}s cannot, strictly speaking, represent γ_∞; however, we are not entirely convinced that the plot of γ versus $1/\sqrt{t}$ is preferable. The rationale for using this extrapolation is tenable only if the processes causing interfacial tension reduction are transport controlled (see section 5). This however is not the case at late stages of the protein adsorption [24]. The difference between the two procedures is so large that a discussion of $d\gamma/dT$ is virtually meaningless. Establishing better or more definitive procedures for determining γ_∞ is therefore a task of considerable importance.

4 Concentration Dependence of the Interfacial Pressure of Human Serum Albumin at the Water-Decane Interface

In this study, axisymmetric drop shape analysis (ADSA-P) was employed to obtain highly accurate measurements of the concentration dependence of the interfacial (surface film) pressure of human serum albumin (HSA) at the water-decane interface. The significance of concentration in the surface activity of proteins has been well documented [7,10,18,25,26]. In general, with an increase in the bulk protein concentration, a decrease in the interfacial tension, i.e., a positive interfacial pressure, has been reported. This reflects increased diffusion of protein to the interface, followed by unfolding and molecular rearrangements of adsorbed molecules [13]. In this study, we report the measurement of negative interfacial pressures at very low concentrations (1×10^{-4} and 1×10^{-3} mg/ml). An interpretation of this finding with respect to the effects of electrical charges and pH is attempted. The variation of the interfacial tension with the change in the surface concentration of adsorbed proteins might be described by the Gibbs adsorption equation. The applicability of Gibbs' law to our adsorption isotherms is also investigated [23].

4.1 Materials

The samples of human serum albumin (Sigma Chemical Corp., USA) and decane (Caledon Laboratories Ltd., Georgetown, ON, Canada) were the same as those described in section 3.1. Fifteen aqueous solutions of the albumin were prepared with distilled water. The protein concentrations ranged from 1×10^{-4} to 5 mg/ml. For a limited number of experiments at a protein concentration of 1×10^{-4} mg/ml, 20 mM Trizma Base buffers (Sigma catalogue No. T-

1503, Sigma, Mississauga, ON, Canada), pH adjusted with HCl, were used to produce aqueous solutions with a pH of 3.5, 4.8 (the isoelectric point of albumin), and 5.6.

4.2 Results

Figure 8 illustrates the time dependence of the interfacial tension of aqueous human serum albumin (HSA) solution of 15 concentrations at a decane interface. In all cases, a reduction of the interfacial tension to a relatively constant value is observed with the passage of time, *t*. In general, the higher the bulk protein concentration, the lower the observed equilibrium value. Also, the rate of decrease in γ at early times increases with increasing bulk protein concentration.

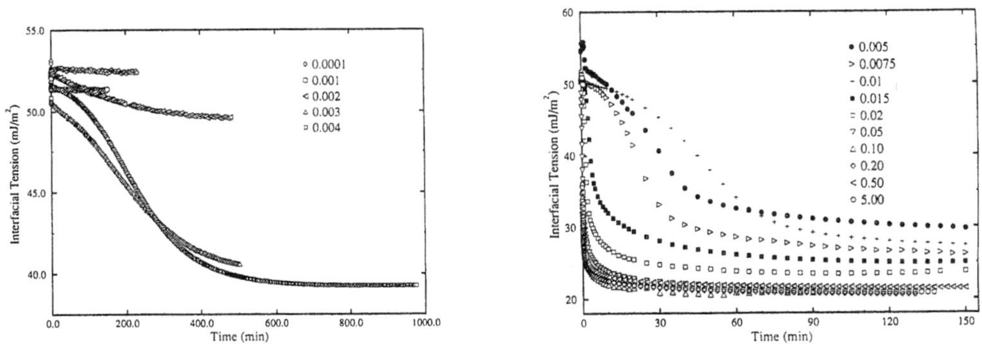

(a) (b)

Fig. 8 The interfacial tension between the aqueous HSA solution and decane versus time for various bulk concentrations of the HSA: (a) 0.0001 to 0.004 mg/ml; (b) 0.005 to 5.0 mg/ml.

As a consequence of this last observation, the duration of each experiment varied with its protein concentration. That is, for low concentrations (from 2×10^{-3} to 4×10^{-3} mg/ml) the time required for an experiment is more than 8 h (Fig. 8a), while for high concentrations (from 5×10^{-3} to 5.00 mg/ml) the time for each experiment is about 2 h (Fig. 8b). An exception is found for the two lowest concentrations (Fig. 8a), where the interfacial tensions are rather constant with the passage of time and they stay above the value of the pure solvent interfacial tension (about 50.4 mJ/m^2 [8]).

In order to establish an "equilibrium" interfacial tension, γ_∞, the "slope method" (see the previous section 3 and Ref. [8]) was adopted. First, for each concentration, the derivative of interfacial tension with respect to time was calculated by the forward difference method for the experimental curve. This results in a calculated slope for each experimental point on the curve. The final, minimum slope was then found for each concentration. Next, the largest value among these minimum slopes was identified and used as the cut-off value for all concentration and the corresponding interfacial tension values were regarded as the equilibrium interfacial tension of the protein solution.

Figure 9 illustrates the "equilibrium" interfacial tension as a function of concentration. In applying the minimum slope criterion, the derivatives of interfacial tension with respect to time for each concentration were calculated and the minimum slope was determined. The largest minimum slope among 15 concentrations was found to be 5×10^{-4} mJ/m^{2}s; this value was used as the cut-off value. The times corresponding to this cut-off point were all above 1 h, with higher concentrations requiring less time to reach this cut-off point. The final interfacial tension value obtained in the time dependent measurement is also given in Fig. 9.

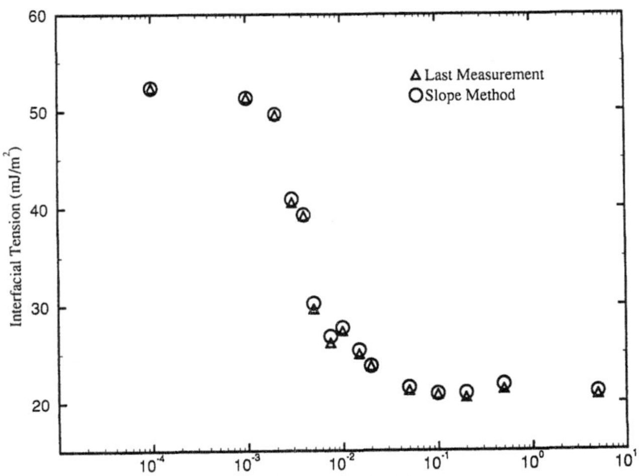

Fig. 9 The "equilibrium" interfacial tension obtained by the minimum slope criterion and the last experimental value for various bulk concentrations of the HSA.

The two types of interfacial tension values agree quite well. The "last measurements" are only slightly lower that the values at $|d\gamma/dt| = 5 \times 10^{-4}$ mJ/m^{2}s. The duration of the kinetic interfacial tension measurement at each concentration had been chosen on the basis of preliminary measurements and was nothing more than a rough optimization of the experimentation. It is therefore significant that the two types of data agree so well. The agreement suggests that we have obtained a reasonable approximation to the true equilibrium interfacial tension. (Note that one may also use the extrapolation method discussed in section 3 to approximate the equilibrium interfacial tension; however, this will not alter the main conclusions given in section 4.4.)

The interfacial tension γ remains almost constant at approximately 52 mJ/m^2 over more than three hours for the two lowest concentrations, i.e., 1×10^{-4} and 1×10^{-3} mg/ml (Fig. 8a). With further increases in the bulk concentration, the interfacial tension declines sharply to approximately 20 mJ/m^2. However, at concentrations above 0.05 mg/ml, γ remains essentially unchanged at that level with an increase in concentration (Fig. 9).

The results may also be interpreted in terms of interfacial pressure, π, as demonstrated in the previous section. The interfacial tension of the pure water-decane interface, γ_0, was measured as 50.4 mJ/m^2. The π values calculated by employing the equilibrium interfacial tension obtained by both the last measured value and the minimum slope criterion are given in Fig. 10. Interestingly, at the two lowest concentrations (1×10^{-4} and 1×10^{-3} mg/ml), a negative interfacial pressure of approximately 2 mJ/m^2 was obtained.

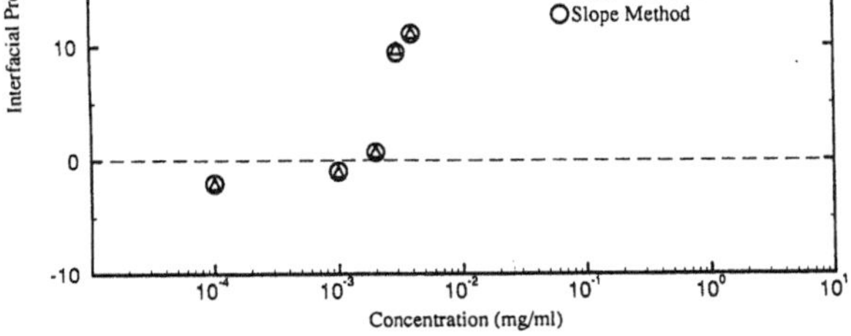

Fig. 10 The interfacial pressure obtained by the minimum slope criterion and the last experimental value for various bulk concentrations of the HSA.

In order to investigate whether the observed negative interfacial pressures are electrostatic in nature, the interfacial tension of the aqueous HSA solution-decane interface was measured at pH values of 3.5, 4.8 and 5.6; a pH of 4.8 represents the isoelectric point of HSA. This was accomplished by preparing the aqueous HSA solution in Trizma buffer. The measurements were performed at an protein concentration of 1×10^{-4} mg/ml. The interfacial tension of the buffer (without HSA) and decane was also measured. The results are given in Fig. 11. At a pH value of 3.5 (Fig. 11a), HSA decreased the interfacial tension approximately from 53.8 to 53.0 mJ/m^2. Therefore, a positive interfacial pressure of 0.8 mJ/m^2 was obtained. A positive interfacial pressure (1.6 mJ/m^2) was also obtained at the pH value of 4.8 (Fig. 11b); the interfacial tension was decreased approximately from 52.4 to 50.8 mJ/m^2. However, at a pH value of 5.6 (Fig. 11c), a negative interfacial pressure (about 0.4 mJ/m^2) was measured. The interfacial tension increased from 53.6 to 54.0 mJ/m^2. The pH value of the protein solutions without buffer was found to be 5.5 (and a negative interfacial pressure was found).

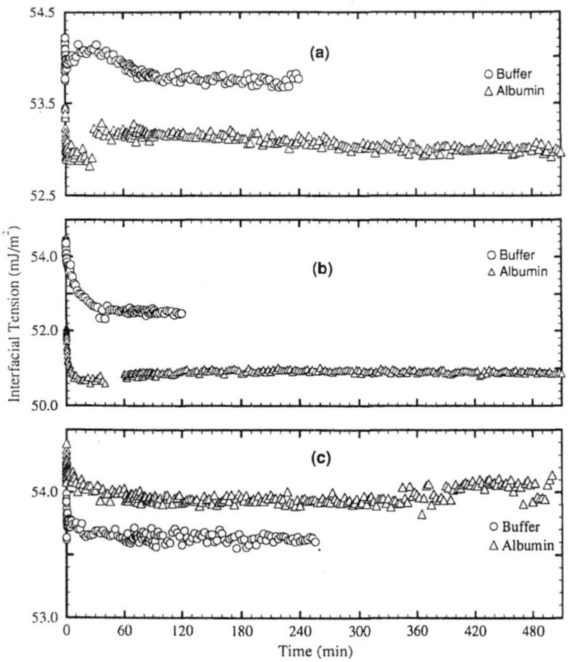

Fig. 11 The interfacial tension versus time at a bulk concentration of 0.0001 mg/ml. Measurements were performed at three pH values: (a) 3.5; (b) 4.8; (c) 5.6.

4.3 Discussion

4.3.1 Concentration Dependence of Interfacial Tension

In view of the concentration dependence of the interfacial tension, three domains may be identified in Fig. 9: (a) the slow change in γ at low bulk concentrations ($C < 10^{-3}$ mg/ml); (b) the sharp decline in γ within the region of intermediate concentrations ($10^{-3} < C < 10^{-2}$ mg/ml); (c) the time independent region of γ ($C > 10^{-2}$ mg/ml). The slow change in the interfacial tension at low concentrations may be indicative of relatively weak interaction between the adsorbed protein molecules in the adsorbed surface layer. As a result, the adsorbed molecules do not affect the interfacial tension strongly. Upon reaching the region of intermediate concentrations, the strong interactions between protein molecules in the adsorbed surface layer induce a sharp decline in the interfacial tension. As the bulk concentration increases further, the surface layer will be saturated with protein molecules forming a close-packed monolayer, resulting in a constant value for the interfacial tension. It can be postulated that the close-packed monolayer has a comparatively stable structure and the interfacial tension does not decrease noticeably at these high protein concentrations.

In order to quantify the concentration dependence of the interfacial tension, Gibbs' adsorption equation [27] may be used

$$\Gamma = -\frac{C}{RT}\frac{d\gamma}{dC} = -\frac{1}{RT}\frac{d\gamma}{d\ln C} \tag{1}$$

where Γ is the surface excess concentration (called surface concentration) of the protein, C is the protein concentration in the bulk phase of the aqueous solution, R is the universal gas constant, T is the temperature, and γ is the interfacial tension. This is the most frequently used adsorption equation in the field of surfactant adsorption; however, use of this relation requires caution. As seen in Fig. 9, the slope $d\gamma/dC$ is close to zero for high bulk concentrations. According to Eq. (1), this would imply that the surface concentration is close to zero, which is obviously incorrect.

Thus, it is apparent that Eq. (1) cannot be applied to the high concentration region. Figure 12 shows the region of Fig. 9 with bulk concentrations between 2×10^{-3} and 0.02 mg/ml; the interfacial tension values are those obtained by the minimum slope criterion. A linear curve-fit is also shown in Fig. 12. From this fit, the derivative $d\gamma/d\ln C$ can be easily calculated to be 57.0

mJ/m^2. Using the gas constant R = 8.31 J/(K∀mole) and the temperature T = 300 K, the resulting surface concentration is approximately 287 mg/m^2. This value is two orders of magnitude greater than the saturated (close-packed) monolayer concentration of bovine serum albumin (BSA) (about 4 mg/m^2 for BSA as measured by radio-tracer techniques [7]). This comparison between the two types of proteins is reasonable since they have similar molecular weights and structures. For BSA at a concentration of 4 mg/m^2, the corresponding surface area per molecule is 2500 Å^2, which is close to the estimated value of 3000 Å^2 for proteins such as BSA and HSA [28]. On the other hand, the surface area per molecule for HSA at a concentration of 287 mg/m^2 is 37 Å^2. Since it is not possible to compress a protein molecule by two orders of magnitude, this calculation indicates the inapplicability of Gibbs' adsorption equation in the region of intermediate protein concentration; similar conclusions have been drawn by others [e.g., Ref. 7].

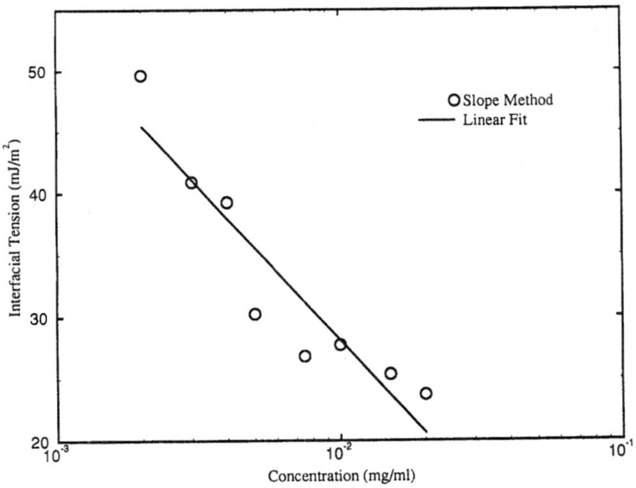

Fig. 12 Linear curve fit to the interfacial tension versus bulk HSA concentrations. The points were obtained by the minimum slope criterion.

In the region of low protein bulk concentrations (below 2×10^{-3} mg/ml), only two data points are available, at concentrations of 1×10^{-4} and 1×10^{-3} mg/ml. Nevertheless, these two points may be used to perform a preliminary evaluation of the applicability of Gibbs' adsorption equation. In a γ-C plot, the slope, $d\gamma/dC$, may be calculated by connecting the two points by a straight line. This slope may be substituted into Eq. (1) to calculate the surface concentration. The resulting surface concentrations are about 2.7 and 27 mg/m^2, which correspond to specific surface molecular areas of 4200 and 420 Å^2, for concentrations 1×10^{-4} and 1×10^{-3} mg/ml, respectively. The molecular area value at 1×10^{-4} mg/ml is of the same order of magnitude as that of the close-packed monolayer (about 3000 Å^2 [28]), but the molecular area value at 1×10^{-3} mg/ml is an order of magnitude smaller.

However, if a γ-log$C$ graph is used, and the slope, $d\gamma/d\log C$, is employed, a surface concentration of 11.6 mg/m^2 is obtained from Eq. (1). This surface concentration corresponds to a surface molecular area of 930 Å^2. This value is of the same order of magnitude as that of the close-packed monolayer. If the surface concentration can be assumed to be close to that of the saturated monolayer at these low protein concentrations in the bulk phase, the above calculation would indicate that Gibbs' adsorption equation may be applicable in the region of low bulk concentrations.

4.3.2 Negative interfacial pressure

Our experiments show a negative interfacial pressure for the aqueous human serum albumin solution-decane interface at low bulk albumin concentrations (i.e., 1×10^{-4} mg/ml and 1×10^{-3} mg/ml). Negative surface pressures have been reported in the past for organic and inorganic solutes in water. For example, the amino acid glycine increases the surface tension of water: For weight percentages (w/v) of 3.62, 6.98, 10.12, and 13.10, surface tensions of 72.54, 73.11, 73.74, and 74.18 mJ/m^2 have been reported, respectively [29]. We have performed measurements at weight percentages of 6.98 and 13.10 of glycine in double-distilled water by ADSA-P and have found close agreement with the literature values. These increases in interfacial tension are thought to have electrostatic origins [30]. When charged particles approach an interface from solution, particles with the same sign of charge will repel one another. This repulsion hinders ions from adsorbing to the interface. Thus, a depletion layer is formed that results in an increase in interfacial tension, and hence in a negative interfacial pressure. In case of the small amino acid glycine, repulsive interactions may occur between the dipolar amino acid molecules [31,32].

In the pH experiments, at the isoelectric point (pH=4.8) and below this point (pH=3.4), a positive interfacial pressure was obtained. However, a negative interfacial pressure was measured for a pH above the isoelectric point. Similarly, in the measurements without the Trizma buffer at a concentration of 1×10^{-4} mg/ml, a surface pressure of -1.7 mJ/m^2 (Fig. 10) and a pH of 5.5 were recorded. It is known that albumin molecules are negatively charged when the pH exceeds the isoelectric point because the side-chains, which have slightly more carboxyl groups than amino groups, are hydrolysed and become negatively charged. A charged albumin molecule at the interface induces a repulsive image potential. The resulting electrostatic repulsion will result in a depletion layer at the interface [30] and an increase in the interfacial tension.

At lower pH values, the albumin molecule exists in a fast-migrating and expanded form where most tyrosines and other hydrophobic residues are exposed to the solvent [33]. The hydrophobicity of the exposed residues provides the driving force for the albumin molecules to adsorb at the interface. Therefore, a surface depletion layer does not form and a rise in interfacial tension does not occur. At pH values above the isoelectric point, the protein has a different expanded form. There exists an increased accessibility of the hydrogen atoms for exchange, an increased mobility of the thiol group, and a slight loss of the helical structure [33]. This variant expanded form exposes less of the hydrophobic residues, and as a result the negative charges of the side-chains play a more dominant role in dictating the behaviour of the molecule at the interface. Hence, negative interfacial pressures are observed at a pH value above the isoelectric point.

The above explanation of charge effects may also be supported or supplemented by observations of the charge properties of hydrocarbon surfaces in aqueous solutions [34-36]. The measurements of the zeta potential at relatively high pH indicate that some hydrocarbons are negatively charged, just as albumin. Hence, there is an electrostatic repulsion between the hydrocarbon and the protein. This is in line with the supposition of the repulsive image potential induced by the protein adsorbed at the interface. It is then reasonable to expect that the observed negative interfacial pressure is electrostatic in nature. However, one has to be cautious about the role that the hydrocarbon plays in the negative interfacial pressure. The existence of hydrocarbon is not essential in obtaining negative interfacial potential. Experiments [37] have shown that, at a water/air interface, human serum albumin of low concentrations also has negative interfacial pressures. For example, in an HSA aqueous solution at concentration of 1×10^{-5} mg/ml, a surface tension of 73.5 mJ/m^2 is observed at 20°C, and this value is steady for more than four hours after an initial equilibration period of about 14 hours. Nevertheless, further experiments must be performed for the system presented here, so that a direct correlation between zeta potentials and negative interfacial pressures may be obtained.

To conclude, negative interfacial pressures were only observed at the two lowest bulk concentrations used in our experiments. There is a concentration region where a transition occurs from negative interfacial pressure to positive interfacial pressure, i.e., zero interfacial pressure. It can be postulated that, with an increase in the bulk albumin concentration, the conformation of the protein at the interface is altered. With increasing concentration, the expansion of the molecules may be restricted due to the close packing of the molecular segments at the interface. Therefore, repulsive electrostatic forces are overcome by the close packing of the protein molecules. As a result, rather than formation of a depletion layer leading to an elevation of the interfacial tension, a reduction in the interfacial tension and a positive surface pressure ensues.

4.4 Conclusions

(1) Three domains were identified in the effect of the bulk protein concentration on the interfacial tension for the aqueous human serum albumin-decane system: (a) a slow change in γ at low bulk concentrations ($C < 10^{-3}$ mg/ml); (b) a sharp decline in γ within the region of intermediate concentrations ($10^{-3} < C < 10^{-2}$ mg/ml): (c) the constant region of γ ($C > 10^{-2}$ mg/ml).

(2) At the two lowest bulk albumin concentrations, 1×10^{-4} mg/ml and 1×10^{-3} mg/ml, negative interfacial pressures were observed that may be attributed to repulsive electrostatic interactions and formation of a surface depletion layer. At all other concentrations, a positive surface pressure was measured.

(3) The Gibbs adsorption equation is unsuccessful in providing a realistic explanation of the variation of the surface concentration, Γ, with the bulk concentration, C, for $C > 10^{-3}$ mg/ml. However, for $C < 10^{-3}$ mg/ml, the possibility of the applicability of the Gibbs adsorption equation exists.

5 Dynamic Surface Tension of Mixed Solutions of a Protein and Small or Medium-Sized Organic Molecules

The preceding two sections study the static interfacial tension of protein solutions as a function of temperature and bulk concentration; this provides a fundamental understanding of protein surface activity and thermodynamic properties at equilibrium. To understand the dynamics of protein adsorption, dynamic surface tension has to be measured. ADSA is an ideal tool to achieve this goal since ADSA calculates surface tension, drop area and volume simultaneously and has recently been equipped with a motorized syringe. It can be used to study the surface tension response to various area changes, through which surface molecular movements and interactions can be revealed.

Transient relaxation experiments have been performed for a human serum albumin solution, where the surface tension response to a trapezoidal area variation was analyzed [18]. In the concentration interval studied there, the relaxation of the protein cannot be modeled by a diffusion controlled mechanism [18, 38-42]. In order to fit the prediction of the diffusion theory to the experimental data, the diffusion coefficient needs to be three or four orders of magnitude higher than the physically expected value [18,41]. This indicates that the actual adsorption is faster than the diffusion controlled process. Therefore, it might be suspected that some other factors are involved in increasing the rate of the adsorption process; presumably, liquid flow due to the rapid change in drop volume would assist the molecular transportation to the interface, and might well outpace diffusion.

In this chapter, we summarize a study in which ADSA-P was employed to measure the dynamic surface tension response to a saw-tooth area variation. A periodical area change was obtained by symmetrical increases and decreases of drop volume through a motorized syringe. The systems used were various aqueous solutions of bovine serum albumin (BSA) and small-medium organic molecules. The purpose of this study was to investigate the molecular interaction between bovine serum albumin and three kinds of small-medium organic molecules: dimethyl sulfoxide (DMSO), ethyl alcohol (ethanol), and a naturally occurring biologically active lipid, hepoxilin A_3, dissolved in DMSO. For recent reviews on the chemistry, biochemistry and pharmacology of the hepoxilins, please see references [43,44].

The interplay and competitive adsorption between proteins and smaller organic molecules, such as surfactants and lipids, at different surfaces is of fundamental importance to many biological processes. It is central to one of the most important functions of proteins, namely the adsorption at biological interfaces, and the structure of biological membranes [27,45-47]. Although a large amount of work has been done in this area [47-51], fundamental understanding of the mechanisms is limited. This is partially because most studies have been focused on the isotherm, i.e., equilibrium behavior of protein and smaller organic molecules at an interface. The dynamic and kinetic processes, which may be more important, especially when the interface itself is undergoing an area variation, have not yet been explored significantly.

5.1 Materials

The sample of bovine serum albumin (BSA) (Sigma Chemical Co., St. Louis, MO, USA.) was essentially fatty acid free and globulin free, with an average molecular weight of 67,000. It was used without further purification. Dimethyl sulfoxide (DMSO) and ethyl alcohol (ethanol) were obtained from Caledon Laboratories Ltd., Georgetown, ON, Canada. Hepoxilin A_3 was prepared by total chemical synthesis [52]. Water used in the experiment was distilled and deionised. Four types of samples were prepared: a) BSA aqueous solution at a concentration of 0.02 mg/ml, b) 1.0 μl DMSO added to 1.0 ml BSA solution, c) 1.0 μl ethanol added to 1.0 ml BSA solution, and d) 1.0 mg hepoxilin A_3 dissolved in 1.0 μl DMSO and added to 1.0 ml BSA solution.

5.2 Results

The above four types of systems were subjected to a symmetric saw-tooth shaped variation in the drop surface area, A. This may produce a similar saw-tooth variation in the surface tension, γ. At least four runs were performed for each sample system in our experiments; good reproducibility was achieved. The results presented below are single but representative of several runs.

In Fig. 13a, for a BSA aqueous solution at a concentration of 0.02 mg/ml, we can see an early transition in the pattern of the γ response, from an initial rather symmetric peak shape to a skewed one in response to the symmetric saw-tooth pattern in the area A. A skewed, asymmetric pattern of the γ response begins to develop after two or three cycles and becomes steady after 60 sec. Figure 13b shows the asymmetric shape more clearly at later times. In general, within each cycle, the dynamic surface tension increases as the surface is expanded (due to the reduction in the surface concentration), and γ decreases when A shrinks. The γ response shows two kinks (see arrows), one each in the branches of the surface expansion and compression.

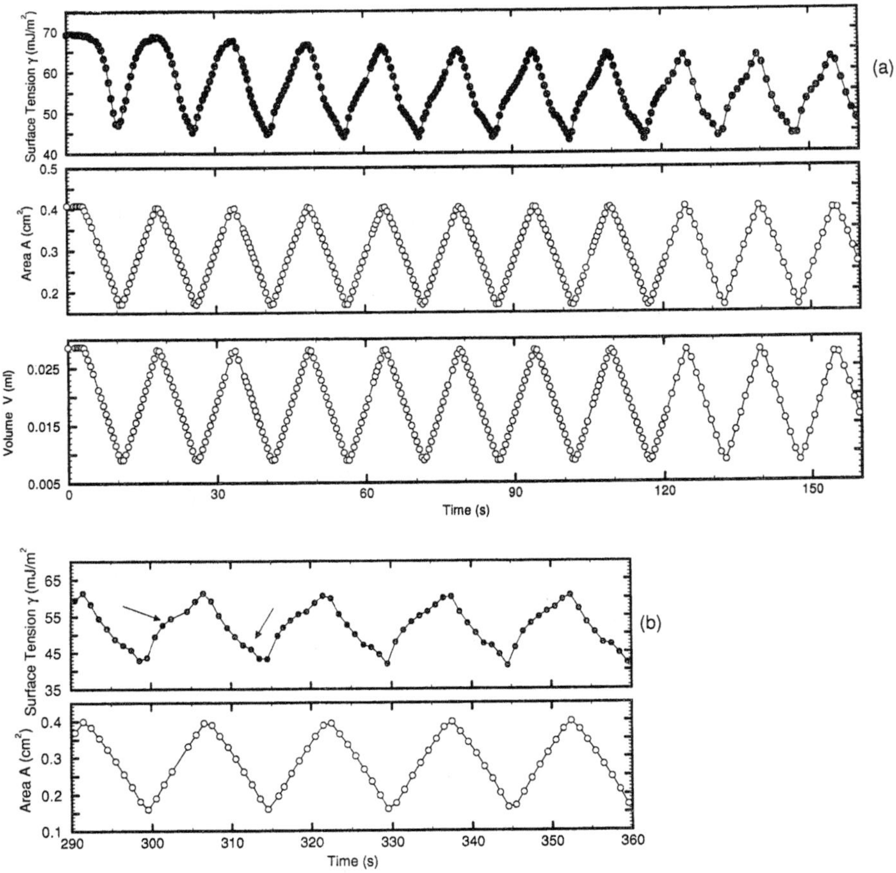

Fig. 13 (a) Dynamic surface tension, γ, response to a saw-tooth change in surface area of a BSA aqueous solution at a concentration of 0.02 mg/ml. (b) The γ response on an expanded scale, in late stages. The skewed pattern in the γ response is revealed clearly by arrows.

Figure 14a illustrates the γ response to the surface area variation of a system in which DMSO was added to the BSA aqueous solution at a concentration of 1.0 μl DMSO to 1.0 ml of 0.02 mg/ml BSA solution. A significant pattern change is observed in the γ response. The surface tension initially does not respond appreciably to the area variation. Then, beginning after 30 or 40 seconds, the surface tension shows cycles which gradually increase in amplitude and have a rather narrow, but symmetric valley. After approximately 180 seconds, the peaks start becoming asymmetric and towards the end of the experiment, after 360 seconds, the shape of the γ curve has become very similar if not identical to the one that was observed in pure BSA aqueous solution, see Fig. 13. Figure 14b provides a detailed picture of the asymmetry in the γ pattern at late stages. In Fig. 15a, the effect of adding ethanol to the BSA aqueous solution at a concentration of 1.0 μl ethanol in 1.0 ml of 0.02 mg/ml BSA solution is illustrated. The observed γ curve is similar to that observed for DMSO, except that the transition to the shape of the pure BSA system occurs earlier. Figure 15b shows the asymmetric tension cycles, on an expanded scale.

Figure 16a illustrates the results of addition of hepoxilin A_3 dissolved in DMSO (1.0 mg/ml) to the BSA aqueous solution at a ratio of 1.0 mg hepoxilin A_3 in 1.0 ml of 0.02 mg/ml BSA solution. A rather different pattern of the γ response is observed. Initially, γ variations have a small amplitude and a rounded shape upwards. With the passage of time, the amplitude of the surface tension oscillation gradually increases; it reaches a constant final value after about 120 s. However, the roundness of the peak remains for a considerable length of time (about 200 s). It is noticed that through the whole range of the experiment, the symmetric shape is maintained from cycle to cycle of the γ curve. At late stages, the γ peaks become symmetric with respect to the time axis as well and resemble closely the shape of the area variation. Figure 16b shows several of the symmetric cycles of the γ response toward the end of the experiment. This feature is different from all the previously mentioned observations where a skewed shape of γ was recorded. It is also observed that the disappearance of the round peaks is accompanied by a slight decrease in the height of the peak in the γ.

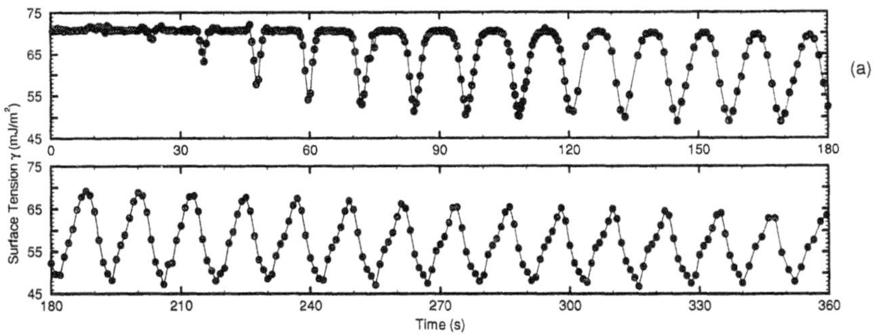

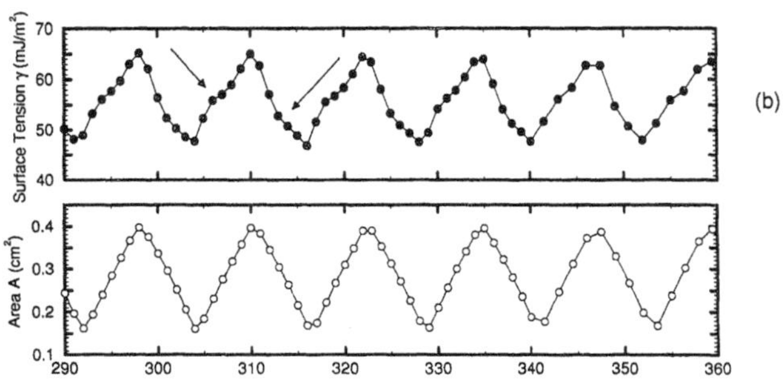

Fig. 14 (a) Dynamic surface tension, γ, response to the area change of a DMSO and BSA solution at a concentration of 1.0 μl DMSO/0.02 mg BSA in 1.0 ml water. The area change is the same as in Fig. 13 (saw-tooth shaped), and for space consideration it is omitted. A transition is shown in the γ response: from an initial symmetric pattern to a later asymmetric one. (b) The asymmetric γ oscillation on an expanded time scale in late stages. The pattern of the γ oscillation is similar to that of the pure BSA solution in Fig. 13. The arrows point to kinks.

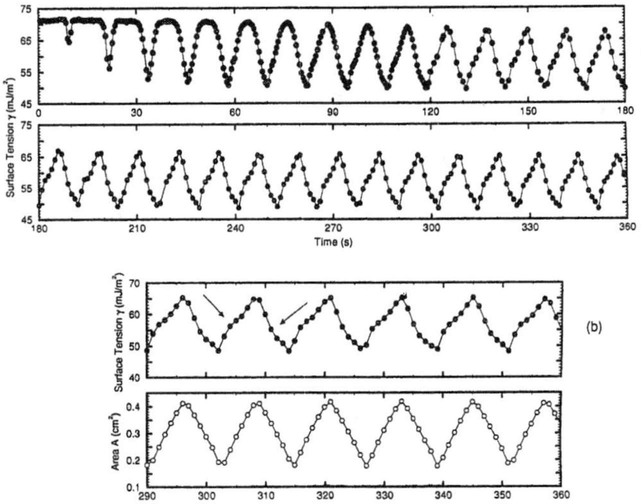

Fig. 15 (a) Dynamic surface tension, γ, response to the area change of an ethanol and BSA solution at a concentration of 1.0 μl ethanol/0.02 mg BSA in 1.0 ml water. The area change is the same as in Fig. (13). A transition in the pattern of the γ response is shown at about 120 s. (b) The skewed shape of the γ response on an expanded scale, in late stages of the experiment. The pattern in the γ response is similar to that of the pure BSA solution in Fig. 13. The arrows point to kinks.

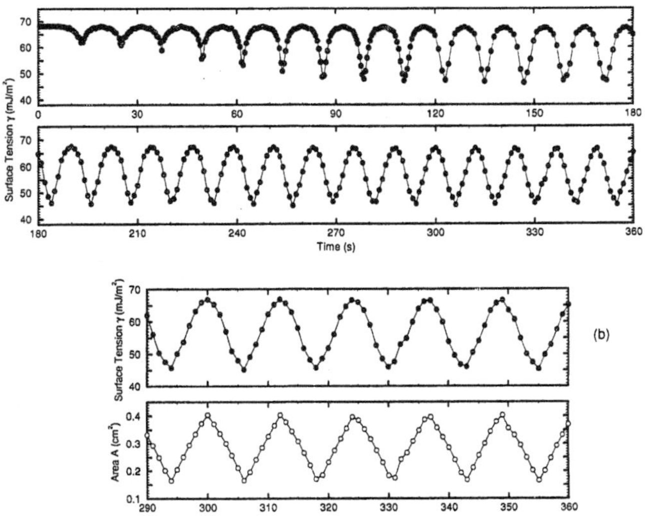

Fig. 16 (a) Dynamic surface tension, γ, response to the area change of hepoxilin A_3 dissolved in DMSO (1.0 μg/μl) and added 1.0 μl to a solution of 0.02 mg BSA in 1.0 ml water. The area change is the same as in Fig. (13). A symmetric pattern in the γ response is observed throughout the experiment. (b) Symmetric shape of the γ response on an expanded scale, in late stages of the experiment. The pattern in the γ response is different from that of the pure BSA solution in Fig. 13.

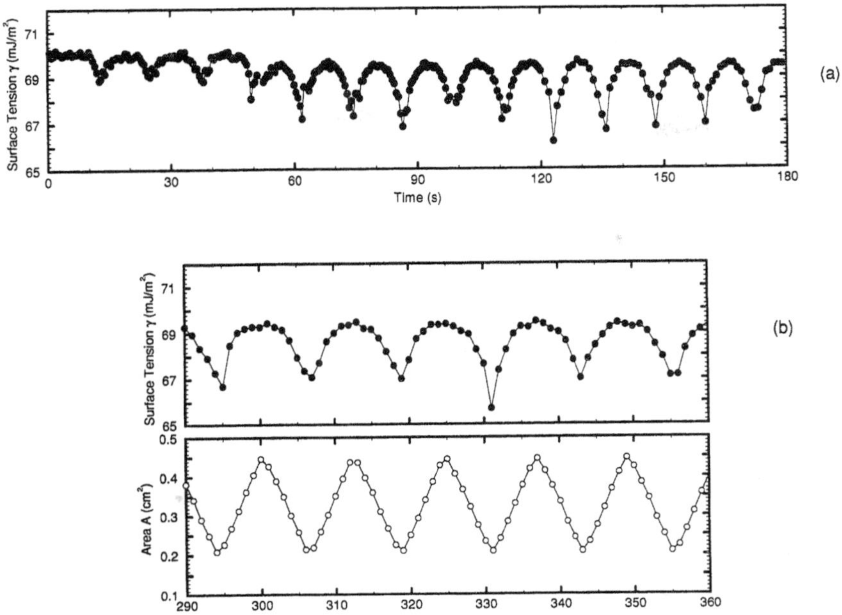

Fig. 17 (a) Dynamic surface tension, γ, response for 1.0 μg of hepoxilin A_3 (1.0 μg/μl DMSO) added to 1.0 ml water. The area change is the same as in Fig. 13. Approximately symmetric pattern in the γ response is observed throughout the experiment. (b) The γ response on an expanded scale, in late stages of the experiment. The amplitude of the response is small in comparison with that of pure BSA aqueous solution as given in Fig. 13.

Since the γ response to the area change for the system with hepoxilin A_3 is different from those for all other systems, it is necessary to study the surface behavior of hepoxilin A_3 on its own in more detail. Additional experiments were performed with hepoxilin A_3 solution alone, in the absence of BSA, at a concentration of 1.0 mg hepoxilin A_3/1.0 μl DMSO/1.0 ml water. The results are shown in Figs. 17a and 17b. Round and symmetric peaks in the γ response are observed while the surface area A varies in a saw-tooth pattern. It is noticed that the amplitude of the γ oscillation is rather small, about 1 mJ/m^2 in the first four or five cycles and increasing to approximately 3 mJ/m^2, as compared with that of BSA of more than 20 mJ/m^2. Throughout the experiment, the pattern of the γ response does not change.

5.3 Discussion

In mixtures of small or medium organic molecules and protein solutions, Figs. 14-16, the dynamic surface tension γ response to the surface area A variation is drastically different from that observed in the pure BSA aqueous solution (Fig. 13) during the early stages. In the presence of small molecules, a small or near zero amplitude is generally observed. During the late stages of the experiment, the γ response in two systems, the DMSO/BSA and the ethanol/BSA solutions, are similar to that observed in pure BSA solution. In contrast, when hepoxilin A_3 is added to the BSA solution, the γ response is quite different from that of pure BSA solution. In the later stages of the experiment, a symmetric pattern, both vertically and horizontally, for the response in γ is observed for the hepoxilin A_3 /BSA system while an asymmetric shape of the γ response is obtained in the case of the pure BSA solution during the same time period.

The key to understand the above observations is as follows: Compared with protein, DMSO and ethanol can adsorb at the interface more readily because of their smaller molecular size and higher diffusivity; they also desorb from the interface more easily, upon compression. But during each expansion, some protein molecules adsorb at the interface; the adsorbed protein molecules do not desorb upon the next compression. As a result, the periodic expansions and compressions lead to increasing surface concentration of protein and decreasing surface concentration of the small molecules; eventually the small molecules may be squeezed out and protein alone remains at the interface.

In Figs. 14 and 15, the small organic molecules of DMSO and ethanol dominate the molecular population at the surface in the early stages. Therefore, in the initial stages of the experiment, the surface properties are governed mainly by the small organic molecules. In general, a change of the surface tension γ is associated with the change in the surface concentration. An increase in the surface area A will tend to induce a decrease in the surface molecular concentration, and hence an increase in the surface tension. Conversely, a decrease in the surface area will cause a decrease in γ. However, the amplitude of the γ oscillation in response to the area change is also related to the desorption behavior of adsorbed molecules, in addition to the area variation. If the surface molecules can desorb from the surface sufficiently fast so that the molecules can quickly adjust their surface concentration to maintain a constant value while the surface is compressed, then the surface tension, which is determined mainly by these molecules, will show little variation with changing surface area [53]. With the continuation of the surface expansion and compression, more and more protein molecules are transported to and adsorbed at the interface. The experiments with the DMSO/BSA and ethanol/BSA solutions show that, in the late stages, the pattern of the surface tension response to the area changes is similar to that of the pure BSA system, Figs. 13-15. This indicates that the surface tension behavior in the late stages is governed mainly by the protein molecules at the surface for these two systems. The skewed shape in the γ response is suggestive of a surface conformational change in the protein molecules [53].

In the late stages of the experiment, the difference in appearance between the DMSO/BSA and ethanol/BSA systems on the one hand (Figs. 14,15), and the hepoxilin A_3/BSA system on the other hand (Fig. 16) has to be seen in this light. The former systems show a very similar γ response to that of the pure BSA aqueous solution, Figs. 13-15. Therefore, one may assume that the adsorption layer is composed of BSA molecules, and the small organic molecules have been squeezed out of the surface. On the other hand, for the hepoxilin A_3/BSA system, a different γ response is observed (compare Figs. 13 and 16). This indicates that the hepoxilin A_3 molecules are not squeezed out of the surface; possibly, binding occurs between the hepoxilin A_3 molecules and the protein molecules at the surface. We expect that this binding occurs both in the bulk and at the surface. The absence of the skewed shape in the γ response in the late stages of the experiment suggests a stabilization of a single conformation of the protein molecules, compared to at least two in the DMSO/BSA, ethanol/BSA and pure BSA systems.

The inference of molecular binding of hepoxilin A_3 to the protein, however, needs to be further justified. One might argue that, without binding, hepoxilin A_3 molecules might nevertheless accumulate at the surface because of the low solubility in water. Then, the physical picture of the surface would be that both protein and hepoxilin A_3 molecules adsorb at the surface, and they both contribute to the dynamic surface tension independently, i.e., additively. Thus, in the later stages of the experiment, the γ response would have to be an addition of the two types of surface tension responses: one from hepoxilin A_3, and the other from the protein. The γ behavior of hepoxilin A_3 alone can be seen in Fig. 17, where the γ peaks are quite symmetric. (Note that the symmetric pattern of the γ curve in Fig. 17 is similar to the γ response in the early stage of the experiment, Fig. 16. This supports our supposition that the small molecules determine the surface tension variation with the area changes in the early stages of the experiment.) The γ response of the pure BSA solution, Fig. 13, shows a rather skewed shape of the peaks. If we add these two γ curves, then an asymmetric shape in the γ peaks would be expected. However, from Fig. 16, we observe a symmetrically shaped surface tension peaks, which are different from the sum of the two independent γ responses of the protein and hepoxilin A_3 [53].

One might also consider that, since the concentration used in the present experiment gives a molecular ratio of 10:1 for hepoxilin A_3:BSA, the γ response from the mixture of hepoxilin A_3/BSA might pre-dominantly result from that of hepoxilin A_3 alone. However, Fig. 17 shows that only small amplitudes (1 to 3 mJ/m^2) of the γ oscillation can be observed from hepoxilin A_3, while the experiment with the mixture of the two shows a nearly ten times larger amplitude in the surface tension response at the same area perturbation. It is thus apparent that the observed response, Fig. 16, cannot result from either of the two elements alone, nor from the simple addition or superposition of the two. The molecular interaction between hepoxilin A_3 and BSA is very likely the main cause for the shape of the γ curve. The experimental data support strongly the notion of binding between protein and hepoxilin A_3 .

From the observations of the DMSO/BSA and ethanol/BSA solutions, the hypothesis of the squeeze-out of small molecules is confirmed. We note that the compression/dilation cycles illustrated here might be useful as a mechanism to purify surface protein layers. That is, by oscillating the surface area, impurities (of small molecular sizes) will be pushed out and only protein molecules left at the surface. This is demonstrated in Fig. 13 of the BSA solution, in which the γ response changes from an initial symmetric pattern to a skewed one (as observed with the pure BSA), as early as about 30 s after starting the experiment. This may indicate that impurities previously existing in the BSA solution may be quickly squeezed out of the surface by oscillating the surface area, resulting in a purified BSA film. Note that, in Fig. 13, the first few γ-peaks are rounded at the top, similar to those of the DMSO/BSA and ethanol/BSA systems at early and intermediate times [53].

5.4 Conclusions

(1) ADSA provides a powerful tool for measuring the dynamic surface tension γ response to a surface area variation for solutions of a protein and of mixtures of protein and other molecular species.

(2) Competitive adsorption between small organic molecules (DMSO and ethanol) and a protein (BSA) has been demonstrated through the γ response. The smaller organic molecules dominate the population at the surface in the initial stages of the experiment, as documented in initial small amplitudes of the surface tension oscillation. In the late stages, a squeeze-out mechanism is operative, and the protein BSA is purified at the surface by the surface area oscillations.

(3) The interaction of hepoxilin A_3 with BSA is very different from that of ethanol and DMSO. The experiments indicate binding of hepoxilin A_3 to the protein.

6 Acknowledgements

The perceptive suggestions of Dr. R. Miller are greatly appreciated. Financial support for this project was provided by the Medical Research Council of Canada (grant No. MT-5462).

7 References

1. J.L. Brash and T.A. Horbett, in "Proteins at Interface II. Fundamentals and Applications," T.A. Horbett and J.L. Brash eds. American Chemical Society, Washington, DC, chapter 1, pp. 1-25, 1995.

2. H.B. Callen, "Thermodynamics and an Introduction to Thermostatics," 2nd. ed. Wiley, New York, 1985.

3. Y. Rotenberg, L. Boruvka and A.W. Neumann, *J. Colloid Interface Sci.* **93**, 169, 1983.

4. S. Lahooti, O.I. del Río, P. Cheng and A.W. Neumann, in "Applied Surface Thermodynamics," A.W. Neumann and J.K. Spelt, eds. Marcel Dekker, New York, chapter 10, pp. 441-507, 1996.

5. O.I. del Río and A.W. Neumann, *J. Colloid Interface Sci.* submitted, 1996.

6. P. Chen, D.Y. Kwok, R.M. Prokop, O.I. del Río, S.S. Susnar and A.W. Neumann, in "Drops and Bubbles in Interfacial Research," D. Möbius and R. Miller eds. Elsevier, Amsterdam, The Netherlands, submitted, 1997.

7. D.E. Graham and M.C. Phillips, *J. Colloid Interface Sci.* **70**, 415, 1979.

8. M.A. Cabrerizo-Vílchez, Z. Policova, D.Y. Kwok, P. Chen and A.W. Neumann, *Colloids Surfaces B: Biointerfaces*, **5**, 1, 1995.

9. M. Paulsson and P. Dejmek, *J. Colloid Interface Sci.* **150**, 394, 1992.

10. D.E. Graham and M.C. Phillips, *J. Colloid Interface Sci.* **70**, 403, 1979.

11, P. Suttiprasit, V. Krisdhasima and J. McGuire, *J. Colloid Interface Sci.* **154**, 316, 1992.

12. E. Tornberg, *J. Colloid Interface Sci.* **64**, 391, 1978.

13. A.J.I. Ward and L.H. Regan, *J. Colloid Interface Sci.* **78**, 389, 1980.

14. A. Voigt, O. Thiel, D. Williams, Z. Policova, W. Zingg and A.W. Neumann, *Colloids Surfaces*, **58**, 315, 1991.

15. H.H.J. Girault, D.J. Schiffrin and B.D.V. Smith, *J. Colloid Interface Sci.* **101**, 257, 1984.

16. C. Maze and G. Burnet, *Surface Sci.* **24**, 335, 1971.

17. P. Cheng and A.W. Neumann, Colloids Surfaces, **62**, 297, 1992.

18. R. Miller, Z. Policova, R. Sedev and A.W. Neumann, *Colloids Surfaces A: Physiochem. Eng. Aspects*, **76**, 179, 1993.

19. S.S. Susnar, H.A. Hamza and A.W. Neumann, *Colloids Surfaces A: Physiochem. Eng. Aspects*, **89**, 169, 1994.

20. G. Gonzalez and F. MacRitchie, *J. Colloid Interface Sci.* **32**, 55, 1970.

21. R. Miller and G. Kretzschmar, *Adv. Colloid Interface Sci.* **37**, 97, 1991.

22. R. Miller and K. Lunkenheimer, *Colloid Polym. Sci.* **261**, 585, 1983.

23. P. Chen, S. Lahooti, Z. Policova, M.A. Cabrerizo-Vílchez and A.W. Neumann, *Colloids Surfaces B: Biointerfaces*, **6**, 279, 1996.

24. R. Miller, S. Treppo, A. Voigt, W. Zingg and A.W. Neumann, *Colloids Surfaces*, **69**, 203, 1993.

25. J.A. de Feijter and J. Benjamin, in "Food Emulsions and Foams," E. Dickinson ed. Royal Society of Chemistry, London, p. 72, 1987.

26. F. MacRitchie, in "Proteins at Interfaces: Physicochemical and Biochemical Studies," J.L. Brash and T.A. Horbett eds. American Chemical Society, Washington, DC, chapter 11, 1987.

27. A.W. Adamson, "Physical Chemistry of Surfaces," 5th edn. John Wiley, New York, 1990.

28. B.R. Malcolm, *Prog. Surf. Membr. Sci.* **7**, 183, 1973.

29. R.C. Weast (Ed.), "Handbook of Chemistry and Physics," The Chemical Rubber Company, Cleveland, Ohio, pp. 28-29, 1971-72.

30. C. Ramachandran, R.A. Pyter, P. Mukerjee, *J. Phys. Chem.* **86**, 3198, 1982.

31. L. Onsager and N.N.T. Samaras, *J. Chem. Phys.* **2**, 528, 1934.

32. F.P. Buff and N.S. Goel, *J. Chem. Phys.* **56**, 2405, 1972.

33. T. Peters Jr. and R.G. Reed, in "Albumin Structure, Biosynthesis, Function," T. Peters and I. Sjoholm Eds. Proceedings of the 11th FEBS Meeting, Copenhagen 1977, Vol. 50, Colloquium B9, Pergamon Press, Oxford, p.11, 1978.

34. K. B. Medrzycka, *Colloid Polym. Sci.* **269**, 85, 1991.

35. H. C. van der Mei, J. de Vries, and H. J. Busscher, *Appl. Environ. Microbiol.* **59**, 4305, 1993.

36. G. I. Geertsema-Doornbusch, H. C. van der Mei, and H. J. Busscher, *J. Microbiol. Meth.* **18**, 61, 1993.

37. P. Chen, Z. Policova and A.W. Neumann, in preparation.

38. G. Loglio, U. Tesei, N. Degli-Innocenti, R. Miller and R. Cini, *Colloids Surfaces*, **57**, 335, 1991.

39. R. Miller, G. Loglio, U. Tesei and K.-H. Schano, *Adv. Colloid Interface Sci.* **37**, 73, 1991.

40. G. Loglio, U. Tesei, R. Miller and R. Cini, *Colloids Surfaces*, **61**, 219, 1991.

41. H.A. Sober, ed. “Handbook of Biochemistry,” The Chemical Rubber Co. Cleveland, OH, 1968.

42. R. Miller, Trends Polym. Sci. 2, 47, 1992.

43. C. R. Pace-Asciak, *Biochim. Biophys. Acta*, **1215**, 1, 1994.

44. C. R. Pace-Asciak, D. Reynaud and P. M. Demin, *Lipids*, **30**, 1, 1995.

45. I. D. Robb, in "Anionic Surfactants," E. H. Lucassen-Reynders, ed. Surfactant Science Series, Marcel Dekker, Inc. New York, vol. 11, p. 109, 1981.

46. G. L. Gaines, Jr. "Insoluble Monolayers at Liquid-Gas Interfaces," Interscience Publishers, New York, 1966.

47. M. J. Rosen, "Surfactants and Interfacial Phenomena," 2nd. edn. John Wiley & Sons, New York, 1989.

48. S. W. H. Eijt, M. M. Wittebrood, M. A. C. Devillers and Th. Rasing, *Langmuir*, **10**, 4498, 1994.

49. M. Wahlgren, S. Welin-Klintstroem, T. Arnebrant, A. Askendal and H. Elwing, *Colloids Surfaces B: Biointerfaces*, **4**, 23, 1995.

50. V. Hlady and J. D. Andrade, *Colloids Surfaces*, **32**, 359, 1988.

51. W. Norde, *Adv. Colloid Interface Sci.* **25**, 267, 1986.

52. P. M. Demin, L. L. Vasilyeva, G. I. Myagkova and K. K. Pivnitsky, *Bioorg. Khim.* **17**, 1133, 1991.

53. P. Chen, Z. Policova, S.S. Susnar, C.R. Pace-Asciak, P.M. Demin and A.W. Neumann, *Colloids Surfaces A: Physiochem. Eng. Aspects*, **114**, 99, 1996.

8 List of Symbols

A	surface area
C	bulk concentration
R	universal gas constant
T	temperature
t	time
Γ	interfacial (surface) concentration
γ	interfacial (surface) tension
γ_0	interfacial (surface) tension of pure solvents
γ_∞	equilibrium interfacial (surface) tension
π	interfacial (surface) pressure

9 List of Abbreviations

ADSA	axisymmetric drop shape analysis
ADSA-P	axisymmetric drop shape analysis-profile
BSA	bovine serum albumin
DMSO	dimethyl sulfoxide
HSA	human serum albumin

Proteins at Liquid Interfaces
D. Möbius and R. Miller (Editors)

SURFACE DILATIONAL RHEOLOGY OF PROTEINS ADSORBED AT AIR/WATER AND OIL/WATER INTERFACES

J. Benjamins and E.H. Lucassen-Reynders*

Unilever Research Laboratorium Vlaardingen, Olivier van Noortlaan 120,
3133 AT Vlaardingen, the Netherlands

* present address: Mathenesselaan 11, 2343 HA Oegstgeest, the Netherlands

Contents

1 INTRODUCTION

Proteins can be extremely effective in producing and stabilising foams and emulsions. To a degree, their role in this is similar to that of low molecular weight surface active molecules: both types of molecules adsorb at air/water (foams) and oil/water (emulsions) interfaces. The primary effect of such molecular adsorption is that it reduces the tension of the interface. However, the reduction of the interfacial tension cannot in itself explain the formation of emulsions and foams with more than transient stability. If this were the case, it should be possible to prepare emulsions and foams in the absence of surface active solutes, from pure low tension liquids. In practice it is impossible to obtain emulsions with any degree of stability in this way.

The essential stabilising function of surface active molecules, including proteins, during emulsification and foaming is that they enable the interface to resist tangential stresses from the adjoining flowing liquids (van den Tempel [1]). A surfactant-covered interface can behave as a two-dimensional body with its own rheological properties, such as elasticity and viscosity. These properties provide the liquid films separating emulsion drops or foam bubbles with a mechanism for dynamic stabilisation, without which any two drops or bubbles just formed would be liable to re-coalescence before the end of the dispersion process.

Rheological interfacial parameters can be defined for both compressional deformation and shearing motion in the interface. The former type of experiment measures the response of the isotropic component of the surface stress tensor to changes in surface area at constant shape of a deformed surface element, while the latter measures the response of the tensor's deviatoric component to changes in shape at constant area. Both types of measurement can be performed at small periodic deformations as well as under continuous expansion or shear (Lucassen and van den Tempel [2]; de Feijter [3]; Dickinson et al. [4]; Miller et al. [5]; Lucassen-Reynders [6]). Surface shear viscosity has enjoyed the greater attention of experimentalists working with small-molecule surfactants (Goodrich et al. [7]; Mohan et al. [8]). Food proteins have also been studied in surface shear by Graham and Phillips [9], Martinez-Mendoza and Sherman [10], Kiosseoglou [11], Benjamins and van Voorst Vader [12] and by Murray and Dickinson [13]. These studies have shown that surface shear viscosities of protein layers are far higher than those of small molecules, and that they keep increasing considerably with increasing age of the interface. While such viscosity may contribute appreciably to the long-term stability of emulsions and foams, it cannot be expected to have much relevance to their short-term stability for two reasons. First, the type of deformation that interfaces undergo during emulsification and foaming is expansion and, to a lesser extent, compression rather than shear. Second, the new interface which is continuously formed during the dispersion process, in time scales which may be as low as 1 ms or less, cannot build up the high shear viscosities found for aged interfaces. For short-term stability, therefore, interfacial rheology in compression/expansion is considered to be far more relevant.

This chapter will focus on the latter type of rheology; the effects of surface shear will be considered only when interfering with the measurement of surface dilational properties. The surface dilational modulus in compression and expansion is defined by the expression originally proposed by Gibbs [14] for the surface elasticity of a soap-stabilised liquid film as the increase in surface tension for a small increase in area of a surface element:

$$\varepsilon = \frac{d\gamma}{d\ln A} \quad (1)$$

where γ is the surface tension and A the area of the surface element. In the simplest case, the modulus is a pure elasticity with a limiting value, ε_0, to be deduced from the surface equation of state, i.e., from the equilibrium relationship between surface tension and surfactant adsorption, Γ:

$$\varepsilon_0 = \left(\frac{-d\gamma}{d\ln\Gamma}\right)_{eq} \quad (2)$$

This limiting value is reached only if there is no exchange of surfactant with the adjoining bulk solution (i.e., $\Gamma \times A$ is constant), and if, moreover, the surface tension adjusts instantaneously to the equilibrium value for the new adsorption. Deviations from this simple limit occur when relaxation processes in or near the surface affect either γ or Γ within the time of the measurement. In such cases, the modulus ε is a surface viscoelasticity, with an elastic part accounting for the recoverable energy stored in the interface and a viscous contribution reflecting the loss of energy through any relaxation processes occurring at or near the surface. Elastic and viscous contributions can be measured separately by subjecting the surface to small periodic compressions and expansions at a given frequency. In such experiments the viscoelastic modulus ε is a complex number, with a real part ε' (the storage modulus) equal to the elasticity, ε_d, and the imaginary part ε'' (the loss modulus) given by the product of the viscosity, η_d, and the imposed angular frequency, ω, of the area variations:

$$\varepsilon = \varepsilon' + i\varepsilon'' = \varepsilon_d + i\omega\eta_d \quad (3)$$

Experimentally, the imaginary contribution to the modulus ε is reflected in a phase difference ϕ between stress ($d\gamma$) and strain (dA), which means that the elastic and viscous contributions are given by:

$$\varepsilon' = |\varepsilon| \cos\phi \quad ; \quad \varepsilon'' = |\varepsilon| \sin\phi \quad (4)$$

where $|\varepsilon|$ is the absolute value of the complex modulus and ϕ is the viscous phase angle.

For low molecular weight surfactants, the dynamic behaviour as reflected in the modulus ε is more or less fully understood and can often be explained from the characteristic parameters of the surfactant, e.g., its surface-equation-of-state and transport parameters. For proteins it has not been possible yet to build a comprehensive model in spite of many recent experimental data (Graham and Phillips [15-17]; Serrien et al. [18]; Gau et al. [19]; van Aken and Merks [20, 21]; Boury et al. [22]). Bottlenecks for building such a model from literature data are that (i) the same proteins from different sources will sometimes give different results, especially if the molecules have been modified, e.g., in radio tracer probing, (ii) the experimental techniques that have been used are not always mutually compatible, and (iii) model building is hampered because it is unclear how the apparent "irreversibility" of protein adsorption works out in the dynamic behaviour.

The aim of the present chapter is to survey the experimental rheological data available for adsorbed protein layers under compression/dilation and, where possible, to relate these data to the molecular properties of the protein through independently measured equilibrium adsorption properties.

2 EXPERIMENTAL METHODS

2.1 PRINCIPLES AND PROBLEMS

Convenient techniques for measuring surface dilational moduli are derived from the longitudinal wave method developed by Lucassen and van den Tempel [2]. In such methods, the surface is periodically expanded and compressed, usually but not invariably by a barrier which oscillates in the plane of the surface, and the response of the surface tension is monitored by a probe, e.g., a Wilhelmy plate, some distance away from the barrier. A problem here is that the amplitude of the area variations generated by the barrier may be substantially damped when the area disturbance reaches the probe. The area variation generated by the barrier travels over the surface as a longitudinal wave, with characteristics derived by Lucassen and van den Tempel [23]. The relevant wave properties, wavelength (λ) and damping coefficient (β), were found to depend far more strongly on the surface dilational modulus than is the case for transverse capillary waves or ripples.

The equations describing small-amplitude surface waves can be obtained by solving the hydrodynamic equations of motion of the solution below the surface, using the boundary condition that the viscous drag exerted by the surface on the solution is compensated by a surface tension gradient, in the absence of appreciable resistance to surface shear. Three possibilities must be distinguished depending on the spatial damping coefficient of the wave and the effective length (L, see Section 2.2) of the trough. Most convenient experimentally is the limiting case where the wavelength (λ) is much greater than L and the damping coefficient is much smaller than 1/L. In this region the wave causes multiple reflections against the walls of the trough and the barrier. As a result, the surface undergoes a practically uniform deformation, without the wave character being apparent from variations in phase and amplitude with distance. This region is characterised by very high values of the wave propagation number W defined by Lucassen and Barnes [24]:

$$W \equiv \frac{1}{\beta L} = \frac{|\varepsilon|^{1/2}}{L\omega^{3/4}\left(\sqrt{\eta\rho} + \sqrt{\eta'\rho'}\right)^{1/2} \sin(\pi/8 + \phi/2)} \tag{5}$$

where η and ρ are liquid viscosity and density respectively, and primed symbols refer to the upper fluid phase. Thus, W represents that fraction of the distance between the wave generator and the end of the trough at which the wave is damped to 1/e of its original amplitude. In cases where W >>1, the viscoelastic modulus simply can be obtained from the surface tension variation (Δγ) measured anywhere on the surface and from the overall area change (ΔA):

$$|\varepsilon| = \left[(\varepsilon')^2 + (\varepsilon'')^2\right]^{1/2} = \frac{A\,\Delta\gamma}{\Delta A} \tag{6}$$

At the other extreme of wave propagation characteristics, the surface wave is fully damped before it has travelled the distance L. In this case the viscoelastic modulus $|\varepsilon|$ and the viscous phase angle φ can be determined by measuring wave number κ (=2π/λ) and damping coefficient β:

$$|\varepsilon| = \frac{\omega^{3/2}(\sqrt{\eta\rho} + \sqrt{\eta'\rho'})}{\kappa^2 + \beta^2} \tag{7}$$

$$\phi = 2\arctan(\beta/\kappa) - \pi/4 \tag{8}$$

(Lucassen and van den Tempel [2]). In this region the wave propagation number W is very small (W <<1, i.e., βL>>1).

In the intermediate region, where W is of order 1, no explicit expressions exist for the calculation of modulus and viscous phase angle φ from measured values of Δγ and phase difference as a function of distance from the oscillating barrier. The only way to determine required data in this region is to compare the curve of the measured Δγ and phase difference with a set of predicted curves [24]. This procedure is laborious; it is advisable to circumvent this region by adapting trough length and/or frequency.

The conventional technique, in which surface waves are generated by a linear barrier oscillating in the surface and the surface tension is monitored some distance from the barrier is particularly suitable for studying the dynamics of small-molecule surfactants at the vapour/water surface [2,25,26]. Difficulties arise when (i) the non-aqueous phase is a viscous liquid rather than vapour and/or (ii) the adsorbed surfactant causes appreciable resistance against surface shear:

(i) *Oil/water interfaces* require smaller effective trough lengths L for the condition of uniform deformation ($W >> 1$) to be met, because the higher viscosity of oil in comparison to vapour increases wave damping and hinders wave propagation. Such small effective lengths L are more easily achieved in the novel method of the Dynamic Drop Tensiometer (Benjamins et al. [27]). This instrument (see Section 2.3) subjects a small drop or bubble, which either hangs from or sits on the tip of a capillary, to sinusoidal oscillations of its volume and, therefore, also of its area. Here the role of the oscillating barrier is played by the rim of the capillary against which the interface is being alternately compressed and expanded. For an external diameter of the capillary of 2 mm, the distance over which the wave can travel over the drop is only a few mm, i.e., much smaller than the effective trough length in the conventional set-up. The drop method, therefore, facilitates rheological measurements at oil/water interfaces.

(ii) *Surface shear resistance* is a complicating factor of particular relevance for proteinaceous surfaces. In contrast to small-molecule surfactants, proteins can produce appreciable resistance to shearing motion. Interference of these properties with the measurement of the dilational modulus ε can be considerable, especially at low values of ε. Instruments designed to obviate such interference include the square-band barrier of Benjamins et al. [28] (described in Section 2.2.), the ring trough of Kokelaar et al. [29], the oscillating bubble method of Wantke et al. [30] and the dynamic drop tensiometer of Benjamins et al. [27]. This last instrument, described in Section 2.3., is suitable for both vapour/water and oil/water interfaces and has the advantage of a small effective trough length. Thus, it can generate a regime of compression/expansion which is both isotropic and uniform.

2.2 CONVENTIONAL AND MODIFIED METHODS: TROUGH WITH BARRIER AND PLATE

The conventional apparatus, schematically drawn in Figure 1, consists of a rectangular shallow trough with one or two movable teflon bars as barrier(s) compressing and expanding the interface, and a Wilhelmy plate for monitoring the response of the interfacial tension. For a vapour/water interface to be investigated, the trough is filled up to the rim with the aqueous solution, and the barriers are placed on top. The interfacial area between the teflon bars is compressed and expanded by a sinusoidal movement of one or both end bars. This sinusoidal movement is generated by an eccentric driven by a constant-speed motor attached to a gear system. The change of the interfacial tension $d\gamma$ produced by the change in the interfacial area dA is measured with a properly positioned Wilhelmy plate attached to a force transducer. The Wilhelmy plate can be situated anywhere on the surface if the deformation of the interface is uniform, i.e., if the wavelength λ of the compression/expansion wave generated by the barrier movement is much larger than the length of the trough ($W \gg 1$; see Section 2.1).

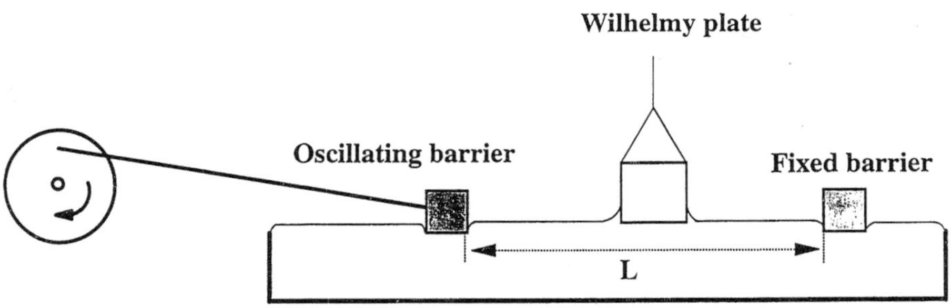

Fig. 1 Conventional barrier-and-plate apparatus used for vapour/water surface (side view).

Oil/water interfaces require not only a smaller trough length (L), but also better precautions against leakage of surface-active material past the movable barrier. In our experiments [27], a reduction of the trough length to 15 cm was enough to satisfy the condition for uniform deformation.

In Fig. 1, the compression/expansion effected by the barrier is unidirectional rather than isotropic. The surface deformation, therefore, contains a shearing component which can interfere with the measurement of the dilational modulus in cases of appreciable resistance against shear. A modified version, shown in Figure 2, avoids such complications by producing isotropic dilational deformation of the surface. The area to be subjected to compression/expansion is isolated from the rest of the surface by a square of elastic bands placed vertically in the surface. The corners of the square are connected to one eccentric driven by an electromotor with gearbox. The construction ensures that the corners move synchronously along the square's diagonals. In this set-up the cycle frequency ω can be varied between 10^{-3} and 1 rad/sec. The amplitude ΔlnA of the sinusoidal change of the surface area A can be varied between 0 and 0.25 by adjusting the eccentricity. Thorough cleaning of the rubber bands is necessary to avoid contamination. This modified version eliminates the anomalous damping of the longitudinal wave found for some protein solutions in the standard set-up.

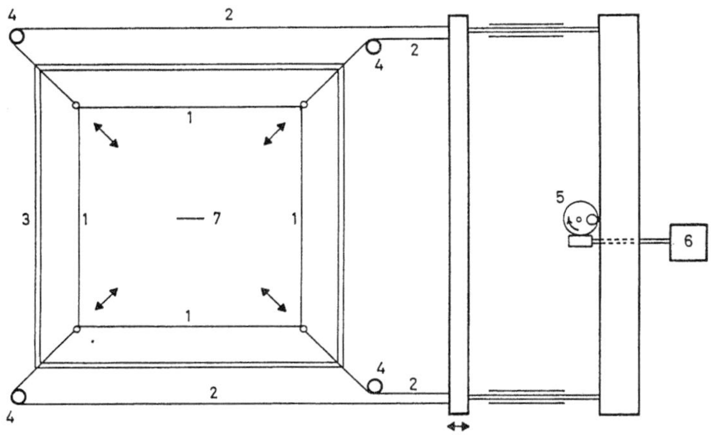

Fig. 2 Modified barrier-and-plate apparatus ensuring isotropic compression/expansion (top view). (1) rubber bands; (2) metal wires; (3) glass vessel; (4) wheels; (5) eccentric driver system; (6) electromotor with gear box; (7) Wilhelmy plate.

In the two versions of the experimental set-up (Figures 1 and 2), both dγ and dA/A are converted into electrical signals recorded on the axes of an xy-recorder. This results in a straight line for purely elastic behaviour of the interface, while viscoelastic interfaces produce an ellipse, illustrated in Figure 3. The phase angle ϕ is calculated from the eccentricity of the ellipse written by the xy-recorder. The elastic component ε' and the viscous component ε'' are calculated from the maximum change of the surface tension, Δγ, and the maximum change of the surface area, ΔA, and from the phase angle ϕ with the aid of Eq (3).

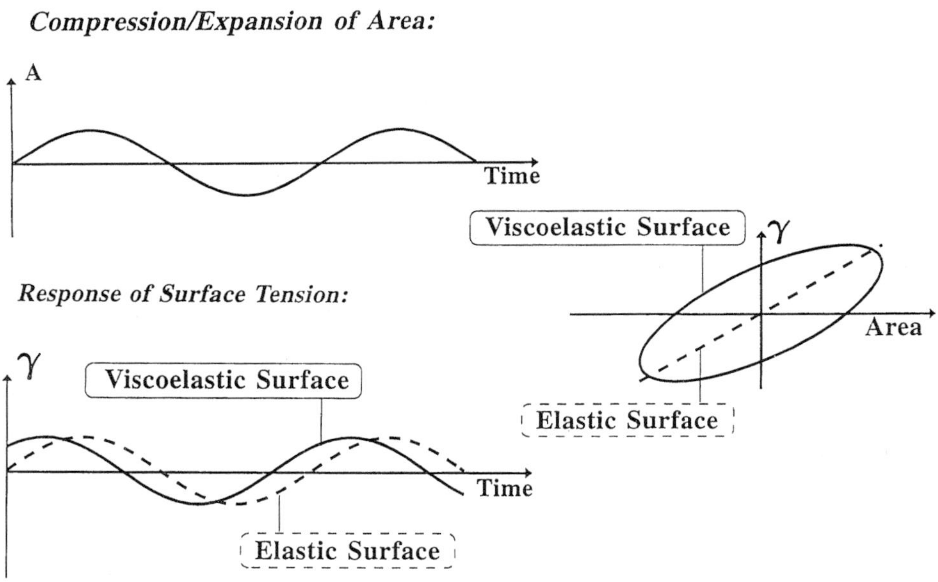

Fig. 3 Schematic illustration of sinusoidal area variations (not drawn to scale) and corresponding tension variations, for purely elastic and viscoelastic surface behaviour.

2.3 Novel method: dynamic drop tensiometer

The Dynamic Drop Tensiometer is a modified version of the Automatic Drop Tensiometer developed by Cagna et al. [31]. The unmodified version of the instrument is reminiscent of recently-described tensiometers using axisymmetric drop shape analysis to study dynamic surface properties *at constant area* (Rotenberg et al [32]; McLeod and Radke [33]; Miller et al. [34]; Nagarajan et al. [35]).

The modifications introduced permit us to determine the interfacial rheological properties in compression/expansion at different rates. The set-up differs from the barrier-and-plate technique in three respects: (i) the oscillations in area and tension are measured on one and the same small interfacial area, (ii) homogeneity of deformation of this area is far more easily ensured and (iii) the tension changes are evaluated by means of the Young-Laplace equation from measurements of the fluctuating drop shape rather than by means of a Wilhelmy plate. Figure 4 shows a diagram of the experimental set-up [27]. The main modification of the original instrument is that the area of the drop can be made to oscillate sinusoidally at a chosen amplitude and frequency. This was achieved by regulating the DC motor driving the piston in the syringe feeding the drop, so as to produce sinusoidal oscillations of the drop volume. The control unit records and plots both the area oscillations (dA) and the resulting interfacial tension oscillations (dγ). As in the barrier-and-plate methods, this produces an ellipse from which the elastic and viscous parts of the modulus ε can be evaluated.

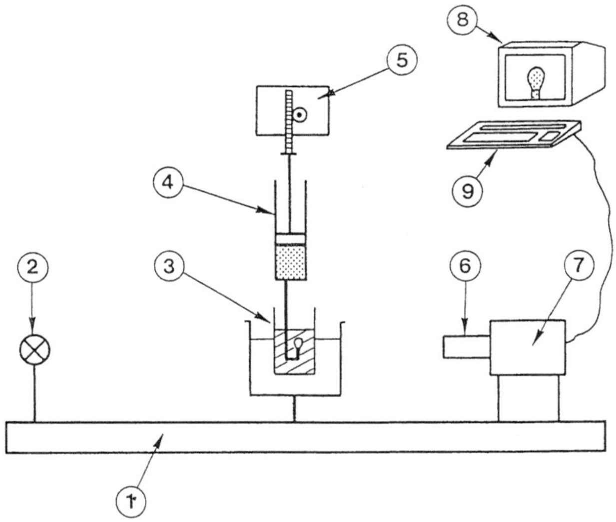

Fig. 4 Dynamic Drop Tensiometer (side view). (1) optical bench; (2) light source; (3) cuvette in which drop is formed; (4) syringe with drop phase; (5) DC motor driving piston of syringe; (6) telecentric gauging lens; (7) CCD camera; (8) video monitor; (9) personal computer.

3 EXPERIMENTAL RESULTS FOR PROTEINS ADSORBED AT AIR/WATER

3.1 Modified method with no-shear barrier

Systematic studies of the surface dilational modulus, in combination with the adsorption and the surface tension on their way to equilibrium, are available for a number of proteins ranging in molecular structure from almost random coil to compact globular. The adsorptions were measured by ellipsometry [28, 36], with the same protein samples and under the same experimental conditions as the surface moduli and the surface pressures. Figures 5 to 9 show some of these results for β-casein, κ-casein, whole casein (sodium caseinate), bovine serum albumin and ovalbumin, with data for a flexible-chain PVA polymer added for comparison.

Figs. 5 to 7 illustrate the slow equilibration of very dilute macromolecular solutions, reflected in a steady increase of the modulus and the surface pressure as a function of the adsorption time, for two concentrations. As expected, the modulus of the higher concentrations starts to increase after a shorter adsorption time. However, the modulus at near-equilibrium (i.e., after 21 hours) does not significantly increase with increasing concentration in most cases. This is illustrated in Table 1, which also gives the corresponding surface pressures and adsorptions. The only exception to this pattern appears to be κ-casein, where the modulus does increase considerably with concentration, an increase apparently linked to the rather steep increase of the adsorption of this protein.

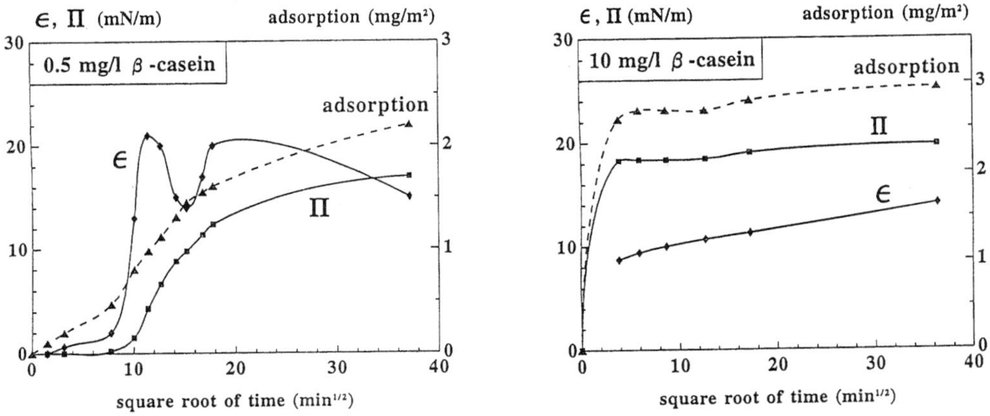

Fig. 5 Surface dilational modulus, adsorption and surface pressure as a function of adsorption time for two concentrations of β-casein. Frequency: 0.84 rad/s; pH=6.7; ΔA/A=0.07.

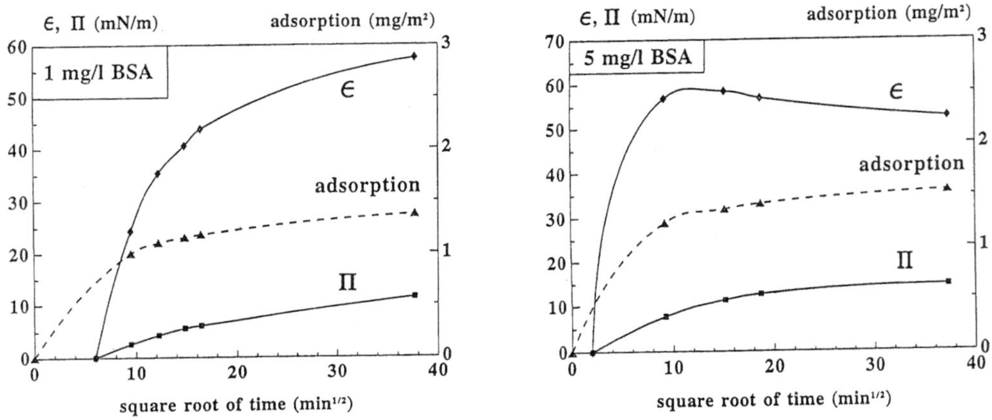

Fig. 6 Surface dilational modulus, adsorption and surface pressure as a function of adsorption time for two concentrations of Bovine Serum Albumin. Frequency: 0.84 rad/s; pH=6.7; A/A=0.07.

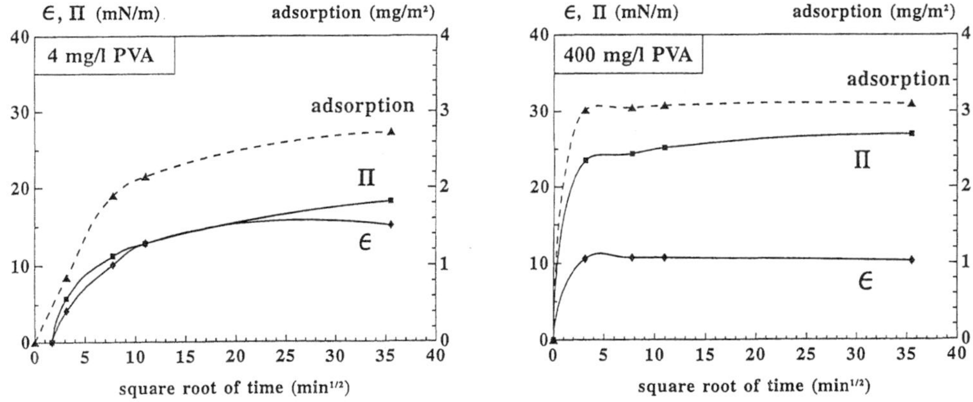

Fig. 7 Surface dilational modulus, adsorption and surface pressure as a function of adsorption time for two concentrations of PVA. Frequency: 0.84 rad/s; pH=6.7; ΔA/A=0.07.

Table 1. Surface dilational modulus at near-equilibrium (after 21 h) for different protein concentrations. Frequency: 0.84 rad/s.

Protein	Molecular Weight (Da)	Concentration (g/l)	Surface Pressure (mN/m)	Adsorption (mg/m^2)	Modulus (mN/m)
β-Casein	24,000	0.0005	17	2.2	17
		0.003	19	2.95	16
		0.01	19.8	2.95	16.4
κ-Casein	19,000	0.0008	13.4	1.95	22
		0.008	17.7	3.1	58
		0.3	18.5	4.6	80
Na Caseinate	23,000	0.0002	12.2	1.55	29
		0.0005	19	2.15	24.2
		0.3	25	3.3	20.6
BSA	69,000	0.001	11.3	1.37	59
		0.005	14.6	1.54	56
		0.1	17.8	1.95	69
Ovalbumin	45,000	0.1	16.3	1.52	75
PVA	42,000	0.001	13.3	2.2	11.1
		0.004	18.2	2.73	15.5
		0.4	27	3.1	11

At all concentrations in this work, modulus values were found to be almost independent of frequency (in the range of $0.01 < \omega < 1$ rad/s) in the region of surface pressures up to around 10 mN/m. At higher surface pressures, i.e., at higher adsorptions, moduli generally did decrease somewhat with decreasing frequency. In this region, viscoelastic surface behaviour was found, indicating the effect of relaxation processes in close-packed protein surfaces. Table 2 summarises viscous phase angles for each of the proteins and the polymer at medium and high surface pressures, at different frequencies.

Table 2. Viscous phase angle at medium and high surface pressure for proteins and PVA, at various frequencies, at pH=6.7.

Protein	Surface pressure (mN/m)	Adsorption (mg/m^2)	Viscous phase angle (degrees)		
	Frequency (rad/s) -->		0.84	0.033	0.0084
β-Casein	9.8	1.45	0	0	
	19	2.95	0	6	8
κ-Casein	9.2	1.65	0	0	
	18.2	3.2	0	16	17
Na Caseinate	9.5	1.3	0	0	
	22	3.3	12	42	
BSA	11.3	1.37	0	0	
	16.9	1.83	7	19	
Ovalbumin	9.4	1.3	0	3	
	14.5	1.45	0	4.5	
PVA 205	11.2	1.9	0	16	
	23.5	3.0	0	21	

Experimental values of the adsorptions and the corresponding surface pressures for all six macromolecules are presented in Figs. 8 and 9. For each protein, all measurements at different bulk concentrations and different surface ages were found approximately to coincide on a single curve characteristic for the protein. This means that equilibrium in the surface is largely established within the time needed for an ellipsometric measurement, which is approximately 5 min. Thus, equilibration within the surface is very much faster than between surface and bulk solution, and each curve in Figs. 8 and 9 can be considered to represent the equilibrium surface equation of state to a fair approximation.

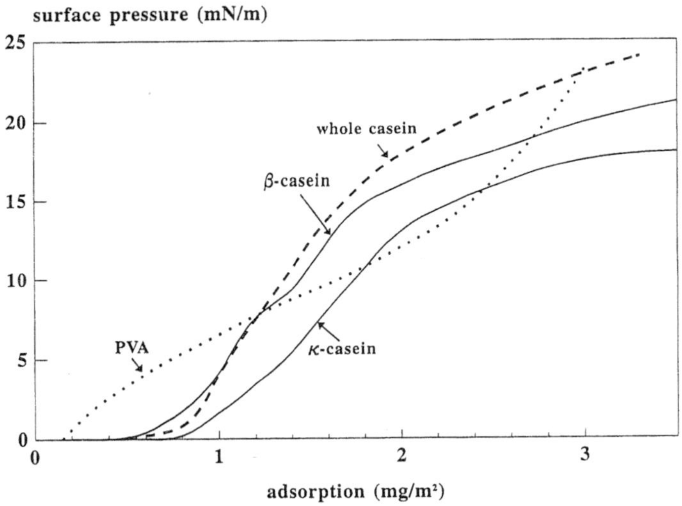

Fig. 8 Surface pressure as a function of adsorption for β-casein, κ-casein, whole casein and PVA 205.

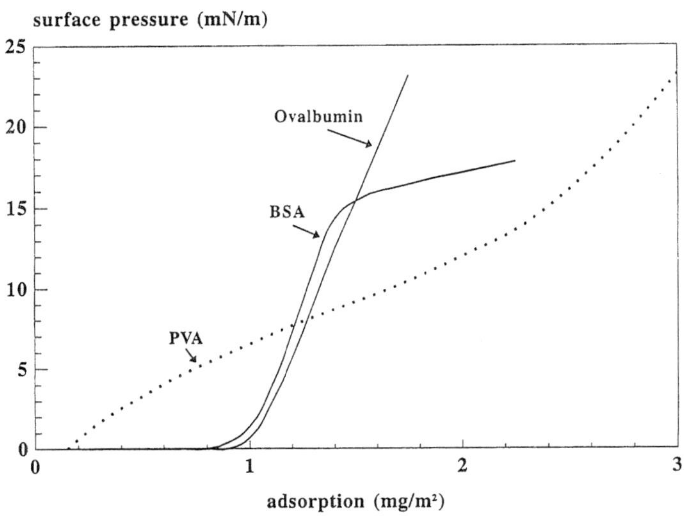

Fig. 9 Surface pressure as a function of adsorption for Bovine Serum Albumin, Ovalbumin and PVA 205.

The curves for the flexible proteins in Fig. 8 show a more gradual increase of the surface pressure and a more pronounced plateau at the highest adsorptions than those of the globular proteins in Fig. 9. The curve for the synthetic polymer PVA shown for comparison is rather different from the caseins in taking off at a lower adsorption and in showing no sign of levelling off at high adsorptions.

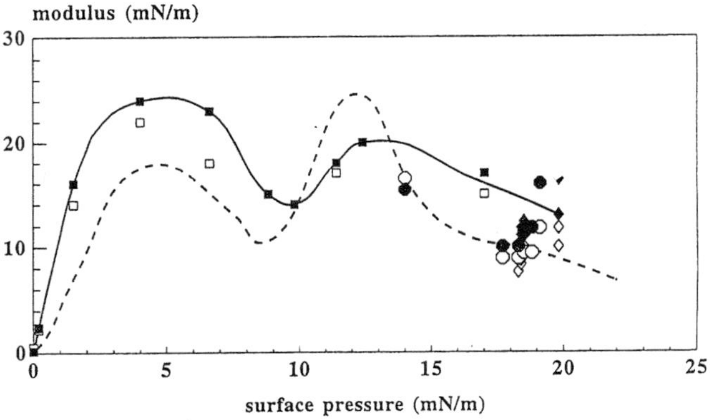

Fig. 10 Surface dilational modulus as a function of surface pressure for β-casein. pH=6.7; ΔA/A=0.07. Closed symbols: ω=0.84 rad/s; open symbols: ω=0.033 rad/s. Squares: c=0.5 mg/l; circles: c=3 mg/l; diamonds; c=10 mg/l. Drawn line: results at highest frequency. Dashed line: limiting modulus, ε_0, from Π *vs* Γ curve in Fig. 8.

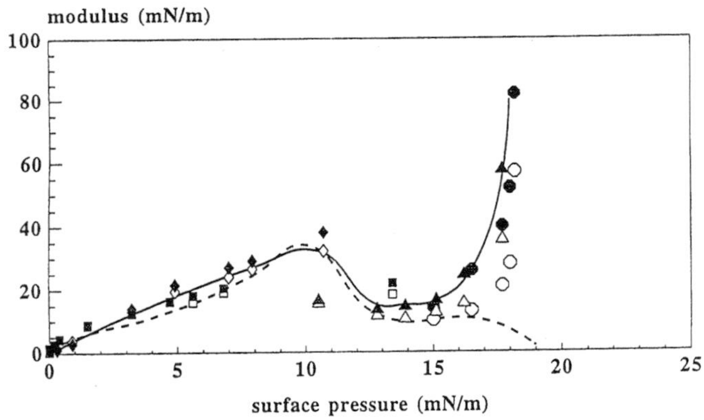

Fig. 11 Surface dilational modulus as a function of surface pressure for κ-casein. pH=6.7; ΔA/A=0.07. Closed symbols: ω=0.84 rad/s; open symbols: ω=0.033 rad/s. Squares: c=0.8 mg/l; diamonds: c=1.5 mg/l; triangles: 8 mg/l; circles: c=30 mg/l. Drawn line: results at highest frequency. Dashed line: limiting modulus, ε_0, from Π *vs* Γ curve in Fig. 8.

The experimental surface pressure *vs* adsorption curves in Figs. 8 and 9 enable us to evaluate the limiting modulus, ε_0, defined in Eq (2) as a function of the adsorption or the surface pressure for each protein. Figs. 10 to 15 show both the measured moduli and the calculated limiting elasticities as a function of the surface pressure, for all six macromolecules. Agreement between measured and calculated moduli appears to be very fair in the range of low surface pressures where surface behaviour is purely elastic (see Table 2.). Beyond this range, viscoelastic surface behaviour is found, with the single curve splitting up into different branches for different frequencies, as illustrated in Figures 12 and 13. In the viscoelastic range, measured moduli exceed the limiting ε_0 values for globular proteins, such as BSA and Ovalbumin.

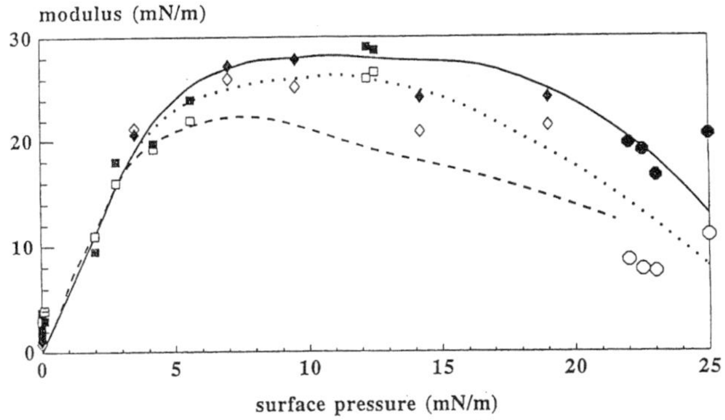

Fig. 12 Surface dilational modulus as a function of surface pressure for whole casein. pH=6.7; $\Delta A/A$=0.07. Closed symbols: ω=0.84 rad/s; open symbols: ω=0.033 rad/s. Squares: c=0.2 mg/l; diamonds: c=0.5 mg/l; circles: c=30 mg/l. Drawn line: results at highest frequency. Dashed line: limiting modulus, ε_0, from Π *vs* Γ curve in Fig. 8.

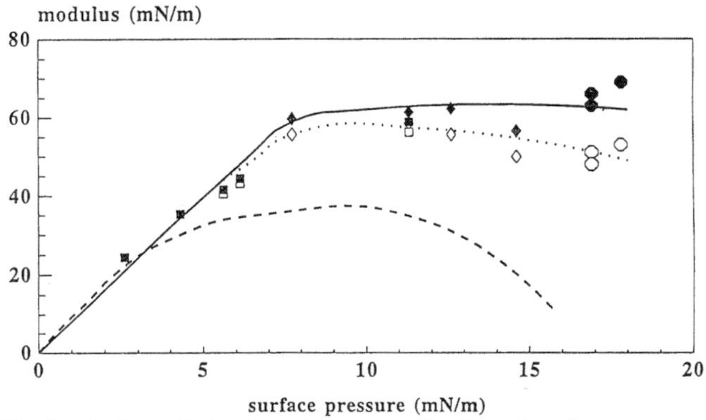

Fig. 13 Surface dilational modulus as a function of surface pressure for Bovine Serum Albumin. pH=6.7; $\Delta A/A$=0.07. Closed symbols: ω=0.84 rad/s; open symbols: ω=0.084 rad/s. Squares: c=1 mg/l; diamonds: c=5 mg/l; circles: c=10 mg/l. Drawn line: results at highest frequency. Dashed line: limiting modulus, ε_0, from Π *vs* Γ curve in Figure 9.

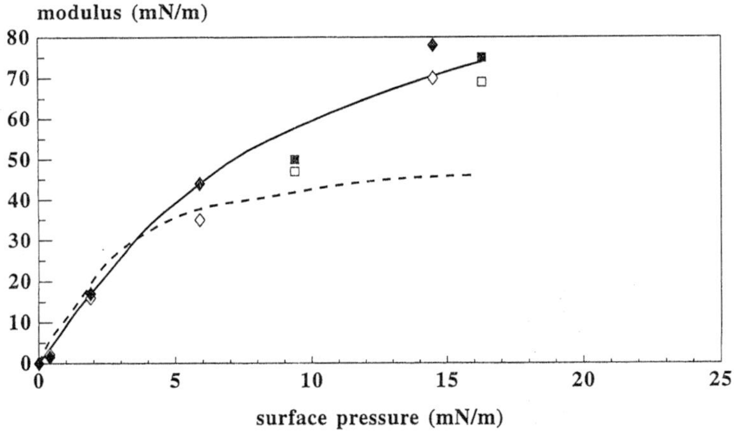

Fig. 14 Surface dilational modulus as a function of surface pressure for Ovalbumin. ΔA/A=0.07; c=100 mg/l. Closed symbols: ω=0.84 rad/s; open symbols: ω=0.084 rad/s. Squares: pH=6.5; diamonds: pH=7.5. Drawn line: results at highest frequency. Dashed line: limiting modulus, ε_0, from Π *vs* Γ curve in Fig. 9.

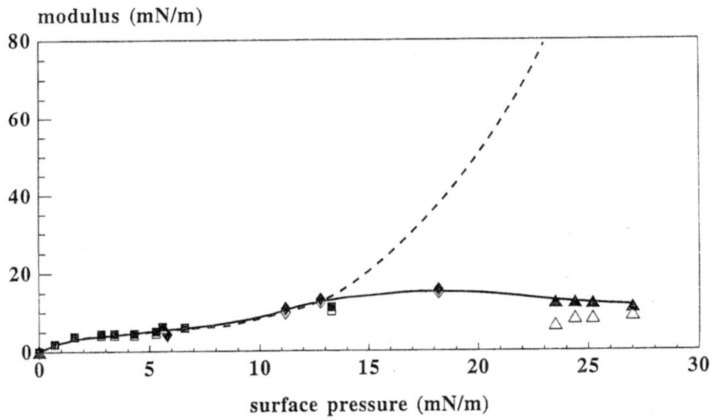

Fig. 15 Surface dilational modulus as a function of surface pressure for PVA. Closed symbols: ω=0.84 rad/s; open symbols: ω=0.084 rad/s. Squares: c=1 mg/l; diamonds: c=4 mg/l; triangles: 400 mg/l. Drawn line: results at highest frequency. Dashed line: limiting modulus, ε_0, from Π *vs* Γ curve in Fig. 8.

3.2 Other results

Reliable published data of dilational moduli for proteins measured after adsorption from solution are relatively scarce compared to results obtained with spread monolayers. Under static conditions, surface behaviour as expressed in the surface pressure *vs* adsorption curve is generally found to be similar for adsorbed and spread layers (Bull [37]; MacRitchie and Alexander [38]; Yamashita and Bull [39]). For this reason, we include results obtained with spread layers, to be compared with the present results. The only surface parameter directly measurable for both adsorbed and spread monolayers is the surface pressure and, therefore, results for the different systems will be compared at given surface pressure. Such comparison is not only necessary in the case of spread layers, where concentrations are unknown, but preferable in any case since it directly reflects what happens in the surface without interference from surface-to-bulk transport phenomena. Therefore, we consider equal surface pressures to reflect equal surface conditions for any given protein. Modulus values for proteins were first reported as a function of surface pressure and compared to the limiting values (ε_0) calculated from the pressure-area curve by Blank et al. [40], followed by Giles and Lucassen [41] and by Joos [42].

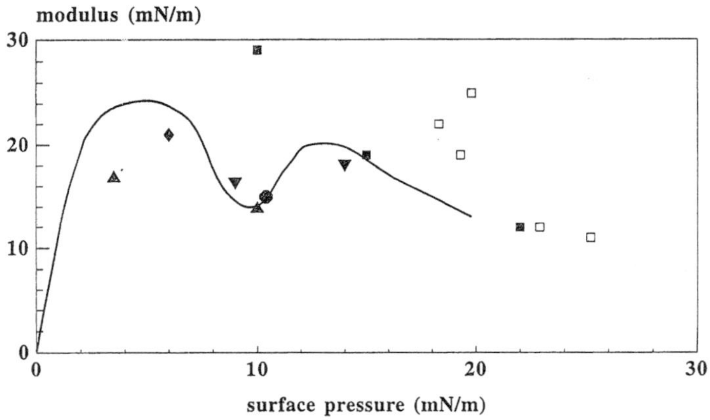

Fig. 16 Surface dilational modulus as a function of surface pressure for β-casein; data from various sources. Drawn line: as in Fig. 10. Points: ●, ▲ [19]; ■ [17]; □ [43]; ▼ [21, 45]; ◆ [44].

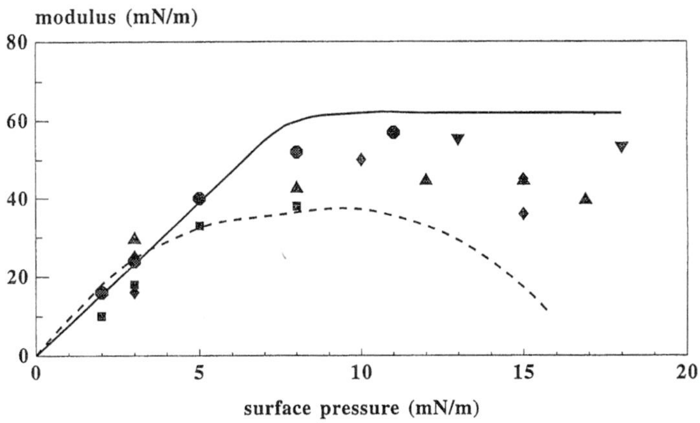

Fig. 17 Surface dilational modulus as a function of surface pressure for BSA; data from various sources. Drawn line: as in Figure 13. Points: ● [41]; ■ [40]; ▲ [17]; ▼ [22]; ◆ [20, 21].

For β-casein, modulus values were reported by Graham and Philips [17], Serrien et al. [18] and Williams and Prins [43] with adsorbed layers, while spread layers were used by Gau et al. [19], Douillard et al. [44] and van Aken and Merks [21,45]. These results are shown in Fig. 16 as a function of the surface pressure, together with the results from Figure 10, represented by the drawn line. For BSA, similarly, Fig. 17 collates data from adsorbed layers [17, 18] and spread layers (Blank et al. [40]; van Aken and Merks [20]; Boury et al. [22]) with the present data from Fig. 13. In some of this work [21, 45] the area variations applied were step-wise rather than sinusoidal; we present only data showing fully elastic behaviour, i.e., no relaxation after the area change.

Inevitably, methods, materials and conditions used in these studies were in no case fully equal to the conditions in the present work. Apart from any differences between adsorbed and spread layers, other possible reasons for discrepancies are:

(i) Most published studies used uniaxial compression and monitored the surface tension changes in the middle of a long and narrow trough, assuming isotropic deformation. A careful analysis of the available information indicates a significant non-isotropy of the deformation in the work of Graham and Phillips [17], as recognised by the authors, and the data of van Aken and Merks [21, 45]. Such non-isotropy causes surface shear effects within the deformed protein layer that adheres to the side-walls. Under these conditions the local area change at the surface tension probe is smaller than that near the barrier, resulting in moduli that are too low at higher surface pressures. Williams and Prins [43] used the ring trough designed to produce purely dilational deformation to study β-casein and β-lactoglobulin as a function of concentration, at a frequency of 0.1 Hz, followed by Boerboom et al. [46] with a study of the same proteins over a range of lower frequencies.

(ii) The proteins used by Graham and Phillips [17] were radio-labelled. For such proteins, the initial increase of the pressure-adsorption curve is less steep, and consequently ε_0 and ε are smaller at low surface pressures compared to the unlabelled proteins.

(iii) The frequencies used by Gau et al. [19] were in the kHz range, i.e., much higher than ours. In this high-frequency range, the authors observed a transition from purely elastic to viscoelastic surface behaviour, indicating a relaxation mechanism with characteristic time scale in this high frequency range. As a result, one would expect their high-frequency moduli to be rather higher than ours. Surprisingly, the agreement is found to be quite good.

In spite of these significant differences in methods and in experimental conditions, the results obtained in this work for the most part agree rather well with earlier studies in which the deformation was non-isotropic. We conclude that interference by shear properties induces only minor errors in the present systems where the dilational modulus exceeds the shear modulus.

3.3 Discussion

3.3.1 EFFECTS OF ADSORPTION TIME AND PROTEIN CONCENTRATION

For the lowest protein concentrations in Fig. 5-7, a time lag is observed during which the modulus is too small to be measured with sufficient accuracy. This time lag corresponds with the time lag ("induction" period) in the surface pressure. In these cases, an appreciable value of the adsorption, characteristic for each protein, is required to produce a measurable non-zero value of the surface pressure and, in consequence, also of the dilational modulus. It is only at very low concentrations that the adsorption, governed by diffusion from the solution, takes a long time to reach this characteristic value. After take-off, both modulus and surface pressure in most cases gradually increase with time, until almost constant values are reached when the system is close to adsorption equilibrium.

During the equilibration process, the modulus ε depends on both time and protein concentration in the solution. Over a large range of surface pressures, however, these two variables combined merely serve to determine the changing values of the really important variable, i.e., the protein adsorption. As a result, the different modulus *vs* time curves in Figures 5-7 tend to correspond to a single modulus *vs* adsorption curve which, for each protein, corresponds to the modulus vs pressure curves in Figs. 10 to 15. The modulus vs adsorption curves are summarised in Fig. 18, where the adsorption values were taken from Figures 8 and 9 for given values of time and concentration. At the highest rate of compression/expansion, where ω=0.8 rad/s, surface behaviour is purely elastic, with the elasticity equal to the limiting value ε_0 shown in Figures 10 to 15, to be discussed in more detail in Section 3.3.2.

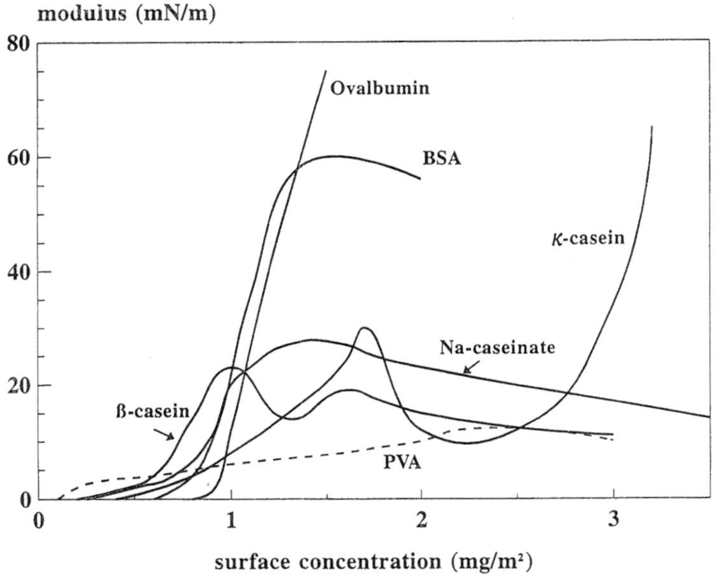

Fig. 18 Surface dilational modulus as a function of adsorption for caseins, BSA, Ovalbumin and PVA.

The negligible values found for the viscous phase angles, implying that relaxation processes do not play any part in a time scale of 1 s, lead to a two-fold conclusion:

(i) diffusional interchange with the solution does not take place because this mechanism requires far longer time scales,

(ii) relaxation processes in the adsorbed layer are largely completed inside this time scale.

Relaxation phenomena do play an important role at higher surface pressures, i.e., higher adsorptions, where modulus values for different frequencies no longer coincide at given values of the adsorption or the surface pressure. As a result, moduli in Figures 10 to 15 all diverge from the surface-equilibrium curve above a certain value of the surface pressure. This range will be further discussed in Section 3.3.3.

Summarising, over a large range of not too high surface pressures, the effects of protein concentration and time on the modulus can be explained quantitatively as the effect of the varying protein adsorption. As a result, experimental data for each protein investigated at different times and concentrations all coincide on a single $|\varepsilon|$ *vs* Γ curve, and also on a single $|\varepsilon|$ *vs* Π curve, characteristic for each protein. In this range the modulus is a pure elasticity, determined by the surface equation of state of the protein. At higher adsorptions and surface pressures, a transition from purely elastic behaviour to viscoelastic behaviour is observed, illustrated by the single curve splitting up into different branches for different frequencies. In this high-adsorption range, mechanisms of surface relaxation become operative.

3.3.2 EFFECTS OF THE SURFACE EQUATION OF STATE

In the pure-elasticity range, where dynamic behaviour is dominated by the surface equation of state, the modulus $|\varepsilon|$ should be given by the limiting value ε_0 defined in Eq (2). Measured moduli in Figures 10 to 15 do follow the curves for the limiting values calculated from the surface pressure *vs* adsorption curves in Figures 8 and 9 quite well for not too high surface pressures. For the most flexible molecules, i.e., the caseins and PVA, reasonable agreement is found at Π values up to 15 mN/m. For instance, the irregular shape of the modulus curve for β-casein, with its two maxima, can be traced back to the two points of inflexion in the surface pressure curve *vs* adsorption curve at surface pressures of about 8 and 15 mN/m (see Figure 8). The inflection at surface pressures between 8 and 10 mN/m has been attributed to a transition from an all-trains configuration, in which flexible polypeptide chains lie fully unfolded in the surface, to a trains-and-loops configuration where some segments protrude as loops into the aqueous phase [17]. Loop formation has also been deduced from enzymatic action on adsorbed β-casein molecules by Leaver and Dalgleish [48]. In the frequency range of our experiments the phase angles are zero, which means that the characteristic time scale of this process of loop formation is much smaller than 1 s (see also Section 3.3.3).

The globular molecules, i.e., BSA and ovalbumin, on the other hand, display a smaller range of about 5 mN/m of limiting elastic behaviour. In this range, both measured and calculated elasticities increase linearly with the surface pressure in all systems. In view of the definition of the limiting elasticity in Eq (2), this means that the surface equation of state in this range follows an equation of the form

$$\Pi = k_1 \Gamma^{k_2} \quad \text{for } 0 < \Pi < 5\ \text{mN/m} \tag{9}$$

where k_1 and k_2, the initial slope, are constants characteristic of the protein. In terms of the adsorption, the simple linear behaviour is observed in the intermediate range, roughly from 0.5 to

1.5 mg/m^2 for the flexible molecules and from 0.8 to 1.5 mg/m^2 for the globular ones. If full monolayer coverage is represented by 3 mg/m^2, as appears to be the case for the present globular proteins, this would mean a range of surface coverage from about 25 to 50%.

Table 3 presents the initial slopes determined from Figures 10 to 15 and from other work referred to in Figures 16 and 17 for the different proteins. It is clear that the slopes from the present work are roughly similar to the literature data. High values for the slope reflect the steep ascent of the surface pressure *vs* adsorption curves in Figures 8 and 9. The surface pressure *vs* adsorption curves do not in themselves provide information on the molecular structure or the flexibility of proteins in particular: essentially similar curves have been measured for the synthetic polymers hydroxyethylcellulose and hydroxypropylcellulose by McNally and Zografi [49]. A highly significant aspect of their data is that for each polymer the surface pressure *vs* adsorption curve is almost independent of molecular weight over a range from 6×10^4 to 10^6; obviously, any differences in structure and flexibility between the larger and the smaller macromolecules have only minor effects on the equation of state.

Table 3. Linear ranges and slopes ($d\varepsilon/d\pi$) for various proteins

Protein	$d\varepsilon/d\pi$	Range (mN/m)	Reference
β-Casein	9	3	Present work
κ-Casein	4	7	Present work
BSA	8	7	Present work
	9	5	Blank et al. [40]
	9		Serrien et al. [18]
	8	6-7	Graham and Phillips [17]
Ovalbumin	12	4	Present work
	8	5	Blank et al. [40]
β-Lactoglobulin	7.5	5	Joos [42]
PVA205	2	2	Present work

Therefore, we cannot expect a direct correlation between the slope of the ε_0 *vs* Π curves in Table 3 and molecular flexibility. Even if such a correlation did exist, the slope for the flexible β-casein should have been lower than for the rigid globular proteins, while in fact it is found to be in the same range. Moreover, similarly high values have been reported for small molecules, e.g., dodecyl triethylene glycol [25], and mixtures of anionic and cationic surfactants [50]. For the present systems, differences between the different proteins become more evident if we consider the range of the linear region. Compared to globular proteins, β-casein and PVA produce linear behaviour over a relatively small range of surface pressures. This suggests that a small initial linear region may be a better measure for the flexibility of the molecule. Support for this suggestion may be derived from Figure 18, in which the same modulus data are given as a function of the adsorption. Initial parts of all curves are no longer linear, but the rigid globular molecules, BSA and Ovalbumin, do show a steeper increase than the more flexible caseins and PVA. As a result, the flexible macromolecules generally produce less cohesive and more compressible films with lower elasticity than the compact proteins.

Existing analytical equations of state are unable to describe important aspects of protein behaviour as they deal neither with complex intermolecular interactions nor with intramolecular rearrangements. At very low adsorptions, all known equations of state reduce to the two-dimensional analogue of the ideal-gas law, which predicts a slope of +RT for the Π *vs* Γ curve and a slope of +1 for the ε_0 *vs* Π curve. A striking feature of Figures 10 to 14 is that there is no observable trace of such a limiting slope in any of the present systems. All exhibit severely-non-ideal behaviour at very low surface pressures, which reduces the range where $d\varepsilon/d\Pi=1$ to the point of invisibility. Such non-ideality, of course, is also apparent from the quite high adsorptions needed to produce any measurable surface pressure. Surface non-ideality of macromolecular systems is caused by a combination of non-ideal entropy and by non-zero enthalpy resulting from molecular interactions. Several theories deal with the entropy aspect, either in statistical-mechanical treatments (Singer [51]; Frisch and Simha [52]; Cohen Stuart et al. [53]) or in a two-dimensional solution approach (Joos [42]; Lucassen-Reynders [54]), but less attention has been paid to the enthalpy. The equation of state proposed by Fainerman et al. [55], in which a protein can occupy a number of configurations with different molecular areas, does account for enthalpy of mixing of the average configuration with the solvent by a Frumkin-type expression, but considers the non-ideal entropy to be negligible. A simple treatment [56] accounting for both entropy and enthalpy, for a protein with only one molecular area, ω_2, is no more than a rough approximation but it can serve to rationalise the steep slopes in Figures 10 to 14. In such a simple version, the surface pressure Π depends on the degree of surface coverage Θ ($=\omega_2\Gamma_2$) according to

$$\frac{\Pi\omega_1}{RT} = -\ln(1-\Theta) - (1-1/S)\Theta - \frac{H}{RT}\Theta^2 \qquad (10)$$

where ω_1 is the molar area of the solvent, S ($=\omega_2/\omega_1$) is the factor by which the protein's molar area exceeds that of the solvent, and H is the partial molar heat of mixing of the Frumkin model. Positive values of H represent attractive interactions between surfactant molecules. The first term in Eq (10) is the surface pressure of an ideal surface mixture of equally sized molecules (also known as the kinetic contribution to Π), the second term is related to the non-ideal entropy of a mixture of small and large molecules and the third term is the enthalpic contribution. Figure 19 illustrates the pressure *vs* adsorption isotherms for three cases representative of ideal mixing, non-ideal entropy and non-ideal entropy combined with heat of mixing. Both the non-ideal entropy and the heat are seen to depress the surface pressure at all surface coverages. The combination of the two effects, in particular, results in very low pressures at low surface coverage: for a value of 30 mN/m for RT/ω_1, the surface pressure is only 0.3 mN/m at a relatively high surface coverage of 25%. The limiting modulus ε_0 according to this model is given by

$$\frac{\varepsilon_0\,\omega_1}{RT} = \frac{\Theta}{1-\Theta} - (1-1/S)\Theta - \frac{2H}{RT}\Theta^2 \qquad (11)$$

The effect of the entropic and enthalpic contributions on the modulus *vs* pressure relationship is shown in Figure 20. Interestingly, it is only the combination of entropy and enthalpy that produces a steep linear ascent of the modulus at moderately high surface coverage. The slope here is very nearly equal to the slope at 50% coverage:

$$\frac{d\varepsilon_0}{d\Pi} = \frac{3 + 1/S - 2H/RT}{1 + 1/S - H/RT} \qquad \text{at } \Theta = 0.5 \tag{12}$$

So, at half coverage, the ideal mixture, with S=1 and H=0, is unable to produce a slope higher than +2, non-ideal entropy on its own can increase the slope to at most +3, but the combination of entropy and heat can produce much higher slopes, e.g., +5 in the example of Figure 20 and even higher values for higher H. Such high values come close to the experimental results in Table 3, and we propose that values of this level are indicative of enthalpy of mixing caused by attractive intermolecular interactions.

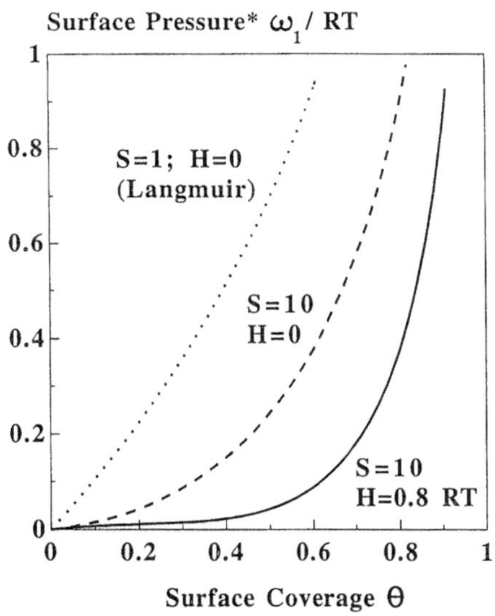

Figure 19. Effects of non-ideal entropy and enthalpy on surface pressure *vs* surface coverage according to Eq (10).

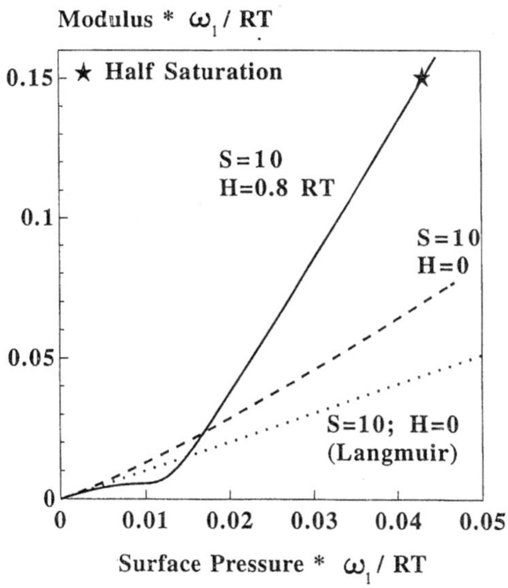

Figure 20. Effects of non-ideal entropy and enthalpy on limiting modulus *vs* surface pressure according to Eq (11).

Finally, at high adsorptions all reasonably simple current models predict that, in the absence of phase separation in the surface, both Π and the slope $d\Pi/d\Gamma$ should steadily increase with increasing surface coverage. In this respect, theory seriously overestimates the surface pressure since nearly all measured Π *vs* Γ curves in Figures 8-13 show a decreasing slope at high Γ. Such flattening off has been attributed qualitatively to the onset of collapse, which is a phase separation phenomenon. Since there are no abrupt changes, this would probably have to be a second-order phase transition rather than the first-order transition known for collapsed monolayers of smaller molecules, e.g., long-chain fatty acids. An alternative explanation is that, in fairly close-packed layers, protein molecules can behave as "soft particles" undergoing a reconformation into a modification with a smaller molecular area (de Feijter and Benjamins [57]; van Aken [58]). This would imply that, at increasing Γ, the area fraction covered by the protein can remain almost constant as the molecules become increasingly more compressed and, as a result, Π can also become almost constant. Both suggestions are attractive but, so far, have not been put into a quantitative framework.

Summarising, the dilational properties of protein monolayers are satisfactorily explained by the equilibrium pressure-area curve for surface coverages from 0 to roughly 1.5 mg/m^2. Surface behaviour is purely elastic at a time scale of 1 s for all proteins considered. The equality of measured and calculated moduli (ε_0) in this range implies that equilibrium in the surface is established within the time scale of the compression/expansion, i.e., within approximately 1 s. Surface pressures and dilational moduli are negligibly small up to a surface coverage of 0.5 mg/m^2; at higher surface coverages, up to 1.5 mg/m^2, the modulus increases sharply and linearly with surface pressure. Such a steep increase of the elasticity points to a severely non-ideal surface equation of state, with an overriding influence of molecular interactions. Existing equations of state have to be extended to cover such interactions in order to account for the measured elasticities.

3.3.3 EFFECTS OF SURFACE RELAXATION PROCESSES

At surface coverages higher than about 2 mg/m^2, viscoelastic behaviour sets in and the modulus curves for different concentrations and frequencies no longer coincide in Figures 10 to 15. Possible processes resulting in viscoelastic behaviour are (i) diffusional interchange with the bulk solution and (ii) relaxation phenomena in the surface layer. Relaxation of surface tension by diffusion is the most common relaxation mechanism in soluble monolayers, and has been modelled quantitatively for single-surfactant solutions [2,23,25]. Purely diffusional relaxation for a single surfactant is characterised by a frequency dependence of both the modulus, $|\varepsilon|$, and the viscous phase angle, ϕ, described by

$$|\varepsilon|/\varepsilon_0 = \left[1 + 2(\omega\tau_{\text{diff}})^{-1/2} + 2(\omega\tau_{\text{diff}})^{-1}\right]^{-1/2} \quad (13)$$

and

$$\tan\phi = \frac{1}{1+\sqrt{\omega\tau_{\text{diff}}}} \quad (14)$$

respectively. The characteristic diffusional time scale τ_{diff} is defined by

$$\tau_{\text{diff}} = \frac{2}{D}\left(\frac{d\Gamma}{dc}\right)^2 \quad (15)$$

where D is the diffusion coefficient of the surfactant, and $d\Gamma/dc$ the slope of the adsorption isotherm.

Eq (13) implies that the frequency spectrum of the viscous phase angle as a function of the dimensionless frequency, $\omega\tau_{\text{diff}}$, is represented by a single curve for any surfactant at any concentration: the characteristics of individual surfactants are reflected in the numerical values of τ_{diff} (and ε_0) but not in the shape of the curve. Therefore, the frequency spectrum of the viscous loss angle (and also that of the dimensionless modulus, $|\varepsilon|/\varepsilon_0$) can be used as a master curve to identify diffusional relaxation. The master curves for diffusional relaxation, shown in Figure 21, so far have been applied to small-molecule surfactants, but they are equally valid for any other surface active agent. Macromolecules have lower values of D and different adsorption isotherms, but changing the values for these two factors merely produces a horizontal shift of the line for tan ϕ in Figure 21, not a change in its shape. If the viscoelasticity of fairly close-packed protein layers is due to diffusional interchange with the solution, modulus and phase angle should show the same frequency dependence as the lines in Figure 21. Figure 22 illustrates that the data of Murray et al for β-Lactoglobulin [59] fail to follow the characteristic diffusional spectrum. In the absence of independently determined values of the characteristic time scale, τ_{diff}, the horizontal position of the curves is undecided but the shape at the lower frequencies is incompatible with the diffusional model. For the caseins, BSA and Ovalbumin, some of the measured moduli exceed the limiting value ε_0 in Figures 10 to 14; this in itself indicates non-diffusional relaxation since diffusion on its own can only lead to modulus values lower than ε_0. We conclude that the dilational viscosity in times scales from 1 to 1000 s cannot be explained by diffusional relaxation. An additional relaxation mechanism is necessary to produce the moduli and phase angles measured. Since the phase angles in Table 2 are fairly low (nearly all at most 26°) the diffusional contribution can only be small. This means that even at fairly high concentrations these protein layers can be regarded as nearly insoluble in the time scale of the dynamic experiments.

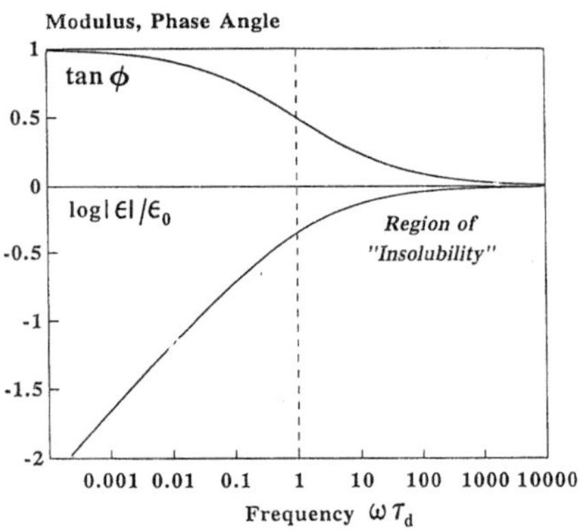

Figure 21. Characteristic spectra for diffusional relaxation. Dimensionless modulus from Eq (13), viscous loss angle from Eq (14).

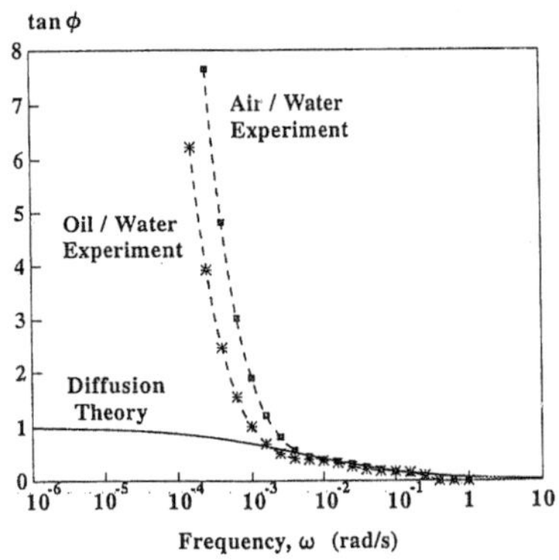

Figure 22. Measured loss angles compared to loss angles for diffusional relaxation. Drawn line: Eq (14) anchored at τ_{diff} = 200 s. Points: 0.01 g/l β–Lactoglobulin (Murray et al [59]).

Relaxation processes other than diffusion in time scales ranging from 10^{-2} to 10^{+3} were reviewed by van den Tempel and Lucassen-Reynders [60] and include (i) retardation of adsorption by an adsorption "barrier"; (ii) slow re-orientation of molecules after adsorption; (iii) complex formation and phase transitions in the surface; (iv) formation or destruction of 3-D structures, either in the surface or in the adjoining solution. In addition, macromolecules may undergo slow internal reconformation processes, such as unfolding of long chains, involving changes in molecular shape, which are generally accepted as important for relaxation of proteinaceous surfaces (Graham and Phillips [15]; MacRitchie [61]; Maksymiw and Nitsch [62]; Serrien et al. [18]). Time scales required for such molecular rearrangements may vary from less than 1 s in not too densely packed surfaces to more than 100 s in the range where viscoelastic behaviour is found in these longer time scales, i.e., at lower frequencies. Another possible conformational change is the change in area requirement described by the soft-particle treatment [57] (see Section 3.3.2.). If such a change requires long times, it will show up as a relaxation process in the modulus measurement [58].

The term "adsorption barrier" is too often used as a blanket term covering any and all time-dependent phenomena that cannot be explained from simple diffusion to an empty surface. In a strict sense the term has been defined for a process which may retard the actual adsorption step, i.e., the equilibration between surface and sub-surface layer after molecules have diffused up from deeper solution layers (Joos et al.[63]). The "induction" periods often found in Π *vs* time curves (as in Figures 5 to 7) do not indicate a "barrier" but a seriously non-ideal surface equation of state, both in the case of small molecules [60] and the case of globular proteins [64]. In both cases, unhindered diffusion fully explains the dynamic adsorption values in the early stages of adsorption.

Comparatively little quantitative evidence is available for specific other mechanisms operative in polymer/protein systems. For flexible proteins, such as the caseins, measured viscous contributions have been ascribed to movement of chain segments from a trains-only to a trains-and-loops configuration [47]. As the relaxation time of such a process has been estimated at about 10^{-8} s, it is difficult to see how it could have an effect in experiments with a much longer time scale of 10^{-3} s [19], let alone in even slower experiments with time scales of 1 s and longer, as in most of the work discussed here.

Furthermore, the various terms in use for macromolecular relaxation phenomena are not very strictly defined, and the phenomena themselves often happen in the same time scale. Moreover, for lack of sufficiently detailed information, a first-order kinetic model is often used as a first approximation to describe any processes by which the originally adsorbed material can undergo any change in the surface. In such a first-order model, the formation of new material, denoted by the subscript S, is assumed to take place sufficiently close to equilibrium for its rate to be proportional to the deviation from equilibrium:

$$-\frac{d\Gamma_S}{dt} = \frac{1}{\tau_{rel}}\left(\Gamma_S - \Gamma_S^{eq}\right) \tag{16}$$

where the superscript (eq) refers to equilibrium at $t=\infty$ when the process has run its full course. The characteristic parameter of the relaxation process described by Eq (13) is the relaxation time τ, which depends on system parameters as prescribed by the specific kinetic model chosen. Such a simple first-order approximation cannot distinguish between particular models for surface reactions, e.g., re-orientation, unfolding, aggregation and collapse. As a result, the same linear model has been used by workers with different relaxation phenomena in mind. For instance, Veer and van den Tempel [65] used Eq (13) to model the effect of exchange of medium-chainlength aliphatic alcohol molecules with collapsed particles on the modulus ε; their model was subsequently applied by Kitching et al. [66] to "reorientation/reconformation" of polymeric surfactant. A first-order model in terms of the surface pressure was first proposed by Rebinder and applied to protein relaxation by Fainerman [67], Trapeznikov et al. [68], Serrien et al. [18] and van Aken and Merks [21]:

$$\frac{\Pi - \Pi^{eq}}{\Pi^0 - \Pi^{eq}} = \exp(-t/\tau_{rel}) \tag{17}$$

where Π^0 is the (extrapolated) value of the surface pressure at the start of the relaxation process. For lack of independent information, this parameter is assumed to be independent of the bulk concentration, and treated as an adjustable parameter. Application of the model of Eq (14) to experimental data is then found to yield a non-constant relaxation time [67]. Alternatively, the model needed to be extended into a bi-exponential expression with two different, constant, relaxation times [21].

Only in the simple case where the surface pressure Π increases linearly with Γ_S is this relationship equivalent to the integrated form of Eq (16):

$$\frac{\Gamma_S - \Gamma_S^{eq}}{\Gamma_S^0 - \Gamma_S^{eq}} = \exp(-t/\tau_{rel}) \tag{18}$$

Like Eqs (16) and (17), Eq (18) is a simple form of a first-order model; therefore it is unable to provide quantitative information on specific relaxation phenomena other than their time scale. Serrien et al. [18], in a wide-ranging study with four different dynamic surface techniques, describe unfolding of proteins as a monomolecular reaction from a native state into a denatured, insoluble, state with dynamic surface pressures given by Eq (17) and the dilational modulus by

$$\frac{\varepsilon - \varepsilon(0)}{\varepsilon(\infty) - \varepsilon(0)} = \frac{1}{\sqrt{1 + (k/\omega)^2}} \tag{19}$$

where, in the virtual absence of diffusion, $\varepsilon(0)$ and $\varepsilon(\infty)$ are the modulus values measured at frequencies much lower and much higher, respectively, than the characteristic relaxation frequency, k (=$1/\tau_{rel}$). At the lower frequencies, the native and denatured molecules relax to their equilibrium values within the time scale of the experiment, while at the higher frequencies the equilibrium is effectively frozen and the denatured state prevails. Values of k are in the order of 0.01 s^{-1} for BSA from the modulus spectrum, in agreement with the phase angles at high surface pressure in Table 2. Separate stress relaxation experiments revealed a second, much slower, process with k_2 in the order of 10^{-4} s^{-1}, which would fit in with the high phase angles found by Murray et al. [59] at very low frequencies (see Figure 22). Even slower relaxation, with a rate constant k in the order of 10^{-5} s^{-1} was observed by Bull [69] for solutions of egg albumin at fairly high surface pressures (Π>10 mN/m). However, as pointed out by Bull, a more detailed model cannot be developed if the molecular areas occupied by the native and the denatured forms are unknown.

The slow conformational changes discussed so far are all described as reactions <u>in</u> the surface. Additionally, exchange with protein layers under the surface is possible in cases of multilayer adsorption. Relaxation by exchange between first and second layer has been proposed by MacRitchie [61], Guzman et al. [70] and Hunter et al [71]. Multilayers provide a reservoir of protein close to the interface, and the protein molecules in it should be able to readily interchange with the interface as the area is varied. Thus one would expect quite fast relaxation from this mechanism, which is most likely to apply to the caseins. The data presented in Figure 10 for β-casein give little indication of exchange with multilayers as phase angles are very small at all frequencies over the whole surface concentration range. However, higher viscous phase angles were reported [43] for high concentrations (> 0.1 g/l) of β-casein in a time scale of 10 s and qualitatively ascribed to such an exchange in combination with diffusion from the bulk of the solution.

Summarising, most relaxation in fairly close-packed protein layers reflects processes by which the adsorbed protein becomes increasingly more insoluble, by either intramolecular or intermolecular interactions, which often proceed simultaneously in time scales from 10^2 to 10^5 s. Current modelling of such relaxation phenomena in specific molecular-kinetic models has not advanced far beyond the stage of curve fitting in terms of simple first-order kinetics with a number of adjustable parameters.

4 EXPERIMENTAL RESULTS FOR PROTEINS ADSORBED AT OIL/WATER

Experimental studies of proteins at oil/water interfaces are few and far between, in comparison with almost countless reports on the air/water surface. Work on dynamic interfacial properties has suffered from the greater experimental difficulties in obtaining reliable values of the dilational modulus, owing to increased wave damping and greater risks of leakage at oil/water interfaces (see Section 2.1). Problems with wave propagation caused by increased damping are easiest to overcome with low-viscosity oils, e.g., n-decane or n-tetradecane. Far heavier damping is to be expected for the highly viscous triacylglycerol oils, e.g., sunflower oil, which are of particular interest for food emulsions.

4.1 Hydrocarbon oil/water interfaces

At the tetradecane/water interface, Murray et al. [13,59] studied BSA and β-lactoglobulin at one concentration (10^{-2} g/l) over a range of frequencies from 10^{-5} to 0.1 Hz, using a novel leak-free enclosure for the interface (Murray and Nelson [72]). Interestingly, their data indicate a relaxation process around $2\text{x}10^{-4}$ Hz, in a longer time scale than at air/water. For a low-viscosity paraffin oil, Williams and Prins [43] used the ring trough to measure moduli and phase angles on solutions of β-casein and β-lactoglobulin with concentrations ranging from 10^{-4} to 1 g/l at one frequency (0.1 Hz). Their results at oil/water are approximately the same as at air/water for both their proteins, but Murray et al. [13,59] find modulus values for β-lactoglobulin at oil/water to be higher by up to 20 mN/m.

4.2 Triacylglycerol oil/water interfaces

Triacylglycerols are more polar than the hydrocarbons and, therefore, may result in surface behaviour distinct from the air/water and the hydrocarbon/water interfaces. Benjamins et al. [27] used the Dynamic Drop Tensiometer at the sunflower oil/water interface to obtain moduli and phase angles for three proteins of different molecular structures over a range of concentrations, interfacial ages and frequencies.

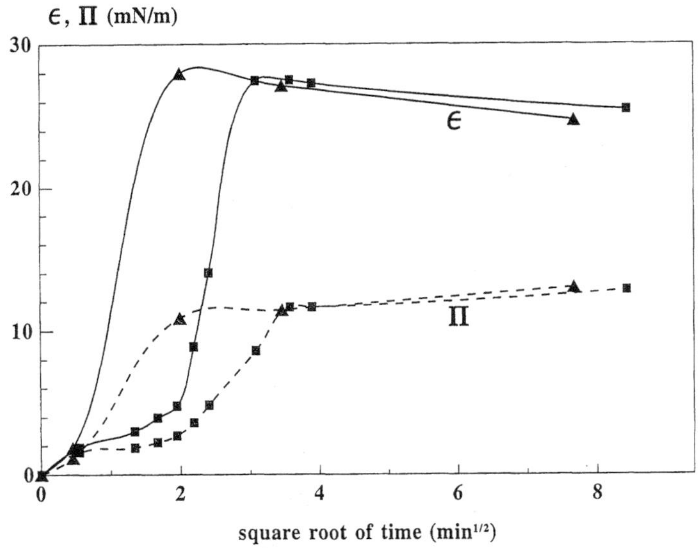

Fig. 23 Example of the time dependence of the viscoelastic modulus, $|\varepsilon|$, and the interfacial pressure Π for two concentrations of Bovine Serum Albumin, adsorbed at the sunflower oil/water interface [27]. ❑ - 5 mg/l; Δ - 10 mg/l. Frequency: 0.1 Hz.

Fig. 23 gives an example of the time dependence of the modulus and the interfacial pressure for two concentrations of BSA. Contrary to what is seen at air/water (see Fig. 6) for these low concentrations, no "induction" time is observed: non-zero values of the interfacial pressure are found immediately after formation of the interface. As at air/water, combining the data of Fig. 23 in a modulus *vs* pressure plot makes them coincide on a single curve, as illustrated in Fig. 24 for Ovalbumin. Similarly characteristic curves were obtained for β-casein, whole casein, Bovine Serum Albumin [27] and β-lactoglobulin; a summary for four proteins is shown in Fig. 25.

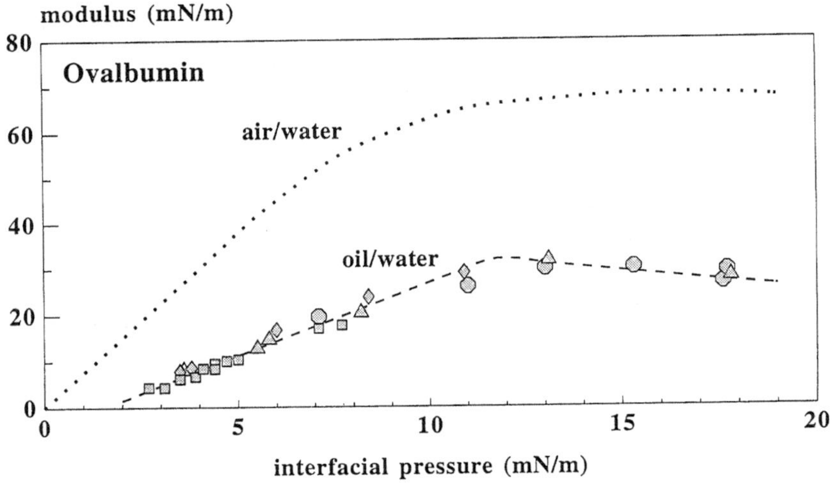

Fig. 24 Viscoelastic modulus, $|\varepsilon|$, measured with the Dynamic Drop Tensiometer, as a function of interfacial pressure, Π, for Ovalbumin, adsorbed at the sunflower oil/water interface [27]. Concentrations: 0.01 g/l (□), 0.1 g/l (Δ) and 1 g/l (○). Frequency: 0.1 Hz. Amplitude of compression/expansion: ΔA/A=0.2. Results for the air/water surface from Figure 14 are given for comparison.

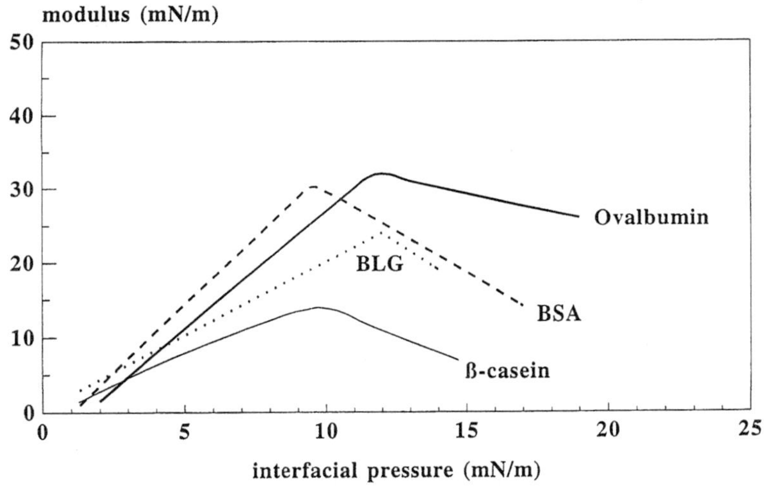

Fig. 25 Summary of the modulus *vs* interfacial pressure results for the different proteins at the sunflower oil/water interface as measured with the Dynamic Drop Tensiometer [27].

Again, as at air/water, surface behaviour at high frequency was found to be purely elastic for all proteins at not too high surface coverage, i.e., at interfacial pressures up to 15 mN/m. At lower frequencies and higher Π, relaxation mechanisms were apparent from the measured non-zero viscous phase angles, ϕ, which increased with decreasing frequency. These phase angles were in the same range as at air/water, although somewhat higher than the values in Table 2 in the case of BSA.

In the early version of the Dynamic Drop Tensiometer, quite high values of the amplitude $\Delta A/A$ were required for the oscillations in tension to be easily measurable. A relative deformation of 0.20 might be expected to be far too high for low-amplitude experiments, but in fact the modulus *vs* interfacial pressure curve was found to be the same at relative deformations of 0.07 and 0.20, within the experimental error [27].

4.3 Discussion

At the hydrocarbon oil/water interface, indications from the work of Williams and Prins [43] are that moduli and viscous phase angles at this interface are very similar to those at air/water for both β-casein and β-lactoglobulin, while the moduli measured by Murray et al. [13,59] for β-lactoglobulin have higher elastic and viscous parts at oil/water than at air/water. However, the limited amount of data does not appear to allow any hard and fast conclusions. Further interpretation is also hampered by difficulties in measuring values for the protein adsorption because the frequently used technique of ellipsometry cannot be applied at oil/water interfaces. Notable exceptions where such data have been obtained for some proteins are in the work of Graham and Phillips [47] at decane/water, reproduced in Figure 26 for β-casein and BSA, and of Murray and Nelson [72] at tetradecane/water. There is a fairly large discrepancy between the two sets of data for BSA at each interface, but there is qualitative agreement that BSA is more expanded at oil/water than at air/water. This is in line with what has often been observed with small surface-active molecules, where it is attributed to reduced van der Waals cohesion of hydrophobic chains in oil, which is a better solvent for the chains than water. Both Graham and Phillips [47] and Murray and Nelson [72] advance this argument in terms of the hydrophobic polypeptide chains of globular proteins, which can unfold into loops in the oil phase. Such reduced cohesion results in a higher pressure at oil/water. However, the flexible protein β-casein exhibits the opposite behaviour, as it was found to be more expanded at air/water [47]. The authors explain this by arguing that the unhindered loop formation which is possible only for flexible macromolecules, results in a smaller number of segments in the surface, i.e., a smaller molecular area and, therefore a smaller degree of coverage (Θ) and a lower surface pressure at the same value of Γ. This second, positive, effect on the surface pressure overrides the negative effect of the van der Waals cohesion of the hydrophobic chains. A quantitative interpretation of the measured elasticities at the oil/water interface in terms of the different pressure *vs* area curves is not available at present.

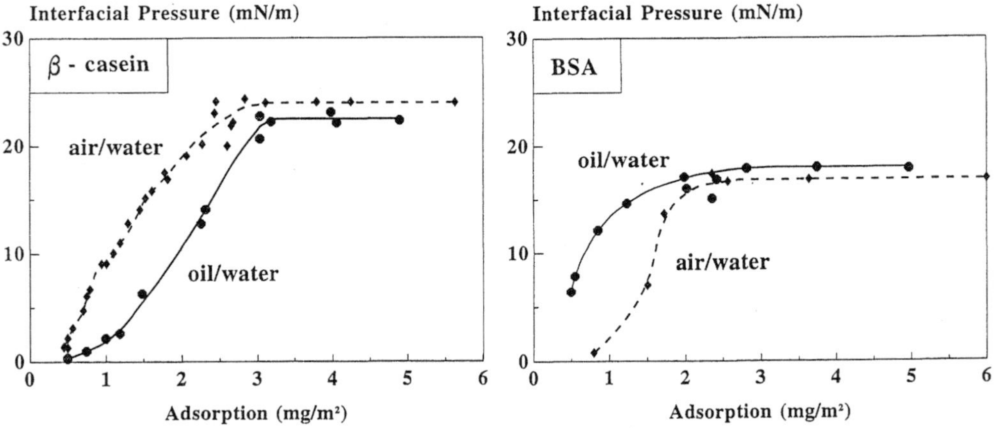

Fig. 26 Interfacial pressure *vs* adsorption at the decane/water interface for β-casein and Bovine Serum Albumin. experimental data from Graham and Phillips [47].

At the triacylglycerol oil/water interface, there is no more than a qualitative similarity to the modulus *vs* pressure curves at air/water in Figure 24. Quantitatively, modulus values for all proteins in Fig. 25 are quite appreciably lower than those at air/water: the modulus at oil/water at its maximum is only half that at air/water, and the ratio is much smaller at all other values of the interfacial pressure. These low values are connected with three characteristic features of the oil/water curves in Fig. 25:

(i) the initial slopes of the $|\varepsilon|$ *vs* Π curves, at pressures $\Pi<1$ mN/m, are very low. Lack of reliable data at these very low pressures probably obscures the +1 slope predicted by the two-dimensional analogue of the ideal-gas law. In contrast to what was found at the air/water surface, here the gaseous region may extend up to pressures of about 1 mN/m.

(ii) at slightly higher pressures, a steeper linear part follows with a slope for all four proteins of approximately +3, i.e., much lower than at air/water (see Table 3). As at air/water, the pressure range of this steeper linear behaviour appears to increase with decreasing flexibility of the protein molecule, as seen from the summarised results in Fig. 25. Linearity holds over only 3 mN/m for the most flexible molecule, Na Caseinate, but over 10 mN/m for the most compact one, Ovalbumin. In this range, the modulus is purely elastic, and should be equal to the value of ε_0 evaluated from the equilibrium pressure *vs* adsorption curve. However, such equality cannot be confirmed for the present oil/water interface as no measured adsorption values are available. If the expansion at the triacylglycerol oil/water interface is similar to that found for BSA by Murray and Nelson [72] at the hydrocarbon oil/water, this would mean that for BSA the modulus at O/W would be a factor of 1.7 lower than at A/W over a large range of pressures, in fair agreement with the measured data in Figure 25.

(iii) after the linear region, viscous phase angles are no longer negligible, all curves pass through a maximum and then decline with increasing pressure. Such decline, even more than the flattening off at air/water, may indicate collapse-type phenomena, although it is by no means clear why these should be facilitated at the oil/water interface.

The fairly high slopes in the linear, elastic, range are indicative of a fair amount of molecular interactions, although these are less pronounced than at air/water. Molecular reconformations, including unfolding of the protein molecules at the interface, are assumed to occur as well but, as at air/water, these take place fairly fast, in a time scale of a few seconds, and they do not affect the modulus at 0.1 Hz. At the higher pressures where viscoelastic behaviour is observed, as at air/water, the measured viscous phase angles cannot be explained from diffusional exchange with the bulk solution, as is clear from Fig. 22. Of the relaxation mechanisms discussed in Section 3.3.3., slow reconformations, collapse phenomena and, in the case of flexible proteins, exchange with multilayers are expected to play a similar part at the oil/water interface.

Summarising, dilational moduli at hydrocarbon oil/water interfaces appear to be fairly similar to - or somewhat higher than - those at air/water, but those at triacylglycerol oil/water are much lower. We attribute this to less severe non-ideality of the latter interface, as indicated by the far shorter "induction" times and by the much lower slopes of the modulus *vs* pressure curves.

5 SUMMARY

All proteins and the synthetic polymer studied show purely elastic surface behaviour at both the air/water and the oil/water interface in a time scale of 1 s over a considerable range of adsorptions and surface pressures. For the caseins and PVA, this is approximately true for almost the entire range of experimental pressures, roughly 20 mN/m, while for BSA and Ovalbumin measurable viscosities are found for pressures higher than 15 mN/m. In the elastic range, the dilational modulus ε is equal to the limiting modulus ε_0 calculated from the equilibrium pressure *vs* adsorption relationship, which is characteristic for each protein. Common features distinguishing this relationship from that for simple small molecules at the air/water surface are (i) a range of adsorptions up to roughly 1 mg/m^2 where the pressure and the modulus are extremely low, followed by (ii) a sudden and steep increase of both pressure and modulus with increasing adsorption at higher adsorptions (up to roughly 2 mg/m^2) and, finally, (iii) a flattening off of both modulus and pressure at the highest adsorptions. In this last region, surface relaxation processes induce viscoelastic behaviour in time scales longer than 1 s.

In the first two regions, where the modulus is a pure elasticity, differences between individual proteins are caused by different degrees of severe non-ideality of the surface equation of state, due to molecular reconformations and intermolecular interactions. The measured steep increase of the elasticity points to an overriding influence of attractive intermolecular interactions. An adequate description of the characteristic features of the elastic region is obtained from a simple equation-of-state model combining non-ideal entropic and enthalpic effects. Such non-ideality is shown to be much less pronounced at a triacylglycerol oil/water interface than at the air/water surface, resulting in appreciably lower moduli at the former interface. As a result of the various intramolecular and intermolecular interactions, the modulus at both interfaces generally increases with decreasing flexibility of the molecules, with lowest levels found for the synthetic polymer PVA.

In the third region, where the protein layers are fairly close-packed and viscoelastic, existing equations of state have to be extended in order to account for the measured viscoelastic moduli. Relaxation processes in this region cause the adsorbed protein to become increasingly more insoluble, by either slow molecular reconformation or collapse-type processes, in time scales from 10^2 to 10^5 s. Reconformation processes, if responsible, clearly proceed at a much slower rate in these close-packed layers than at the lower adsorptions of the elastic region. Current modelling of such relaxation phenomena is still in the stage of curve fitting in terms of simple first-order kinetics with a number of adjustable parameters; more specific molecular-kinetic models will have to be developed.

6 REFERENCES

1. M. van den Tempel, 3rd Int. Congress Surface Active Agents (Cologne), Volume II 345 (1960)
2. J. Lucassen and M. van den Tempel, J. Chem. Sci., 27 (1972) 1283
3. J.A. de Feijter, J. Colloid Interface Sci., 69 (1979) 375
4. E. Dickinson, B.S. Murray and G. Stainsby, J. Chem. Soc. Faraday Trans. I, 84(1988) 871
5. R. Miller, R. Wüstneck, J. Krägel and G. Kretzschmar, Colloids Surfaces, 111 (1996) 75
6. E.H. Lucassen-Reynders, Food Structure 12 (1993) 1
7. F.C. Goodrich, L.H. Allen and A. Poskanzer, J. Colloid Interface Sci., 52 (1975) 201, 213
8. V. Mohan, B.K. Malvija and D.T Wasan, Can. J. Chem Eng., 54 (1976) 515
9. D.E. Graham and M.C. Phillips, J. Colloid Interface Sci., 76 (1980) 240
10. A. Martinez-Mendoza and P. Sherman, J. Dispersion Sci. Technology, 11 (1990) 347
11. V. Kiosseoglou, J. Dispersion Sci. Technology, 13 (1992) 135
12. J. Benjamins and F. van Voorst Vader, Colloids and Surfaces, 65 (1992) 161
13. B.S. Murray and E. Dickinson, Food Sci.Technol. Int. 2 (1996) 31
14. J.W. Gibbs, Collected Works (Longmans, Green, New York), 1 (1928) 302
15. D.E. Graham and M.C. Phillips, J. Colloid Interface Sci., 70 (1979) 403
16. D.E. Graham and M.C. Phillips, J. Colloid Interface Sci., 70 (1979) 415
17. D.E. Graham and M.C. Phillips, J. Colloid Interface Sci., 76 (1980) 227
18. G. Serrien, G. Geeraerts, L. Ghosh and P. Joos, Colloids Surfaces, 68 (1992) 219
19. C-S. Gau, H. Yu and G. Zografi, J. Colloid Interface Sci., 162 (1994) 214
20. G.A. van Aken and M.T.E. Merks, Spec. Publ. Roy. Soc. Chem., (E. Dickinson, P. Walstra, Eds.) 113 (1993) 402
21. G.A. van Aken and M.T.E. Merks, Progr. Colloid Polym. Sci., 97 (1994) 281
22. F. Boury, Tz. Ivanova, I. Panaiotov and J.E. Proust, Langmuir 11 (1995) 2131
23. J. Lucassen and M. van den Tempel, J. Colloid Interface Sci., 41 (1972) 491

24. G.T. Barnes and J. Lucassen, J. Chem. Soc. Faraday Trans. I, 68 (1972) 2129
25. J. Lucassen and D. Giles, J. Chem. Soc. Faraday Trans. I, 171 (1975) 217
26. L. Ting, D.T. Wasan and K. Miyano, J. Colloid Interface Sci., 107 (1985) 345
27. J. Benjamins, A. Cagna and E.H. Lucassen-Reynders, Colloids Surfaces, 114 (1996) 245
28. J. Benjamins, J.A. de Feijter, M.T.A. Evans, D.E. Graham and M.C. Phillips, Faraday Discussions of the Chemical Society, 59 (1975) 218
29. J.J. Kokelaar, A.Prins and M. de Gee, J. Colloid Interface Sci. 146 (1991) 507
30. K.D. Wantke, K. Lunkenheimer and C. Hempt, J. Colloid Interface Science., 159(1993)28
31. A. Cagna, G. Esposito, C. Rivière, S. Housset and R. Verger, 33rd Intern. Conf. on Biochemistry of Lipids, Lyon, France (1992).
32. Y. Rotenberg, L. Boruvka and A.W. Neumann, J. Colloid Interface Sci., 93 (1983) 169
33. C.A. McLeod and C.J. Radke, J. Colloid Interface Sci., 160 (1993) 435
34. R. Miller, Z. Policova, R. Sedev and A.W. Neumann, Colloids Surfaces, 76 (1993) 179
35. R. Nagarajan, K. Koczo, E. Erdos and D.T. Wasan, AIChE Journal, 41 (1995) 915
36. J.A. de Feijter, J. Benjamins and F.A. Veer, Biopolymers, 17 (1978) 1759
37. H.B. Bull, Adv. Protein Chem., 3 (1947) 95
38. F. MacRitchie and A.E. Alexander, J. Colloid Interface Sci., 18 (1963) 453
39. T. Yamashita and H.B. Bull, J. Colloid Interface Sci., 27 (1968) 19
40. M. Blank, J. Lucassen and M. van den Tempel, J. Colloid Interface Sci., 33 (1970) 94
41. D. Giles, J. Lucassen, unpublished data (1971)
42. P. Joos, Biochemica et Biophysica Acta, 375 (1975) 1
43. A. Williams and A. Prins, Colloids Surfaces, 114 (1996) 267
44. R. Douillard, M. Daoud, J. Lefebvre, C. Minier, G. Lecannu and J. Coutret, J. Colloid Interface Sci., 163 (1994) 277
45. G.A. van Aken and M.T.E. Merks, Voedingsmiddelen Technologie, 27 (1994) 11
46. F.J.G. Boerboom, A.E.A. de Groot-Mostert, A. Prins and T. van Vliet, Netherlands Milk & Dairy J., 50 (1996) 183
47. D.E. Graham and M.C. Phillips, J. Colloid Interface Sci., 70 (1979) 427
48. J. Leaver and D.G. Dalgleish, J. Colloid Interface Sci. 149 (1992) 49
49. E.J. McNally and G. Zografi, J. Colloid Interface Sci., 138 (1990) 61
50. E.H. Lucassen-Reynders, J. Lucassen, P.R. Garrett, D. Giles and F. Hollway, Advances in Chemistry Series, 144 (1975) 272
51. S.J. Singer, J. Chem. Phys., 16 (1948) 872

52. H.L. Frisch and R. Simha, J. Chem. Phys., 27 (1957) 702
53. M.A. Cohen Stuart, G.J. Fleer and J.M.H.M. Scheutjens, J. Colloid Interface Sci., 97 (1984) 515
54. E.H. Lucassen-Reynders, Colloids Surfaces, 91 (1994) 79
55. V.B. Fainerman, R. Miller and R. Wüstneck, J. Colloid Interface Sci., 183 (1996) 26
56. E.H. Lucassen-Reynders, 11th Intern. Symposium on Surfactants in Solution, Jerusalem, Israel (1996)
57. J.A. de Feijter and J. Benjamins, J. Colloid Interface Sci., 90 (1982) 289
58. G.A. van Aken, Spec. Publ. Roy. Soc. Chem. (E. Dickinson, D. Lorient, Eds.), 156(1995) 43
59. B.S. Murray, A. Ventura, C. Lallement, Colloids Surfaces, in press
60. M. van den Tempel and E.H. Lucassen-Reynders, Adv. Colloid Interface Sci., 18 (1983)281
61. F. MacRitchie, Adv. Colloid Interface Sci., 25 341 (1986)
62. R. Maksymiw and W. Nitsch, J. Colloid Interface Sci., 147 (1991) 67
63. P. Joos, G. Bleys, G. Pétré, J. chim. phys., 79 (1982) 387
64. J.A. de Feijter and J. Benjamins, in Food Emulsions and Foams (E. Dickinson, Ed.), Special Publication Roy. Soc. Chem., 58 (1987) 72
65. F.A. Veer and M. van den Tempel, J. Colloid Interface Sci., 42 (1973) 418
66. S. Kitching, G.D.W. Johnson, B.R. Ridmore and T.M. Herrington, J. Colloid Interface Sci., 177 (1996) 58
67. V.B. Fainerman, Kolloidn. Zh., 40 (1978) 530
68. A.A. Trapeznikov, V.G. Vins and T.Y. Shirokova, Colloid J. USSR, 43 (1981) 262
69. H.B. Bull, J. Colloid Interface Sci., 41 (1972) 305
70. R.Z. Guzman, R.G. Carbonell and P.K. Kilpatrick, J. Colloid Interface Sci. 14 (1986) 536
71. J.R. Hunter, P.K. Kilpatrick and R.G. Carbonell, J. Colloid Interface Sci., 142 429 (1991)
72. B.S. Murray and P.V. Nelson, Langmuir 12 (1996) 5973

7 LIST OF SYMBOLS

A - interfacial area

c - protein concentration in solution

D - diffusion coefficient of protein or other solute

L - effective length of trough

H - molar enthalpy of mixing in Frumkin model

S - size factor of protein ($=\omega_2/\omega_1$)

β - damping coefficient of surface wave

Γ - adsorption of protein

γ - interfacial tension

ε - surface dilational modulus

ε_0 - limiting value defined in Eq (2)

$\varepsilon^{/}$ - elastic part of surface dilational modulus (storage modulus)

$\varepsilon^{//}$ - viscous part of surface dilational modulus (loss modulus)

ε_d - surface dilational elasticity ($=\varepsilon^{/}$)

η - viscosity of aqueous phase

$\eta^{/}$ - viscosity of non-aqueous phase

η_d - surface dilational viscosity ($=\varepsilon^{//}/\omega$)

Θ - fraction of interface covered by protein

κ - wave number of surface wave ($=2\pi/\lambda$)

λ - wavelength of surface wave

ν - frequency of surface wave

Π - interfacial pressure

ρ - density of aqueous phase

$\rho^{/}$ - density of non-aqueous phase

τ_{rel} - time scale of relaxation process

τ_{diff} - time scale of diffusional relaxation

ϕ - viscous phase angle

ω - angular frequency of surface wave ($=2\pi\nu$)

ω_1 - partial molar area of solvent

ω_2 - partial molar area of protein or other solute

Proteins at Liquid Interfaces
D. Möbius and R. Miller (Editors)

PROTEIN-LIPID INTERACTIONS

Tommy Nylander

Physical Chemistry 1, Center for Chemistry and Chemical Engineering,

Lund University, P. O. Box 124, S-221 00 Lund, Sweden

Contents

1 INTRODUCTION

The membrane of the living cell, which separates one micro-environment from another, gives the most prominent example of a liquid interface, where proteins, whether membrane bound or extracellular, interact with lipids. The complexity of these interactions has triggered a vast number of publications (c.f. the book edited by Watts (1)). The degradation of lipids by lipases is an example of a biological process which occurs at the lipid-aqueous interface (2, 3). The properties of the interface formed, strongly affect the enzyme activity. Furthermore, the interface is not constant, but as the enzymatic process proceeds the inclusion of degradation products (e. g. fatty acids) at the interface will change its properties. The pulmonary surfactant system is another example where the interfacial properties of a complex mixture of proteins and lipids are essential for the function of a biological system - the lungs (4-6). Alterations of the interfacial behaviour of the pulmonary surfactants might affect their ability to promote lung expansion on inspiration, prevent lung collapse during expiration, balance the pulmonary fluids and stabilise small airways (6).

The biological role of some proteins is to act as carriers of hydrophobic molecules with low aqueous solubility, which bind to hydrophobic pockets on the proteins. For instance, the central role of fatty acid binding to albumin in the mammalian lipid transport system is now well established (7). Another example is β-lactoglobulin, which has been thought to take part in the transport of retinol (8-10). This protein has also been shown to have a high affinity for phospholipids, fatty acids and triglycerides (11-14). An important application of this interaction was reported by Kurihara and Katsuragi (15), who found that a complex formed between β-lactoglobulin and phosphatidic acid, could mask the taste of bitter substances. This property was suggested to be specific for phosphatidic acid as no effect was observed for mixtures of β-lactoglobulin and phosphatidylcholine, triacylglycerol or diacylglycerol.

The colloidal stability of a dispersed phase found in emulsions and foams is controlled by the combination of properties of the emulsifying and foaming agents. Proteins and lipids are probably the most frequently used emulsifying and foaming agents in applications ranging from food, pharmaceuticals and cosmetics to paint, paper and petrochemical products. The interaction between proteins and lipids can play a crucial role in the stability of these system, whether they are used in combination as stabilisers or one of the components is used as stabiliser and the other is already present in the system. Their common amphiphilic nature makes them orient at an interface. This property provides a driving force for the formation of associated structures of lipids as well as for the folding of a polypeptide chain into the unique conformation of the native protein. However, the mechanisms by which they stabilise emulsions and foams can be quite different. Many proteins stabilise foams and emulsions by intermolecular interactions between the adsorbed protein molecules which encapsulates the dispersed phase. Such a mechanically rigid layer immobilises the proteins and prevents perturbations, droplet coalescence or flocculation. In contrast, lipids usually stabilise the dispersed droplet or bubble by forming a densely packed and much less rigid monolayer which is stabilised by dynamic processes, e. g. the Gibbs-Marangoni effect. These aspects are discussed more in detail elsewhere in this book. Although these stabilisation processes are quite well understood for the individual components, the effect of protein-lipid interaction at the interfaces is less known.

In this review the interaction involving lipids will be the main theme. What, then, is a lipid? I will use the definition adopted by my teacher, Prof. Kåre Larsson (16), who quotes the formulation proposed by Christie "Lipids are fatty acids and their derivatives, and substances related biosynthetically or functionally to those compounds"(17). Lipids can be divided into non-polar lipids, e. g. triglycerides, which hardly interact with water, and polar lipids, e. g. phospholipids, which readily interact with water and associate into lipid-water phases (16). The structures of some of the main lipids discussed in this review are shown in Fig. 1. Surfactants, e. g. sodium dodecylsulphate (SDS) can be regarded as synthetic analogues to polar lipids. The water soluble polar lipids (e. g., ionised fatty acids, lysophospholipids and bile salts) and synthetic surfactants, ionic and non-ionic, have a considerable monomeric solubility (millimoles) and form micelles at higher concentrations, starting at the critical micelle concentration (cmc) (18). The interaction between water soluble lipids and proteins usually takes place via monomers (19, 20) and occurs around or above cmc for the lipid/surfactant. Furthermore the surfactant can associate on the protein and thereby decorating the polypeptide chain of the unfolded protein to form a structure resembling a necklace (19, 21, 22). Polar lipids with very low water solubility usually swell into liquid crystalline or gel phases in the presence of water. Thus, the lipid aggregates (dispersed particles, liposomes or vesicles, liquid crystalline or gel phases) are generally already formed when the interaction with the protein takes place. Knowledge of the phase behaviour of these lipids is therefore a prerequisite to be able to understand the interaction with proteins.

The focus in this review will be on the interaction between proteins and polar lipid with low aqueous solubility. Some of the possible mixed protein-lipid structures, formed when protein molecules interact with a lipid monolayer, are shown in Fig. 2. The diversity of these interactions will be further demonstrated as the discussion proceeds from "simple" monolayers at the air/aqueous and oil/aqueous interface to the curved interfaces of lipid aggregate and finally the protein encapsulation in such intriguing structures as the cubic phases formed in lipid-aqueous systems.

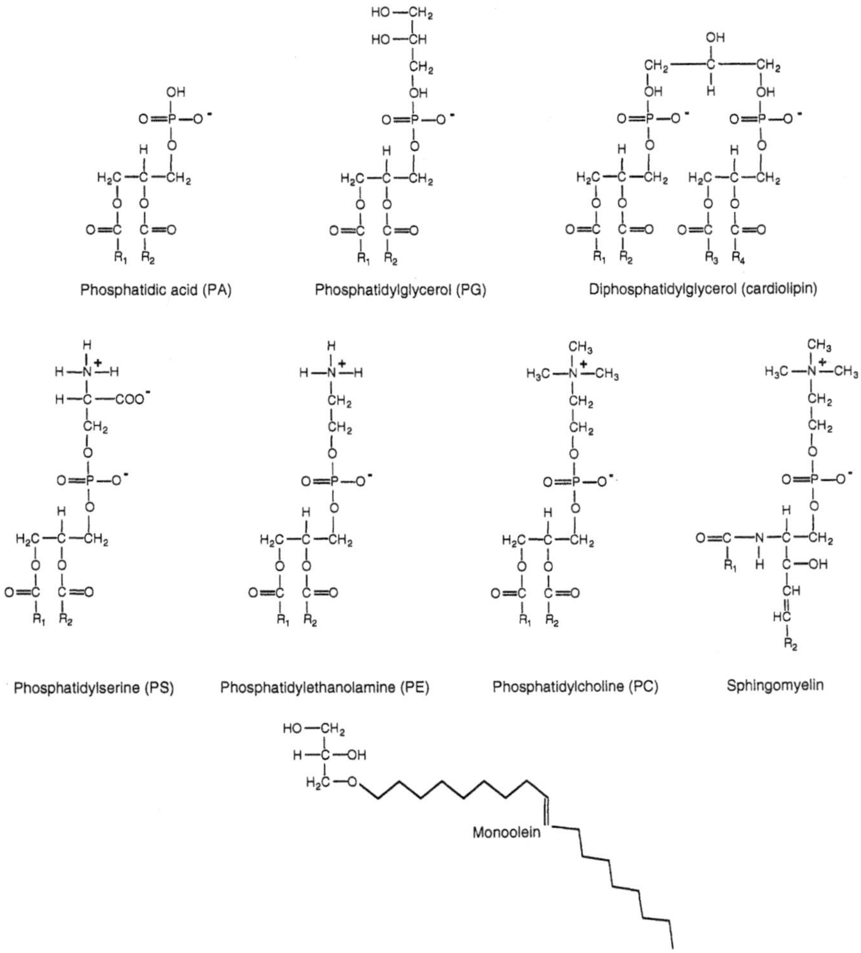

Fig. 1 The molecular structure of some of the lipids discussed. The charge is given as at pH 7. R1, R2, R3, and R3 denote the acyl chains of the lipids.

Although similar driving forces for the protein-lipid interaction prevails, it is important to stress the difference in nature between the two types of liquid interfaces, the liquid/air interface and the one between two condensed media. The oil/water interface, as well as the one formed between an aggregated lipid and the aqueous surrounding medium, allows hydrophobic residues to become dissolved in, and interact favourably, with the oil phase or the lipidic medium, which is not possible at the pure air/water interface. This is important as the unfolding of protein, induced by lipid/surfactant binding or by the presence of an interface, generally leads to exposure of hydrophobic residues, that is the unfolded protein can be substantially more "oil soluble" than the native one. Furthermore, Ninham and Yaminsky recently showed that dispersion interactions are significant for specific ion adsorption, also at a charged interface like a protein or a lipid layer (23). Thus the effect can be qualitatively different at air-water, and oil-water surfaces.

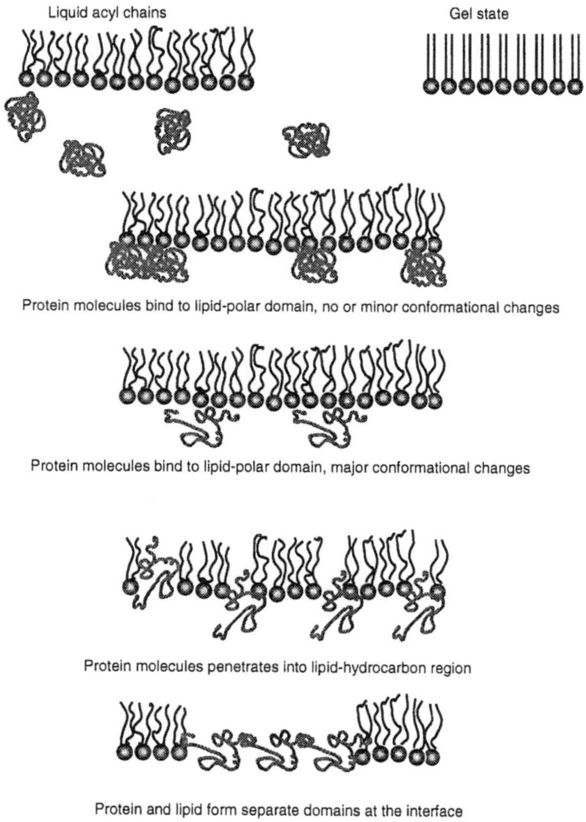

Fig. 2 Schematic representation of protein molecules interacting with a lipid monolayer. Some of the possible structures of a mixed protein-lipid monolayer are represented, which are also applicable for the protein interacting with a lipid bilayer.

A substantial part of the review will be dealing with the interaction between proteins and lipid aggregates such as vesicles, micelles, gels and liquid crystals. It should be realised that these systems are dynamic and flexible and that the interfaces themselves are prone to change as a consequence of the interaction with the protein. In turn it can completely change the structure of the aggregate. The interest for these interactions have also increased considerably in the last few years, not only due to their significance in biological systems, but also for the development of new applications in such as foods, drug delivery and analytical systems.

2 PROTEINS AND LIPIDS AT INTERFACES

A delicate balance of intermolecular forces, including electrostatic forces, hydrogen bonding, van der Waals forces, conformational entropy and so-called hydrophobic interactions (cf. (24-27)), as well as cross-links (e. g. disulphide bridges), gives a protein the unique conformation which determines its function. The core of the protein molecule consists mainly of nonpolar amino acid residue, while the polar amino acid residues (uncharged and charged) are predominant on the surface of the protein molecule. It should be borne in mind that even though the net charge of a protein is, say, negative, it also carries positively charged groups which by proper orientation of the protein at an interface can bind to negative groups at the interface.

The interior of a globular protein is very densely packed, with values of packing density (typical values ≈ 0.74 ml/g) close to those observed in crystals of small organic molecules (28). However, proteins are only marginally stable at room temperature. Thus only slight changes of the protein structure, like the exchange of one or two amino acid residues or a change in the conditions like the binding of a lipid molecule to the protein, or the presence of an interface, can considerably affect the stability as well as the structure of the protein. The formation of a globular protein structure is mainly counteracted by the loss of entropy on protein folding (26). Thus, the unfolding of a protein on adsorption is an entropically favoured process and can in fact in many cases be the driving force for adsorption (29, 30). In addition, at a lipid-aqueous interface the exposure of unfolded hydrophobic domains of the protein to the aqueous environment can be minimal if these domains interpenetrate into the lipid mono- or bilayer (Fig. 2).

During the transfer from the native to the completely unfolded state, the protein can adopt an intermediate structure, the molten globular state (31-37). This state is characterised by a retained secondary structure, but with a fluctuating tertiary structure, and consequently hydrophobic domains from the interior are exposed. The translocation of proteins across biological membranes has been proposed to occur with the protein in the molten globular state (38, 39). The molten globular state can be achieved by for instance pH-changes, increase of temperature, the use of denaturation agents, breaking of disulphide bridges and removal of ligands or co-factors bound to the protein as well as when the protein interacts with an interface (37). The latter has been demonstrated for α-lactalbumin, which was found to be more surface active when the solution conditions favoured the molten globular state.

Polar lipids are more capable of reducing the interfacial tension than proteins. Proteins can, on the other hand, form kinetically stabilised layers, where the mobility is substantially lower than for a corresponding film of polar lipids. This is due to the fact that protein molecules are usually anchored at multiple sites at the interface and that the unfolding process can entropically stabilise the protein at the interface. Inter-molecular interaction between protein molecules can further increase the stability of the film.

Interfacial phenomena in mixed protein-lipid systems are often strongly linked to the association state of the lipid. An example is the interaction between β-lactoglobulin and phosphatidic acid which was found too occur only when the lipid was present as a dispersion, but not when in the gel state (14). Discrepancies between results from different studies can often be related to differences in the state of the lipid.

3 AIR/AQUEOUS INTERFACE

Much of the knowledge concerning the interaction between proteins and lipids with low aqueous solubility has been gained from monolayer studies at the air/aqueous interface (2, 40). For these studies pure substances are needed, as any impurity with high enough surface activity will be preferentially adsorbed at the interface and reduce the surface tension. The presence of impurities in commercial surfactants, e. g. sodium dodecylsulphate, (41-43), and proteins, e.g. fatty acids, bound to β-lactoglobulin (44) can be substantial, and anomalies in the interfacial behaviour can be removed by careful purification of the sample.

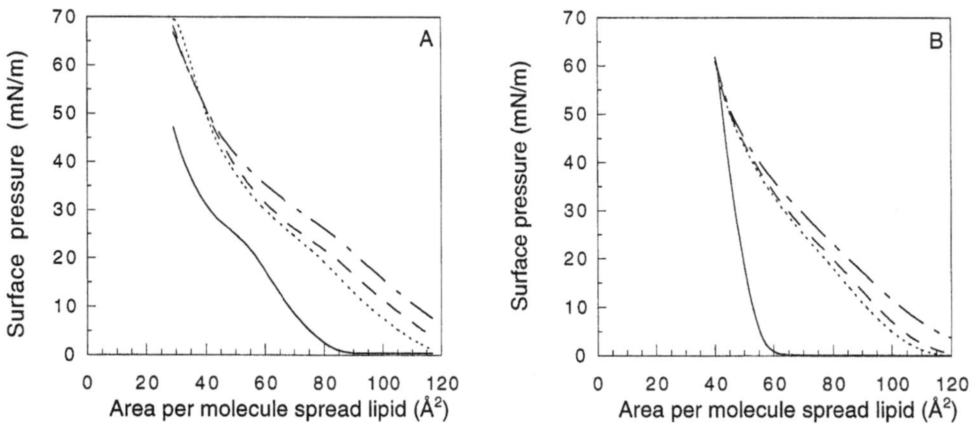

Fig. 3 Dynamic surface pressure (Π) as a function of the molecular area of the spread amount lipid for compression of (A) sphingomyelin and (B) distearoylphosphatidylcholine (DSPC) monolayers on a phosphate buffered subphase (40 mM phosphate containing 0.1 M sodium chloride, pH = 7.4) with or without xanthine oxidase (5 mg/ml). The isotherms recorded for the lipid spread on pure buffer (——)

and at 5 (- - - - -), 10 (— — — —), 20 (— - — -) min. elapsed between spreading and compression. The lipid (25 μg) was spread from a chloroform/methanol (2:1, v/v) solution on a maximum area of 50 × 450 mm^2 and a compression speed of 12.5 mm/min. was used. Data adopted from Kristensen *et al.* (49), where also the experimental details are given.

3.1 Effects of proteins on the monolayer stability

It is important to bear in mind that even though the lipid that makes up the monolayer has very low solubility, the monolayer formed at the air-aqueous interface can be metastable. Apart from processes such as rearrangement of the amphiphiles or dissolution into the subphase, a transformation to a three dimensional phase can occur at pressures above the equilibrium spreading pressure (45, 46). Furthermore, the stability of the monolayers is very much dependent on the solvent and the techniques used for spreading the lipid (47). For example, different surface pressure - area isotherms of digalactosyldiacylglycerol were recorded, depending on whether the lipid was spread from chloroform or from a liposomal aqueous dispersion (48). In the latter case no increase in surface pressure was observed upon compression. The stability of the monolayer can also be considerably changed by the type of counterion, where the stability of an arachidic (n-eicosanoic, C20:0) acid monolayer was found to increase in the order $H^+ < Li^+ < Na^+ < Ca^{2+} < Mg^{2+}$ (45).

As was demonstrated by Kristensen *et al*, the presence of a protein can increase the stability of a lipid monolayer (49). They investigated the interaction between one of the major proteins, xanthine oxidase, and the major lipids, sphingomyelin and phosphatidylcholine, in the milk fat globule membrane at the air /aqueous interface by using the monolayer technique. Both lipids have a similar phosphorylcholine headgroup, which is zwitterionic in the neutral pH range, although the belt regions linking the phosphorylcholine group with the acyl chains are different (see Fig. 1). The Π-A isotherms of sphingomyelin and phosphatidylcholine are shown in Fig. 3A and B, respectively. The isotherms for sphingomyelin monolayers spread on pure buffer and a xanthine oxidase solution are shown. As is obvious from the figure, the slope of isotherm and the area of the compressed monolayer for pure sphingomyelin is smaller than expected for these type of lipids. This indicates that the sphingomyelin monolayer is metastable, which is also confirmed by the large hysteresis and the dependence on the compression speed not observed for distearoylphosphatidylcholine. The difference in stability of monolayers can probably be related to the different conformation of choline groups in the two types of lipids, where intra molecular hydrogen bonding is possible between the phosphate group and the amide and hydroxyl groups in the belt region of sphingomyelin (50). An increase in π at maximum compression of the sphingomyelin monolayer, which reflects an increase in the monolayer stability, was observed in the presence of xanthine oxidase. Furthermore, the area per sphingomyelin molecule increases in the presence of xanthine oxidase even at high Π-values. This is in contrast to the results from the parallel study of the phosphatidylcholine monolayers with and without xanthin oxidase, where the interacting protein could be completely squeezed out from the lipid monolayer at high enough surface pressures without affecting the collapse pressure. This indicates that interaction between xanthine oxidase and sphingomyelin is much stronger than that between the protein and phosphatidylcholine. Although it is difficult to make a direct comparison between the simplified "model membrane" used in their study and such a complex system as the milk fat globule membrane, the results indicate that this type of interaction can take place and might even be of significance for the stabilisation of the milk fat globule membrane.

Vikholm and Teleman found that the presence of a protein (anti-CRP), which is the antibody to C-reactive protein, in the subphase, stabilised a monolayer of octadecylamine (51). They found that the antibodies had a fluidising and expanding effect on the octadecylamine monolayer. The area per molecule at the collapse of the film was found to be independent of the protein concentration, which indicates that the protein is expelled from the monolayer at high surface pressures. The collapse pressure on the other hand, was found to increase with the subphase protein concentration. This suggests that the expelled protein molecules remained adjacent to the octadecylamine monolayer thereby stabilising the monolayer by interacting with the polar head groups.

3.2 Driving force for lipid-protein interactions at air/aqueous interfaces

What are then the driving forces for the lipid-protein interaction to take place? As shown previously, the interaction between xanthine oxidase and sphingomyelin takes place at pH 7.4, although the lipid is zwitterionic and the protein is slightly positive under these conditions (49). Thus, the interaction seems not to be governed by simple electrostatic attraction. In this context it is rewarding to refer to the work of Ninham and Yaminsky concerning ion binding and ion specificity (23). They conclude that specific ion adsorption due to dispersion interactions can be dominant even at charged interfaces, and especially at high salt concentrations. It is thus not always possible to separate between the electrostatic and dispersion forces.

Some of our own experimental data from a monolayer study of the phospholipid - β-lactoglobulin interaction (52) will be used in the following to illustrate the diverse nature of protein-lipid interaction. In this study, different phospholipids were spread on a β-lactoglobulin solution of different pH and ionic strength. From the Π-A isotherms it was found that the protein was completely expelled from the monolayer at high enough pressure (≈40 mN/m) and the presence of protein did not affect the collapse pressure. The following procedure was therefore used to estimate the rate of penetration/adsorption of the protein in the lipid layer: Before each penetration/adsorption measurement, the spread monolayer was compressed to ≈40 mN/m to remove any protein from the lipid layer (Fig. 4B). The monolayer was then expanded to the desired surface pressure, and the pressure was then held constant while the area expansion, due to the protein interaction, was recorded (Fig. 4B and 4A, respectively). As expected the rate of area increase and hence the rate of penetration/adsorption decreases with surface pressure.

In order to compare the kinetics of this interaction for the different systems, a simple first order kinetics model (40), where only the surface pressure barrier is taken into account, was applied. The obtained rates of incorporation of β-lactoglobulin into monolayers of distearoylphosphatidic acid (DSPA), distearoylphosphatidylcholine (DSPC) and dipalmitoylphosphatidic acid (DPPA) versus surface pressure (Π) at pH 7 and different ionic strengths are summarised in Fig. 5.

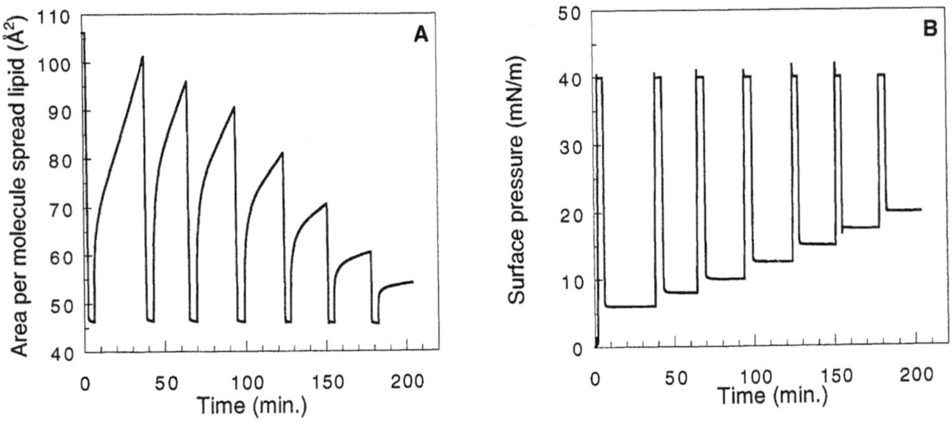

Fig. 4 Examples of the adsorption of β-lactoglobulin into a DSPC monolayer at various surface pressures (Π). The surface pressure is changed in the sequence shown in (B) and the corresponding response in area per molecule of the spread lipid is shown in (A). The subphase was a 10 mM phosphate buffer, pH 7, containing 150 mM sodium chloride, and the protein concentration was 1.15 mg/l. The lipid (125 μg) was

spread from a chloroform/methanol (2:1, v/v) solution on a maximum area of 150 × 675 mm^2. Data adopted from Bos and Nylander (52), where also the experimental details are given.

Most notable is that significant incorporation of the protein occurred at a higher surface pressure into a DSPA monolayer than into monolayers of the other lipids. The calculated rate of β-lactoglobulin incorporation was also significantly higher. Differential scanning calorimetry (DSC) measurements confirm the presence of a specific interaction between phosphatidic acid and β-lactoglobulin also with the lipid dispersed in solution (14). The presence of distearoylphosphatidic acid (DSPA) as well as dipalmitoylphosphatidic acid (DPPA) thermally stabilised the protein, which was not observed when the protein was mixed with phosphatidylcholine, phosphatidylethanolamine or phosphatidylglycerol.

As the protein has zero net charge at pH ≈ 5.2 (53), simple electrostatic binding can only take place at pH 7 if the protein is oriented with its positive groups towards the lipid layer. One would expect a stronger interaction at pH 4 since the protein then has a net positive charged. However, almost the same rates where found at this pH (52), indicating that not just simple ion binding takes place. This is somewhat contradictory to the findings of Cornell and Patterson, who studied the adsorption of β-lactoglobulin to a negatively charged lipid monolayer, composed of a mixture of palmitoyloleoylphosphatidylcholine (POPC) and palmitoyl oleoylphosphatidyl-glycerol (POPG) (65/35 mol %) (54). They only observed a substantial binding of β-Lactoglobulin at pH 4.4, which is when the protein carries a net positive charge, but not at higher pH (pH 7). Similar findings where reported for other milk proteins, where the highest amounts of α-lactalbumin or BSA bound to mixed monolayers of POPC and POPG were observed below the isoelectric point of the proteins (55). Less was adsorbed around, and almost nothing above, the isoelectric point. The interaction was also found to be reduced with increasing ionic strength as well as in the presence of calcium. It was concluded that the interaction could be explained by simple electrostatic attractive forces. This is also in line with the results from an earlier study by Cornell, where he investigated the interaction in mixed spread monolayers of β-lactoglobulin and egg yolk phosphatidic acid (e-PA) or egg yolk phosphatidylcholine (e-PC) (56). He found that the interaction with the anionic e-PA took place at pH's where the protein carries a net positive charge (pH 1.3 and 4). No interaction was observed for e-PC or for e-PA when the pH was increased to values in the neutral pH range. The different conclusion from our study compared to the results presented by Cornell *et al.* probably arises from the use of different lipids and methodology (54-56). Cornell *et al.* measured the amounts of protein adsorbed to the lipid layer by transferring the layer to a solid support (54-55). During the transfer, the surface pressure was kept at 30-35 mN/m, thus preventing insertion of portions of the protein in the lipid monolayer (55) (compare Fig. 2).

Only protein molecules which interact strongly with the lipid headgroups are transferred to the solid supported. Another difference is that their surface pressure data of the protein penetration is recorded under constant area, not at constant pressure as in our study. In addition Cornell *et al.* used lipids with their chains in the liquid state, which, as discussed below, can influence the interaction (54-56).

The highest rate of incorporation of β-lactoglobulin in a DSPA monolayer at pH 7 was observed when the sodium chloride concentration was increased to 150 mM (Fig. 5A). Under these conditions the screening of the anionic lipid headgroup and the positively charged residues on the protein would be expected to decrease the rate due to the weaker interaction. Instead the reduced repulsion within the phosphatidic acid– protein monolayer at a higher ionic strength seems to favour inclusion of the protein. Consequently as observed, the incorporation into the zwitterionic DSPC monolayers is almost independent of the salt concentration (Fig. 5B). These results imply that the incorporation is not only determined by electrostatic interactions, but also that other types of interactions are important. This is confirmed by the fact that the incorporation of β-lactoglobulin into the lipid monolayers was significantly reduced when a corresponding lipid with shorter acyl chains, DPPA, was used (Fig. 5C). This indicates that the interaction between the protein and the hydrophobic part of the lipid is significant.

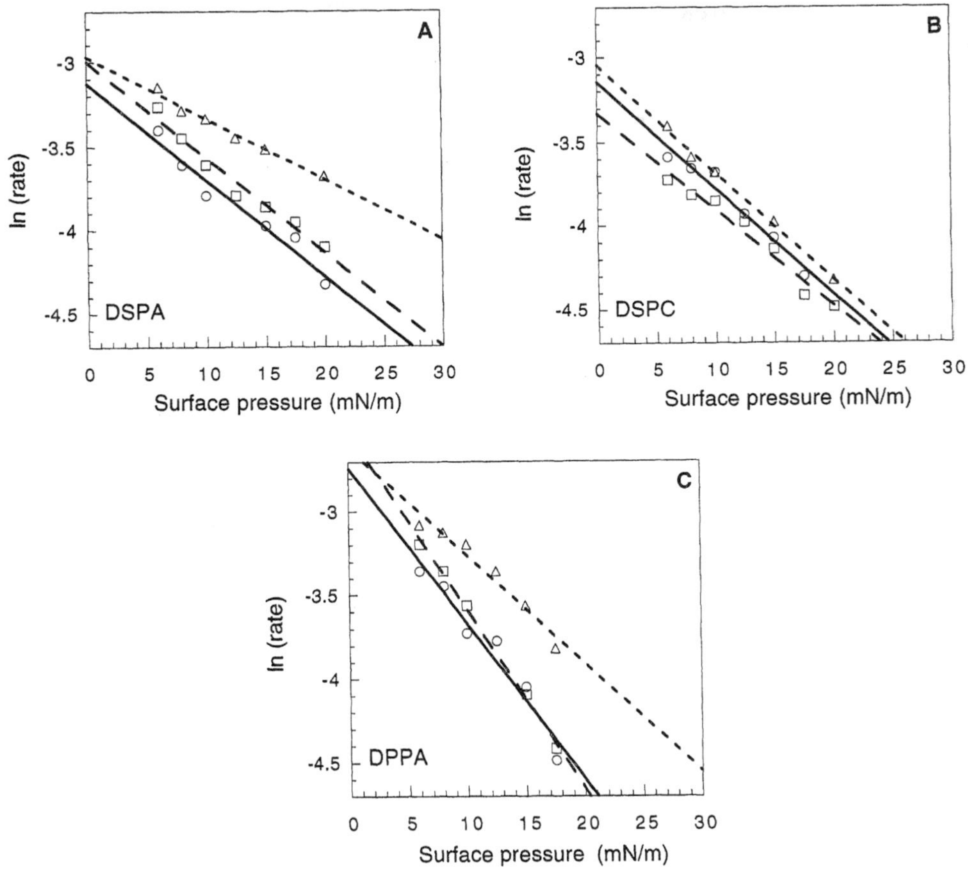

Fig. 5 The rate of incorporation of β-lactoglobulin into monolayers of distearoylphosphatidic acid (DSPA) (A), distearoylphosphatidylcholine (DSPC) (B) and dipalmitoylphosphatidic acid (DPPA) (C), versus surface pressure (Π). The data was recorded at constant surface pressure by measuring the area increase of the lipid monolayer spread on a protein solution contain 1.15 mg/l in 10 mM phosphate buffer of pH 7, with 0 mM (—O—), 50 mM (— —□— —), 150 mM (- - -Δ - - -) sodium chloride as depicted in Fig. 4. The rate in mg/m² was calculated from the area increase by using the Π-area isotherm of spread monolayers of β-lactoglobulin. Data adopted from Bos and Nylander (52), where also the experimental details are given.

Quinn and Dawson arrives at a similar conclusion in a study of the interaction between cytochrome C (net positive charge below pH 10) and monolayers of phospholipids from egg yolk and cardiolipin from heart, although electrostatic factors prevail (57, 58). They measured the pressure increase caused by the penetration/adsorption of the protein to the lipid monolayers as well as the amount adsorbed by using ^{14}C-labelled protein. Their results show that the limiting surface pressure for penetration of the protein into the lipid monolayer is 20 and 24 mN/m for e-PC and e-PE, respectively. The slightly higher value for e-PE was attributed to its partly net negative charge under the unbuffered conditions used in their study. In contrast, the protein penetration into the anionic e-PA and diphosphatidylglycerol (cardiolipin) monolayers occurred up to pressures close to the collapse pressure of the film (<40 mN/m). Furthermore, the penetration into the e-PC monolayers was not affected by increasing the sodium chloride concentration to 1 M. Cytochrome C bound to the e-PC monolayers could not be removed by increasing the ionic strength. This is in contrast to the cardiolipin and e-PA monolayers where the penetration was reduced when the sodium chloride content was increased to 1 M. It was also possible to partly desorb some cytochrome C from e-PA monolayers. However, the pH dependence of the interaction was found to be quite complex, which suggests that subtle changes in the protein conformation also affects the interaction.

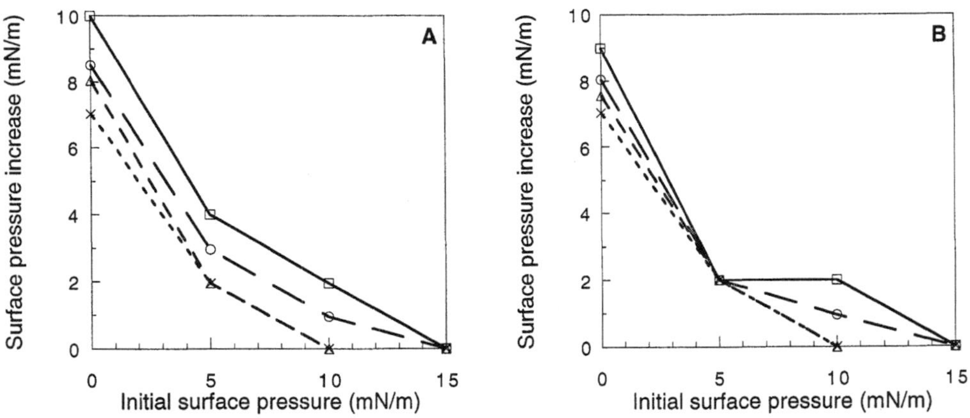

Fig. 6 The increase in surface pressure (ΔΠ) versus initial pressure for monolayers of glucolipids with various headgroups and chain lengths spread on a glucose oxidase solution. The results for dialkyl glycerylether-β-glucosides and dialkyl glycerylether-β-D-maltosides are shown in (A) and (B), respectively, with alkyl chain lengths of 18 (—□—), 16 (— — ○ — —), 14 (– – – Δ – – –) and 12 (- - -×- - -). Data adopted from Du *et al.* (59), where also the experimental details are given.

Quinn and Dawson (58) found that the threshold surface pressure, above which no penetration of cytochrome C took place in phosphatidylcholine monolayers, was considerably lower when DPPA was used instead of hydrogenated egg yolk phosphatidylcholine (e-PC).The latter lipid contained fatty acid with a longer chain length, about 60% C18:0 and 30% C16:0. A study of the influence of the alkyl chain length of glucolipids on the interaction between lipid monolayers and glucose oxidase was presented by Du *et al* (59). Some of their data for the dialkyl glycerylether-β-D-glucosides and dialkyl glycerylether-β-D-maltosides are presented in Fig. 6, as the increase in surface pressure versus initial surface pressure. The interaction, as shown by an increase in surface pressure, was found to increase with increasing lipid chain lengths for both types of lipids. These results suggest that the hydrophobic interaction is the predominant force. Furthermore it is interesting to note that the interactions were not so strong with the lipids having the more bulky head group, that is the dialkyl glycerylether-β-D-maltosides (Fig. 6B), although the Π-A isotherms for the corresponding dialkyl glycerylether-β-D-glucosides was similar. This illustrate that a bulky head group can sterically hamper the protein-lipid interaction.

3.3 Effect of lipid fluidity on lipid-protein interactions

The lipid-protein interaction is also certain to be affected by the state of the lipid acyl-chains, that is whether they are fluid or in the solid (gel) state (Fig. 2). Calorimetric studies of the interactions between β-lactoglobulin and phospholipids dispersed in solution has revealed that the presence of distearoyl- and dipalmitoylphosphatidic acid increases the temperature at which the thermally induced unfolding takes place (14). However, with phosphatidic acid from egg yolk, where a considerable proportion of the acyl chains are unsaturated, no such effect was observed. The experimental data presented could not give a definite explanation, but if the binding of the lipid is assumed to take place in a hydrophobic pocket on the protein, the binding of a lipid with a bulkier unsaturated acyl chain should be more difficult. The bulkiness of an unsaturated lipid also shows up at the liquid interface, where at a given surface pressure of the lipid monolayer, the number of saturated lipids per area will be larger than the number of corresponding unsaturated ones.

The complete hydrogenation of e-PC was found not to affect the surface pressure threshold for penetration of cytochrome C compared to the native e-PC (58). However, the plot of the change in surface pressure versus initial surface pressure was less steep for the saturated one. A similar trend was observed for the e-PE samples (57). The conclusion was that the limiting pressure for penetration to take place is likely to be determined by the work necessary for the penetration, that is $\int \Pi dA$, where an area of interface, A, has to be created for the protein to penetrate. Once the penetration is feasible the magnitude will depend on the space between the molecules and thus the degree of penetration is expected to be lower for the hydrogenated sample (58). The surface pressure threshold below which penetration of cytochrome C into the anionic diphosphatidylglycerol (cardiolipin) monolayer took place was also found to decrease when the lipid was fully hydrogenated (58). Ibdah and Phillips found the same trend in their study of the effect of lipid composition and packing on penetration of apolipoprotein A-I into lipid monolayers (60). In the biological system this protein interacts with the phospholipid membrane of the serum high density lipoprotein (HDL) particles (see discussion in oil/aqueous interface section). Their results show that this protein adsorbs to a larger extent on expanded monolayers than on condensed monolayers, that is, protein adsorption decrease in the order e-PC > egg sphingomyelin > DSPC. Furthermore it was found that protein adsorption generally decreased with increasing amount of cholesterol in the lipid monolayer. It was suggested this was due to the condensing effect of cholesterol.

3.4 Influence of the protein structure on the interaction

The importance of the protein structure was demonstrated by Hanssens and Van Cauwelaert (61), who studied the penetration of α-lactalbumin in monolayers of DPPC and cardiolipin at physiological pH (pH 7.4) and at pH 4.6 with and without calcium. Reports have shown that the α-lactalbumin is transferred to an intermediate, partially unfolded (molten globule) state at low pH and by depletion of calcium (32, 34, 35). The calcium depleted form of the protein is also known to be more hydrophobic, which actually can be used for the purification of the protein (62). The more hydrophobic character of the protein at low pH was found to increase the rate as well as the capacity of the protein to penetrate the lipid monolayer compared to the effect at physiological pH (61). Consequently, penetration was also prevented if the protein was adsorbed from a calcium solution.

One technique which can be used to monitor the structural changes of the protein on binding to a lipid monolayer is to transfer the mixed protein-lipid film to a quartz plate and analyse the circular dichroism (CD) spectra. This was done for β-lactoglobulin, α-lactalbumin or BSA bound to mixed monolayers of POPC and POPG (54, 55). The recorded CD-spectra was similar to the protein solution spectra, which indicates that the protein conformation did not change significantly when interacting with the lipid monolayer.

3.5 The structure of the protein-lipid layer

The transfer of a film from the air/aqueous interface to a solid support, can be used to reveal the structure within the mixed protein-lipid film. The structure of the film on the solid support can then be imaged using electron microscopy or atomic force microscopy. The former technique was used by Cornell and Carroll to study the miscibility of proteins and lipids at the air/water interface (63). They found that only lipids with the chains in liquid state, e-PA, DOPC and DOPE, formed homogenous films with β-lactoglobulin, whereas DPPA and DSPC formed heterogeneous layers. The used of AFM to reveal the structure of mixed films of glucose oxidase and behenic acid was demonstrated by Sommer *et al.* (64). They spread a behenic acid layer on top of a glucose oxidase solution, which after equilibration was compressed to 30 mN/m and transferred to a graphite surface by the Langmuir-Blodgett technique. The images obtained featured randomly organised enzyme aggregates composed of 1000-5000 enzyme molecules. By using the so called "Tapping Mode" they were able to resolve the individual molecules and found they where organised as quite ordered aligned structures within the aggregates. Lateral force measurements were used to analyse the composition of the different domains as the friction of the protein part and lipid part were found to be quite different. The glucose oxidase aggregates were found to be partly covered by a behenic acid monolayer.

From the study of the penetration of protein versus surface pressure it is also possible get some hints about the structure of the mixed layer. Indeed Cornell *et al.* (54, 55) observed penetration of β-lactoglobulin, α-lactalbumin or BSA into mixed monolayers of POPC and POPG at a higher surface pressure than it is possible to obtain with a pure protein solution, that is when the protein penetrates into a pure protein layer. Thus, they concluded that the formation of pure protein patches is unlikely and portions of the protein probably intercalated into the lipid monolayer. A similar observation was made by Bos and Nylander for the interaction between β-lactoglobulin and DSPC and DSPA monolayers (52).

There are a number of direct techniques that can be applied to image the structure of the film at the air/aqueous interface, e.g. fluorescence microscopy and Brewster angle microscopy (BAM). Their lateral resolution is, however, limited by the resolution of the optical microscope. The structure of mixed phospholipid cytochrome C and b film was studied by Heckl *et al* [(65), using fluorescence microscopy and the surface film balance technique. Cytochrome c (positively charged) was found to interact with dimyristoylphosphatidic acid (DMPA) monolayers but not with dipalmitoylphosphatidylcholine (DPPC) layers, suggesting that the interaction was electrostatically controlled. As expected, the penetration into the lipid monolayer was reduced with increasing pressure. The mixed films were found to consist of solid lipid domains without protein, surrounded by a fluid membrane phase, containing the proteins. Schönhoff *et al* concluded from their study of the incorporation of membrane proteins into DPPA/DOPA monolayers that incorporation mainly takes place in the fluid phases of the matrix (66).

The mutual interaction between proteins and a lipid monolayer is nicely demonstrated in the work of Diederich *et al*, who studied the interaction between bacterial surface layer proteins (S-layer proteins) and phosphatidylethanolamine (DMPE and DPPE) monolayers using dual label fluorescence microscopy, FTIR spectroscopy, and electronmicroscopy (67). The phase state of the lipid monolayer was found to affect the interaction with the protein. When the monolayer is in the two phase region, with one isotropic and one anisotropic fluid phase, the S-layer protein is preferentially adsorbed to the isotropic phase. However, 2D crystallisation, nucleated in the boundaries between the two phases, proceeds mainly underneath the anisotropic phase and is much slower underneath the isotropic fluid phase. The FTIR-measurements clearly indicate that the protein crystallisation leads to increased ordering of the lipid acyl chains. Their data suggests that the protein interacts with the lipid headgroups and do not intercalate into the lipid monolayer. In the section dealing with the interactions between proteins and lipid-aqueous liquid crystalline phases, similar types of ordering of lipids and proteins will be discussed.

4 OIL/AQUEOUS INTERFACES

Compared to the vast number of studies of events at the air/aqueous interface, less results discussing the mechanism of protein-lipid interactions at oil/aqueous interfaces have surfaced. The reasons for this have been experimental difficulties and a lack of experimental techniques. For instance, the surface film balances for the oil/aqueous interface are quite complex and hard to handle. However, Murray and Nelson recently presented a novel and simple Langmuir trough for equilibrium and dynamic measurements on both air/water and oil/water monolayers (68). The use of optical techniques like ellipsometry is also hampered by the lower optical contrast at the oil/aqueous interfaces. Furthermore, it is more difficult to bring surface active material, in particular lipids with low aqueous solubility, to this interface, by say spreading, without disturbing it.

Most studies of lipid-protein interaction have been carried out using model oil in water emulsions. Aynié *et al* used a model food emulsion to study the interaction between nitroxide homologues of fatty acids and milk proteins, using electron spin resonance (ESR) to measure the change in mobility of the nitroxide radicals (69). They investigated the interaction of proteins with both unordered, α_{s1}-casein and β-casein, and globular structure, β-lactoglobulin, at pH 7 and found no correlation with the protein structure. Instead the strength of the interaction increased with the number of positive charges on the protein, in the order α_{s1}-casein > β-lactoglobulin > β-casein. This suggested that the interaction was of electrostatic nature and was achieved by proper orientation of the protein at the interface. They also found that α_{s1}-casein in contrast to β-lactoglobulin and β-casein, induces an ordering of a monolayer of nitroxide fatty acids on the surface of an emulsion droplet. This is likely to be a consequence of the stronger interaction between α_{s1}-casein and the lipid.

It is important to bear in mind that the composition of the oil phase, in particular the amount of surface active substances, can largely affect the interaction at the oil/aqueous interface. The variation of binding of β-casein to oil-water interfaces was studied by following the trypsin-catalysed hydrolysis of the protein at the interface (70). The structure of the β-casein layer adsorbed on the emulsion droplets was found to depend on whether tetradecane or a triglyceride (soy) oil was used. The triglycerides are more surface active than a pure alkane and thus give an oil/aqueous interface with different properties. The lipoproteins are emulsion droplets with a core of mainly triacylglycerols and cholesterylesters enveloped mainly by phospholipids with associated cholesterol and apolipoprotein (71). To mimic the protein-lipid interactions in the very-low-density lipoproteins and chylomicrons, Badr *et al* used e-PC/cholesterol stabilised emulsions of methyloleate, ethyloleate, glyceryltrioleate, and erythrityltetraoleate, which were allowed to interact with plasma or plasma fractions (72). The glyceroltrioleate and erythrityltetraoleate emulsions bound the plasma apolipoproteins AI, AIV, CII, CIII, and E as well as albumin to the same extent and composition as lymph chylomicrons. However, the emulsions based on methyl and ethyl oleate bound much less protein and the composition of the bound proteins were different. As concluded in the paper, the methyl and ethyloleate are more polar and are likely to be more surface active than glycerol trioleate and erythrityltetraoleate. Thus the partition of them at the aqueous/oil interface is likely to be larger leading to changes in the surface properties of the emulsion droplets. Heertje *et al* used confocal laser microscopy to monitor the displacement of fluorescent-labelled caseinate from a planar oil/aqueous interface by monoglycerides, monooleoylglycerol and monostearoylglycerol, dissolved in the oil phase (73). They found that the displacement was directly correlated with the adsorption of the monoglycerides at the oil-water interface, which differed substantially for the two lipids. The amount of monooleoylglycerol increased gradually with concentration and reached a plateau when approaching an oil phase concentration of 1 wt%. Under these conditions all of the caseinate was displaced from the interface. The saturated lipid, monostearoylglycerol, was much more efficient in displacing the protein. Already, at a concentration in the oil phase of between 0.2 and 0.3 wt% the adsorbed amount of monostearoylglycerol increases sharply and reaches much higher surface concentrations than monooleoylglycerol. At 0.3 wt% all of the caseinate is removed from the interface.

Ellipsometry can be used to directly measure the protein lipid interactions at interfaces. As mentioned above this technique depends on the optical contrast between a reflecting substrate, the film and the medium. Thus the highest resolution is generally achieved for measurements at the solid/aqueous interface and one way to overcome the low resolution at the oil/aqueous interface is therefore to make a lipid surface on a solid support. Malmsten developed a technique to prepare phospholipid layers by spin-coating them from an organic solvent onto methylated silica surfaces (74). These surfaces were found to have similar properties to corresponding Langmuir-Blodgett deposited layers. He studied the adsorption of human serum albumin, IgG, and fibronectin from human plasma, to lipid layers from aqueous solutions containing 0.15 M sodium chloride at pH 7.2 (75). In general, the protein adsorption was found to be very low to lipid surfaces with no net charge (phosphatidylcholine, phosphatidylethanolamine, sphingomyelin) or when the charges are shielded like for phosphatidylinositol and ganglioside G_{M1}. The values of the amount of protein adsorbed were substantially lower than to the bare hydrophobised silica. In contrast, when the lipid surfaces contained unprotected charges as in lipid surfaces composed of phosphatidic acid, diphosphatidylglycerol and phosphatidylserine, the protein adsorption was high, in particular for fibrinogen which even showed a higher adsorption than to bare hydrophobised silica.

5 INTERFACES OF LIPID AGGREGATES

When the protein interacts with the interfaces of lipid structures, like a liquid crystalline phase, an additional factor has to be considered, that is the curvature of the interface. On the scale of the molecular dimensions of a protein, the surface of an emulsion droplet appears to be almost flat. However this is not the case for the curved interfaces in lipid structures, where the protein molecule will experience a curved interface. The role of curvature in biological systems has recently been discussed in the book by Hyde *et al* (76). McCallum and Epand demonstrated the effect of curvature of biological membranes in their study of the membrane physical properties, which modify insulin receptor autophosphorylation and signalling (77). They found that adding compounds which raise the bilayer to reverse hexagonal (H_{II}) transition temperature of model membranes, that is promotes a decrease of curvature, also inhibited the insulin stimulation of the receptor phosphorylation.

As for polar lipids and synthetic analogues, which are water soluble in monomeric and micellar form (e. g. lyso-phospholipids), insoluble polar lipids do interact with water and swell into liquid crystalline phases. The self association of polar lipids in an aqueous environment leads to the formation of different aggregates depending on the structure of the lipids and the composition of the aqueous solvent. Some of the most common type of aggregates are shown in Fig. 7. These aggregates will have a polar interface, which separate the hydrocarbon and aqueous regions.

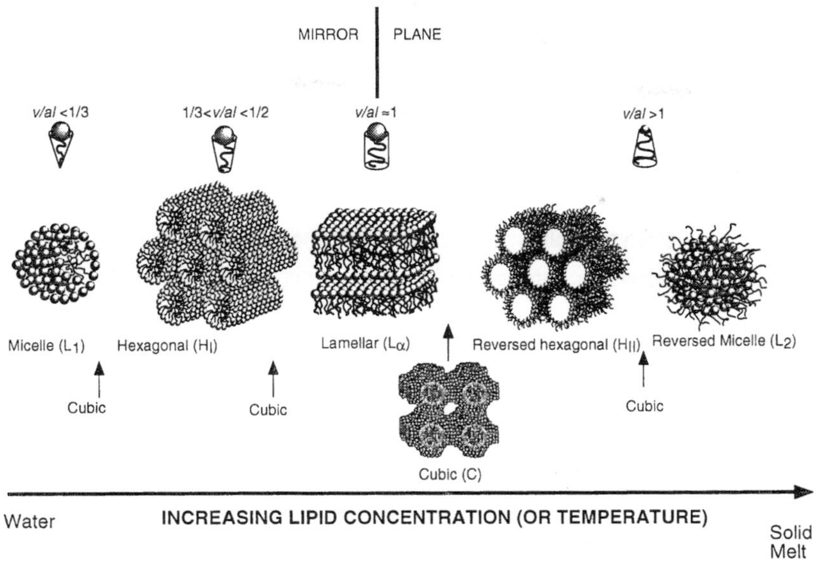

Fig. 7 Some of the most common structures formed by polar lipids. Phase transitions can be induced by changes in water content, temperature or by interaction with other solution components, like proteins. The lamellar liquid crystalline phase (L_α) can be regarded as the mirror plane, where the aggregates are of the "oil-in-water" type on the water rich side and of "water-in-oil" type on the water poor side (83). There are two possible locations for cubic phases on both the water rich and water poor sides of the L_α-phase. Other "intermediate phases" may also occur. The formation of a particular phase can in many cases be understood by considering the geometric packing of the amphiphilic molecule, that is its shape, in the particular environment (86). This property can be expressed by the so called packing parameter (v/al), which is defined as the ratio between the volume of the hydrophobic chain (v) and the product of the head group area (a) and the chain length (l).

The structure of the common mesophases was determined by Luzzati and co-workers in the early 1960s by X-ray diffraction (reviewed by Luzzati in 1968 (78)). The application of new spectroscopic techniques, e. g. NMR, and the development of theoretical models have largely increased understanding of the dynamic nature of these phases. The lamellar phase (L_α) consists of stacked infinite lipid bilayers separated by water layers, while the hexagonal phases consists of infinite cylinders, having either a hydrocarbon core in the hexagonal phase (H_I) or a water core in the reversed hexagonal phase (H_{II}) (Fig. 7). Other phases might occur, most notable cubic liquid crystalline phases (Q) which can exist in several locations in the phase diagram as indicated in the figure.

The cubic phases, which exist in a number of lipid systems (79), have the most intriguing structure of the lipid-aqueous liquid crystalline phases. They are isotropic and highly viscoelastic. Several types of structures with cubic symmetry has been described for different lipid-aqueous systems (80-83). The structure proposed by Luzzati *et al.* consists of networks of rods, formed of identical rod-like elements, linked three by three into two three-dimensional networks (80). These networks are interwoven, but unconnected. In some cases the rods are filled by the hydrocarbon chains, forming *two unconnected "oil" continuous systems*, and in others by polar moieties. The other main type of cubic phases, observed in systems of water insoluble lipids like monoglycerides, phospholipids and glyceroglucolipids (16, 81), are bicontinuous and based on curved non-intersecting lipid bilayers. These are generally organised in such a way that *two unconnected "water" continuous systems* are formed (cf. (81, 84)). This is illustrated in Fig. 8, where the curved surface is located in the gap between the methyl end groups of the lipid in the bilayer. The curvature of any surface is given by the two principal radii of curvature, R_1 and R_2. The average curvature $1/2(1/R_1 + 1/R_2)$ is by definition zero at any point for a minimal surface, that is, in all points the surface is as concave as it is convex. The surface described by the mid plane of the bilayer in the cubic lipid-aqueous phase can be described as a minimal surface (81, 85) or actually as it exhibits periodicity it is an *Infinite Periodic Minimal Surface* (IPMS).

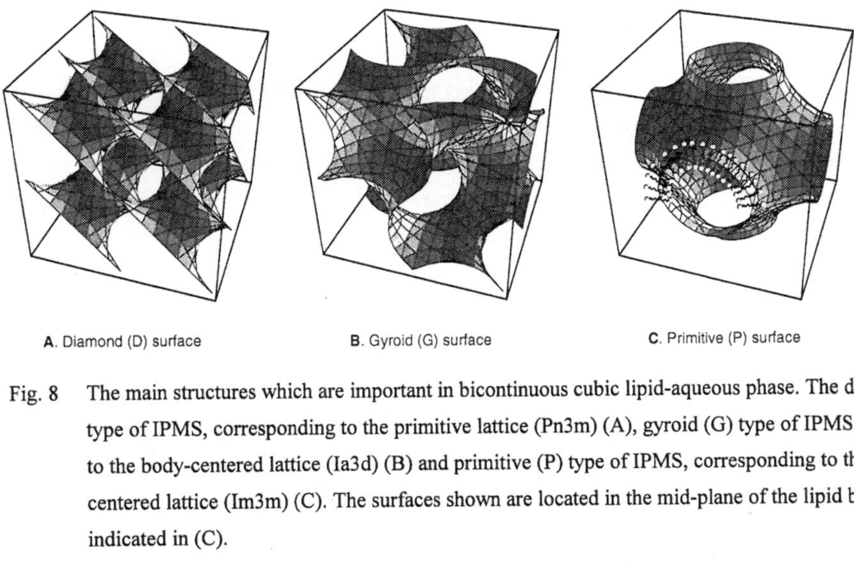

Fig. 8 The main structures which are important in bicontinuous cubic lipid-aqueous phase. The diamond (D) type of IPMS, corresponding to the primitive lattice (Pn3m) (A), gyroid (G) type of IPMS, corresponding to the body-centered lattice (Ia3d) (B) and primitive (P) type of IPMS, corresponding to the body-centered lattice (Im3m) (C). The surfaces shown are located in the mid-plane of the lipid bilayer as indicated in (C).

Three types of IPMS, corresponding to different cubic space groups, have been shown to be important in lipid systems (81, 85):

Diamond (D) type of IPMS, corresponding to the primitive lattice (Pn3m) (Fig. 8A).

Gyroid (G) type of IPMS, corresponding to the body-centered lattice (Ia3d) (Fig. 8B).

Primitive (P) type of IPMS, corresponding to the body-centered lattice (Im3m) (Fig. 8C).

The shape of the amphiphilic molecule, that is if it is cone shaped, cylindrical or shaped like a reversed cone, in its specific environment indicates which particular phase the amphiphile tends to form (86, 87). This property can be expressed by the so called packing parameter, v/al, where v is the volume of the hydrophobic chain, a is the head group area and l is the chain length (86). A value of the packing parameter lower than unity (cone shaped amphiphile) facilitates the formation of structures where the polar interface is curved towards the hydrocarbon phase, i. e. structures of "oil-in-water" type (L_1, H_I), whereas a value larger than 1 (reversed cone shaped amphiphile) will reverse the curvature and favour "water-in-oil" structures like H_{II} and L_2. The packing parameter does not only depend on the molecular structure of the amphiphile, but it is also affected by for instance, water content, ionic strength, temperature or the addition of other molecules like proteins. Decreased hydration will decrease the head group repulsion, resulting in a decreased head group area and thus in an increase of the packing parameter. As indicated in Fig. 7, the phase transitions observed for water soluble polar lipids often follows the sequence: micellar phase (L_1) $\rightarrow$ hexagonal phase (H_I) $\rightarrow$ lamellar phase (L_α) with decreasing water content. For lipids with low water solubility the sequence is usually lamellar phase (L_α) $\rightarrow$ reversed hexagonal phase (H_{II}) $\rightarrow$ reversed micellar phase (L_2). An increase in temperature will increase chain mobility and thereby increase the volume of the lipophilic part of the molecules. Thus the same sequence of phase transitions are usually observed when increasing the temperature as when the water content decreases. Referring to the structure of the bicontinuous cubic phases discussed above, the packing parameter for a lipid in such a curved bilayer can be related to the Gaussian curvature ($1/(R_1 R_2)$) of the IPMS and is larger than 1 (88).

In nature most of the polar lipids occur with the hydrocarbon chains in a fluid state at room temperature. One interesting exception is the sphingolipid fraction from the milk fat globule membrane, discussed earlier, which contains lipids with long saturated acyl chains with chain melting transitions around 35-85 °C, depending on the water content (c.f. (89, 90)). A lipid lamellar phase changes to a lipid gel phase at temperatures below the chain melting transition under which the basic structure of the lamellar phase is retained, but with the acyl-chains in the solid state (16). The chain melting transition in such a gel phase usually occurs at a much lower temperature than the melting of the pure lipid.

The phase behaviour described above is, in nature and in many technical applications, more complex as the aggregates are generally composed of a mixture of different lipids, which either exist in a homogenous mixture or separate into domains. The lateral distribution in these mixed aggregates is influenced by a number of factors like ionic strength, presence of polymers/proteins as well as the composition of the lipids. It is thus hard to give any general rules to predict under which conditions phase separation will occur (91). In some cases the phase behaviour in mixed lipid systems can be predicted by estimating a mean packing parameter. However, this assumes that the lipids form a homogenous mixture, which is definitely not always the case. Lateral phase separation can, as discussed by Raudino (91), occur when a protein binds to heterogeneous bilayers. In turn this can lead to the formation of defective lamellar phases, so called ribbon phases, where the broken lamellae are limited by curved rims. The occurrence of these phases can be explained by differences in packing shapes of the lipids in the separated domains, which will evoke different curvatures. This is demonstrated in the work by Caboi *et al.*, where large effects on the phase diagram of the monoolein-water system was observed by only small amounts of the oil soluble vitamin K_1 (92). The transition of cubic phase to H_{II}, which in the binary system only occurs at 80 °C, was found to take place in the presence of a few wt% of vitamin K_1 at room temperature. These small amounts were found not be sufficient to change the packing parameter enough to induce this transition. Instead it was suggested that the vitamin was not evenly distributed in the cubic phase. This could eventually produce a local change of the bilayer curvature in the cubic phase, which in turn induces the phase transition.

6 LIPID VESICLES

When a lamellar phase is dispersed in a large excess of water, the bilayers, which make up the structure separate and become unstable. They can then curve to form lipid vesicles, liposomes, with either one lipid bilayer or where the lipid bilayers are organised in concentric multi-bilayer aggregates (93). Depending on the lipid and the mode of preparation several types of vesicles can be formed and are usually separated into large multilamellar vesicles (MLV), and large (LUV) and small (SUV) unilamellar vesicles (93). The term liposomes is usually reserved for multilamellar vesicles (16). Much of the understanding of bilayer structures (such as vesicles), including lipid-protein interactions, has been derived from monolayer studies (94). As MacDonald points out, the forces operating in bilayers are essentially the same as those in monolayers, except, which is important to stress, that an oil/air interfacial tension is not present (94).

6.1 Driving force for lipid-protein interactions at vesicle/aqueous interfaces

As in the studies by Quinn and Dawson (57, 58) mentioned above, Rytömaa *et al.* studied the interaction between cytochrome C and a model membrane (95). Instead of using monolayers and the Langmuir trough, they studied the interaction with cardiolipin-phosphatidylcholine liposomes by monitoring the fluorescence resonance energy transfer from a pyrene-fatty acid containing phospholipid derivative to the heme group in cytochrome C. In parallel to the Quinn and Dawson study, they found a strong electrostatic contribution when cytochrome C binds to the lipid surface, which only took place if the negatively charge lipid cardiolipin was present in the membrane. Accordingly, the protein was dissociated from the vesicle in the presence of 2 mM $MgCl_2$ and 80 mM NaCl at pH 7. The presence of adenine nucleotides at mM concentrations had a similar effect, indicating that they compete with cardiolipin for the same binding sites on the protein. The apparent affinity of cytochrome C to the vesicles increased when the pH was dropped to 4 and the protein could no longer be dissociated by the addition of either salt or nucleotides. This indicated that the interaction was no longer ionic and it was suggested that the interaction was now mediated by hydrogen bonds. The interaction was found to be completely reversible for pH changes, that is if the pH was increased to 7, the protein could be dissociated from the vesicle by adding salt or nucleotides.

The thickness of adsorbed layers of the milk proteins, β-casein, κ-casein, α_{s1}-casein and β-lactoglobulin, on negatively charged phosphatidylglycerol (PG) and zwitterionic phosphatidylcholine (PC) vesicles was determined by Brooksbank *et al.*, using photon correlation spectroscopy (96). In general, the proteins were found to give a thicker layer on the negatively charged vesicles, in spite of the fact that all of the proteins studied carried a negative net charge under the conditions used (pH 6.2). However, it was necessary to perform the adsorption experiments in the presence of sodium chloride (160 mM) in order to see any increase in layer thickness. In most cases the difference in thickness of the adsorbed layers on the two types of vesicles could be explained in terms of the charge distribution on the proteins. The binding to the vesicle surface was then assumed to take place mainly through hydrophobic interactions. For instance the hydrophilic, N-terminal, part of β-casein has a net charge of -11, whereas the remainder of the molecule carries almost no net charge. Thus, on the negatively charge vesicle surface, the molecules adopt a more extended configuration as the N-terminal part is likely to be pushed away from the surface. This explains the thicker layers on this surface. A similar reasoning can be applied for κ-casein, while the charge distribution of the globular protein, β-lactoglobulin is more intriguing. The apparent very thick adsorbed layer of α_{s1}-casein was explained by bridging flocculation of the vesicles induced by the protein. Again, this could be rationalised from the charge distribution on the protein. The middle section of α_{s1}-casein carries a negative net charge, while the two ends have no net charge. One of the uncharged ends protrudes into the vesicle bilayer, while the middle section is repelled from the vesicle surface. This leaves the other uncharged end of the peptide chain free to interact with another vesicle.

6.2 Influence of the protein structure on the interaction with the bilayer

Protein interaction with bilayers can also be favoured by conformational changes of the protein, e. g. transition to molten globular state. As discussed above α-lactalbumin is transformed to the molten globular state at low pH. Kim and Kim studied the interaction between α-lactalbumin and phosphatidylserine/phosphatidylethanolamine vesicles (1:1 molar ratio) versus pH [(97). They found that the interaction, which almost did not exist at neutral pH, increased with decreasing pH (Fig. 9). What is interesting to note (Fig. 9), is that vesicle fusion, as estimated from increase of the initial rate of Tb fluorescence increase, correlates with the binding of the protein to the vesicles. Once the protein was bound to the vesicle, it was impossible to dissociate it by increasing the pH. The binding was suggested to be due to hydrophobic interaction via protein segments penetrating into the lipid bilayer. This was further confirmed by using proteolytic enzymes on the protein bound to the vesicle surface. Both ends of the polypeptide chain was digested and only a loop, which was likely to penetrate into the bilayer, was left intact. The penetration of this protein loop into the lipid bilayer was believed to be the reason for the fusion of the vesicles.

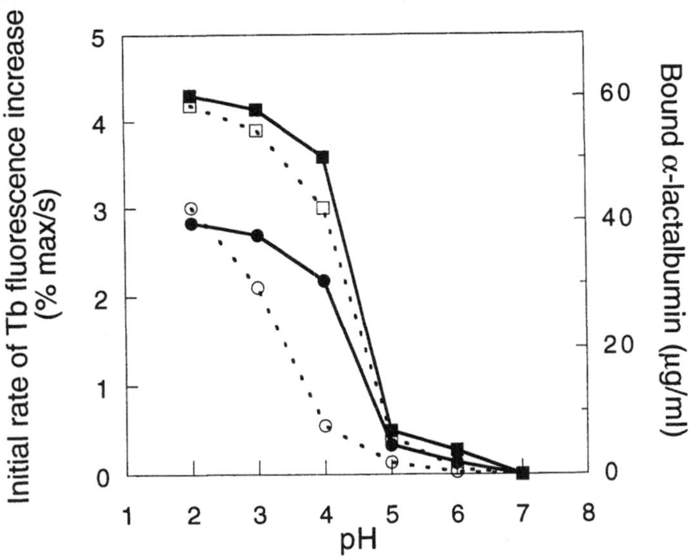

Fig. 9 Dependence of initial rate of Tb fluorescence increase (- - -○- - -, - - -□- - -) upon α-lactalbumin induced fusion of phophatidylserine/phosphatidylcholine (1:1 molar ratio) vesicles as a function of pH. The pH-dependent binding of α-lactalbumin is shown as the amount of protein bound per ml vesicle suspension (●——, ——■——), which contained 1 mM lipid molecules (determined from the phosphor content) per ml suspension. The results for initial protein concentrations of 50 (○,●) and 100 (□,■) μg/ml are presented. As the curves for the fusion process represents kinetic data and the binding studies represent equilibrium data when the fusion process is over, only qualitative comparison is possible. Data adopted from Kim and Kim. (97), where also the experimental details are given.

Conformational changes of the protein can also be provoked by dissolving them in an organic solvent. This is illustrated in the study of Brown *et al.* (98), where they initially found no interaction between native β-lactoglobulin and DPPC vesicles. However, when β-lactoglobulin was pre-treated by diluting an aqueous protein solution 1:5 with a 2:1 mixture of chloroform and methanol and redissolving the resulting precipitate in aqueous solution, the modified protein did interact with the vesicles. Moreover, the lipid-protein complex formed had an α-helix content of at least 25-30% larger than for the native protein. The interaction was found to lead to aggregation of the vesicles at pH 7.2, while no aggregates were observed at 3.7. This was explained by the larger net charge at pH 3.7 (+20) compared to pH 7.2 (-10). Brown suggested a possible binding mechanism, where the acyl chains of the lipid interact with the hydrophobic interior of the α-helix and the polar head group is more likely to interact with the hydrophilic exterior of the protein (99). The technical implication is that protein modification either during processing or by special treatment, can increase the helix content, which in turn can be boosted by lipid interaction. These lipid-protein complexes can be used to improve the emulsification processes (99, 100).

The interaction between PhoE signal peptide, which is the synthetic signal sequence analogue of the *E. coli* outer membrane protein PhoE, and anionic phospholipid vesicles of dioleoylphosphatidylglycerol (DOPG) and dioleoylphosphatidylserine was found to increase the α-helix content of the peptide as observed from circular dichroism measurements (101). This was not observed when using the zwitterionic dioleoylphosphatidylcholine vesicle and at least 50% DOPG in mixed DOPG/DOPC vesicle was needed in order to keep the induced α-helix content on a constant level. This suggested that the interaction was partly of electrostatic nature and consequently a decrease in the ionic strength was found to lower the amount of DOPG needed to give the same α-helix content. The effect of curvature was studied by comparing the interaction with SUV and LUV DOPG vesicles. It was found that the less curved LUV vesicles gave a smaller increase in the α-helix content. The PhoE signal peptide probably interacts with the membrane by first binding to anionic phospholipids upon which the peptide adopts a helical structure (101). In turn this domain inserts into the hydrophobic core of the lipid bilayer. This mechanism was suggested to play a role in the initial step by which the precursor-protein is transported across the membrane from the cytoplasm.

It is clear that the α-helix structure is particular in terms of interacting with lipid bilayers, where this type of structure seems to be able to penetrate into the bilayer. This was demonstrated by Subbarao *et al*, who studied pH dependent lipid bilayer destabilisation by using a synthetic 30 residue amphiphatic peptide which had a pH induced random coil to α-helix transition (102). They monitored the tryptophan fluorescence of the peptide, circular dichroism spectra and vesicle leakage in mixtures of the peptide and unilamellar egg phosphatidylcholine vesicles. No interaction or leakage was observed when the peptide adopted a random coil conformation, that is at pH 7.4 and above. However interaction and leakage was observed when the pH was brought down to 6 and below, which correlated with the appearance of the helical structure. As pointed out by Subbaro *et al*, the hydrophilic part of the lipids also seemed to play a role as divalent cation-peptide complexes were much less efficient in inducing leakage from the vesicles. This is in spite of the fact that divalent ions like Ca^{2+}, Mg^{2+} and Zn^{2+}, strongly promoted the helical structure.

6.3 Lateral phase separation in the lipid-bilayer

The lateral lysozyme induced phase separation in vesicle bilayers composed of a mixture of phosphatidic acid and phosphatidylcholine was studied by Raudino and Castelli (103). They used differential scanning calorimetry to follow the excess heat capacity peak originating from lipid chain melting transition. Without lysozyme, good mixing was evident as only one peak occurred, where the peak was shifted towards higher temperatures as the phosphatidic acid content increased. In the presence of lysozyme however, a lateral phase separation did occur as the lipid peak was split into two peaks. This was observed starting at a mole fraction of charged lipid of 0.15. The effect on the unfolding temperature of the protein was also followed and no effect was found in the presence of pure phosphatidylcholine vesicles compared to unfolding in the absence of lipid. However, when increasing the amount of phosphatidic acid the protein unfolding occurred at a higher temperature, suggesting a stabilisation of the protein due to the lipid-protein interaction.

It is important to bear in mind that microheterogeneity of the bilayer does not only occur for mixtures of different lipids, but also close to the gel-to-fluid phase transition of the lipid. This was demonstrated in the work by Hønger *et al*, who studied the relation between phospholipase A_2 catalysed hydrolysis of one component phosphatidylcholine vesicles and the microheterogeneity of the lipid bilayer (104). They varied the microheterogeneity by changing the temperature in the vicinity of the gel-to-fluid phase transition and by using lipid chain lengths between C14 to C18. The results gave a strong correlation between the maximal lipase-lipid interaction, as interpreted from a minimum in the lag time before the onset of the lipid hydrolyses after lipase addition, and the maxima in interfacial area between gel and fluid domains, as obtained by Monte Carlo computer simulations.

7 LIQUID CRYSTALLINE PHASE/GELPHASES

So far only isolated lipid surfaces, like monolayers, the surface of an emulsion droplet and a bilayer in a vesicle, have been considered. However, in liquid crystalline phases and gel phases the dimensions of the aqueous cavity generally approaches the dimensions of the interacting protein. This means that not only does the protein molecule experience the curvature of the particular interface, but the penetration of the macromolecule into the lipid structure can be obstructed. Furthermore, when the protein is incorporated into the structure it can be surrounded and in close contact with lipid interfaces in all directions. This means the reciprocal protein-lipid interaction is generally strong and hence the impact on the protein and lipid structure is profound. Much of the work on the lipid phase transitions induced by soluble proteins has been focused on intrinsic and peripheral membrane proteins. Most of these proteins have clusters of positive charges and often carry a positive net-charge at physiological pH, while most biological membranes are neutral, zwitterionic, or negative under the same conditions (105). In general, the number of bound protein molecules can be correlated to the positive net charge or the number of positively charged amino acid residues (105). This implies that binding of these proteins generally takes place via direct electrostatic interaction between basic residues, like lysine and arginine, and the opposite charges on the lipids. However, as discussed above, other effects occur, like hydrophobic interaction through penetration of hydrophobic protein segments into the lipid membrane. These segments are either already present at the surface of the protein or exposed as a result of conformational changes of the protein during the adsorption process. In addition the protein-lipid interactions can be hampered by steric factors. If any penetration into the lipid layer occurs it is likely to have a larger impact on the lipid structure than if the protein-lipid interaction is restricted to the polar region of the lipid. Some examples of this have already been given, but more will follow below.

In the absence of any specific interaction, proteins can also have an impact due to the limited space of the aqueous cavity. This was demonstrated in the work by Minami *et al*, where the incorporation of lysozyme, β-lactoglobulin and α-lactalbumin in a sphingomyelin gelphase containing 0.6 wt% sodium palmitate and 80 wt% aqueous phase was investigated (90). The amount of protein which could be incorporation was suggested to be limited by the dimension of the aqueous layer in the gel phase. Above this limit, phase separation will occur with a gelphase and an "outside" protein rich solution. As discussed in the paper, at high enough protein concentration, the protein will probably also compete for the water in the interlamellar spacing. This will eventually lead to a reduction of the aqueous layer thickness. This effect was demonstrated for high molecular weight polymers in equilibrium with the phosphatidylcholine lamellar phase (106). The polymer was unable to enter the aqueous layer, but still exerted an osmotic stress which was large enough to compress the lamellar lattice as shown by X-ray diffraction data. This method has been used to measure the interaction between the lipid bilayers (106, 107).

Due to their diverse properties, protein can affect the phase behaviour of the liquid crystalline lipid aqueous phase in a number of ways. The significance of electrostatic and other polar interactions between protein and lipid polar region, hydrophobic interactions by penetrating into the hydrocarbon region of the lipid structure, osmotic stress imposed by the protein or effects due to the size of the macromolecule, is thus determined by the properties of the particular system. The changes in the phase behaviour in the lipid-protein-aqueous system relative to the two component lipid-aqueous system can be explained in terms of changes of geometrical packing parameters of the lipid imposed by the proteins, that is changes in effective head group area and the volume of the acyl chains. However, the whole truth can only be obtained by studying the entire three component system. Examples of such studies, including amphiphiles and proteins, have been reported by Morén *et al* (108, 109). They determined the phase diagrams of the ternary systems β-lactoglobulin – dodecyltrimethyl ammonium chloride (DOTAC) -water and lysozyme - sodium dodecylsulphate - water up to protein and surfactant concentrations of 20 wt%, respectively. In both systems the surfactant and the protein carry opposite net charges and precipitation occurs and reaches its maxima when the molar ratio between protein and surfactant corresponds to the net charge of the protein. Further addition of surfactant lead to a redissolution of the precipitate. However at higher protein concentrations, above 6-7 wt%, a narrow strip of gel phase occurred, which dissolved into an isotropic solution at higher surfactant concentrations.

7.1 Protein interactions that increase the curvature of the lipid-aqueous interfaces

Gramicidin A is a hydrophobic polypeptide, which forms channels in phospholipid membranes, that are specific for monovalent cations. The conformation of gramicidin A in these channels is as an amino terminal-to-amino terminal helical dimer (110). The peptide was found to favour the transition lamellar phase → reversed hexagonal (H_{II}) phase in dioleoylphosphatidylcholine (DOPC) and dioleoylphosphatidylethanolamine (DOPE) systems in an excess of water, as observed by NMR-studies (111). The effect was found to be less strong for DOPE compared with DSPC, which was suggested to partly depend on the stronger intermolecular attractive forces between the PE headgroups. At low water content gramicidin was not able to induce the H_{II} -phase, while upon increasing the water content, the lamellar phase separated into one gramicidin poor and one gramicidin rich lamellar phase. The gramicidin rich phase had a significant decreased order of the entire lipid molecule. Further increase of the water content lead to an increase of the disordered lamellar phase, which if sufficient water was available transformed into the reversed hexagonal phase. The ability of the peptide, which can form β-type of helices, to promote cylindrical (curved) structures, like the hexagonal phase, was suggested to be due to its conical shape and strong intermolecular attractive forces.

Not only proteins or peptides that penetrate into the lipid bilayer can induce phase transitions, but also proteins that are certain to interact with the headgroups of the phospholipid bilayer can give rise to similar effects. This has been demonstrated for cytochrome C, which has a positive net-charge and has been shown to interact with negatively charged phospholipids (112). The consequences of this interaction was reported by de Kruijff and Cullis, who found that binding of cytochrome C to anionic cardiolipin liposomes induced the formation to an inverted hexagonal, H_{II}, structure (112). No interaction and hence no phase transition was observed in the presence of liposomes composed of neutral zwitterionic lipids like PC and PE. However, if a sufficient fraction of these lipids was replaced for cardiolipin, the phase transition to the H_{II}-phase was observed. Interestingly, if cardiolipin was replaced with another anionic lipid like phosphatidylserine (PS) the protein was found to interact with the lipid, but did not induce any phase transition, which shows the specificity of the cardiolipin - cytochrome C interaction. The interaction between cardiolipin and cytochrome C was also studied by Spooner and Watts, using deuterium and phosphorus 31 NMR measurements (113). They likewise found that the interaction can, depending on the lipid stoichiometry, cause a transition from a lamellar to a non-bilayer structure. However, their NMR data did not show the characteristics of a well defined inverted hexagonal phase (H_{II}). The binding of the protein with the liquid-crystalline bilayers of cardiolipin was also found to cause extensive derangement of the cytochrome C secondary structure (113, 114).

Heimburg *et al* (115) reported results from phosphorous 31 NMR and resonance Raman spectroscopy studies of the interaction between cytochrome C and suspensions of DMPG or admixtures of dioleoylglycerol (DOG) or DOPC with DOPG. These results also indicate that binding of cytochrome C could promote an increase in surface curvature of the lipid aggregates from a bilayer structure. This is deduced from the NMR-data where an isotropic peak occurs in the presence of cytochrome c, indicating cubic lipid phases, small spherical vesicles or extended bilayers with high local curvature.

The structure of cytochrome C was found to change on binding to the lipid, and two forms, depending on the lipid composition, were identified:

I. close to the native conformation in solution

II. unfolded with the heme crevice opened.

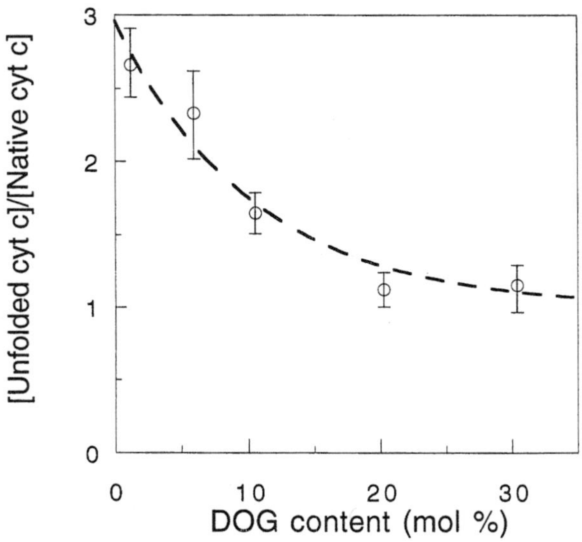

Fig. 10 Concentration of unfolded (II) and native (I) cytochrome C (cyt c) in dioleoylphosphatidylcholine (DOPC) / dioleoylglycerol (DOG) dispersions versus DOG mol % determined from Raman resonance spectra. The concentrations of lipid and cytochrome C were 300 and 20 μM, respectively, in an aqueous buffer (1 mM Hepes, 1 mM EDTA) of pH 7.5. Data adopted from Heimburg *et al.* (115), where also the experimental details are given.

They could in fact correlate the changes in protein structure with the curvature of the lipid bilayer. Some of their findings from resonance Raman measurements are illustrated in Fig. 10 as the ratio between the unfolded (II) and native (I) cytochrome C (cyt c) in DOPC/DOG dispersions versus DOG mol %. The presence of DOG was found to induce spontaneous curvature in the DOPG lipid bilayer in the pure lipid system. At high enough DOG content (□50%) this leads to the transition to a reversed hexagonal (H_{II}) phase. In the absence of DOG, that is a strict bilayer structure, the binding of the more unfolded form (II) of cytochrome is favoured, whereas the fraction of the more native globular protein structure (I) increases with the amount of DOG (Fig. 10) and thus with curvature of the surface. The physical state of the lipid was also found to affect the proportions of the two structural forms of cytochrome C. In the fluid state of pure DMPG, the fraction of the more unfolded form (II) was larger (85%) than when the lipid was in the gel state (80%). It is noteworthy that they found that the bound fraction of the more unfolded form (II) to the fluid DOPG bilayer structure was substantially lower (75%). Thus it was concluded that not only the fluidity of the bilayer matters, but other effects, such as if the lipid charge distribution matches that of the protein, have to be considered.

The interaction between cytochrome C and monoolein in the cubic phase was studied by Razumas *et al* by differential scanning calorimetry (DSC) and optical microscopy (116). In line with the studies reported above they also found that the presence of cytochrome C at high enough concentrations favoured lipid aggregates with a larger curvature. Thus they observed that the phase transition cubic $\rightarrow H_{II} \rightarrow L_2$ in the monoolein - cytochrome C - water system took place at a lower temperature than in the binary monoolein - water system (116).

7.2 Protein interactions that decrease the curvature of the lipid-aqueous interfaces

Fraser *et al* investigated the ability of a range of basic proteins to convert a reversed hexagonal (H_{II}) phase, consisting of dioleoylphosphatidylethanolamine (DOPE) and mixtures of DOPE and phosphatidylserine (PS), to stable lamellar (L_α) phases at pH 9 where DOPE is anionic and at pH 7 when it is zwitterionic (117). The proteins investigated where all capable of binding to the H_{II} -phase at pH 9, but only myelin basic protein and polylysine did induce transition to the L_α -phase. Lysozyme formed a new H_{II} -phase where the protein was included. A lowering of the pH seemed to release the proteins, except for mellittin which also seemed to penetrate into the hydrophobic core of the lipid aggregates. The introduction of PS into the H_{II} -phase at pH 7 increased the protein binding, but only myelin basic protein was able to induce the formation of a lamellar phase. Based on earlier studies, Fraser *et al.* suggested that the stabilisation of the lamellar phase by myelin basic protein was achieved by interaction in the polar portion of DOPE and thereby increasing the effective size of the lipid headgroup (117). They concluded that the properties of myelin basic protein in terms of stabilising the lamellar structure could be related to the role of the protein to stabilise the myelin sheath multilayers.

The impact on the swelling of lecithin/cardiolipin bilayers in the presence of bovine serum albumin was studied at pH 3.3, using X-ray diffraction by Rand (118). At this pH, the protein carries a positive net-charge and it is also likely to adopt a more expanded structure, thus exposing more hydrophobic segments. He found that the inter-lamellar spacing of the lamellar phase which was retained, decreased with increasing cardiolipin/bovine serum albumin ratio. This is due to a reduction of the negative charge of the lipid layer as the amount of bound protein increases. However, as the fraction of charged lipid (cardiolipin) increased, no further increase in the lamellar spacing was observed, suggesting that the interlamellar spacing in this system was not only controlled by simple electrostatics. Based on his X-ray data, Rand offers two plausible models for the interaction between the protein and the lipid bilayer, which are depicted in Fig. 11. The lipid bilayer is continuous in the first model (B), but the polar groups are spread apart and 23% of the exposed surface is constituted by the non-polar parts of the lipid giving a thinning of the lipid layer. The protein can interact with these hydropobic patches in addition to the electrostatic interaction with the charge cardiolipin molecules. In the second model (C) the thickness of the lipid bilayer is the same as the one in the pure lipid aqueous system (B), but the bilayer is no longer continuous as the more hydrophobic parts of the protein penetrate into the non-polar domain of the layer. The same type of forces are likely to be important and the two models give the same X-ray repeat distance.

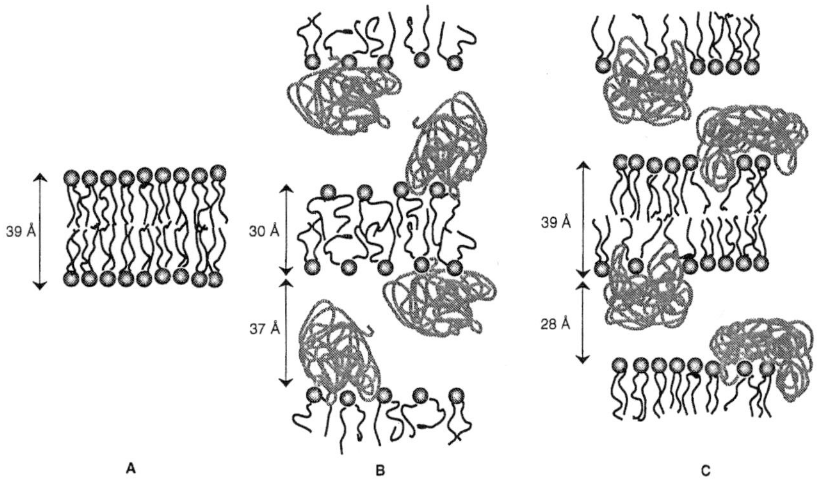

Fig. 11 Tentative structures of lamellar phases without protein (A), or with protein (B, C). The figure is adopted from Rand (118) and refers to X-ray data concerning the inclusion of bovine serum albumin into a phosphatidylcholine/cardiolipin lamellar phase (1:1). In (B), the bilayer is continuous, but the polar groups are spread apart, exposing some 20% of the surface to hydrophobic parts of the lipid. In (C), the protein penetrates into the lipid bilayer. Both models imply hydrophobic interaction with the protein and the gives the same repeat distance (67 Å).

8 LIPID-PROTEIN-AQUEOUS CUBIC PHASES

Lipid bilayers can also be folded into such intriguing structures as cubic liquid crystalline lipid aqueous phases. In these structure not only the surface properties and the curvature, but also the dimensions of the aqueous space affect protein-lipid interactions. Examples of the bicontinuous type of these structures are outlined in Fig. 8. The monoolein-aqueous system is a thoroughly studied example of such a system, where two types of cubic phases have been observed on the water-rich side of the lamellar phase (119-122). The cubic phase in excess of water (□ 40%) was identified to have a primitive lattice, corresponding to a diamond type of IPMS (Fig. 8A).

As the water content decreases a narrow opaque two phase region appears and is replaced by the other type of cubic phase observed in this system when the amount of water is further reduced. The cubic phase formed has a body-centered lattice, which corresponds to the gyroid type of IPMS (Fig. 8B).

One of the important features of the bicontinuous cubic lipid-aqueous phases is that they contain two water continuous systems. Thus aqueous soluble molecules, provided they fit into the structure, move more or less freely through the structure. This is illustrated in the work of Mattisson *et al* (123) where the diffusion of a glucose from a cubic monoolein-aqueous phase was studied by using holographic laser interferometry. Some of the concentration profiles of glucose in the cubic phase they obtained are shown in Fig. 12. These profiles could be fitted to Ficks 2nd law, which gave a diffusion coefficient 4 times lower than the value in aqueous solution. The mobility of the molecules in the aqueous channels of the cubic phase is certain to be affected by the dimensions of the channels and the size of the solute. Thus, electrochemical studies of the transport of cytochrome C in the monoolein-aqueous cubic phase shows that the mobility of the entrapped enzyme is limited in the cubic phase compared to bulk solution, with values of diffusion coefficients that were found to be about 70 times lower than the bulk values (116).

As the knowledge of the structure and the formation of these lipid-aqueous cubic phases has grown, they have increasingly been recognised as important in many biological systems (16, 76, 81, 82, 124, 125). In fact Landh compared calculated electron density maps of given periodic cubic surfaces with numerous published transmission electron micrographs of biological specimens showing cubic membrane morphologies (126). He identified cubic structures formed in conjunction with a number of cell organelles, e. g. endoplasmatic reticulum, inner nuclear envelope, mitochondria, trans-Golgi apparatus, chloroplasts, plasma membranes and lysosomes.

One possible process where the structure of bicontinuous phase can be advantageous is fusion of biological membranes. In fact studies of the fusion process have revealed that the rate is larger in systems showing the lamellar phase $\rightarrow$ inverted hexagonal phase transition (127-129). The discovery of lipidic particles during the fusion event (130) indicates that additional liquid crystalline phases occur during the fusion process. It should be pointed out that not only a certain lipid composition or distribution promote fusion, but the process can also be trigged by proteins and polypeptides interacting with the lipidic membrane (91).

Another process where the cubic lipid phases are sure to take part is the lipolyses of triglycerides catalysed by lipases. Patton and Carey directly monitored this process in an intestinal-like environment in a (polarised) light microscope, where an oil droplet was placed on the microscope slide and exposed for human lipases (131). A viscous isotropic phase composed of monoglycerides and fatty acids, was observed in addition to the initially occurring lamellar liquid crystalline phase. These type of isotropic phases, e. g. cubic phases, are also formed in monoglyceride aqueous systems as discussed above. In the body, the lipolysis products are rapidly solubilised in mixed micelles with bile salts, which generally are in excess in the intestine. However, after a meal rich in fats, the bile acid amounts in vivo are not sufficient to solubilise all lipids (132), which implies that liquid crystalline phases appear in vivo. The diminishing fat droplet is surrounded by these phases formed by the lipolysis products, which must be penetrated by lipase and water to continue the lipolytic process. The bicontinuity as well as the ability to incorporate other molecules, are important features of the cubic monoglyceride phases which favour the lipolysis process (133). In addition a cubic phase creates a large effective area compared to, for instance, an oil-aqueous interface, which can lead to a much faster lipase catalysed lipolytic process (134).

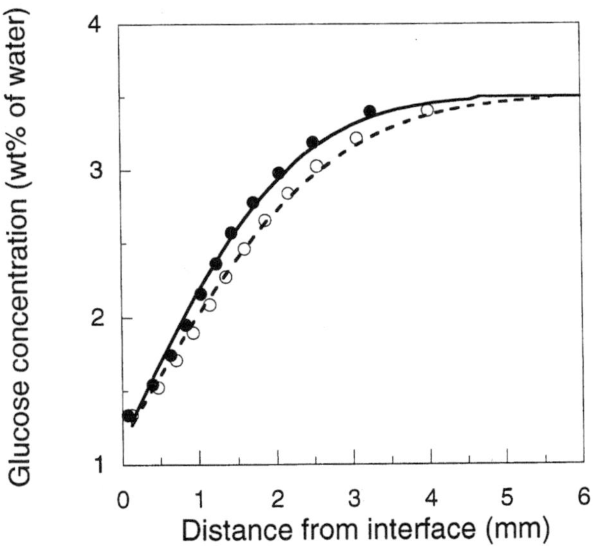

Fig. 12 Glucose concentration profiles in a monoolein - aqueous cubic phase (62:38 wt%), where the aqueous solution initially contained 3.5 wt% glucose, after 3 h (●) and 4 h (○) equilibration against pure water. The concentration is given as the wt% glucose in the aqueous solution of the cubic phase. The solid and

broken lines represent the best theoretical fit of Fick's law, giving diffusion coefficients of $1.39 \cdot 10^{-10} m^2 s^{-1}$ and $1.47 \cdot 10^{-10} m^2 s^{-1}$ after 3 and 4 hours, respectively. The corresponding bulk value is $6.7 \cdot 10^{-10} m^2 s^{-1}$. The data, obtained by holographic laser interferometry, are adopted from Mattisson *et al.* (123, 144), where also the experimental details are given.

Apart from their significance in biological systems, the structural features of the lipid-aqueous cubic phases, such as flexibility, bicontinuity and the presence of aqueous cavities with dimensions similar to proteins, have triggered a number of studies of protein entrapment in these phases (116, 124, 135-146).

A variety of hydrophilic proteins with molecular weights up to 590 kD can be entrapped in the aqueous cavity of the monoolein-aqueous cubic phases (116, 136, 142-145). The entrapped proteins have been found to be protected in the cubic phase as they can be kept for a very long time (months in some cases), with retained activity, which is not possible in aqueous solution (142, 144). The cubic monoglyceride phases have also the ability to solubilise lipophilic proteins like A-gliadin from wheat (135) and bacteriorhodopsin (146) as well as relatively large amounts of membrane lipids (143, 144, 147) and other hydrophobic compounds as vitamin K_1 (92). These compounds are most probably dispersed in the lipid bilayer region of the cubic phase.

It seems that properties like visco-elasticity, microscopic appearance and the nature of the X-ray diffraction pattern at low protein concentrations are basically the same for the ternary lipid-protein-aqueous cubic phases as for the corresponding binary lipid-aqueous cubic phase (116, 136, 139, 140, 142). However, there are some important differences in particular at higher protein concentrations. The data for the ternary monoolein-lysozyme-aqueous system presented by Ericsson (148) (recalculated from earlier work (136) to take into account the existence of two cubic phases, gyroid and diamond IPMS) indicate that the amount of protein as well as the amount of water determines the phase behaviour. She found that at very high protein concentrations (34.2 wt%) and a water content of 34.2 wt%, the preferred cubic structure seems to be of the diamond type. At lower protein concentration (11.5 wt%) and 45.9 wt% water the cubic phase formed is of the gyroid type. It has also been suggested that the third type of cubic phase, that is the primitive IPMS (corresponding to the body centered lattice (Im3m), is present in this system (138). This was confirmed by the data presented by Razumas *et al*, for the phase behaviour in the presence of 5, 7 and 8 wt% lysozyme concentrations at constant water content of 38-39 wt% (143). The encapsulation of haemoglobin in the monoolein-aqueous cubic phase was also found to affect the type cubic phase formed (145). At low haemoglobin concentration (0.5-2.5 wt%) and constant lipid content of 50% the diamond type was formed, while the protein concentration lead to a two phase region where the diamond and the gyroid type coexisted. Above a protein concentration of 5 wt%, the gyroid type prevailed. The formation of the different type of cubic phases can be understood when considering the differences in structure between the three types of cubic phases (Fig. 8) (81). The primitive and diamond types are more restricted than the gyroid type of structure. As shown in Fig. 8 the primitive and diamond feature circular necks, which impose geometrical constraints when packing the lipid bilayers in these types of structures. The necks are the most narrow for the primitive type of cubic phase. Consequently, this is usually not seen in the binary monoolein system where the maximum swelled cubic phase has a lipid content of about 60 wt% (81). However, as discussed above it does appear in ternary monoolein-protein-aqueous system where the bilayer takes up a smaller portion of the unit volume (81). The primitive cubic structure has also been reported for monoolein-aqueous systems, which has been swollen by introducing an anionic phospholipid (DSPG) (143). The gyroid type can, because of fewer geometrical constraints, occur over a larger concentration range down to low water content. This type of cubic phase was also found when a zwitterionic phospholipid (PC) was introduced in the cubic phase (144, 147). Furthermore it was found that the gyroid type of monoolein-aqueous phase could accommodate a larger content of vitamin K_1 in its bilayer than the diamond type before the transition to the reversed hexagonal phase occurred (92).

Generally it has been observed that the cubic phases can swell (considerable increase in unit cell dimension of the cubic lattice) to a higher water content when protein is present (136). The amount of protein also affects the swelling. For instance, the unit cell dimensions of the primitive cubic phase increased from 123 to 130 Å, when the lysozyme concentration increased from 5 to 8 wt% in the monoolein-lysozyme-aqueous system containing 38-39 wt% water (143). Furthermore it has been observed that serum albumin, having a considerably larger hydrodynamic dimension than lysozyme, also gives a larger unit cell dimension (148). However, large proteins as glucose oxidase (150 kD) have a limited capability to become incorporated in the cubic phase (136, 142). It is not only the size of the protein which determines the incorporation in the cubic phase but also solution conditions like the ionic strength. Razumas *et al* found that no protein was released when a cubic monoolein-cytochrome C aqueous phase was exposed to pure water (116). Instead a swelling of the cubic phase was observed, suggesting an uptake of water. However, if a similar sample was exposed to a buffer solution of physiological ionic strength, almost all of the protein was released. The corresponding unit cell dimension decreased down to the value for the pure monoolein-buffer phase. This suggests that the ions and the proteins compete to interact with the lipid structure and according to our experience it is often harder to incorporate proteins into the cubic phase at high ionic strength.

Spectroscopic data have revealed changes in the molecular organisation of the lipids evoked by the presence of the protein. FT-IR measurements on the monoolein - cytochrome C aqueous system showed an increased conformational order of the monoolein acyl chain as well as structural rearrangements of the polar head group region in the presence of cytochrome C (116). This was in line with X-ray data, where the increase in unit cell dimension suggests a decrease of the packing parameter on incorporation of cytochrome c. The Raman scattering studies on the monoolein - lysozyme - aqueous system demonstrated an increase in the number of hydrogen -bond C=O groups of monoolein, but no increase in the acyl chain order relative the binary lipid-aqueous system (143). Similar increase in the hydrogen bonding, caused by the presence of a protein, was observed in the monoolein - haemoglobin - aqueous system using FT-IR spectroscopy (145). However, in this case the protein incorporation also caused a decrease in the acyl-chain order.

The properties of the lipid polar interface in the cubic phase can be altered by dispersing various lipids in the bilayer. As discussed above, this can cause a intercubic phase transition. Both properties are sure to affect the protein encapsulation. The introduction of zwitterionic phosphatidylcholine into a monoolein based cubic phase was found to considerably increase the time the enzyme activity of encapsulated glucose oxidase was preserved (144). However it was not possible to introduce the cationic protein lysozyme into a cubic phase formed from monoolein and anionic DSPG, in spite of considerable swelling of the cubic structure (143). Razumas *et al.* suggested that the initial electrostatic interaction between DSPG and lysozyme and subsequent penetration of the protein into the mixed lipid bilayer distorted the cubic phase (143).

Several investigations on the structure of proteins entrapped in the cubic phase indicate that the native structure is retained. For instance, differential scanning calorimetry and enzyme activity measurements show that lysozyme retains its conformation and activity in the cubic phase (136, 143). Similarly it has been observed by circular dichroism measurements that bacteriorhodopsin and melittin (140) as well as α-chymotrypsin (139) retain their native conformation on incorporation in lysolecithin-aqueous cubic phases. FT-IR measurements on the monoolein - haemoglobin aqueous cubic phase gave no evidence for conformational changes of the protein on incorporation (145). On the other hand, differential scanning calorimetry data and other observations on cytochrome C entrapped in the cubic phase, suggested that some interactions, which affect the thermal stability of the protein as well as of the cubic phase, take place (116).

Apart from the biological significance of cubic lipid-aqueous phases, Razumas *et al* demonstrated that cubic monoolein-aqueous phases containing enzymes can be used as the biocatalytic layer in amperometric and potentiometric biosensors (142). Their results for biosensors, based on a variety of enzymes, show that the long-term stability decreases in the order lactate oxidase > creatinine deiminase > glucose oxidase > urease, that is basically in the order of increasing molecular weight. Also the cubic phases of other amphiphiles like ethoxylated fatty alcohols can be used to entrap glucose oxidase, to construct a simple glucose monitor (141). The bicontinuous cubic structures have by virtue of their well defined porosity also a large potential in drug delivery systems (16). Stable particles of lipid-aqueous cubic phases, cubosomes, can also be produced (16, 149).

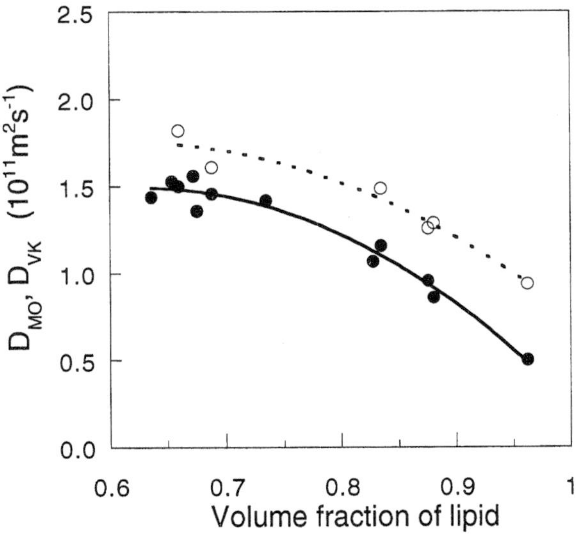

Fig. 13. NMR self-diffusion coefficients at 25 □C in monoolein-aqueous cubic phases containing 0-5 wt % vitamin K_1, are shown as a function of the lipid volume fraction (including vitamin K_1). The self-diffusion coefficients were measured in the cubic (both gyroid and diamond type) and in the reversed micelle, L_2, phases. Self-diffusion coefficients of monoolein (D_{MO}) (●) and vitamin K_1 (D_{VK}) (○) are shown. The lines are arbitrary fits to demonstrate the similar trends. The data are adopted from Caboi *et al.* (92), where also the experimental details are given.

It should be born in mind that the cubic lipid-aqueous phases are flexible structures. Not only aqueous soluble molecules are movable in the water channels, but also the lipids that constitutes the bicontinuous structure are highly mobile. Any molecules that are introduced in the lipid region will in general also be mobile. This is demonstrated in Fig. 13, where the mobility, in terms of the NMR self diffusion coefficients, of monoolein and vitamin K_1 dispersed in the lipid bilayer is plotted versus lipid volume fraction in the cubic phases. As shown in the figure the mobility of the introduced vitamin K_1 follows that of monoolein, indicating complete dispersion of vitamin K_1. The mobility is of large importance in biological systems as well as in applications. Recently Landau and Rosenbusch demonstrated that the bicontinuous phases based on monoolein and monopalmitolein could provide matrices for the crystallisation of membrane proteins like bacteriorhodopsin (146). They pointed out that the use of these types of cubic phases is advantageous as they provide both nucleation sites, as the membrane proteins can be dissolved in the lipid bilayer, and support growth by allowing lateral diffusion of the protein molecules in the membrane.

9 ACKNOWLEDGEMENTS

The fruitful discussions with Dr. Andrew Fogden, Prof. Kåre Larsson, Dr. Jane Morris, Prof. Barry Ninham, and Prof. Valdemaras Razumas are acknowledged. This work has benefited from the financial support from the European Commission - FAIR programme (MADGELAS - Concerted Action CT96 1202) and from The Swedish Research Council for Engineering Sciences.

10 REFERENCES

1. Watts, E., *Protein-Lipid Interactions*; Elsevier: Amsterdam, 1993.
2. Verger, R.; Pattus, F., *Chem. Phys. Lipids*, 30 (1982) 189.
3. Piéroni, G.; Gargouri, Y.; Sarda, L.; Verger, R., *Adv. Colloid Interf. Sci.*, 32 (1990) 341.
4. Hills, B. A. *The Biology of Surfactants*; Cambridge University Press: Cambridge, 1988.
5. Harwood, J. L.; Cryer, A.; Gurr, M. I.; Dodds, P. In *The lipid handbook.* ; F. D. Gunstone, J. L. Harwood, F. B. Padley, Eds.; Chapman & Hall: London, 1994 p. 665.
6. Pison, U.; Herold, R.; Schürch, S., *Colloids Surfaces A: Physicochem. Eng. Aspects*, 114 (1996) 165.
7. Spector, A. A., *J. Lipid Res.*, 16 (1975) 165.
8. Papiz, M.J.; Sawyer, L.; Eliopoulos, E.E.; North, A.C.T.; Findlay, B.C.; Sivaprasadarao, R.; Jones, T.A.; Newcomer, M.E.; Kraulis, P.J., *Nature*, 324(1986)383.
9. Sawyer, L., *Nature*, 327 (1987) 659.
10. North, A. C. T., *Int. J. Biol. Macromol.*, 81 (1989) 56.
11. Diaz de Villegas,M.C.; Oria,R.; Salva, F.J.; Calvo,M., *Milchwissenschaft*, 42(1987)357.

12. Sarker, D.K.; Wilde, P.J.; Clark, D.C., *Colloids Surfaces B: Biointerfaces*, 3 (1995)349.

13. Creamer, L. K., *Biochemistry*, 34 (1995) 7170.

14. Kristensen, A.; Nylander, T.; Paulsson, M.; Carlsson, A., *Int. Dairy J.*, 7 (1997) 87.

15. Kurihara, K.; Katsuragi, Y., *Nature*, 365 (1993) 213.

16. Larsson, K. *Lipids — Molecular Organization, Physical Functions and Technical Applications*; The Oily Press Ltd.: Dundee, 1994.

17. Christie, W. W. *High-Performance Liquid Chromatography and Lipids*; Pergamon Books: Oxford, 1987.

18. Lindman, B.; Wennerström, H., *Topics Current Chem.*, 87 (1980) 1.

19. Mattice, W. L.; Riser, J. M.; Clark, D. S., *Biochemistry*, 15 (1976) 4264.

20. Ananthapadmanabhan, K. P. In *Interactions of Surfactants with Polymers and Proteins*; E.D.Goddard, K.P.Ananthapadmanabhan, Eds.; CRC Press: Boca Raton; Florida 1993 p.319.

21. Guo, X. H.; Zhao, N. M.; Chen, S. H.; Teixeira, J., *Biopolymers*, 29 (1990) 335.

22. Guo, X. H.; Chen, S. H., *Chem. Phys.*, 149 (1990) 129.

23. Ninham, B. W.; Yaminsky, V., *Langmuir*, 13 (1997) 2097.

24. Pace, C. N.; Fisher, L. M.; Cupo, J. F., *Acta Biol. Med. Germ.*, 40 (1981) 1385.

25. Privalov, P. L.; Gill, S. J., *Adv. Protein Chem.*, 39 (1988) 191.

26. Dill, K. A., *Biochemistry*, 29 (1990) 7133.

27. Ponnuswamy, P. K., *Prog. Biophys. Molec. Biol.*, 59 (1993) 57.

28. Richards, F. M., *Ann. Rev. Biophys. Bioeng.*, 6 (1977) 151.

29. Norde, W., *Adv. Colloid Interface Sci.*, 25 (1986) 267.

30. Haynes, C. A.; Norde, W., *Colloids Surfaces B: Biointerfaces*, 2 (1994) 517.

31. Dalgleish, D.G.; Dickinson, E.; Whyman, R.H., *J. Colloid Interface Sci.*, 108(1985)174.

32. Dolgikh, D. A.; Gilmanshin, R. I.; Brazhnikov, E. V.; Bychkova, V. E.; Semisotnov, G. V.; Venyaminov, S. Y.; Ptitsyn, O. B., *FEBS Lett.*, 136 (1981) 311.

33. Ohgushi, M.; Wada, A., *FEBS Lett.*, 164 (1983) 21.

34. Dolgikh, D. A.; Abaturov, L. V.; Bolotina, I. A.; Brazhnikov, E. V.; Bychkova, V. E.; Gilmanshin, R. I.; Lebedev, Y. O.; Semisotnov, G. V.; Tiktopulo, E. I.; Ptitsyn, O. B., *Eur. Biophys. J.*, 13 (1985) 109.

35. Kuwajima, K., *Proteins: Struct. Funct. Genet.*, 6 (1989) 87.

36. Ptitsyn, O. B.; Pain, R. H.; Semisotnov, G. V.; Zerovnik, E.; Razgulyaev, O. I., *FEBS Lett.*, 262 (1990) 20.

37. Dickinson, E.; Matsumura, Y., *Colloids Surfaces B: Biointerfaces*, 3 (1994) 1.

38. Bychokova, V. E.; Pain, R. H.; Ptitsyn, O. B., *FEBS Lett.*, 238 (1988) 231.

39. van der Goot, F.G.; González-Mañas, J.M.; Lakey, J.H.; Pattus, F., *Nature*, 354(1991)408.

40. MacRitchie, F. *Chemistry at Interfaces* ed.; Academic Press: San Diego, 1990.

41. Miller, R.; Lunkenheimer, K., *Colloid Polymer Sci.*, 264 (1986) 273.

42. Lunkenheimer, K.; Miller, R., *J. Colloid Interface Sci.*, 120 (1987) 176.

43. Lunkenheimer, K.; Czichocki, G., *J. Colloid Interface Sci.*, 160 (1993) 509.

44. Clark, D. C.; Husband, F.; Wilde, P. J.; Cornec, M.; Miller, R.; Krägel, J.; Wüstneck,R., *J. Chem. Soc. Faraday Trans.*, 91 (1995) 1991.

45. Vollhardt, D., *Adv. Colloid Interface Sci.*, 47 (1993) 1.

46. Vollhardt, D.; Kato, T.; Kawano, M., *J. Phys. Chem.*, 100 (1996) 4141.

47. Gericke, A.; Simon-Kutscher, J.; Hühnerfuss, H., *Langmuir*, 9 (1993) 2119.

48. Carlsson, A.; Bergqvist, M.; Nilsson, R.; Nylander, T., *Pro. Drug Delivery System (Tokyo)*, IV (1995) 105.

49. Kristensen, D.; Nylander, T.; Rasmussen, J. T.; Paulsson, M.; Carlsson, A., *Langmuir*, 12 (1996) 5856.

50. Siminovitch, D. J.; Jeffrey, K. R., *Biochim. Biophys. Acta*, 645 (1981) 270.

51. Vikholm, I.; Teleman, O., *J. Colloid Interface Sci.*, 168 (1994) 125.

52. Bos, M.; Nylander, T., *Langmuir*, 12 (1996) 2791.

53. Hambling, S. G.; McAlpine, A. S.; Sawyer, L. In *Advanced dairy chemistry. Vol. 1: Proteins* ; P. Fox, Ed.; Elsevier Applied Science Publishers Ltd.: London, 1992 p. 141.

54. Cornell, D. G.; Patterson, D. L., *J. Agric. Food Chem.*, 37 (1989) 1455.

55. Cornell, D. G.; Patterson, D. L.; Hoban, N., *J. Colloid Interface Sci.*, 140 (1990) 428.

56. Cornell, D. G., *J. Colloid Interface Sci.*, 88 (1982) 536.

57. Quinn, P. J.; Dawson, R. M. C., *Biochem. J.*, 113 (1969) 791.

58. Quinn, P. J.; Dawson, R. M. C., *Biochem. J.*, 115 (1969) 65.

59. Du, Y.-K.; An, J.-Y.; Tang, J.; Li, Y.; Jiang, L., *Colloid Surfaces B: Biointerfaces*, 7 (1996) 129.

60. Ibdah, J. A.; Phillips, M. C., *Biochemistry*, 27 (1988) 7155.

61. Hanssens, I.; Van Cauwelaert, F. H., *Biochem. Biophys. Res. Commun.*, 84 (1978) 1088.

62. Lindahl, L.; Vogel, H. J., *Anal. Biochem.*, 140 (1984) 394.

63. Cornell, D. G.; Caroll, R. J., *J. Colloid Interface Sci.*, 108 (1985) 226.

64. Sommer, F.; Alexandre, S.; Dubreuil, N.; Lair, D.; Duc, T.-m.; Valleton, J. M., *Langmuir*, 13 (1997) 791.

65. Heckl, W. M.; Zaba, B. N.; Möhwald, H., *Biochim. Biophys. Acta*, 903 (1987) 166.

66. Schönhoff, M.; Lösche, M.; Meyer, M.; Wilhelm, C., *Progr. Colloid Polym. Sci.*, 89 (1992) 243.

67. Diederich, A.; Sponer, C.; Pum, D.; Sleytr, U. B.; Lösche, M., *Colloids Surfaces B: Biointerfaces*, 6 (1996) 335.

68. Murray, B. S.; Nelson, P. V., *Langmuir*, 12 (1996) 5973.

69. Aynié, S.; Le Meste, M.; Colas, B.; Lorient, D., *J. Food Sci.*, 57 (1992) 883.

70. Leaver, J.; Dalgleish, D. G., *J. Colloid Interface Sci.*, 149 (1992) 49.

71. Miller, K. W.; Small, D. M., *J. Biol. Chem.*, 258 (1983) 13772.

72. Badr, M.; Kodali, D. R.; Redgrave, T. G., *J. Colloid Interface Sci.*, 113 (1986) 414.

73. Heertje, I.; Nederlof, J.; Hendrickx, H. A. C. M.; Lucassen-Reynders, E. H., *Food Structure*, 9 (1990) 305.

74. Malmsten, M., *J. colloid Interface Sci.*, 168 (1994) 247.

75. Malmsten, M., *J. Colloid Interface Sci.*, 172 (1995) 106.

76. Hyde, S.; Andersson, S.; Larsson, K.; Blum, Z.; Landh, T.; Lidin, S.; Ninham, B. W. *The language of shape. The role of curvature in condensed matter: Physics, chemistry and biology*; Elsevier: Amsterdam, 1997.

77. McCallum, C. D.; Epand, R. M., *Biochemistry*, 34 (1995) 1815.

78. Luzzati, V. In *Biological Membranes*; D. Chapman, Ed.; Academic Press: New York, 1968 p. 77.

79. Fontell, K., *Colloid Polym. Sci.*, 268 (1990) 264.

80. Luzzati, V.; Tardieu, A.; Gulik-Krzywicki, T.; Rivas, E.; Reiss-Husson, F., *Nature*, 220 (1968) 485.

81. Larsson, K., *J. Phys. Chem.*, 93 (1989) 7304.

82. Lindblom, G.; Rilfors, L., *Biochim. Biophys. Acta*, 988 (1989) 221.

83. Fontell, K., *Adv. Colloid Interface Sci.*, 41 (1992) 127.

84. Lindblom, G.; Larsson, K.; Johansson, L.; Fontell, K.; Forsén, S., *J. Am. Chem. Soc.*, 101 (1979) 5465.

85. Andersson, S.; Hyde, S. T.; Larsson, K.; Lidin, S., *Chem. Rev.*, 88 (1988) 221.

86. Israelachvili, J. N.; Mitchell, D. J.; Ninham, B. W., *J. Chem. Soc. Faraday Trans., II*, 72 (1976) 1525.

87. Mitchell, D. J.; Ninham, B. W., *J. Chem. Soc. Faraday Trans. 2*, 77 (1981) 601.

88. Hyde, S. T., *J. Phys. Chem.*, 93 (1989) 1458.

89. Malmsten, M.; Bergenståhl, B.; Nyberg, L.; Odham, G., *J. Am. Oil Chem. Soc.*, 71 (1994) 1021.

90. Minami, H.; Nylander, T.; Carlsson, A.; Larsson, K., *Chem. Phys. Lipids*, 79 (1996) 65.

91. Raudino, A., *Adv. Colloid Interface Sci.*, 57 (1995) 229.

92. Caboi, F.; Nylander, T.; Razumas, V.; Talaikyté, Z., T.; Monduzzi, M.; Larsson, K., *Langmuir*, 13 (1997) 5476.

93. Lasic, D. D. *Liposomes: From Physics to Applications*; Elsevier: Amsterdam, 1993.

94. MacDonald, R. C. In *Vesicles* ; M. Rosoff, Ed.; Marcel Dekker: New York, 1996; Vol. 62, p. 3.

95. Rytömaa, M.; Mustonen, P.; Kinnunen, P. K. J., *J. Biol. Chem.*, 267 (1992) 22243.

96. Brooksbank, D. V.; Leaver, J.; Horne, D. S., *J. Colloid Interface Sci.*, 161 (1993) 38.

97. Kim, J.; Kim, H., *Biochemistry*, 25 (1986) 7867.

98. Brown, E. M.; Caroll, R. J.; Pfeffer, P. E.; Sampugna, J., *Lipids*, 18 (1983) 111.

99. Brown, E. M., *J. Dairy Sci.*, 67 (1984) 713.

100. de Wit, J. N. In *Developments in Dairy Chemistry - 4*; P. F. Fox, Ed.; Elsevier Applied Science: London, 1989 p. 285.

101. Keller, R. A. C.; Killian, J. A.; De Kruiff, B., *Biochemistry*, 31 (1992) 1672.

102. Subbarao, N. K.; Parente, R. A.; Szoka Jr, F. C.; Nadasdi, L.; Pongracz, K., *Biochemistry*, 26 (1987) 2964.

103. Raudino, A.; Castelli, F., *Colloid Polymer Sci.*, 270 (1992) 1116.

104. Hønger, T.; Jørgensen, K.; Biltonen, R.L.; Mouritsen, O.G., *Biochemistry*, 35(1996)9003.

105. Sankaram, B.; Marsh, D. In *Protein-Lipid Interactions*; A. Watts, Ed.; Elsevier: Amsterdam, 1993; Vol. 25, p. 127.

106. LeNeveu, D. M.; Rand, R. P.; Parsegian, V. A., *Biophys. J.*, 18 (1977) 209.

107. Cowley, A.C.; Fuller, N.L.; Rand, R.P.; Parsegian, V.A., *Biochemistry*, 17 (1978) 3163.

108. Morén, A. K.; Khan, A., *Langmuir*, 11 (1995) 3636.

109. Morén, A. K.; Eskilsson, K.; Khan, A., *Colloids Surfaces B: Biointerfaces*, 9(1997)305.

110. Wallis, B. A., *Biophys. J.*, 49 (1986) 295.

111. Chupin, V.; Killian, J. A.; De Kruijff, B., *Biophys. J.*, 51 (1987) 395.

112. De Kruijff, B.; Cullis, P. R., *Biochim. Biophys. Acta*, 602 (1980) 477.

113. Spooner, P. J. R.; Watts, A., *Biochemistry*, 30 (1991) 3880.

114. Spooner, P. J. R.; Watts, A., *Biochemistry*, 30 (1991) 3871.

115. Heimburg, T.; Hildebrandt, P.; Marsh, D., *Biochemistry*, 30 (1991) 9084.

116. Razumas, V.; Larsson, K.; Miezes, Y.; Nylander, T., *J. Phys. Chem.*, 100 (1996) 11766.

117. Fraser, P. E.; Rand, R. P.; Deber, C. M., *Biochim. Biophys. Acta*, 983 (1989) 23.

118. Rand, R. P., *Biochim. Biophys. Acta*, 241 (1971) 823.

119. Larsson, K., *Nature*, 304 (1983) 664.

120. Hyde, S. T.; Andersson, S.; Ericsson, B.; Larsson, K., *Z. Kristallogr.*, 168 (1984) 213.

121. Landh, T., *J. Phys. Chem.*, 98 (1994) 8453.

122. Briggs, J.; Chung, H.; Caffrey, M., *J. Phys. II France*, 6 (1996) 723.

123. Mattisson, C.; Nylander, T.; Axelsson, A.; Zacchi, G., *Chem. Phys. Lipids*, 84 (1996) 1.

124. Mariani, P.; Luzzati, V.; Delacroix, H., *J. Mol. Biol.*, 204 (1988) 165.

125. Seddon, J. M., *Biochim. Biophys. Acta*, 1031 (1990) 1.

126. Landh, T., *Cubic cell membrane architectures. Taking another look at membrane bound cell spaces.* ; Thesis, University of Lund: Lund, Sweden, 1996.

127. Verkleij, A. J.; Mombers, C.; Gerritsen, W. J.; Leunissen-Bijvelt, L.; Cullis, P. R., *Biochim. Biophys. Acta*, 555 (1979) 358.

128. Mandersloot, J. G.; Gerritsen, W. J.; Leunissen-Bijvelt, L.; Van Echteld, A.; Noordam, P. C.; De Gier, J., *Biochim. Biophys. Acta*, 640 (1981) 106.

129. Hope, M. J.; Cullis, P. R., *Biochim. Biophys. Acta*, 640 (1981) 82.

130. Cullis, P. R.; Van Dijck, P. W. M.; De Kruijff, B.; De Gier, J., *Biochim. Biophys. Acta*, 513 (1978) 21.

131. Patton, J. S.; Carey, M. C., *Science*, 204 (1979) 145.

132. Borgström, B., *Scand. J. Gastroenterol.*, 20 (1985) 389.

133. Patton, J. S.; Vetter, R. D.; Hamosh, M.; Borgström, B.; Lindström, M.; Carey, M. C., *Food Microstructure*, 4 (1985) 29.

134. Wallin, R.; Arnebrant, T., *J. Colloid Interface Sci.*, 164 (1994) 16.

135. Larsson, K.; Lindblom, G., *J. Disp. Sci. Technol.*, 3 (1982) 61.

136. Ericsson, B.; Larsson, K.; Fontell, K., *Biochim. Biophys. Acta*, 729 (1983) 23.

137. Gulik-Krzywicki, T.; Aggerbeck, L. P.; Larsson, K. In *Surfactants in solution.* ; K.L.Mittal, B. Lindman, Eds.; Plenum Press: New York, 1984; Vol. 1, p. 237.

138. Buchheim, W.; Larsson, K., *J. Colloid Interface Sci.*, 117 (1987) 582.

139. Portmann, M.; Landau, E. M.; Luisi, P. L., *J. Phys. Chem.*, 95 (1991) 8437.

140. Landau, E. M.; Luisi, P. L., *J. Am. Chem. Soc.*, 115 (1993) 2102.

141. Wallin, R.; Engström, S.; Mandenius, C. F., *Biocatalysis*, 8 (1993) 73.

142. Razumas, V.; Kanapieniené, J.; Nylander, T.; Engström, S.; Larsson, K., *Anal. Chim. Acta*, 289 (1994) 155.

143. Razumas, V.; Talaikyté, Z.; Barauskas, J.; Larsson, K.; Miezis, Y.; Nylander, T., *Chem. Phys. Lipids*, 84 (1996) 123.

144. Nylander, T.; Mattisson, C.; Razumas, V.; Miezes, Y.; Håkansson, B., *Colloids Surfaces A: Physicochem. Eng. Aspects*, 114 (1996) 311.

145. Leslie, S. B.; Puvvada, S.; Ratna, B. R.; Rudolph, A. S., *Biochim. Biophys. Acta*, 1285 (1996) 246.

146. Landau, E. M.; Rosenbusch, J. P., *Proc. Natl. Acad. Sci.*, 93 (1996) 14532.

147. Gutman, H.; Arvidson, G.; Fontell, K.; Lindblom, G. In *Surfactants in Solution* ; K. L. Mittal, B. Lindman, Eds.; Plenum: New York, 1984; Vol. 1, p. 143.

148. Ericsson, B. *Interactions between globular proteins and lipids.*, Thesis, University of Lund: Lund, Sweden, 1986.

149. Gustafsson, J.; Ljusberg-Wahren, H.; Almgren, M.; Larsson, K., *Langmuir*, 12(1996)4611.

Proteins at Liquid Interfaces
D. Möbius and R. Miller (Editors)

CHARACTERISATION OF GELATIN/SURFACTANT INTERACTION AND ITS RELEVANCE TO LIQUID FILM COATING

R. Wüstneck[§] and J. Krägel[*]

[§] Universität Potsdam, Institut für Physik, Am Neuen Palais 10, D-14415 Potsdam, Germany

[*] Max-Planck-Institut für Kolloid- und Grenzflächenforschung, Rudower Chaussee 5, D-12489 Berlin, Germany

Contents

1. INTRODUCTION

This chapter focuses on the gelatin/surfactant interaction, and in particular on their influence on liquid film coating processes in photo industry. Basic aspects of gelatin/surfactant interaction and their consequence for coating processes will be reviewed. The interfacial chemical fundamentals of the related mechanisms are described.

Gelatin is a polyampholytic biopolymer. It is applied in food, pharmaceutical, and photographic industries. Gelatin is a natural surface-active substance with high molecular mass. Until now it is on the first place of binders in photographic emulsions although many attempts were made to substitute gelatin partially or completely. The use of gelatin in photographic emulsions dates back to 1870 when it was used to replace the colloidon wet process. Gelatin emulsions have been continually improved in quality, and it is still the best medium known for photographic film material. The term "photographic emulsion" is used even though it contains dispersed solid microcrystalline particles of silver halide, so that the term "suspension" would be more correct.

Uncontested gelatin significantly influences each phase of manufacturing, application, and processing of light sensitive materials. It plays an important role in the formation and growth of silver halide crystals and guarantees colloidal stability by adsorption. In addition to the thermo-reversible gelling, it acts as a binding, peptising, emulsifying, or stabilising agent.

Often it is modified and applied with other additives, for instance one or more surfactants [1 - 7]. Therefore the understanding of gelatin/surfactant interaction and knowledge of the dynamic interfacial properties of aqueous gelatin/surfactant systems are extremely important in development of new photographic products.

Coating is a process of replacing air contacting a substrate by another material. It is complex and contains aspects of colloid and surface sciences, wetting, spreading, adhesion, fluid mechanics, rheology and others. It comprises applications such as painting and protective films, processes in publishing industry, production of magnetic storage media, and of course photo material. A photographic film is a coated final product used to expose plates or flexible films and to prepare hard copy pictures or movies. A colour film consists of many layers. The most widely used methods for producing a multilayer film are the flow on a moving web slide and the curtain coating. In both methods a free fluid surface is formed by pushing the coating solution through a slot on an inclined plane. Surface active substances affect the free film surface. Phenomena controlling this process are diffusion, adsorption, aggregation, bulk and interfacial rheology, and the interaction of different components in the coating solution.

Up to now not all aspects of the coating process are completely understood. Studies by several authors [7 –20] show that the coating process with different techniques can be differently affected by one and the same surface-active material.

Gelatin is prepared under carefully controlled conditions in order to guarantee a product with desired photographic properties. Gelatin serves many useful purposes in the preparation of the silver emulsion: it acts as a protective colloid during the precipitation of the silver halides, is important in controlling the size of the silver halide grains, protects the halide grains in the developer so that their reduction to metallic silver is directly proportional to the amount of exposed light.

The added lubricants, wetting and, antistatic agents have to be chosen carefully to exclude altering of the original properties of the silver halide microcrystals. On the contrary, the additives have to ensure uniform and faultless film formation on various backings such as photographic films or paper.

The oldest and best known photographic wetting agent is saponin, a natural product which despite variations in the quality of the starting material (quillaia bark) has not yet been completely displaced. The advantage of saponin is its good wetting property and photographic neutrality. In addition synthetic surfactants are in use, such as anionics containing sulphate, sulphonate, or carboxylic groups (e.g. fatty acid sulphates, alkylarylsulphonates, sulphosuccinic acid derivatives), nonionics with alkyl or alkylaryl groups (often with several alkyl chains containing a total of 8-18 carbon atoms), and ester-ether-amide or polyethylene-glycol groups (e.g. polyhydric alcohols, glycerol, sorbitol, and sugar). Perfluorinated surfactants such as Bayer FT 248 have been used successfully with modern multilayer pourers and backing speeds of more than 100 m/min.

Antistatic agents in photographic manufacturing are to prevent disadvantages associated with electrical charging, such as inadvertent exposure or attraction of dust. Readily hydratable ionic surfactants are used, such as sulphated alcohols or sulphosuccinic acid esters, betaines, amine oxides, or short-chain alkyl polyglycol ethers.

Lubricants ensure satisfactory film running in cameras and projectors. They consist of dispersions, for example, of stearic acid, short-chain alkyl polyglycol ethers, wax, polyethylene, or silicon oil, which are added as oil-in-water emulsions to the uppermost layer of the material. Since a number of antistatic agents and lubricants causes faults in the uniformity of the pouring, anionic wetting agents have to be added.

In addition, colour components that couple to the dyes during photographic development must also be dispersed because they are generally insoluble in gelatin solutions. Alkan sulphonates, condensed sulphonates, sulphated fatty acid amides, or alkylnaphthalene sulphonates are applied, with additions of saponin and nonionics. This demonstrates that photographic emulsions are very complex multicomponent systems and a control of their bulk and interfacial behaviour requires a deep fundamental knowledge.

2. GELATIN SURFACTANT INTERACTION IN AQUEOUS BULK PHASES

2.1 STRUCTURE OF GELATIN

Gelatin is a product obtained by partial hydrolysis of collagen derived from skin, white connective tissue, and bones of animals [21]. Contrary to popular belief, gelatin cannot be recovered from horns, hoofs, and other non-collagen-containing parts of animals.

Collagen is an insoluble fibrous protein that occurs in vertebrates. It is distinctive among proteins because it contains an unusually high level of two cyclic amino acids, hydroxyproline and proline. Collagen consists of three helical polypeptide chains wound around each other and connected by intermolecular cross-links. This structure acts as a rigid rod and is the major stress-bearing element for the vertebrate connective tissues.

Gelatin as recovered from collagen by hydrolysis, i.e. cleavage of the cross-linked structure into single chains, is soluble in hot water. It is a heterogeneous mixture of different components depending on the way of collagen decomposition (alkaline, acidic, or enzymatic). Usually it shows a wide distribution of molecular mass, which can contain several maxima. Many of the gelatin properties are determined by the structural particularities of the collagen. As they are essential for the understanding of gelatin surfactant interaction, these properties are reviewed here briefly.

In the amino acids balance of collagen 2/3 are hydrophobic, half of them are glycines while cystine is present only in small amounts (Fig. 1). Therefore disulphide binding is not characteristic for collagen [22]. The differences in the amino acid balance of collagen and gelatin are only small and depend on the origin [23]. Exceptions are collagens of fishes and invertebrates. For collagen fibrils a length of 285 nm was found by electron microscopy [24, 25]. The collagen molecule is extremely anisotropic because its diameter is only 1.4 nm, and has a molecular mass of about 300.000, in good agreement with results of birefrigent measurements [26].

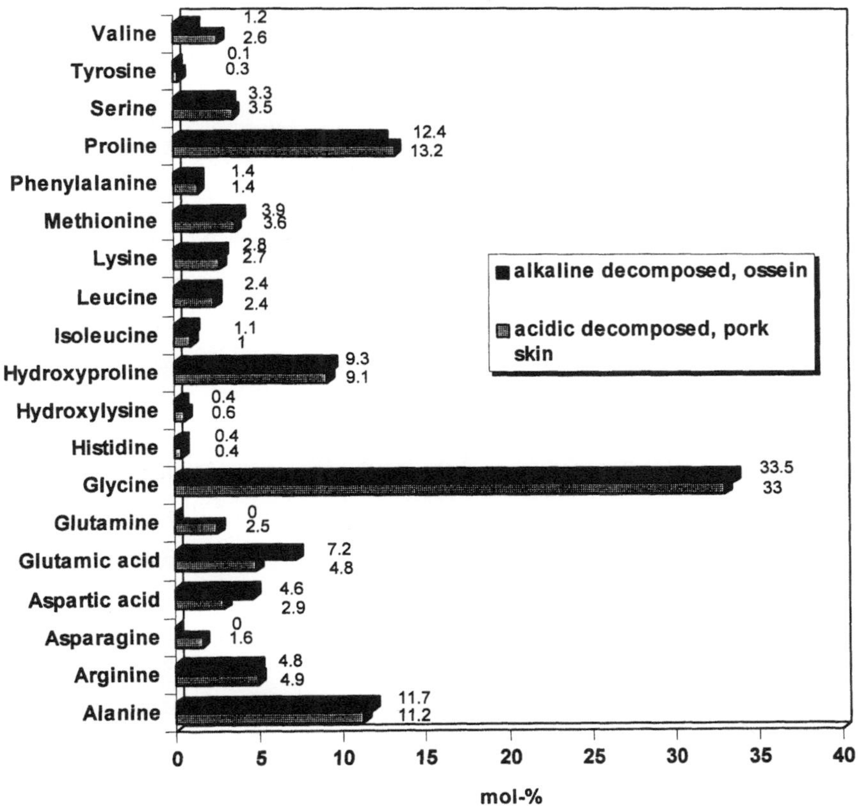

Fig. 1 Amino acid content of acidic and alkaline decomposed gelatin [43].

The generally excepted model for the collagen molecule, first suggested by Ramachandran and Kartha [27], is the triple helix consisting of 3 parallel strands (α_1, α_1, α_2) with a molecular mass of about 95 000 g/M for every strand. The amino acid sequence of the two α_1 strands are identical. The sequence of α_2 is yet unknown. The triple helix is stabilised by H-bonds between the NH-groups of the backbones of one chain and the C=O groups of a neighbouring chain and water as an intermediary in interchain and intrachain hydrogen bonding [28 – 33]. For gelatin associates of triple helix components (γ), associates of two strands (β), α-strands, and fragments were established [34]. Silver and Trelstad [35] determined the hydrodynamic radii for the α, β, and γ components as 13.8 nm, 21.5, and 25.7 nm.

Glycine (Gly) is found over a wide range of the amino acid sequence at every third position of α_1 [36]. Therefore collagen can be assumed as $(\text{Gly-X-Y-})_n$. Proline is found nearly without exception in the X-position. Hydroxyproline is frequently found in Y-position. Furthermore the amino acid sequences show that extensive apolar ranges alternate with polar ranges. For the polar ranges frequently COOH-groups are concentrated, basic groups, and also parts where acid and basic groups alternate. Basic groups of bovine and calf collagen are often terminally concentrated [37]. This discontinuous structure of collagen can be demonstrated by bright and dark crosswise stripes [38], which occur by addition of phosphor-tungsten acid or chrome(III) salts, which leads to a multivalent binding of polar collagen parts [39].

The native collage shows an isoelectric point (IEP) of 9 [40]. By hydrolysis of the amides of glutamic and aspartic acid the IEP is decreased due to the increasing amount of carboxylic groups. This happens by alkaline decomposition. Therefore the IEP of alkaline decomposed gelatin is in the range of 4.7-5.2 [41]. In contrast the hydrolysis is not complete in acidic decomposition so that an IEP in the range of 7.5-9.3 results [42].

Beside collagen the precursor of gelatin contains different parts of connective tissue, acidic mukopolysaccharides, DNA. These components are mostly excluded in gelatin production. Small parts however remain in the gelatin and cause specific photographic properties [44]. The role of these components is understood only in few cases.

The alkaline decomposition is more time-consuming, however, the specific active components are the reason that it is indispensable in photo industry and acidic decomposed gelatin is only used for covering layers as additive to alkaline decomposed gelatin, although mixed alkaline and acid decomposed gelatin may form associates, which is often undesirable. In the following mainly results of alkaline decomposed gelatin are presented, however, most of them can be qualitatively extended to acid decomposed gelatin [42].

Gelatin denatures when heated in aqueous solutions above the melting point and forms statistic coils. By cooling the triple helix structure is reestablished [45 – 47] . The kind and completeness of refolding depends on the solvent, temperature, and concentration. It determines the properties of the solution, the viscosity, the rheology of the gel, the volume change and heat absorption [48]. The process is comparable to the refolding of soluble collagen [34], although recently Djabourov et al. [49] stated that despite the almost identical molecular compositions of collagen and gelatin, the gelation leads to different types of molecular assemblies.

In thin gelatin layers fibrillar structures are formed. The dimension of these fibrils may be even larger than the expected dimension of an isolated gelatin triple helix [50]. Helical rods and fibrils orient to a great extent parallel to the layer surface [51] and determine in turn the rheological properties of the gel. Qualitatively, the non-linear elastic properties can be explained by assuming that the gelatin chains are partially in a crystalline triple helix state (the cross-links) and partially in a random coil state (the network bonds). The more extensive the rigid cross-link regions are, the shorter and more stretched the network bonds become when an external deformation is applied [52].

Concerning experimental physico-chemical data of gelatin it should be taken into account that many of them relate to a stationary metastable equilibrium and depend therefore on the prehistory and the pre-treatment of the system. Very strongly controlled conditions have to be kept when gelatin systems are prepared.

On the other hand it is surprising that some data in literature on mixed gelatin/surfactants systems are repeatable even for classical experiments carried out nearly 50 years ago [53 - 56]. Obviously some of the properties are not dominantly influenced by the gelatin structure, the molecular distribution etc. but are rather determined by the amino acid sequence, the amount and the kind of ionic groups, and the accessibility of these groups, which seems to be comparable for different gelatin blends. This is a hint that for gelatin/surfactant systems the ionic interaction predominates and that the hydrophobic interaction is more influenced by the primary structure of the gelatin strands. Fortunately many data are universally valid and reliable in the same way as those for other proteins.

In respect to the interfacial behaviour, however, it should be taken into account that gelatin is a mixture of components. These components differ in the molecular weight and thus in their surface activity. As a result of gelatin/surfactant interaction such differences may even become still more pronounced, when different components bind different amount of surfactant. Although such differences may not influence the general binding balance, they can be dominant as even small amounts of a component can strongly influence interfacial properties. Nevertheless, most papers deal with gelatin as a simple protein. In the following some points connected with the molecular distribution are considered.

The reactivity of gelatin to surfactants is mainly determined by the functional groups of the side chains. Because of the high average molecular weight the amount of terminal groups and their influence is small except for the formation of networks [44]. Usually protonation of the nitrogen atom in the peptide binding is neglected when discussing the reactivity [57, 58].

The amount of alkaline and acidic groups can be determined by titration or chromatography [36, 39, 59]. Derivation of the titration curve leads to a distribution of the amount of groups with equal dissociation constants over the pH (Fig. 2). An interpretation of these curve is possible by following the criterion of Cannan [59]. The maximum at pH 4 corresponds to the dissociated acidic groups. The small shoulder (pH 6-7.5) shows the relatively small amount of imidazolyl groups. The maximum at pH 8-11.5 corresponds to the dissociated ε-amino groups, but also to dissociated OH-groups at high alkaline pH [44]. The guanidyl groups cannot be separately titrated but determined from the total amount of basic groups. The totally amount of negatively charged polar groups of gelatin is about 0.122-0.131 M/100g, this of positively charged 0.096-0.103 M/100g [61-64].

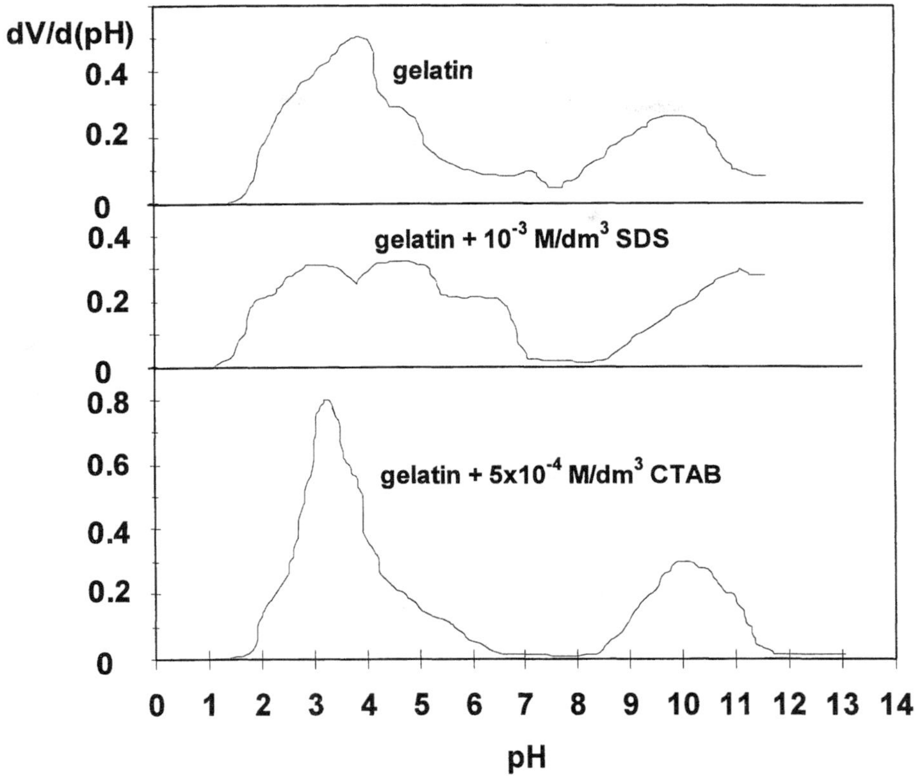

Fig. 2 Derived titration curve of an alkaline decomposed bone gelatin, V is the volume used up for titration [60].

2.2 *TYPES OF INTERACTION*

Gelatin is known to form complexes with some surfactants, in particular with anionics. In the photographic industry the formation of gelatin/surfactant complexes is particularly relevant in different applications since surfactants are commonly to gelatin to promote emulsification and to control surface tension during coating operations. Therefore different techniques have been used to study complex formation, including precipitate formation [65]: determination of binding isotherms by titration [66], bulk rheology [66 - 70], surface tensions [71 - 74], interfacial rheology [16, 18, 20, 75 - 85], fluorescence quenching [86], equilibrium dialysis [87–89], film thickness measurements [90, 91], ion-selective electrode [88, 89, 92-100], small-angle neutron scattering [101, 102], ^{13}C NMR spectroscopy [103], and pulsed-gradient spin-echo NMR spectroscopy [104].

The interaction between non-ionic polymers and anionic surfactants has theoretically and experimentally been established and give some crude ideas to understand the peculiarities of interaction. Lange [105, 106] and others [107-109] observed that above a critical surfactant concentration, known as critical aggregation concentration (CAC), some ionic surfactants can bind co-operatively to non-ionic polymers. Analogous results were reported also for some surfactants and proteins [110-115]. For co-operative transformation the steps of interaction are not independent. After the formation of a certain „germ“ the following steps are relieved. [116-118]. The CAC is usually much smaller than the critical micelle concentration (CMC) which signals the onset of micelle formation in the corresponding polymer-free surfactant solution. In dilute solutions, the polymer/surfactant complex is assumed to be composed by spherical micelles with their surfaces covered by polymer segments and connected by strands of the polymer molecule resembling a necklace of beads (so-called necklace model) [119].

The structure of complexes between SDS and polymers in general has been studied using small-angle neutron scattering [120-123] and NMR studies [124, 125]. These results also provide ample evidence that the water-soluble polymer strands do not penetrate into the hydrophobic micellar core, but interact with the micelle surface and remain in close contact to the surfactant hydrophilic headgroups. The structure of these polymer/surfactant complexes can again be seen as a polymer necklace decorated with micelles.

Nikas and Blanckschtein [126] proposed a theoretical description of the complexation of non-ionic polymers and surfactants in dilute aqueous solutions. The theoretical approach involves a thermodynamic description of polymer/surfactant solution and a molecular model of polymer/surfactant complexation. This model can predict the number of micelles bound per polymer chain, the aggregation number of these micelles, the average distance between polymer-bound micelles, and the mean square and end-to-end distance of the complex.

In the case of biopolymers the situation is more complex [127]. For proteins additional interactions are found, arising from the properties of the amino acids' side groups which may be hydrophobic or hydrophilic and/or ionic depending on pH. The net charge of a protein is zero at its isoelectric point (IEP), and positively or negatively charged at lower or higher pH respectively. Because of the ionic groups the interaction between ionic surfactants and proteins is different to that of uncharged polymers. The interaction between a biopolymer and an ionic surfactant starts at concentrations lower than the CMC, and lower than the CAC of the surfactant in a polymer solution.

Punkhurst et.al.[53-55], Tamaki and Tamamushi [56], Knox and Wright [65] were the first characterising the interaction between gelatin and different surfactants. Kragh [71] studied the effect of gelatin on the dynamic and static surface tension of Aerosol OT solutions (sodium di(isooctyl) succin 1 sulphonate) using the dynamic bubble-pressure method. Knox and Parshall studied the interaction with sodium dodecyl sulphate (SDS) above and below the IEP [68] and Aerosol OT [69] also by surface tension measurements. Interestingly, they found no interaction with the non-ionic Triton X-100. The complexes of SDS or Aerosol OT with gelatin appeared to be more surface-active than the surfactants alone.

Many attempts had been done to clarify the structure of gelatin/surfactant complexes. Reynolds and Tanford [116] assumed a rode-like structure for the gelatin/surfactant complex, where its length is proportional to the molecular mass. This could explain the electrophoretic mobility of the complexes in polyacrylamid gels, which also depends on the molecular mass. Wright et al. [128] and Rowe et al.[129] concluded from birefrigence measurements that the complexes are deformable rode-like ellipsoids, as proposed by Collins and Haller [130]. Shirahama [131] assumed from electrophoreses measurements a chain-like model. Tokiwa [132] investigated the activity of Na^+ ions in polyvinylpyrolidon/SDS solutions. The interaction decreases the activity of free Na^+ and it was concluded that the complexes behave similar to micelles. These results are supported by results of other authors [133, 134].

A deeper insight into the structure of gelatin/SDS was given by Whitesides and Miller [86] using a fluorescence quenching technique. They detect the onset of the formation of hydrophobic aggregates using the fluorescent probe anilinonaphthalene sulphonate [135, 136]. Free surfactant micelles, which occur at sufficiently high SDS concentration, were shown to be similar in size and shape to those formed in absence of gelatin.

^{13}C-NMR spectroscopy and the pulsed gradient spin-echo NMR (PGSE-NMR) technique have been used [137] to further probe into the structure and mobility of gelatin/SDS complexes. Many of the carboxylic acid resonances of the gelatin amino acids (anionic residues such as aspartic and glutamic acid or polar uncharged residues such as hydroxyproline) were unaffected by the surfactant. There was however a significant broadening of the cationic residues (arginine and lysine) in the presence of SDS. Also the hydrophobic residues leucine, isoleucine, valine, and methionine are broadened so much that they disappear into the baseline noise of the spectrum. This demonstrates that in turn of surfactant binding the conformation of the gelatin strands is changed and only the first carbon atom of the SDS (those closest to the sulphate headgroup) is affected by the presence of gelatin [138]. The translational motion of gelatin/SDS complexes was also studied by PGSE-NMR [137]. Combining self-diffusion coefficients for all proton-bearing species and the fluorescence quenching results of Whitesides et al. [86] a picture was proposed in which a gelatin chain wraps itself around the outer surface of surfactant micelles. The binding mechanism is ultimately limited by the repulsion between adjacent micelles. In the presence of gelatin the diffusion of larger non-ionic surfactant micelles was faster than that of SDS. Miller et al. [137] concluded that the retarded motion of SDS in gelatin is due to micellar complexation to the biopolymer chains. The SDS diffuses with gelatin as a unit. The effect of micellar dynamics in increasing the rate of diffusion (via complex formation or dissolution) cannot be excluded.

Gelatin/SDS gels at 25°C have been studied by Cosgrove et al. [101] using small-angle neutron scattering. Strong interaction with SDS changes the structure of gelatin dramatically. At a surfactant concentration just above the CAC, the SDS adsorbing onto the gelatin strands disrupts the gelatin network structure. The correlation length of the network structure was found to decrease with increasing SDS concentration. Adsorbed micelles at the gelatin strands act as bridges between the cationic and hydrophobic sites inducing the gelatin network. The SDS micelles adsorbed onto the gelatin strands were found to be slightly larger than those formed in absence of gelatin.

Furthermore the effects of temperature and pH on gelatin/SDS interaction in gels was studied [102]. Extreme pH on both sides of the IEP strongly disturb the gelatin structure, and the surfactant binding is reduced but for rather different reasons. Although the physical properties of the system change dramatically with temperature, from a gel at 25°C to a fluid at 65°C, the effects on the structure are rather weak over the dimensions probed by small-angle neutron scattering. The mobility of the gelatin network does not play an important role in the adsorption process and the dynamic exchange of monomers between the bulk, adsorbed micelles and free micelles is strongly influenced by temperature. Using PGSE-NMR Griffiths et al. [139] investigated the individual dynamic environments of gelatin and surfactant over a wider range of concentration. Adding SDS slows the diffusion of the gelatin and passes through a minimum with increasing surfactant concentration. The diffusion data are compared to light scattering, small-angle neutron scattering and viscosity data from similar systems. The diffusion data show that SDS induces finite size clustering of the gelatin strands. At higher concentrations the SDS diffusion is dominated by the presence of SDS micelles. An binding isotherm of SDS to gelatin has been derived from the self-diffusion data. Furthermore the effect of alkyl chain length (C_8 – C_{14} sulphate) on the diffusion behaviour of gelatin and surfactant was studied by PGSE-NMR spectroscopy [140]. Changes in the surfactant diffusivity can be rationalised in terms of a two-state model consisting of gelatin-bound micelles in equilibrium with freely diffusing surfactant monomers. A minimum in the gelatin diffusivity is observed when the binding of surfactants amounts to about 1 micelle/strand. The depth of this minimum increases with the surfactant's chain length. These effects are explained in terms of micelle-mediated transient cross-linking as proposed by Greener et al. [68]. The effective strength of the cross-links is a decreasing function of the number of micelles/strand because of the electrostatic repulsion between the micelles. The strength increases with the increase in micelle size.

The interaction between surfactants and different kinds of gelatin have been compared very rarely [141, 142]. The main results of such studies can be summarised as follows. The binding between gelatin and ionic surfactants is effected by the affinity between the hydrocarbon chains of the surfactant and the non-polar ranges of the gelatin. Below a certain surfactant concentration there is no interaction (co-operative binding [65]). At first, the surfactant interaction causes conformational changes favouring electrostatic interaction with further surfactant molecules [114]. The second step of interaction is ion-ion binding, leading to a neutralisation of polar groups. The solubility of these complexes depend on pH, the electrolyte concentration and the gelatin/surfactant ratio, i.e. the degree of charge neutralisation. The charge neutralisation causes the hydrophobic properties of the complexes which even may precipitate. A third step is hydrophobic interaction. This leads to a solubilisation of the hydrophobic complex. Precipitated complexes become soluble again. As the hydrophilic and hydrophobic sections of the gelatin strands are separated the resulting complex is often compared with surfactant micelles bound to a gelatin strand (necklace model). At sufficiently high surfactant concentration free surfactant micelles are formed. All steps of interaction causes more or less pronounced changes of the gelatin structure.

Several critical concentrations can be formulated. There is a critical gelatin/surfactant aggregation concentration CAC which is usually lower than the CMC of the individual surfactant indicating the formation of free surfactant micelles. Due to the surfactant amount bound to gelatin this concentration however is higher than the CMC of the same surfactant in absence of gelatin.

The ability of surfactants to interact with gelatin is different. Usually anionic surfactant interact stronger than cationics [143, 144], and these stronger than non-ionics [69]. Makino [145] assumes that the interaction between non-ionic surfactants and gelatin is not only of hydrophobic nature but involves polar groups of the gelatin too. The ability of a surfactant to interact with gelatin depends on how this surfactant is able to unfold the gelatin in the first step of interaction. The second step of interaction is limited by the accessibility of the polar groups, i.e. the kind of the surfactant polar head group and the accessibility of the oppositely charged groups of the gelatin.

Addition of electrolyte narrows the range of complex precipitation [65]. Such effects can also increase protein/surfactant interaction [96]. Nevertheless there are studies in which no increase of interaction in the range of low surfactant concentrations was found [146, 147].

2.3 SURFACTANT INFLUENCE ON THE PROTON-ACCEPTOR-DONATOR RATIO

The addition of ionic surfactant to a gelatin solution changes the pH until above a certain surfactant concentration it remains constant. This is attributed to the complex formation [148], which causes a proton uptake for an anionic, or a release for a cationic surfactant. Fig. 3 shows the change in pH caused by the addition of two surfactants to a gelatin solution. The addition of SDS does not further change the pH above a concentration of 10^{-2} M/dm³ [60].

The pH remains constant when the addition of surfactant neither increase the surfactant amount bound to the gelatin, nor does the conformation of gelatin change, which in turn may cause an increase of the polar groups accessibility. The CMC of SDS is $8 \cdot 10^{-3}$ M/dm³. At this concentration the formation of complexes is finished and all positively charged gelatin groups are occupied by SD^- ions [65].

In contrast the cationic surfactant CTAB decreases the pH of a gelatin solution. A constant pH is obtained above a concentration of $(5\text{-}7) \cdot 10^{-3}$ M/dm³. This concentration exceeds the CMC of CTAB at about $1.5 \cdot 10^{-3}$ M/dm³.

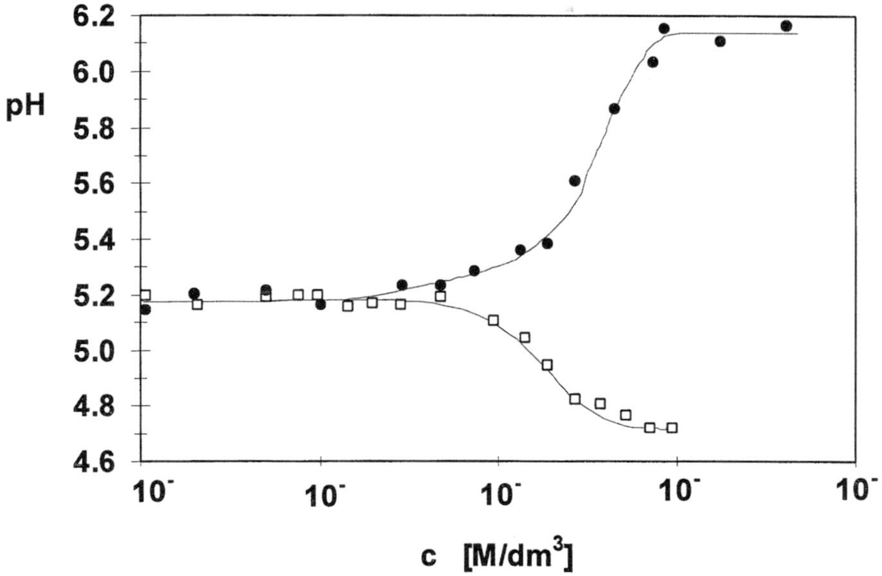

Fig. 3 Change of pH of a 0.2% solution of an alkaline decomposed gelatin caused by addition of SDS (•) and CTAB (□) [60].

The actual surfactant amount bound by gelatin can be determined by gelatin back-titration [66], and determining the actual concentration of the free surfactant molecules in solution [65]. The amount of acid or alkaline for back-titration depends on the gelatin concentration. It was established for instance, that the amount required for a 0.2% gelatin solution is comparably smaller than for a 0.1% solution [117, 149]. This may be explained by the more unfolded structure of gelatin in the more diluted solution [150].

A vary clear picture may be obtained by deriving the titration curves (Fig. 2). Using the criterion of Cannan [59] the pK values of the polar groups can be identified. The amount of α-amino and imidazolyl groups is small (pK ≈ 6-8.5). Therefore there is no change of dV/d [pH] within the pK range of these groups. One can clearly see that SDS binding shifts the maximum of the basic groups to higher pH still at a concentration much lower than the CMC. Some of the basic groups in the range of pH 8 to 11 obviously are shielded. Therefore the amount needed to titrate the groups in this range is remarkably lower. The influence of SDS on the dissociation of carboxyl groups is also strongly pronounced, although the surfactant ions do not bound to them (pK ≈ 1.5-6). This result clearly shows that the anionic surfactant obviously changes the gelatin structure, thus increasing the accessibility of the polar groups.

The addition of CTAB causes a small shift of the spectrum to lower pH. The maximum of the carboxyl groups is strongly increased. This is rather caused by a gelatin unfolding than by a strong surfactant binding, because a strong binding would decrease the titration volume. The dV/d [pH] values in the range of pH 4 to 6 however are only slightly decreased. Analogous curves for different ionic and non-ionic surfactants are given by Izmailova et al. [151].

2.4 INFLUENCE OF SURFACTANT ON THE TRIPLE HELIX STRUCTURE OF GELATIN IN SOLUTION

Gelatin is known to denature when heated in aqueous solution. The triple helix structure is supposed to change into random coil [45-47]. When cooled off gelatin restores some of the collagen-fold-structure again [152]. The refolding extend depends on solvent, temperature, and concentration [45, 48]. CD technique can characterise such structure transfer as the triple helical content of collagen and gelatin solutions is related to a CD peak at 222 nm [153-155]. The amplitude depends on the raw material, the way of collagen decomposition, the preparation of the solution, and the solvent itself [48]. The conformation of a single chain in the triple stranded structure corresponds to the ploy proline II helix [156, 157] with a similar spectrum [156, 158-162].

As only residues in trans configuration can be integrated into the triple helix, the refolding stops when in the peptide chain a proline and hydroxyproline segment are in cis configuration. The cis-trans isomerism becomes rate determining. According to Wetzel et al. [163] a CD minimum at λ>230 nm can be correlated to the cis configuration of the peptide bonds. Negative CD values at 222 nm are obtained after triple helix destruction.

The influence of surfactants on the secondary structure of gelatin was also characterised by CD measurements using the criteria mentioned above. The maximum triple helix content is usually found at the IEP which is slightly decreased to pH 7. Figs. 4 and 5 show some representative

CD spectra found for gelatin surfactant mixtures [152, 164, 165]. These spectra are similar for gelatin samples of different origin and in contact with other surfactants [83].

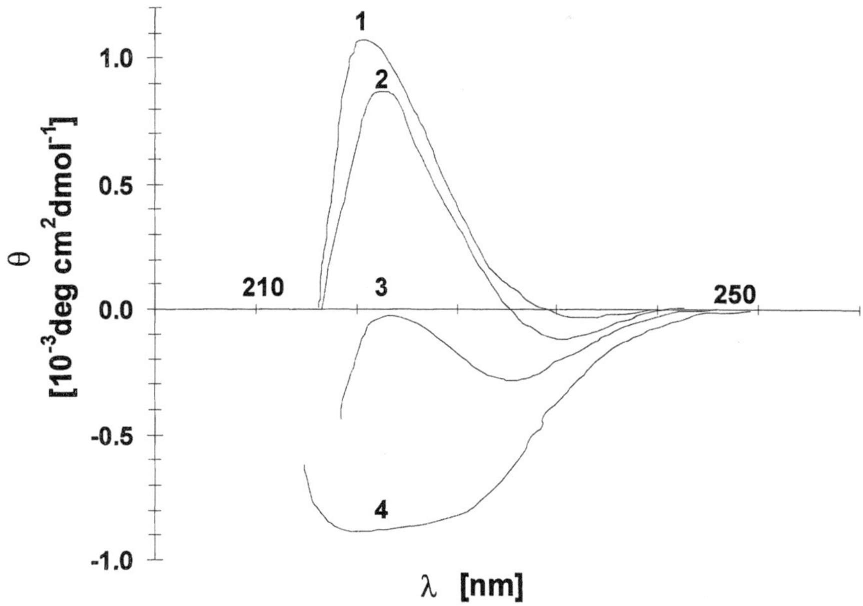

Fig. 4 CD spectra of 0.2% gelatin/SDS solutions at pH 7 and 283 K. (1) 10^{-3} M/dm^3 SDS, (2) $5 \cdot 10^{-3}$ M/dm^3 SDS, (3) 10^{-2} M/dm^3 SDS, (4) $2 \cdot 10^{-2}$ M/dm^3 SDS [152].

All effects are temperature and pH dependent. The refolding process was monitored by keeping the gelatin/surfactant mixtures at 298 K, where the triple helical content is smaller than at 283 K. Cooling down the solution from 298 K to 283 K allows to check the refolding ability.

SDS is only weakly affecting the triple helical structure of gelatin below a SDS concentration of 10^{-4} M/dm^3. Above this concentration however the refolding is remarkably influenced and at $2 \cdot 10^{-2}$ M/dm^3 the triple helical structure is destroyed and ß-sheets are formed with similar spectra. The concentrations were chosen in consideration also of surface aspects [100] which are discussed below.

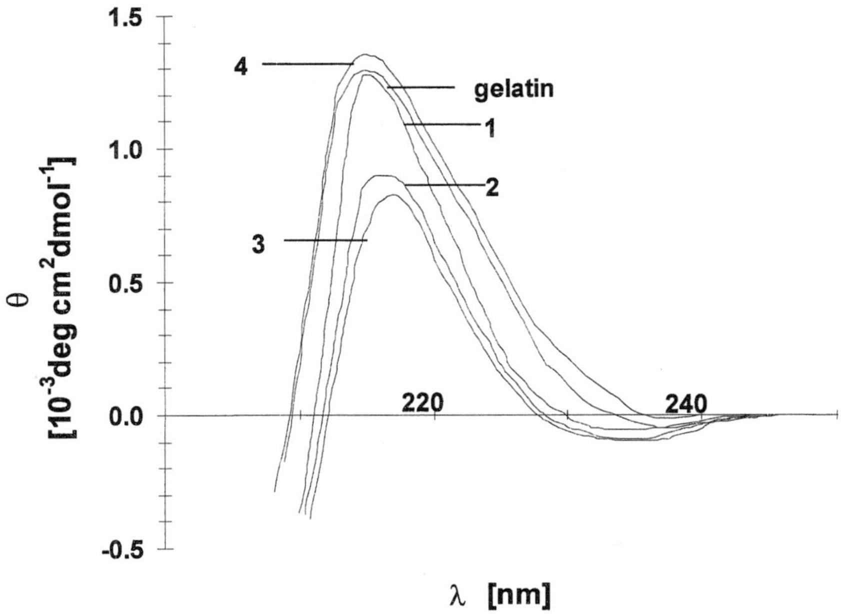

Fig. 5 CD spectra of a 0.2% gelatin solution at pH 7 and 283 K, and with addition of CTAB and the non-ionic surfactant ethoxylated dodecanol 5EO. (1) $5 \cdot 10^{-5}$ M/dm³ CTAB, (2) $5 \cdot 10^{-3}$ M/dm³ CTAB, (3) 10^{-2} M/dm³ CTAB, (4) 10^{-4} M/dm³ ethoxylated dodecanol 5EO [164, 165].

At high SDS concentrations these conformational changes were shown to be thermally irreversible. The rotations of the C_α-C- and N-C_α- bonds are restricted because of the partial double bond character [166]. The rate determining step in unfolding and refolding of proteins is the isomerisation around the gly-pro bonds (glycine, proline) [167] which can only rotate around the C_α-C-bonds because of the pyrrolidine rings. In acidic environment protonation of proline facilitates the rotation around the peptide bonds, which promotes the cis-trans isomerisation [57, 168], i.e. the unfolding. Unfolding increases the rotation of the single strands, which in turn improves the accessibility of the polar side chains for anionic surfactants at the IEP. Therefore the strong interaction at high surfactant concentrations in acidic environment causes a stronger influence of the triple helical structure. In alkaline bulk phases a more rigid conformation of the gelatin strands has to be expected. This restricts the influence of anionic surfactants on the triple helical structure of gelatin at pH 4.9 and 7.

In contrast the influence of CTAB is weak even at pH values, which are optimal for a gelatin/cationic interaction. The pK values of the carboxyl groups of the gelatin, which are expected to interact electrostatically with the cationic surfactant, are in the range of 3.0 to 4.0 [44]. There are two aspects, which have to be considered:

the interaction of the cationic may be enhanced by the partial positive charged aminogroups being in a short distance from the carboxyl groups;

the rigid conformation of the gelatin strands in alkaline environment has to be considered [169] which restricts the accessibility of the gelatin side chains. At pH 10 histidine, the N-terminal groups, and the NH_2 groups of lysine and hydroxylysine carry positive charges.

For different non-ionic surfactants the triple helix content of gelatin increases at small surfactant concentrations [165]. The reason for this increase, however, is not clear up to now. Only at concentrations exceeding the CMC and at pH<IEP the triple helical content decreases (Fig. 5). The decrease of the triple helical structure is caused by dipol-dipol interaction and refolding at cooling off the solution is not restricted.

The use of the polymer structure is of practical interest to create a preferred orientation and to influence the properties of embedded material in thin gelatin films [170-174]. Studies on the tertiary structure of gelatin in thin cold-dried transparent films show that the influence of surfactants on the ordered gelatin structure turns up in a similar way. In opposite to ionic surfactants which lead to a reduction of the ordered structure, the addition of non-ionic and amphoteric surfactants causes an increase of the ordered structure. The influence is independent of the gelatin/surfactant ratio and can be explained by gelatin/gelatin interaction competing with the gelatin/surfactant interaction at sufficiently high gelatin concentration [175].

3. INTERFACIAL BEHAVIOUR OF GELATIN/SURFACTANT ADSORPTION LAYERS

3.1 DYNAMIC INTERFACIAL TENSION

Interfacial tensions and surface energies are important in coating operations. The main role of surface-active compounds in a liquid film coating process is to control the wetting behaviour of the coating solution. In order to wet the moving web, the interfacial tension of this solution has to be lower than the surface energy of the web. In multilayer coating, the interfacial tension of the top layer has to be lower than that of any other layer [176]. During the adsorption process surface-active molecules replace high-energy surface water molecules, which reduces the interfacial tension. The liquid film coating process is highly dynamic and involves generation and extension of surfaces at a high rate. Therefore dynamic surface tensions are relevant for fast coating processes. Further information about the concept of dynamic interfacial tensions and its relevance for practical applications can be found elsewhere [177-179].

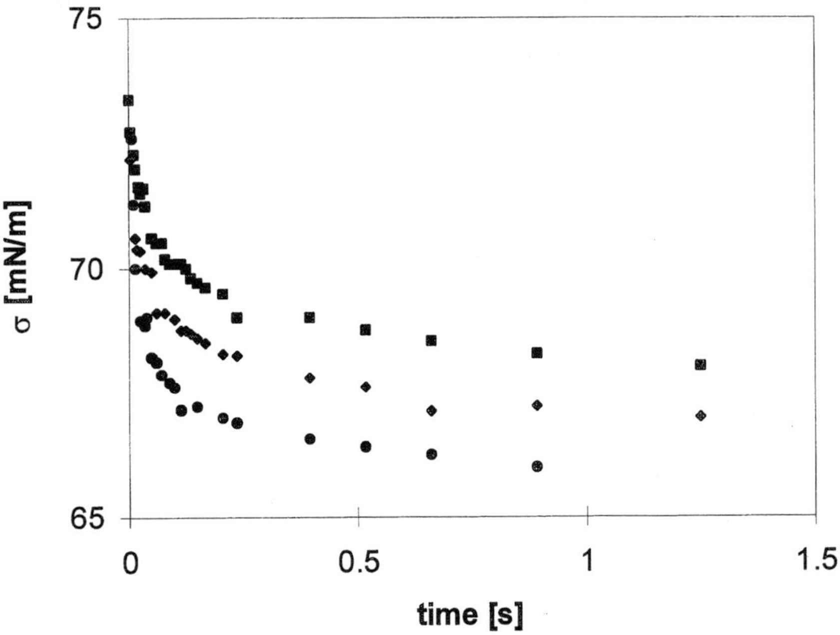

Fig. 6 Dynamic surface tension of an aqueous gelatin solution (10 g/l) at different temperatures: 25°C (■), 30°C (◆) and 35°C (●) [183].

There are only few methods to measure interfacial tensions in a time range below one second. The oscillation jet, the inclined plane, and the maximum bubble pressure method are most frequently used. Schreiter et al [180, 191, 192] applied the bubble pressure technique according to Schwen [181] to measure the dynamic interfacial tension of different gelatin, gelatin/surfactant, and gelatin/surfactant/dye coupling agent systems. They obtained correlations between the dynamic interfacial tension and a critical coating speed. The role of the dynamic interfacial tension in slide coating has been recently underlined by Valentini et al. [16, 20], Tricot [74], and Schunk and Scriven [182]. Miller et al. [183] determined the dynamic interfacial tension depending on temperature for aqueous gelatin solution using a maximum bubble pressure device (Fig. 6). The advantage of this method is the possibility to characterise the dynamic behaviour in the time regime that corresponds to the dynamics of the coating process, i.e. in the milliseconds range. Unfortunately no further systematic investigation are available from literature using this method to characterise coating solutions.

To predict the quality of a coating process by dynamic interfacial tension measurements is not trivial. The reality of coating involves many interaction parameters and their applicability has to be supported additionally by coating experiments. The examples for including dynamic surface tension into the optimisation of film coating are given below in paragraph 4.

3.2 STATIC INTERFACIAL TENSION

In contrast to gelatin/surfactant binding isotherms surface properties are remarkably influenced by the molecular weight distribution [100, 184, 185]. Fig. 7 shows the molecular weight distribution for three different alkaline decomposed gelatin samples while Fig. 8 gives the corresponding σ/log c isotherms for different mixtures of these gelatin samples with SDS and CTAB. The figures show remarkable changes of the surface tensions with increasing surfactant content. Even at small surfactant concentrations the surface behaviour cannot be explained as a sum of the surface tension depression of the individual components. In the shape of the σ/log c isotherms several kinds of plateau regions are recognised.

Usually the appearance of a plateau indicates the saturation of the surface, in the present case by a definite surface active gelatin/surfactant complex. The surfactant concentration at the beginning of a plateau is the CAC. As some curves in Figs. 8 and 9 show more than one plateau it should be concluded that more than one surface active complex can be formed, which in turn is able to saturate the surface. The curves, which do show more than one plateau also exhibit more than one peak in the molecular weight distribution (Fig. 7). Therefore these irregularities are possibly caused by the gelatin molecular weight distribution, i.e. different components with distinct surface activity. In the present system however we do not see only the consequence of the gelatin decomposition, but also the result of interaction with the surfactants, i.e. gelatin components differing in the molecular weight form different complexes. These complexes have different surface activity and CAC. Such effects are usually interpreted by multiple equilibria [117, 149].

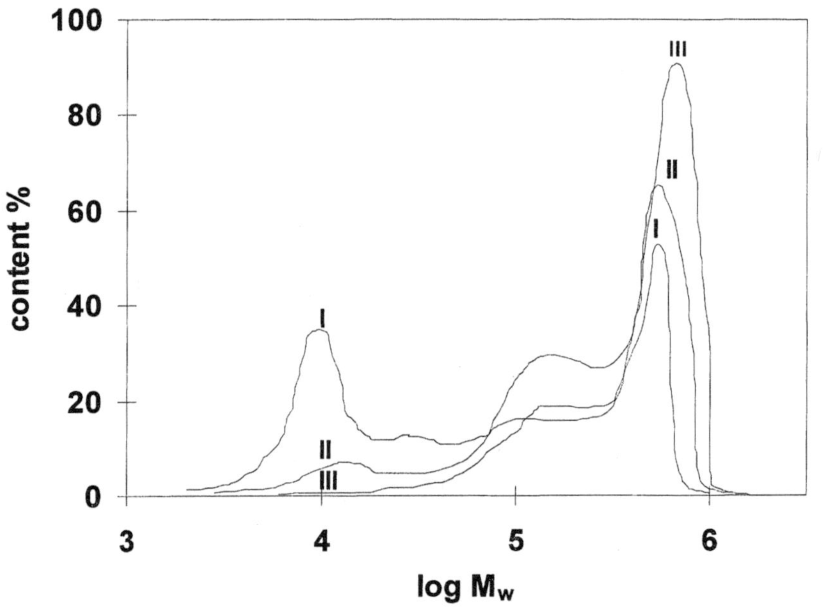

Fig. 7 Molecular weight distribution of different alkaline decomposed gelatin samples. I - skin, M_W= 207 000, M_n= 23 100, 0.1% ash content. II - bone, M_W= 343 700, M_n= 85 400, 1.6% ash content. III - bone, M_W= 511 700, M_n= 204 100, 0.08% ash content.

At a certain surfactant concentration (>CAC) the surface tension starts to decrease again. The composition of the complexes is changed, i.e. they loos their surface activity and are replaced by the higher surface-active surfactant molecules. Finally at a certain surfactant concentration σ levels off at a value that nearly coincides with the maximum surface tension depression of the surfactant alone. This concentration is the CMC of the present system. The difference between this CMC and that of the surfactant in absence of gelatin yield the surfactant amount bound by gelatin.

The isotherms found for the gelatin III having a small content of low molecular weight components is in good agreement with results usually reported in literature [73]. The maximum surfactant binding capacity of the different gelatin samples however seems to be comparable.

A general agreement between the appearance of CAC and a precipitation of gelatin/surfactant complexes, as claimed in [65, 72], could not be proved [184]. This is not surprising because the formation of a precipitate depends not only on the gelatin/surfactant ratio, but also on other conditions, for instance the presence of electrolyte.

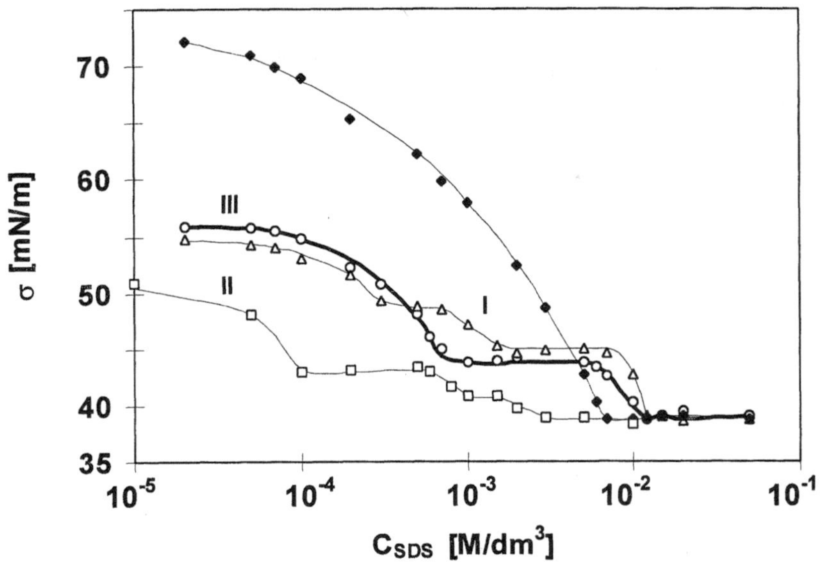

Fig. 8 σ/log C isotherms of SDS (◆) and systems of different 0.2% gelatin solutions + SDS, sample numbers of gelatin indicated in Fig. 7, pH 7, 293 K.

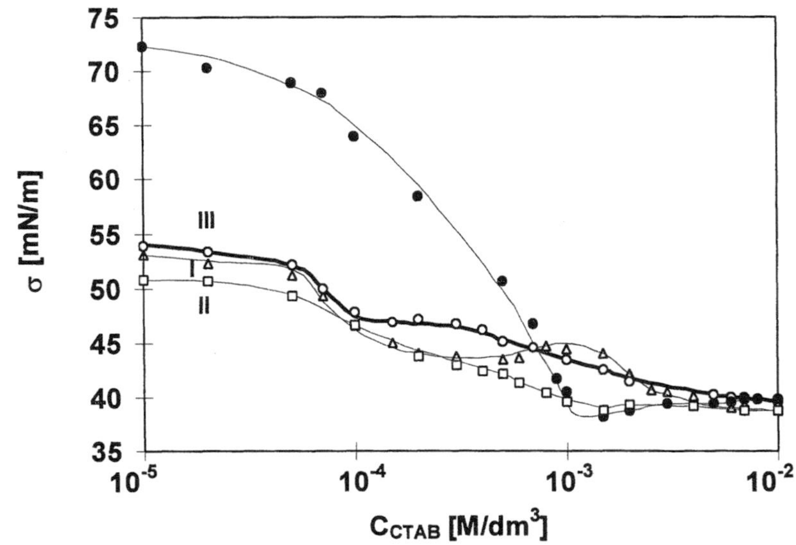

Fig. 9 σ/log C isotherms of CTAB (●) and systems of different 0.2% gelatin solutions + CTAB, sample numbers of gelatin indicated in Fig. 7, pH 7, 293 K.

It should be noted that in contrast to the system gelatin III/CTAB which did not show any precipitation, for the system gelatin I/CTAB a precipitation was observed in the range $(1\div2)\cdot10^{-3}$M/dm³. This slight precipitation is in the range of the CMC of CTAB. Obviously it is caused by the low molecular weight components. A precipitation of gelatin II/CTAB complexes appears in the range $(1.5\div3)\cdot10^{-4}$M/dm³. Gelatin II contains a high ash content. It is well known that a high inorganic content (electrolyte) shifts the precipitation region to lower surfactant concentrations [73, 186], an effect which can be interpreted in the same way as for the system gelatin II/SDS. This system reaches the maximum surface tension depression at lower surfactant concentrations than the other gelatin/SDS mixtures.

Analogous results were also obtained at liquid/liquid interfaces [187]. The same principal behaviour is also found for other ionic surfactants and even for some hydrophilic dye-coupling agents, which are amphiphilic. Pitt et al. [188] reported σ/log c isotherms for gelatin mixtures with a large number of different surfactants. The principal picture obviously holds for all these systems. Fig. 10 shows a peculiarity of a system gelatin/hydrophilic colour coupling agent. The coupler contains 3 hydrophilic COO^-Na^+ groups and an aliphatic residue $C_{17}H_{35}$ [189].

In this special case the maximum surface tension depression in the plateau region is lower than the maximum depression of the individual dye-coupling agent which indicates the formation of a very high surface-active complex saturating the surface. In subsequent steps of interaction however all these complexes are completely solubilised by further interacting coupler molecules. The resulting complexes are more hydrophilic, i.e. less surface active, and are displaced by competitive surfactant molecules. This is the same principle usually found for ionic surfactants. Analogous results are reported by Meguro et al. [190] for other protein/surfactant systems.

The behaviour of gelatin/non-ionic mixtures is different. Because of a very low interaction usually no plateau in the σ/log c isotherm can be detected. Fig. 11 shows the σ/log c isotherm of an ethoxylated surfactant. Obviously there is no difference between the CMC of the surfactant solution with and without gelatin. The surface tension depression of the gelatin/non-ionic system can be easily explained by a superposition of the depression of the individual components. Analogous results can be provoked in gelatin/ionic surfactant systems by deteriorated conditions of interaction; for instance there is no plateau in gelatin/CTAB systems at pH 1.5.

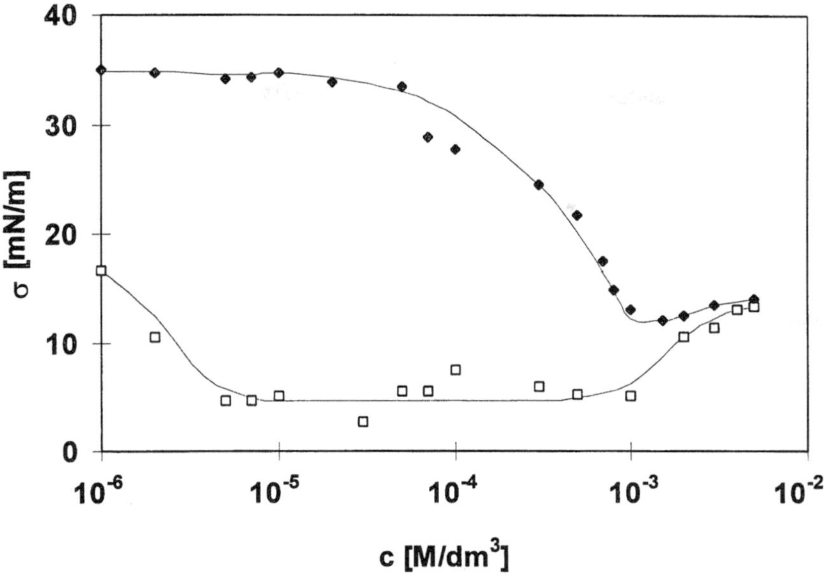

Fig.10 Surface tension isotherm of a dye coupling agent containing 3 hydrophilic COO^-Na^+ groups and an aliphatic residue $C_{17}H_{35}$ in water (●) and in presence of 0.2% gelatin (□) [190].

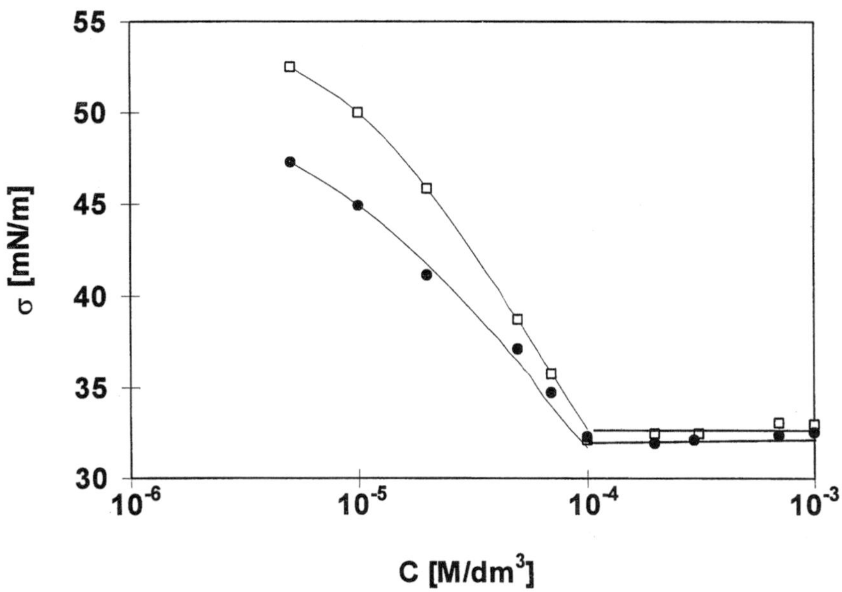

Fig. 11 Surface tension isotherm of an ethoylated octylpheol containing 10.5 EO units and a mixture with 0.2% gelatin, ethoylated octylpheol (□), 0.2%gelatin+ethoylated octylpheol (●), pH 7, 298 K [83].

The principal behaviour described is true for a large range of gelatin concentration and temperature [191, 192], whereas the surfactant amount bound by gelatin seems to decreases at gelatin concentration > 1g/100 ml.

3.3 INTERFACIAL SHEAR RHEOLOGY

Interfacial rheological properties are often used to quantify the interaction at surfaces, because these parameters are very sensitive to gelatin/surfactant interactions. For technical processes, for instance in film coating, different films flow on top of one another these rheological properties are of key interest.

For interfaces, in contrast to bulk phases, we be strongly distinguish between dilational and shear properties [193]. In Chapter 9 of this book Benjamin and Lucassen-Reynders discuss the concept of interfacial rheology in more detail.

Vollhardt and Kretzschmar [76] studied the shear rheological behaviour of aqueous gelatin and mixed gelatin/surfactant solutions at the air/water interface using a rotation pendulum. They observed that the surface age, the pH-value of the bulk phase and the gelatin type have a decisive influence on the surface shear viscosity, while a direct relationship to the bulk viscosity does not exist. In presence of surfactants the surface viscosity was found to decrease strongly in all systems. More detailed investigation however show that gelatin/surfactant complexes may even produce enormously mechanically stable adsorption layers [77, 194-196].

The σ/log c isotherms can be correlated with the surface rheological behaviour [185, 186, 197]. From creep compliance measurements it was obtained that the surface shear elasticity and viscosity exhibit maxima at surfactant concentrations of the plateau region (Figs. 12 and 13). To describe the creep compliance a Schoffield-Scott-Blair model was used in [79, 80, 100]. For simplicity the following figures show only the spontaneous modulus G^s and the Newtonian viscosity η^s, which describe the elasticity in the beginning and the viscosity in the stationary state of creep compliance, and the compliance limit f^s. Below f^s no creep can be forced.

The surface shear rheological parameters are very sensitive and can change by some orders of magnitude. The figures show only the behaviour of the systems containing the gelatin III

(Fig. 7). The differences to the systems containing other gelatin samples are only qualitative. Irregularities were not observed, for instance not more than one maximum was observed. For the results presented below the adsorption layer age was 2 h.

The results of Figs. 8 and 9 show that the σ/log c plateaus agree with the region where mechanically very stabile gelatin networks at the interface were formed with pronounced viscoelastic properties. These layers even show a compliance limit. Contrary the surface shear rheological behaviour of a surfactant solution is usually Newtonian and η^s is by some orders of magnitude lower. The surface shear rheology of gelatin without surfactant is viscoelastic, but again the parameters are much lower than for the mixtures when gelatin/surfactant complexes are adsorbed.

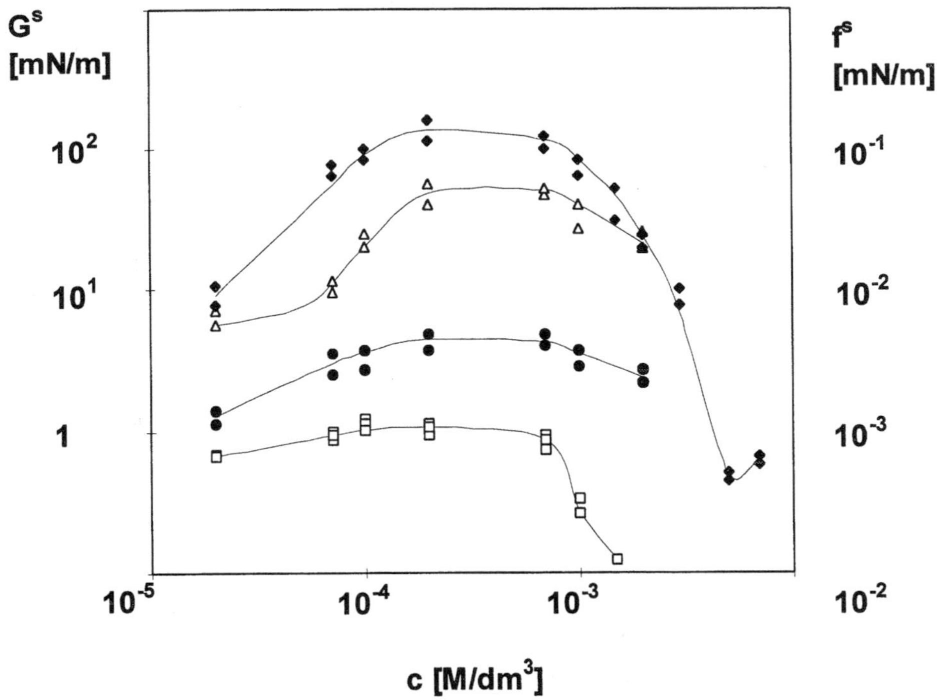

Fig. 12 Modulus of spontaneous surface shear elasticity G^s (□◆, Δ), and the compliance limit f^s (□, ●) of a 0.2% gelatin solution (gelatin sample III) + SDS (Δ, □), and CTAB (◆, ●), pH=7, 293 K.

For high surfactant concentrations the surface shear rheological parameters dramatically decrease and η^s almost agrees with values found for surfactant solutions without gelatin. Also all the parameters for gelatin/surfactant mixtures exhibit a strongly pronounced time dependence. Even at very high surfactant concentrations the time dependence of η^s remarkable differs from that of a pure surfactant. This, however, is not surprising when taking into account the multitude of ingredients present in gelatin samples.

Using a wide spectrum of surface rheological equipment and other surfactants the results reported above have been confirmed [151, 195-198].

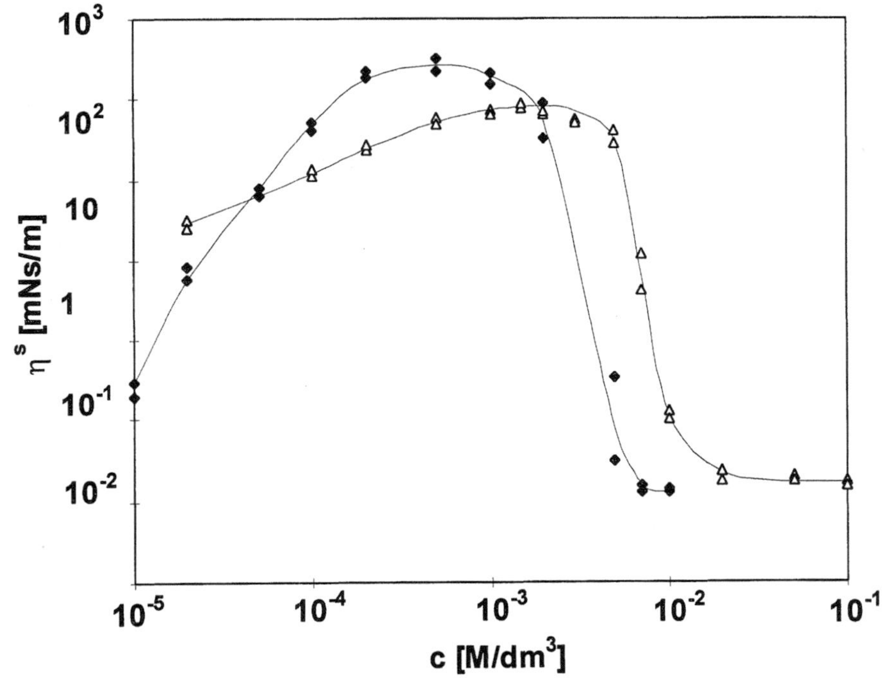

Fig. 13 Newtonian surface shear viscosity η^s of a 0.2% gelatin solution (gelatin sample III) + SDS (Δ), and CTAB (◆), pH=7, 293 K.

The surface shear rheological parameters in Fig. 14 show the influence of a non-ionic surfactant (Fig. 11). There is only a more or less pronounced transition between the behaviour of the individual components without the formation of a mechanically stable surface structure.

Nevertheless even for non-ionic surfactant concentrations exceeding the CMC the surface rheological parameters are still orders of magnitude higher than those characteristic for the pure surfactant, although the σ/log c isotherms could be interpreted by assuming a total displacement of gelatin by surfactant molecules. Furthermore this result strongly supports that gelatin/ionic surfactant complexes are responsible for the formation of viscoelastic and mechanically highly stable adsorption layers.

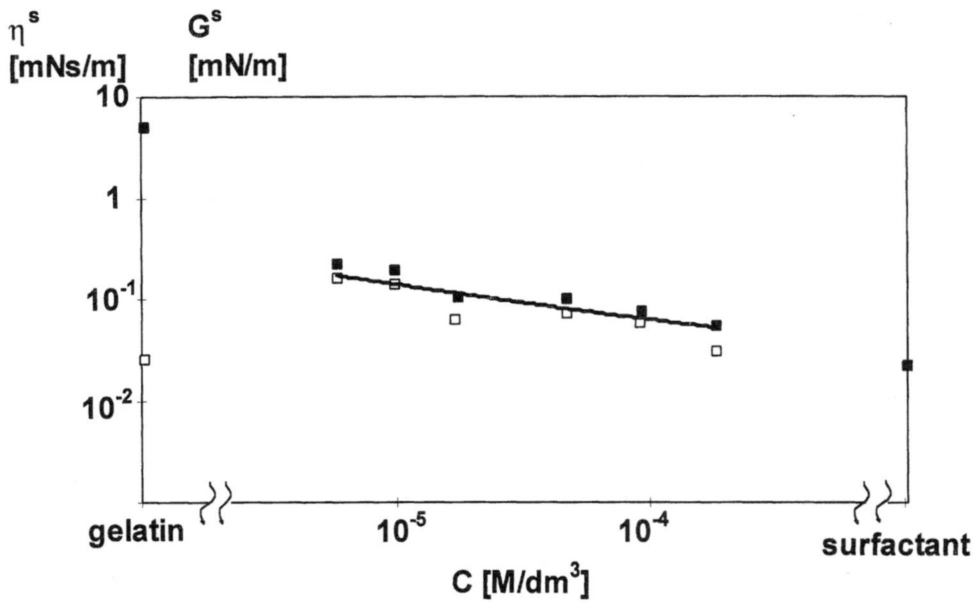

Fig. 14 Dependence of the Newton surface shear viscosity η^s (■) and the spontaneous surface shear elasticity G^s (□) on the concentration of an ethoylated octylpheol containing 10.5 EO units in a mixture with 0.2% gelatin, pH 7, 298 K [83].

The formation of mechanically highly stable layers in the range of gelatin/surfactant complex adsorption presented here are determined most of all at low gelatin concentrations and low temperatures. The surface concentration in these cases is obviously high enough to cause sol-gel-transitions at the interface. With increasing temperature the sol-gel transition is restricted and the surface shear rheological parameters decrease. The surface shear behaviour becomes

Newtonian with $\eta^s \approx 10^{-2}$ mNs/m [83]. When gelatin is cooled down however the earlier properties are restored and become important again.

The stress-deformation behaviour in a frequency range of 0.1 to 1 Hz of gelatin mixed with the non-ionic surfactant Triton X-100 (phenoxy polyethoxy ethanol) was investigated by an oscillatory shearing flow by Lee et al. [199]. This non-ionic interacts with gelatin and from the temperature dependence of the elasticity and viscosity it is possible to distinguish between two structures of gelatin, i.e. β-sheets at 35°C and α-helical conformations at 40°C and 45°C. In contrast to the data reported above Lee et al. used higher gelatin concentrations and higher temperature. Therefore the obtained surface shear viscosities are low in comparison to the effects mentioned above. Analogous results are reported in [83].

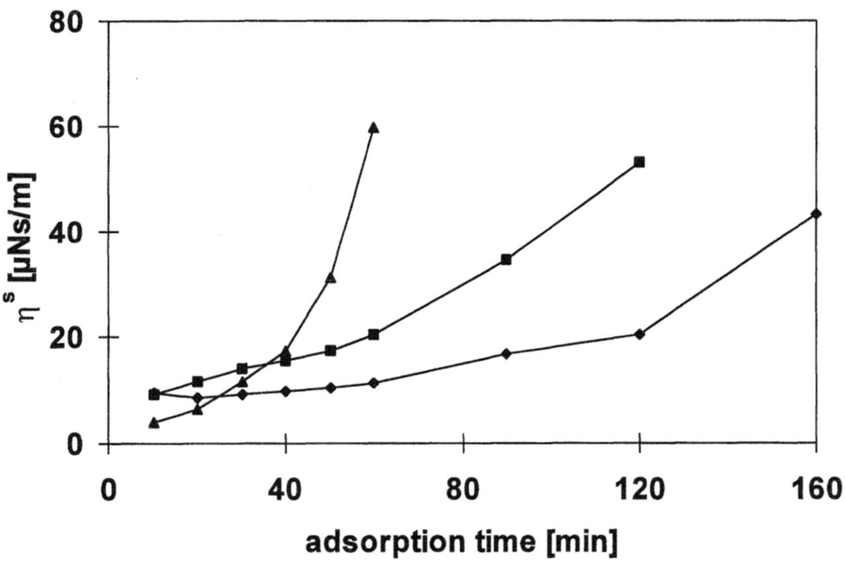

Fig. 15 Surface shear viscosity of gelatin/SDS mixtures: (▲) 0.5 wt% gelatin without SDS, (■) 0.5 wt% gelatin with $9 \cdot 10^{-4}$ M/dm³ SDS, (◆) 0.5 wt% gelatin with $4 \cdot 10^{-3}$ M/dm³ SDS [200].

Krägel et al. [200] studied the influence of the surface age on the viscoelastic properties of mixed gelatin/SDS adsorption layers at the air/liquid surface using a torsion pendulum technique. The results are shown in Fig. 15. While pure gelatin solutions form stiff adsorption layers very fast, the mixed systems show initially higher viscosities and elasticities, which increase more slowly than that of the pure gelatin solution. The higher the surfactant concentration, the smaller the slope of the viscosity and elasticity changes with increasing adsorption time. The small values of both rheological parameters at long adsorption times can be explained by a partial displacement of the adsorbed protein by surfactant molecules with increasing surfactant concentration.

3.4 INTERFACIAL DILATION RHEOLOGY

Another aspect of interfacial rheology is the dilational rheology. Contrary to the shear rheology which yields information about rearrangement processes and flow at the interface, the dilation rheology can provide information about exchange of matter between the bulk and the interface [194, 201] depending on the time scale of surface compression/expansion. The dilational rheology also yields the surface elasticity defined by $E = \frac{d\sigma}{d\ln A}$, were $d\sigma$ is the surface tension change and dA is the relative area change causing $d\sigma$.. For example using an oscillating bubble tensiometer working at frequencies between 10 and 150 Hz the effective surface elasticity can be determined.

The expansion of the interfacial area leads to an increase of surface tension $d\sigma$. During expansion surfactant molecules adsorb (diffusion controlled) and decrease $d\sigma$, so that the measured elasticity E is typically smaller than the thermodynamic value obtained from the adsorption isotherm. When the frequency of oscillation gets too fast E levels off at a certain value, i.e. the adsorption layer becomes quasi insoluble. For a pure surfactant E increases with increasing concentration up to the CMC.

In contrast macromolecules, e.g. gelatin, show a constant E over the whole range of oscillation frequencies 10-150 Hz, and increases with increasing gelatin concentration [82]. This result corresponds to a negligible exchange of matter, i.e. the frequency of surface area oscillation is always to fast for the macromolecules to desorb and rearrangement processes within the range of measurement were not detected.

For gelatin/ionic surfactant mixtures it can be stated that for a constant gelatin concentration the addition of surfactant initiates obviously an exchange of matter at low frequencies. $d\sigma$ however levels off to constant values at lower frequencies than the pure surfactant does. The values of E especially reported for mixtures of CTAB and gelatin [82] are abnormally high and in contrast to values for CTAB reported recently [202].

Valentini et al. [20] investigated the effect of surface dilation elasticity of coating solutions on the stability of the liquid bridge in a slide coater and the curtain in a curtain coater. In this study the stability was defined as the minimum flow required to establish and maintain a stable liquid bridge or curtain. Different technical surfactants were tested. The dilation modulus passes a maximum the localisation of which depending on the surfactant concentration. Higher dilation moduli are favourable to produce thinner films. Therefore the dilation modulus is recognised to be one important parameter in optimising the coating process.

Avramidis and Jiang [203] used the growing bubble technique to determine the interfacial dilational rheology of aqueous gelatin/surfactant mixtures (non-ionic Triton X-100). The interfacial dilational viscosity exceeds the shear elasticity [200]. Over the entire range of surfactant concentration the dilational viscosity increases with temperature and reaches a minimum at the CMC of the surfactant.

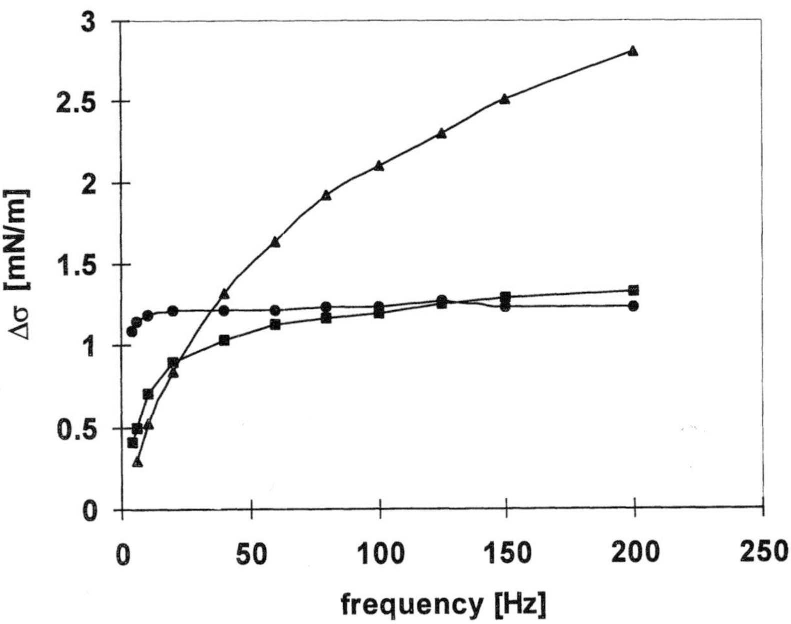

Fig. 16 Dynamic surface tension amplitudes $\Delta\sigma$ as a function of frequency f at a relative area change of 10.2%; (●) 0.7 wt% gelatin, (■) $8 10^{-7}$ mol/cm^3 sodium dioctyl sulfosuccinate, (▲) mixture of 0.7 wt% gelatin and $8 \cdot 10^{-7}$ mol/cm^3 sodium dioctyl sulphosuccinate [204].

A new oscillating bubble apparatus has been developed recently and used for measuring the dilational properties of mixed gelatin/surfactant adsorption layers in a frequency range of 1 to 300 Hz [202, 204]. Experimental dependencies of the dilational elasticity as a function of oscillation frequency obtained for some surfactant solutions (CTAB, Aerosol OT) are in good agreement with a diffusional exchange of matter theory. The relaxation behaviour of gelatin adsorption layers shows no significant dependence in the studied frequency interval as remarkable effects can be expected only at much lower frequencies. However, in presence of a surfactant the mixed gelatin/surfactant adsorption layer shows a linear frequency dependence on the elasticity, which would be in line with a dilational viscosity effect. The pure surfactant solution shows elasticity values levelling off at higher frequencies while the gelatin alone does not show a measurable frequency dependence. The presence of a surfactant changes the elastic and relaxation behaviour dramatically as shown in Fig. 16 for gelatin mixed with the anionic surfactant Aerosol OT. Unfortunately there are only few experiments reported in literature and the results reported in [209] remarkably differ from those given in [82], so that general conclusions cannot be drawn at the moment.

3.5 COMPARISON OF RHEOLOGICAL PROPERTIES IN BULK AND AT INTERFACES

The surface rheological properties of gelatin/surfactant adsorption layers may differ significantly from those of the bulk phase. As shown here these properties are based on completely different phenomena. Therefore it is impossible to predict one of these from knowing the other behaviour.

The viscosity of dilute gelatin solution was studied as a function of pH [205] and qualitatively interpreted on the basis of changes in molecular shape caused by charged groups. At the IEP the gelatin chain configuration is contracted by attractive forces between the balanced charges along the molecule and internal hydrogen bonding so that the viscosity becomes minimum. The addition of acid or alkali to an iso-ionic solution of gelatin changes the dissociation and yields a positive or negative net charge. The excess charges repel each other and cause a molecular extension. At the same time the amount of free ions in the solution increases. These free ions may reduce the repulsion between equally charged pairs and can even result in an attraction [206, 207]. Therefore apart from the IEP different maxima are obtained from the interplay of these two effects.

The influence of the interaction of gelatin and surface active ions was studied first by Tamaki and Tamamushi [56]. They reported that the rheological behaviour of dilute gelatin solution is non-Newtonian characterised by a viscosity and a yield value. The influence of anionic and cationic surfactants on the rheology of dilute gelatin solution differs in the range of the IEP and outside. At pH above and below the IEP the addition of surfactants remarkably reduces the viscosity and the yield value of the gelatin solution until precipitation occurs. After solubilisation of the precipitate the rheological behaviour becomes almost Newtonian and the viscosity slightly increasing with increasing surfactant concentration. This behaviour is completely different from that reported for gelatin/surfactant adsorption layers. At the IEP the viscosity and the yield value pass through a maximum. The viscosity is maximum when 1:1 complexes are formed, which precipitate above and below the IEP. The viscosity at the IEP is somewhat higher than the that above and below the IEP [68]. These effects were explained by folding and refolding of the gelatin molecules by interaction with the ionic surfactants. Recently some of these results could be confirmed for acidic decomposed gelatin [93].

The principal rheological behaviour at an interface is different. The elasticity and the viscosity are higher by nearly one order of magnitude at the IEP in comparison to the pH above or below the IEP [83, 100, 186, 198]. Furthermore the changes in surface rheology caused by ionic surfactants are generally stronger than those reported for the bulk phase of dilute gelatin solutions. The principal differences of the influence of ionic surfactants on the rheological behaviour in bulk and at surfaces of diluted gelatin solutions can be explained by the sol-gel transition and network formation [83, 208]. These processes take place only at complex-saturated surfaces and they are not present in the bulk phase.

3.6 ADSORPTION LAYER THICKNESS

The structure of a gelatin/surfactant adsorption layer is also reflected in the adsorption layer thickness. The thickness can be determined by micro-interferometry [209] from foam film lamellae [92, 93] where the electrical double layer repulsion was suppressed by addition of electrolyte. Usually gelatin forms adsorption layers of a thickness of about 40 nm (about one half the foam film thickness) at the IEP and 313 K. This thickness increases with increasing gelatin concentration yielding foam film thicknesses >100nm for 0.5% gelatin solutions. The results are in good agreement with [210, 211] which were found for adsorption layers adsorbed at solid interfaces. For comparison, the foam film thickness of surfactants is usually about 4-5 nm. The increasing film thickness at very high gelatin concentrations was explained by some authors on the basis of a multilayer formation [212, 213].

The addition of ionic surfactants to gelatin results in a decrease of the adsorption layer thickness at sufficiently high electrolyte content. In the region of complex adsorption, i.e. the formation of mechanically highly stable viscoelastic adsorption layers, common black films are formed with a thickness of 8-12 nm. At surfactant concentrations exceeding the binding capacity of the gelatin foam films are formed with a thickness comparable to that of surfactants alone (Newton black films). This however requires a high content of electrolyte and therefore it does not prove the picture of a total displacement of gelatin or complexes at sufficiently high surfactant concentration.

3.7 INFLUENCE OF SURFACTANT ON THE TRIPLE HELIX STRUCTURE OF GELATIN IN ADSORPTION LAYERS

From adsorption layer thickness measurements one can conclude that the adsorption layer of gelatin/surfactant mixtures contains more unfolded components. A direct method to prove this model is the determination of CD spectra of adsorption layers [83]. To get such spectra 10 quartz sheets were dipped into gelatin solutions permitting measurements through 20 adsorption layers. The transferred adsorption layers were cold-dried. As the related surface concentrations are not exactly known for comparison the amplitude ΔA of the instrument was monitored only. Fig. 17 shows the dependence of ΔA on the wavelength for different pH. The content of triple helix structure in the region of IEP characterised by the amplitude at 222 nm is maximum, i.e. the same principal shape of the graph which was found for gelatin solutions. The minimum at 238 nm is only weekly pronounced.

The following figures show the influence of SDS, CTAB and ethoxylated paratertiar octylphenol (non-ionic) on the adsorption layer structure. The triple helical structure is strongly depressed by SDS, and less pronounced by CTAB. Addition of non-ionic surfactant decreases the triple helical content in the adsorption layer continuously (Fig. 20).

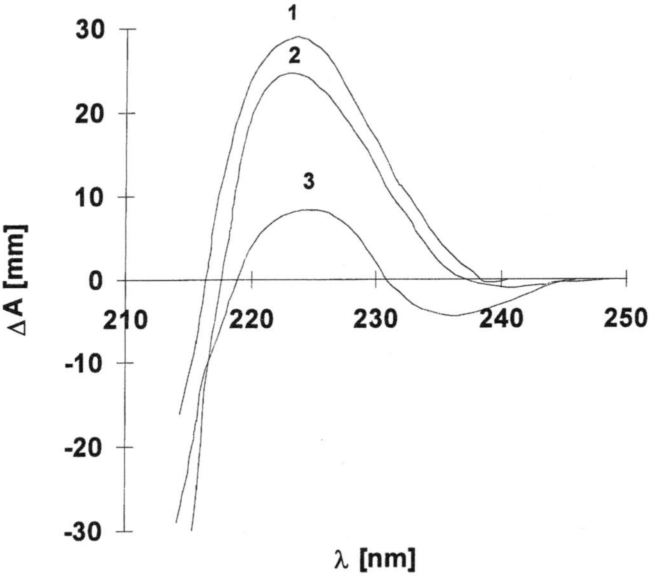

Fig. 17 CD spectra of 20 adsorption layers of a 0.2% gelatin solution on quartz sheets at different pH, pH 4.9 (1), pH 10 (2), and pH 2 (3), 283 K [83].

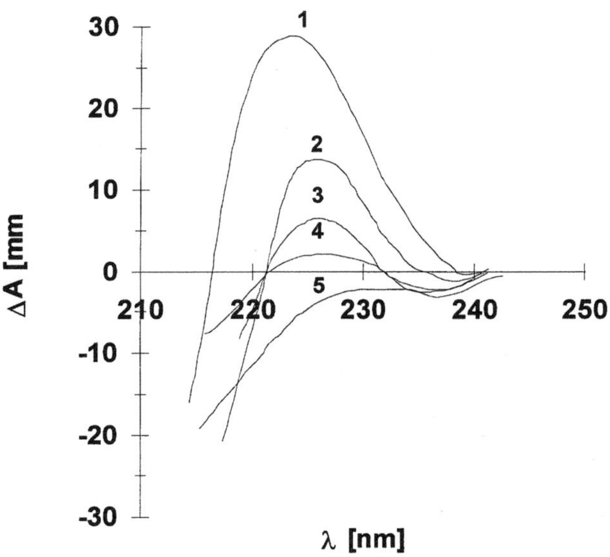

Fig. 18 CD spectra of 20 adsorption layers of a 0.2% gelatin solution+SDS on quartz sheets depending on the content of SDS. 0.2% gelatin (1, 2), gelatin + 10^{-3} M/dm³ (3), 5·10^{-3} M/dm³ (4), 5.10^{-2} M/dm³ SDS (5), pH 4.9 (IEP of gelatin), measurements 1 h after sheet coating and keeping at 283K (1, 3, 4, 5) and 293K (2) [83].

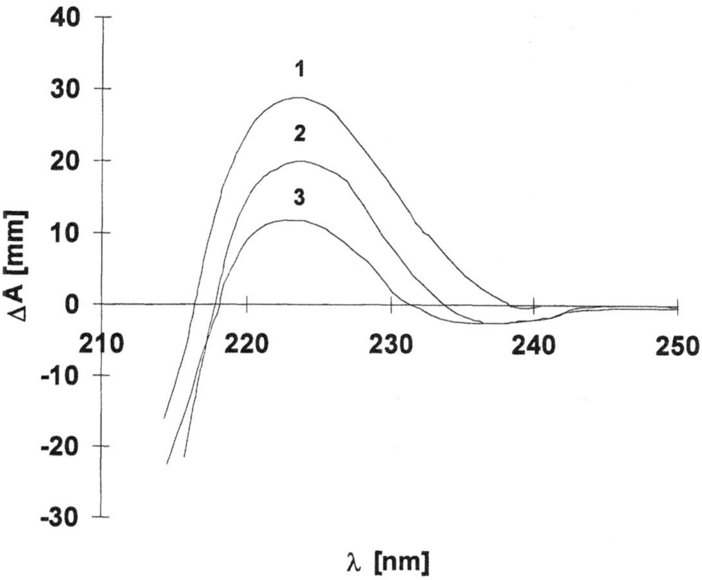

Fig. 19 CD spectra of 20 adsorption layers of a 0.2% gelatin solution+CTAB on quartz sheets depending on the content of CTAB. 0.2% gelatin (1), gelatin + $5 \cdot 10^{-4}$ M/dm³ (2), 10^{-2} M/dm³ CTAB (3), pH 4.9, 283K [83].

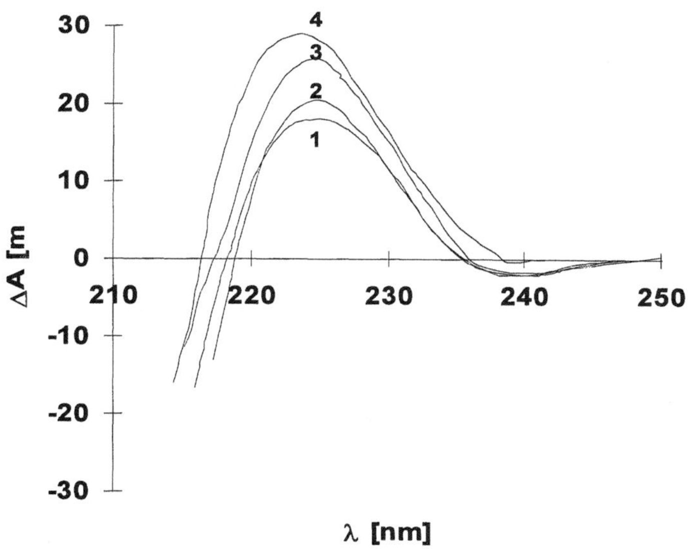

Fig. 20 CD spectra of 20 adsorption layers of a 0.2% gelatin solution+ethoxylated octylphenol (10.5 EO units) on quartz sheets depending on the content of octylphenol. Gelatin (1), gelatin + 10^{-4} M/dm³ (2), $5 \cdot 10^{-4}$ M/dm³ (3), 10^{-3} M/dm³ octylphenol (4), pH 4.9, 283K [83].

This is obviously caused by the predominant adsorption of more unfolded components which however show only partially depressed refolding ability. In the range of gelatin/surfactant complex adsorption which are characterised by highly mechanically stable networks the adsorption layers exhibit a high triple helix content. At high SDS concentrations the triple helical structure strongly decreases (Fig. 18).

It should be noted however that even at very high SDS concentrations, which change the gelatin structure to a great extent into ß-sheets (Fig. 4), there is still a shoulder between 220 and 230 nm. The ß-sheets formed in solution are obviously not dominant at the surface. In contrast for CTAB and non-ionic surfactants at concentration even much above the CMC of the surfactants a remarkable triple helical content is still detected (Figs. 19 and 20). There are no results to exclude the presence of gelatin/surfactant complexes even for surfactants strongly interacting with gelatin and at very high surfactant concentrations, because surfactants do not yield peaks in a CD spectra.

The CD results are in good agreement with the surface rheological properties of these systems but not with the results of adsorption layer thickness measurements. It cannot be excluded that the results are not exactly comparable with those found at the air water interface caused by specific gelatin or complex adsorption at the quartz surface and the high electrolyte concentration, which was necessary to add in the foam film measurements. Due to better sensitivity new CD instruments should allow direct measurements of the protein structure at liquid interfaces. However, no systematic data are available so far.

3.8 MIXED GELATIN/SURFACTANT ADSORPTION AT LIQUID-LIQUID INTERFACES

Interfacial tension isotherms of gelatin/surfactant mixtures show that the adsorption behaviour is similar to that principally found at the water/air interface [188]. Therefore it was concluded that analogous gelatin/surfactant complexes determine the interfacial behaviour. Interfacial adsorption layers of gelatin at the benzene/gelatin D_2O-solution interface were studied by NMR using oil-in-water emulsions prepared by ultrasonic dispersion. The result shows that the fraction of mobile gelatin segments in the interfacial adsorption layers decreases virtually to zero. The presence of two different signals in the NMR spectrum suggests that benzene is solubilised by gelatin macromolecules and constitutes one component of the interfacial adsorption layer. The results are similar to that of collagen-like helices and gels of gelatin [214].

In microemulsions gelatin is frequently used to control the viscosity and to create a carrier material with specific properties [215-220]. Beside surfactant microemulsions may contain co-surfactants, most of all alcohols. These alcohols also influence the structure of gelatin in aqueous solution in a specific way. Bianchi et al. [221] claimed an additional folding of collagen by additional interpeptide bonding. Other authors reported an acceleration of gelatin refolding [222]. Furthermore a decrease of the dielectric permeability of the environment and decrease of the ionisation degree of ε-amino groups of lysine and the guanidyl groups of arginine [223]. Especially, the addition of polyols was shown to increase the number of cross-linking junctions in the gel network. On the contrary, the effect of carbon atoms of hydroxy compounds disturbs the formation of the helices and the aggregation among the helices [224]. Much more details on properties of proteins at liquid/liquid interfaces was already given in Chapter 3 by Izmailova and Yampolskaya.

4. THIN LIQUID FILM COATING

The purpose of a coating process is to deposit a thin and uniform liquid film – in some cases a multilayer film – on a flexible sheet or solid substrate. Generally coating methods can be divided into two categories, self-metering and pre-metered coating processes. To the first category belong processes where the coating thickness is determined by the physical properties of the coating solutions, the geometry of the device, and the operating conditions. Dip and roll coating are examples of this category. In processes of the second category the coating thickness is pre-set and it is theoretically independent of any variables. Extrusion or slot, slide, and curtain coating are considered as typical pre-metered coatings. The fluid mechanics and the scientific principles of such coating methods have recently been reviewed in [7].

From the interfacial chemistry point of view all technologies for the production of thin liquid layers or films on various substrates are founded on the same basic principles. The rheological properties of the coating liquids, the dynamics of the three phase contact line (substrate/liquid/air), and the regularity of the layer or film formation, respectively, are often drastically influenced by the interfaces involved in these processes. Some coating technologies are only applicable when the interfacial properties are systematically modified by surfactants.

Adhesion energy, dynamic surface tensions, dynamics of wetting, and interfacial rheology are parameters which can be modified remarkably depending on the chemical structure of the applied surfactants.

The dynamic behaviour of surfactant/protein mixtures is of great importance for the coating process of photographic films where mixtures of gelatin with surfactants adsorb at the interface simultaneously. Extremely small thicknesses, down to 0.1 μm (dry thickness), must be very precisely reproduced. Unevenness in the thickness of a single layer can lead, for example, to a deviation of the dye density in a colour film. An error of a few percent in the coating thickness can cause a colour imbalance. Nowadays colour films can consist of up to 20 different layers. Furthermore there is a trend to higher coating speeds and larger support widths. For economy reasons the number of passes must be kept as small as possible, thus it is necessary to apply many coatings in a single pass. Coatings on a support consist of light-sensitive layers and a number of auxiliary layers. Together or separately they perform the following functions: light filtration, separation of differently sensitised emulsions containing different dye couplers, antihalation, mechanical protection, non-curling, subbing, antistatic, hardening, matt finish, and many others. The number of layers is steadily increasing, especially in colour materials, because of more and more stringent requirements. Generally all coatings are one-sided. Certain materials, however, such as X-ray films, have coatings on both sides. Photographic films have the most complex structure. A finished Polaroid instant colour film, for example, contains supports on both the top and bottom. Between them are active layers sensitised to the three primary colours (blue, green, and red), timing layers, and a receiving layer to display the image through the clear top support. The film polyester supports are also coated to increase adhesion to the sensitised layers. A black coating on the back of the bottom support prevents light from penetrating into the interior.

Coatings are usually applied in form of aqueous gelatin solutions, emulsions, or suspensions. In some cases solutions in organic solvents are employed. The coating material has to be prepared before application. This procedure may include melting, filtration, degassing, tempering, and the addition of finals to adjust surface tension and rheological behaviour. Coating solution can be prepared batch-wise or continuously. The design and sizing of the tanks and metering equipment depend on the lot size and on the shelf life of the photographic emulsion, including the finals. Equipment for coating of long film or paper webs includes devices similar to those employed in the paper converting or printing industry. Special machinery is used for the coating of plates made of glass or other materials, i.e. optical filters, photographic plates for scientific purposes, or plates serving as the initial stage in the manufacture of printed circuits and microelectronic devices.

The advantage of pre-metered coating devices is the predetermined coating thickness. Furthermore the uniformity of coating is excellent, the coating speed can be very high and simultaneous multilayer coating is possible. Even though the coating speed for a pre-metered coating device is high, it is still limited. For a given coating thickness, the coating speed will go up until a critical value is reached (called - critical coating speed). A further increase would create coating defects. Analogously for a given coating speed the coating thickness can be reduced to a critical value below which also coating defects can appear. These two limits are called the low-flow limit of coatability.

Slot or extrusion coating are versatile operations for coating single layers onto a web. A wide range of application with the viscosity of the coating solution varying between less than 1mPa·s up to several thousand Pa·s can be handled with a single design. An important limitation of the slot coater is the difficulty of designing devices for more than two layers simultaneously. As for colour film production multilayer superposition is required, the slide coating technique was developed in the photographic industry. Mercier et al. [225] first described a slide coater for a single and multilayer flow. A sketch of this technique is shown in Fig. 21.

One or more solutions are dispensed through distribution cavities and slots, thus forming a multilayer stack onto an inclined plane without convective mixing. The liquids are bounded sideways by special shaped edge guides. These prevent the sucking out of liquid from the multilayer stack by the meniscus that is formed at the three phase contact line (air/coating liquid/coater material). The slide coater is placed at a small distance (0.1-0.5 mm) from the moving web, which is supported by a coating roll. At the end of the inclined plane the multilayer stack bridges the narrow gap between the coater and the web. If the flow in this liquid bridge is stable, steady, uniform and without vortices, the slide coating operation is satisfactory. A liquid bridge or bead, where maximum surface stretching occurs, is formed between the moving web and the coater die face. An upper and a lower free surface, which is often denoted as meniscus, bound the bead. The latter is maintained at subatmospheric pressure. Both menisci extend across the gap, which is the distance between the face and the moving web. The web has a previously applied subcoating to ensure proper wettability. The coating roll supporting the web ensures that the latter one remains parallel to the die face at the application point. In principle the number of layers that can be coated simultaneously should be unlimited, but the stability of the multilayer flow on the inclined plane may become a problem upon increasing their length. To avoid vortices in the multilayer flow it is necessary to design a special shape of the distribution cavities and slot exits depending on the bulk rheology. In such a coater geometry the top layer tends to flow back, wetting the upstream coater surface. This wetting line must be straight to avoid coating defects. A sketch of the curtain coater technique is given in Fig. 22.

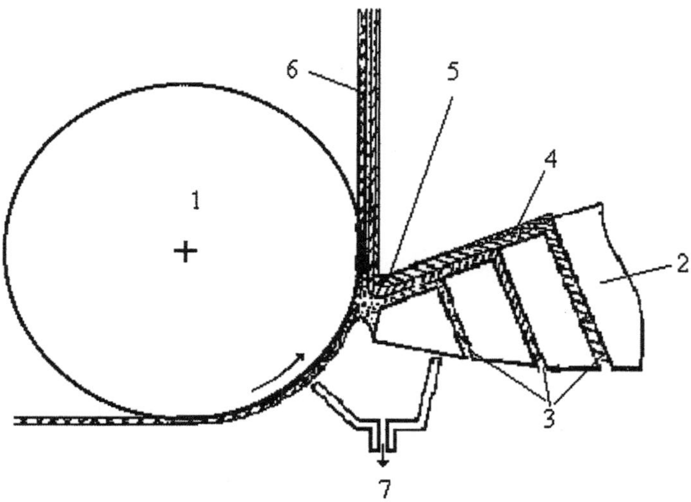

Fig. 21 Multilayer slide coating: (1) supporting roll, (2) cascade coater, (3) coating solution feed, (4) multilayer film flow, (5) liquid bridge or meniscus area, (6) moving web, (7) vacuum box.

In this coating technique which was previously described by Hughes [226], solutions are delivered from the slots onto the slide to form a multilayer flow. These solutions form a free falling vertical curtain that impinges on the moving web and leads to a wet coated product. On each side of the slide, two vertical rods are mounted to adjust the edges of the curtain, thereby avoiding any receding of the edges or necking in. Both edge guides extend from the slide down to 2 - 3 mm above the moving web. The web movement is horizontal with respect to the floor and perpendicular to the edge guides. The distance between the slide and the web has to be covered in order to prevent air circulation. The properties of the liquid curtain, the geometry of the slide and the flow characteristics largely determine the coating quality. Surfactants affect the stability of the film at the edges. Edge stability is sensitive not only to the geometry of the edge guides but also to the type and flow of liquid used as wetting agent.

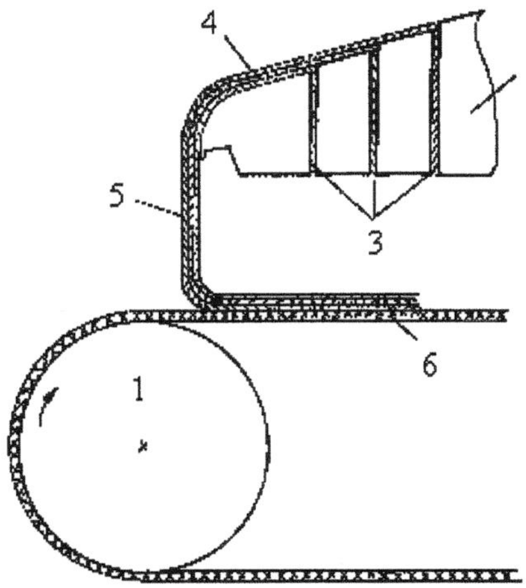

Fig. 22 Multilayer curtain coating: (1) supporting roll, (2) cascade coater, (3) coating solution feed, (4) multilayer film flow, (5) free falling curtain, (6) moving web.

Slide and curtain coating are versatile methods of pre-metering and assembling multiple layers on a support. They allow to transfer multiple layers to a moving web with an excellent uniformity and within an industrially interesting operation range. In these procedures the coating liquids flow trough separate channels so that the fluids exit the slots and flow down an inclined plane. In slide coating the fluid layers form a meniscus (or a liquid bridge) between the narrow gap (a few hundred microns) of the die face of the slide coater and the moving web which in turn carries off the fluid to form the thin liquid layer. A low vacuum applied to the liquid bridge stabilises the coating. The most distinguishing feature in curtain coating is that the fluid layers form a curtain that falls freely and laminarly over a certain height (several centimetres) before it impinges onto the moving web. A successful coating flow is steady and two-dimensional. In both techniques surfaces are created, stretched and/or compressed. Therefore, the effective surface age, or the residence time of a particular element of the liquid/air interface, can vary widely. For example, the residence time of the layer surface on the inclined plan (up to a few seconds) is long enough for the surfactants to adsorb remarkably. A proper choice of surfactant ensures stable interlayer wetting, sufficient wetting of the coater surface itself, and guards against the formation of surface waves. In slide coating the age of the upper surface of the liquid bridge can be very short (in the range of parts of a millisecond). Therefore, surfactants have only little influence because the time is too short to form an adsorption layer. The surfactants affect also the coating homogeneity after deposition on the moving web, until the layers have solidified. All of these phenomena are influenced by the gelatin/surfactant interaction and therefore play an important role in the coating quality.

The transition from dynamic interfacial tension measurements to a high quality coating is not trivial. For example, other important factors, such as bulk rheology and flow characteristics may cause coating failure. Contaminants may cause local coating defects, but have no effect on the measured dynamic interfacial tension. In reality coating involves a lot of different parameters. Therefore, the usefulness of interfacial tension data must be proven by practical coating tests. A good illustration of this concept has been given by Fruhner et al [13-15, 19] and Valentini et al. [16, 20]. While the former focused on investigation of slide coating, the latter performed experiments with slot, slide, and curtain coaters.

Fruhner et al. [12, 227] introduced a technique to observe and record the liquid bridge and the neighbouring coating flow during the coating process. They performed model coating experiments with aqueous gelatin solutions of low bulk viscosity (1-20 mPa·s) to study the influence of the shape and position of the liquid bridge on the coating quality. This method is applicable to investigate the dynamic contact angle formed by the liquid bridge on the moving web. The dependence of the dynamic contact angle on important coating parameters was examined. In a liquid film coating process, the main role of a surfactant is to control the wetting behaviour of the coating solutions. It was found that irregularities of the contact angle on the web cause defects especially at high speeds of 50-100 m/min so that thin layers of 15-20 μm thickness in wet state are formed on the web. In another paper Fruhner et al. [13] examined the influence of defined changes of the contact angle on the coating quality. The effect of different conditions, and the possibilities to stabilise the liquid bridge is also discussed. Disturbances caused by oscillations of the liquid bridge produce periodical irregularities in the thickness of the coated layer. The stabilising effect of a suction applied to the liquid bridge and the influence of other parameters on the coating quality were determined. To quantify the contact angle changes on the web an optical method was introduced to record the movement of the three phase contact line [228, 229].

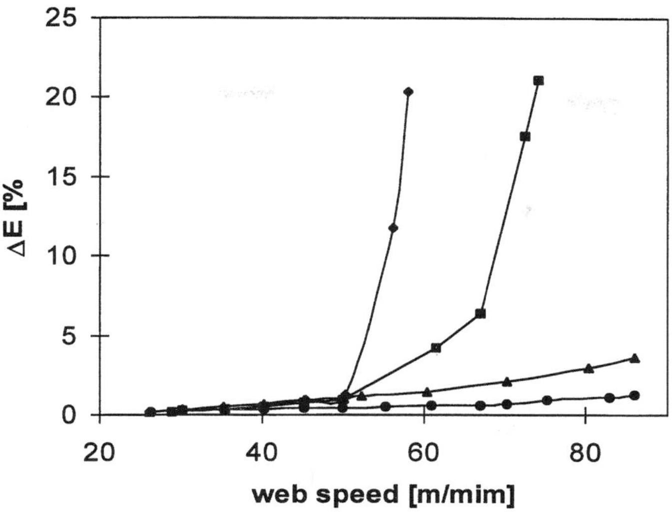

Fig. 23 The effect of different surfactants on the dependence of the web speed on the extinction fluctuations. The specific flow rate is 1 l/h·cm, the bulk viscosity 6 mPa·s, distance between web and coater die face 150 μm, vacuum 200 Pa, surfactant concentration 0.5 g/l.: (◆) CTAB; (■) dodecyl saccharose urethane; (▲)

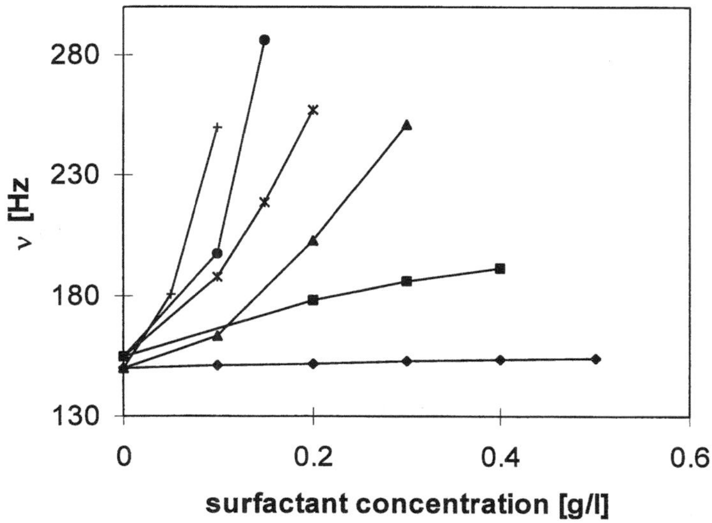

Fig. 24 The influence of the concentration of different surfactants on the oscillation frequency. The specific flow rate is 1 l/h.cm, bulk viscosity 6 mPas, distance between web and coater die face 150 μm, vacuum 200 Pa.: (◆) CTAB, dodecyl pyridinium bromide, tridecenylsuccinic acid polyglyceride, (■) dodecyl saccharoseurethane, (▲) sulfosuccinates, (✱) sodiumalkylsulfonate (E30), (●) Aerosol OT, (+) SDS [15].

Fruhner et al. [14, 15] also examined the effect of several surfactants on the stability of the bridge between the die face and the web and demonstrated a clear improvement in coating quality at higher web speeds. The degree of improvement, however, was highly dependent on the type of surfactant and its interaction with gelatin. The influence of different surfactants on the coating quality is shown in Fig. 23. The extinction fluctuations are used as a measure of the evenness of coating. This figure clearly shows that only anionic surfactant are able to stabilise the liquid bridge by shifting the oscillation frequency to higher values.

In slide coating the surfactants must be added to the coating solution to ensure the stability of the liquid bridge. Experiments show that the stabilising action of an ionic surfactant is caused by interfacial rheological properties of the adsorption layers at the upper surface of the liquid bridge. The influence of the concentration of different surfactants on the oscillation frequency of the liquid bridge is shown in Fig. 24.

The stability of the liquid bridge against oscillations will be attained by a value of its resonance frequency ($\upsilon > 400\ s^{-1}$). The frequency of oscillation of the liquid bridge is determined by the value of the reduced pressure applied to the lower surface of liquid bridge and by the dilational properties of the gelatin/surfactant adsorption layers. The influence of the subathmospheric pressure in the vacuum box on the oscillation frequency of different model systems are shown in Fig. 25. By comparing interfacial rheological data obtained by different methods with results of coating experiments it was shown, that the interfacial dilational properties are of essential importance to the stabilisation of the liquid bridge against oscillations [15].

Fruhner et al. [16] performed coating experiments to investigate the additional stabilising effect of the liquid bridge by viscoelastic flow properties of the bulk phase measured using an oscillatory capillary rheometer. The experiments show that already at a loss angle of viscoelasticity Θ of about 3-5° the stability of the liquid bridge is increased remarkably. The degree of viscoelasticity is modified by adding small amounts of a water soluble polyacrylamide [230]. To measure the influence of different components on the stability of a multilayered film flow on an inclined plane, an optical method was proposed [231].

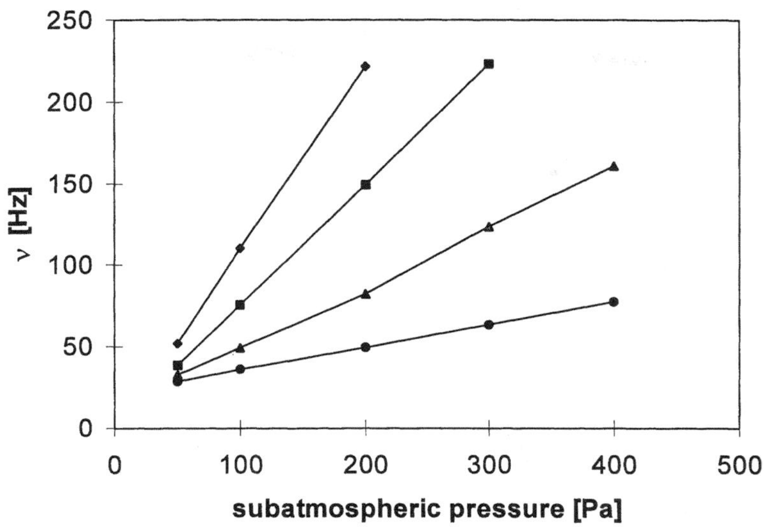

Fig. 25 The influence of vacuum on the oscillation frequency of different model systems. The specific flow rate is 1 l/h·cm, bulk viscosity 6 mPas, distance between web and coater die face 150 μm, web speed 40-60 m/min.: (◆) gelatin solution + 0.1 g/l sodium alkansulfonate, (■) gelatin solution, (▲) glycerol/water mixture + sulfobernsteinsäureester, (●) glycerol/water mixture[15].

5. SUMMARY

The manufacturing of coating emulsions is a complex problem and has to meet the requirements of the coating process. Beside gelatin coating emulsions for photographic materials contain up to more than hundred other ingredients, and many of them are surface active. In particular the multilayer coating requires a detailed balance of surface and bulk properties. The present chapter summarises the main physico-chemical properties of gelatin solutions and gelatin/surfactant mixtures, i.e. the principles of their interaction, the influence of surfactants on the secondary structure of gelatin in the bulk phase and at the interface, the modification of the interfacial tension, the adsorption layer thickness, and the changes of the rheological properties in bulk and interfaces. The dynamic properties are of special interest.

The strength of interaction between gelatin and surfactant depends on the charge and the hydrophilic/lypophilic balance of the surfactant. It is strong for anionics, weaker for cationics, and very weak for non-ionics. As a consequence of the interaction the secondary structure of the gelatin is changed starting from a certain surfactant concentration. At high surfactant concentration the triple helical structure is changed into a ß-sheet structure. In contrast the adsorption of triple helical structured complexes is favoured and a considerable amount of triple helix structures is found at the interface. Gelatin/surfactant complexes in a wide range of surfactant concentration govern the surface behaviour. When the binding capacity of gelatin is exhausted surfactant micelles are formed, the mixed complexes are solubilised and replaced from the interface by competing surfactant molecules. The adsorption layer thickness is much lower for the mixed complex adsorption (common black films for foam lamellae) as compared with the thickness of a pure gelatin adsorption layer. The rheological properties of a gelatin/surfactant adsorption layer differ remarkably from those of the bulk phase. This is caused by the comparably high surface concentration and the lower temperature that favours the sol-gel transition and leads to a strongly pronounced viscoelastic behaviour of a network formed after complex adsorption.

Gelatin fractions which differ in molecular weight distribution and ash content were compared. The main differences between the fractions can be explained by the formation of different complexes having different surface activity.

For a slide hopper coater the influence of ionic surfactants on the stability of the liquid bridge between the edge of the hopper and the moving web was demonstrated. The results of coating experiments show that the interfacial dilational properties are of essential importance for the stabilisation of the liquid bridge against oscillations, thus controlling the coating quality.

6. REFERENCES

1. H. Frieser, G. Haase and E. Klein, Grundlagen der photografischen Prozesse mit Silberhalogeniden, Akademische Verlagsgesellschaft, Frankfurt a. M. 1968
2. W. Walther (ed.), Fotografische Verfahren mit Silberhalogeniden, VEB Fotokinoverlag Leipzig 1983
3. H. Böttcher and J. Epperlein, Moderne photographische Systeme, Wirkprinzipien, Prozesse, Materialien, 2nd ed., VEB Verlag für Grundstoffindustrie, Leipzig 1988
4. B. Elvers, S. Hawkins and G. Schulz (eds.), Ullmanns Encyclopedia of Industrial Chemistry, 5th Completely Revised Edition, Vol. A20, 1-159, VCH Verlagsgesellschaft, Weinheim 1995
5. E.D. Cohen and E.B. Gutoff, Modern Coating and Drying Technology, VCH Verlagsgesellschaft, Weinheim, New York, Cambridge 1992
6. O. Lapp, W. Krafft, R. Winiker, H. Frenken, and H.Lühring, Manufacture of Photographic Materials, In: Science and Technology of Photography, K. Keller (ed.), VCH Verlagsgesellschaft, Weinheim 1993, pp.43-82
7. S.F. Kistler and P.M. Schweizer (eds.), Thin Liquid Film Coating – Scientific principles and their technological implications, Chapman & Hall, London 1997
8. R. Lipp, PhD thesis, TH Leuna-Merseburg 1977
9. J. Krägel, PhD thesis, Academy of Sciences of GDR, Berlin 1984
10. L. Grader, PhD thesis, TH Leuna-Merseburg 1985
11. L. Grader, Colloid and Polymer Sci., 264(1986)719
12. J. Krägel, H. Fruhner and G. Kretzschmar, J. Inf. Rec. Mater., 15(1987)251
13. H. Fruhner, J. Krägel and G. Kretzschmar, J. Inf. Rec. Mater., 17(1989)107
14. H. Fruhner, J. Krägel and G. Kretzschmar, J. Inf. Rec. Mater., 17(1989)273
15. H. Fruhner, J. Krägel and G. Kretzschmar, J. Inf. Rec. Mater., 19(1991)45
16. J.E. Valentini, W.R. Thomas, P. Sevenhuysen, T.S. Jiang, H.O. Lee, Y. Liu and S.C. Yen, Ind. Eng. Chem. Res., 30(1991)453
17. K.Y. Lee, L.D. Liu and T.J. Liu, Chem. Eng. Sci., 47(1992)1703
18. G. Kretzschmar, H. Fruhner and J. Krägel, Tenside, Surf. Det., 30(1993)110
19. H. Fruhner, D. Wantke, J. Krägel and G. Kretzschmar, J. Inf. Rec. Mats., 22(1994)29

20. J.E. Valentini, J.T. Chandler, Q. Jiang, Y.C. Chiew and L.J. Fina, Ind. Eng. Chem. Res., 35(1996)434
21. National Formulary XVI, United States Pharmacopeial Convention, Inc. Rockville, Maryland 1985, pp. 1563
22. V. Hopp, Leder, 15(1964)45
23. J.E. Eastoe and A.A. Leach, in „Food Sci. and Technology of Gelatin" ed. A.G. Ward, A. Courts, Academic Press, L., N-Y., S-F, 1977 pp.78
24. G. Balian and J.H. Bowes, in „Food Sci. and Technology of Gelatin" ed. A.G. Ward, A. Courts, Academic Press, L., N-Y., S-F, 1977 p.10
25. G.N. Ramachandran, „Treatise on Collagen", Vol.1 Academic Press, N-Y. 1967 p.187
26. P. Boedtker and J. Doty, J. Phys.Chem., 58(1954)968
27. G.N. Ramachandran and G. Kartha, Proc. Indian Acad. Sci., 42A(1955)215
28. D.W. James and L. Rintoul, Aust. J. Chem., 35(1982)1157
29. C.A.J. Hoeve and P.C. Lue, Biopolymers, 21(1982)1661
30. S.F.B. Petrie and R. Becker, „Analytical Calorimetry" ed. R.S.Porter, J.F. Johnson, Plenum Press, N-Y, vol.2 (1970), p. 225
31. G.N. Ramachandran and R. Chandrasekharan, Biopolymers, 6(1968. B)1649
32. M. Lüscher-Mattli, Biopolymers, 21(1982)403
33. K. Okuyama, S. Arnott, M. Takayanagi and M. Kakudo, J. Mol. Biol., 152(1981)427
34. W. Borchard, K. Bergmann, A. Emberger and G. Rehage, Progr. Colloid Polymer Sci., 60(1976)120
35. F.C. Silver and R.L. Trelstad, Biopolymers, 20(1981)359
36. T.C. James, „Teorija fotograficseskogo processa" L., Chimija, 1980, p.58
37. A. Nordwig, H. Hörmann, K. Kühn, W. Grassmann and W. Hoppe-Seylers, Z. phys. Chem., 325(1961)242
38. K. Kühn, Naturwissenschaften, 54(1967)101
39. H.D. Jakubke and H. Jeschkeit, „Aminosäuren, Peptide, Proteine" Akademie-Verlag, Berlin, 1973, p.325
40. C.D. Warburton and E.P. Przybylowicz, J. phot. Sci., 15(1967)201
41. J.W. Janus and D.R. Nellist, J. phot. Sci., 15(1967)207
42. G. Reich, S. Walther and P. Stather, „Ges. Abhandlungen des deutschen Lederinstituts Freiberg/Saale", 17(1962)127

43. J.E. Eastoe, A.A. Leach Rec. Adv. in Gelatin and Glue Res., Pergamon Press, N-Y, 1958, p. 173
44. A. Narath and D. Gundlach, Photografische Korrespondenz, 105(1969)95
45. W.F. Harrington and N.V. Rao, Biochem., 9(1970)3714
46. W.F. Harrington and G.M. Karr, Biochem., 9(1970)3725
47. P.V. Hauschka and W.F. Harrington, Biochem., 9(1970)3734
48. D.D. Mascuga, Biopolymers, 11(1972)2521
49. M. Djabourov, J-P. Lechaire and F. Gaill, Biorheology, 30(1993)191
50. J.M. Chabala, R. Levisetti and T.J. Maternaghan, Journal of Imaging Science & Technology, 39(1995)222
51. J. Crespo, M.A. Satorre, JA. Ania and F. Quintana, Journal of Materials Science , 30(1995)6145
52. A. Bot, I.A. Vanamerongen , R.D. Groot, N.L. Hoekstra and G.M. Agterof, Polymer Gels & Networks, 4(1996)189
53. K.G.A. Pankhurst and R.C.M. Smith, Trans. Faraday Soc., 40(1944)565
54. I.I. Harris, K.G.A., Pankhurst and R.C.M. Smith, Trans. Faraday Soc., 43(1947)506
55. K.G.A. Pankhurst, Discussions Faraday Soc., 6(1949)52
56. K. Tamaki and B. Tamamushi, Bull. Chem. Soc. Japan, 28(1955)555
57. V.A. Pcelin and R.A. Kul'man, Vysocomolejularnye soed., 3(1961)768
58. M. Sen, S.P. Mitra and D.K. Chattoraj, Indian J. Biochem. Biophys., 17(1980)405
59. R.K. Cannan, Chem. Rev., 30(1942)395
60. R. Wüstneck, N.P. Wüstneck, H. Hermel and L. Zastrow, Kolloid. Z. USSR, 49(1987)244
61. A.W. Kenchington and A.G.Ward., Biochem. J., 58(1954)202
62. J.E. Eastoe, Biochem. J., 61(1955)597
63. H.M. Barchet, „Chemie photographischer Prozesse“, Akademie Verlag, Berlin 1967
64. K. Heyns, G. Legler, in „Recent Adv. in Gelatin and Glue Research“, Pergamon Press, London 1958, p.186
65. W.J. Knox and J.F. Wright, J. Colloid Interface Sci., 20(1965)177
66. B.H. Tavernier, J. Colloid Interface Sci., 93(1983)419
67. J.P.S. Arora, C. Pal and D. Dutt, Tenside Surf. Det., 28(1991)215
68. J. Greener, B.A. Contestable and M.D. Bale, Macromolecules, 20(1987)2490

69. A.M. Howe, A.G Wilkins and J.W. Goodwin, J. Photogr. Sci., 40(1992)234
70. A.M. Howe, A. Clarke and T.H. Whitesides, Langmuir, 13(1997)2617
71. A.M. Kragh, Trans. Faraday Soc., 60(1964)225
72. W.J. Knox and T.O. Parshall, J. Colloid and Interface Sci., 33(1970)16
73. W.J. Knox and T.O. Parshall, J. Coll. Interf. Sci., 40(1972)290
74. Y.M. Tricot, In: Liquid Film Coating, S.F. Kistler and P.M. Schweizer (eds.), Chapman and Hall, London 1997, Chapter 4, pp. 99-136
75. G. Kretzschmar and K. Lunkenheimer, Ber. Bunsenges. Phys. Chem., 74(1970)1064
76. D. Vollhardt and G. Kretzschmar, J. Signal AM, 4(1976)333
77. R. Wüstneck and H. Fruhner, Colloid Polymer Sci., 259(1981)1228
78. K. Lunkenheimer, Habilitation, Academy of Sciences of GDR, Berlin 1983
79. R. Wüstneck, V.V. Krotov and M. Ziller, Colloid Polymer Sci., 262(1984)67
80. R. Wüstneck, Colloid Polymer Sci., 262(1984)821
81. K. Lunkenheimer, C. Hartenstein, R. Miller, and K.D. Wantke, Colloids and Surfaces A, 8(1984)271
82. C. Hempt, K. Lunkenheimer and R. Miller, Z. phys. Chem. Leipzig, 266(1985)713
83. R. Wüstneck, Habilitation, Akademie der Wissenschaften der DDR, Berlin, 1988
84. K.S. Avramides and T.S. Jiang, J. Colloid and Interface Sci., 147(1991)262
85. H. Fruhner and K.D. Wantke, Colloid and Surfaces A, 114(1996)53
86. T.H. Whitesides and D.D. Miller, Langmuir, 10(1994)2899
87. T. Isemura, F. Tokiwa and S. Ikeda, Bull. Chem. Soc. Japan, 35(1962)240
88. J.P.S. Arora, S.P. Singh and V.K. Singhal, Tenside Surf. Det., 21(1984)197
89. J.P.S. Arora, D. Soam, S.P. Singh and R. Kumar, Tenside Surf. Det., 21(1984)87
90. R. Wüstneck and H.J. Müller, Abh. Akad. Wiss. DDR, Abt. Math. Naturwiss. Tech., 1N(1986)491
91. R. Wüstneck and H.J. Müller, Colloid Polymer Sci., 264(1986)97
92. H. Fruhner, Abh. Akad. Wiss. DDR, Abt. Math. Naturwiss. Tech., 1N(1986)239
93. H. Fruhner and G. Kretzschmar, Colloid Polym. Sci., 267(1989)839
94. H. Fruhner and G. Kretzschmar, Colloid Polym. Sci., 270(1992)177
95. K. Fukushima, Y. Murata, G. Sugihara and M. Tanaka, Bull. Chem. Soc. Japan, 55(1982)1376

96. K. Fukushima, G. Sugihara, Y. Murata and M. Tanaka, Bull. Chem. Soc. Japan, 55(1982)3113
97. K.L. Mittal and E.J. Fendler, in „Solution behaviour of surfactants", ed. K.L.Mittal and B. Lindmann, Vol.1, Plenum Press, New York 1983
98. B.J. Birch, D.E. Clarke, R.S. Lee and J. Oakes, Analytica Chimica Acta, 70(1974)417
99. K. Hayakawa, A.L. Ayub and J.C.T. Kawa, Colloid and Surfactants, 4(1982)389
100. R. Wüstneck, L. Zastrow and G. Kretzschmar, Kolloid. Z. USSR, 47(1985)462
101. T. Cosgrove, S.J. White, A. Zarbakhsh, R.K. Heenan and A.M. Howe, Langmuir, 11(1995)744
102. T. Cosgrove, S.J. White, A. Zarbakhsh, R.K. Heenan and A.M. Howe, J. Chem. Soc., Faraday Trans., 92(1996)595
103. D.D. Miller, W. Lenhart, B.J. Antalek, A.J. Williams and J.M. Hewitt, Langmuir, 10(1994)68
104. P.C. Griffiths, P. Stilbs, A.M. Howe and T. Cosgrove, Langmuir, 12(1996)2884
105. H. Lange, Kolloid Z. Z. Polym., 243(1971)101
106. H. Lange, Chemiker Zeitung, 96(1972)192
107. H. Arai, M. Murata and K. Shinoda, J. Colloid Interface Sci., 37(1971)223
108. M. Murata and H. Arai, J. Colloid Interface Sci., 44(1972)475
109. E.D. Goddard, In: Interactions of Surfactants with Polymers and Proteins, E.D. Goddard and K.P. Ananthapadmanabhan (Eds.), CRC, Boca Raton, Fl, 1993, p. 203
110. J. Oakes and M.C. Cafe, Eur. J. Biochem., 36(1973)553
111. J. J. Oakes, Chem. Soc., Faraday Trans., I 70(1974)2200
112. M.L. Smith and N. Muller, Biochem. Piophys. Res. Commun., 62(1975)723
113. M.J. Schwuger, Kolloid-Z. Z. Polym., 246(1971)626
114. R.K. Burkhard and G.E. Stolzenberg, Biochemistry, 11(1972)1672
115. A. Malovikova, K. Hayakawa and J.C.T. Twak, J. Phys. Chem., 88(1984)1930
116. J.A. Reynolds and C. Tanford, J. biol. Chem., 245(1970)5161
117. J. Steinhardt and J.A. Reynolds, „Multiple Equilibria in Proteins", Academic Press, New York 1969
118. J.A. Reynolds, S. Herbert, H. Polet and J. Steinhardt, Biochemistry, 6(1967)937
119. I.P. Purcell, R.K. Thomas, J. Penfold and A.M. Howe, Colloids and Surfaces A, 94(1995)125

120. B. Cabane and R. Duplessix, J. Physique, 43(1982)1529
121. B. Cabane, R. Duplessix and T. Zemb, J. Physique, 46(1985)2161
122. B. Cabane and R. Duplessix, Colloid and Surfaces, 13(1985)19
123. K. Chari, B. Antalek, M.Y. Lin and S.K. Sinha, J. Chem. Phys., 100(1994)5294
124. B. Cabane, J. Phys. Chem., 81(1977)1639
125. K. Chari and W.C. Lenhart, J. Colloid Interface Sci., 137(1990)204
126. Y.J. Nikas and D. Blanckschtein, Langmuir, 10(1994)3512
127. E. Dickinson, In: Interactions of Surfactants with Polymers and Proteins, E.D. Goddard and K.P. Ananthapadmanabhan (Eds.), CRC, Boca Raton, Fl, 1993, p. 295
128. A.K. Wright, M.R. Tompson and R.L. Miller, Biochemistry, 14(1975)2579
129. E.S. Rowe and J. Steinhardt, Biochemistry, 15(1976)2579
130. R.C. Collins and W. Haller, Anal. Biochem., 54(1973)47
131. K. Shirahama, K. Tssuji and T. Takagi, J. Biochem (Tokyo), 75(1974)309
132. F. Tokiwa, Yukagaku, 19(1970)236
133. C. Tanford, „Physical Chemistry of Macromolecules“, Wiley, New York 1969, p. 573
134. K. Sumida, H. Kishimoto, S. Shimabayashi and M. Nakagaki, Yakugaku Zasshi, 95(1975)853
135. K.S. Birdi, T. Krag and J. Klausen, J. Colloid Interface Sci., 62(1977)562
136. K.S. Bridi, H.H. Singh and S.U. Dalsager, J. Phys. Chem., 83(1979)2723
137. D.D. Miller, W. Lenhart, B.J. Antalek, A.J. Williams and J.M. Hewitt, Langmuir, 10(1994)68
138. K. Chari, J. Colloid Interface Sci., 151(1992)294
139. P.C. Griffiths, P. Stilbs, A.M. Howe and T. Cosgrove, Langmuir, 12(1996)2884
140. P.C. Griffiths, P. Stilbs, A.M. Howe and T.H. Whitesides, Langmuir, 12(1996)5302
141. A. Küntzel and J.C. Le Nénaon, Leder, 12(1961)9
142. D. Vollhardt and G. Kretzschmar, J. Signal AM, 4(1976)333
143. A.V. Few, R.H. Ottewill and H.C. Parreira, Biochim. Biophys. Acta, 18(1955)136
144. Y. Nozaki, J.A. Reynolds and C. Tanford, J. Bio. Chem., 249(1974)4452
145. S. Makino, J.H. Reynolds and C. Tanford, J. Biol. Chem., 248(1973)4926
146. H.M. Rendall, J. Chem. Soc. Faraday Trans., I. 72(1976)481
147. R.V. Decker and J.F. Forster, Biochemistry, 5(1966)1242
148. M.N. Jones, J. Colloid Interface Sci., 23(1967)36

149. D. Robb, in „Anionic surfactants", Lucassen-Reynders E.H. (ed.), Marcel Dekker, Vol. 11, p. 109, New York 1981
150. A.A. Trapeznikov, V.G. Vins and T.J. Shirokova, Kolloid. Z. USSR, 43(1981)322
151. V.N. Izmailova, S.R. Derkach, K.V. Zotova and R.G. Danilova, Colloid Journal of the Russian Academy of Sci., 55(1993)54
152. R. Wüstneck, R. Wetzel, E. Buder and H. Hermel, Colloid Polymer Sci., 266(1988)1061
153. A. Gardi, H.S. Nischmann and K. Rieder, Chimia, 27(1973)116
154. R. Mandel and G. Holzwarth, Biopolymers, 12(1973)655
155. T. Nishio, R. Hayashi, Agric. Biol. Chem., 49(1985)1675
156. K. Sutch and H. Noda, Biopolymers, 13(1973)2391
157. J.W. Caldwell and J. Applequist, Biopolymers, 20(1984)1891
158. O.D. Jennes, C. Sprecher, W. Curtis and W.C. Johnson, Biopolymers, 15(1976)513
159. E.S. Pysh, Biopolymers, 13(1974)1563
160. W. Ronish and S. Krimm, Biopolymers, 13(1974)1635
161. V. Madison and J. Schellman, Biopolymers, 9(1970)65
162. J. Appliequist, Biopolymers, 20(1981)2311
163. R. Wetzel, E. Buder, H. Hermel and A. Hüttner, Colloid Polymer Sci., 265(1987)1036
164. R. Wüstneck, E. Buder, R. Wetzel and H. Hermel, Colloid Polymer Sci., 267(1989)429
165. R. Wüstneck, E. Buder, R. Wetzel and H. Hermel, Colloid Polymer Sci., 267(1989)516
166. H. Tschesche, in „Biophysik" ed. W. Hoppe, W. Lohmann, H. Markl, H. Ziegler, Springer-Verlag, Berlin, 1982, p.35
167. J.F. Brandts, H.R. Halvorson and M. Brennan, Biochemistry, 14(1975)4953
168. I.Z. Steinberg, W.F. Harrington, A. Berger, M. Selas and E.J. Katchalski, JACS, 82(1960)5263
169. Z.N. Izmailova and P.A. Rehbinder, in „Structuroobrazovanie v belkovych sistemach", Nauka, Moskow 1974, p.179
170. A. Seeboth and H. Hermel, Thin solid films, 119(1989)173
171. A. Seeboth, H. Hermel and G. Kretzschmar, Thin solid films, 201(1990)197
172. A. Seeboth, H. Hermel, O. Zaplo and J. Stumpe, Mol. Cryst. And Liq. Cryst., 213(1992)153
173. A. Seeboth, Thin solid films, 223(1992)140

174. A. Seeboth, A.R. Mackie, D.C. Clark and S.A. Clark, Colloid Polym. Sci., 272(1994)348
175. A. Seeboth, R. Wüstneck and K. Otto, Colloid Polymer Sci., 268(1990)286
176. H. Fruhner, J. Krägel and G. Kretzschmar, DD-Patent 214304, 1984
177. S.S. Dukhin, G. Kretzschmar and R. Miller, In: Dynamics of Adsorption at Liquid Interfaces, D. Möbius and R. Miller (eds.), Elsevier, Amsterdam 1995
178. G. Kretzschmar and R. Miller, Adv. Colloid Interface Sci., 36(1991)65
179. M. Ziller, R. Miller and G. Kretzschmar, Z. phys. Chem., 266(1985)721
180. W. Schreiter, PhD Thesis, Universität Halle-Wittenberg 1976
181. G. Schwen, Tenside, Surf. Det., 1(1964)46
182. P.R. Schunk and L.E. Scriven, In: Liquid Film Coating, S.F. Kistler and P.M. Schweizer (eds.), Chapman and Hall, London 1997, Chapter 11d, pp. 495-536
183. R. Miller, V.B. Fainerman, K.H. Schano, W. Heyer, A. Hofmann and R, Hartmann, Labor Praxis, 18(1994)56
184. R. Wüstneck, H. Hermel. and G. Kretzschmar, Colloid Polymer Sci., 262(1984)827
185. R. Wüstneck, L. Zastrow and G. Kretzschmar, Kolloid. Z. USSR, 49(1987)10
186. K. Fukushima, G. Sugihara, Y. Murata and M. Tanaka, Bull. Chem. Soc. Japan, 54(1981)3122
187. R. Wüstneck and T. Wärnheim, Colloid Polym. Sci., 266(1988)926
188. A.R. Pitt, S.D. Morley, N.J. Burbidge and E.L. Quickenden, Colloids and Surfaces A, 114(1996)321
189. R. Wüstneck and G. Kretzschmar, Kolloid Z. USSR, 49(1987)555
190. K. Meguro, T. Kanbe and K. Esumi, JAOCS, 60(1983)1050
191. W. Schreiter, F. Wolf and W. Walther, J. Signal AM, 8(1980)451
192. W. Schreiter, F. Wolf and W. Walther, J. Signal AM, 9(1981)63
193. R. Miller, R. Wüstneck, J. Krägel and G. Kretzschmar, Colloids and Surfaces A, 111(1996)75
194. A.A. Trapeznikov, G.V. Fedotov and T.I. Korotina, Kolloid Z. USSR, 41(1979)103
195. V.G. Vins and A.A. Trapeznikow, Kolloid Z. USSR, 43(1981)424
196. V.G. Vins and A.A. Trapeznikow, Kolloid Z. USSR, 41(1979)392
197. R. Wüstneck, L. Zastrow and G. Kretzschmar, Kolloid. Z. USSR, 49(1987)239

198. V.N. Izmailova, S.R. Derkach, K.V. Zotova and R.G. Danilova, Colloid Journal of the Russian Academy of Sci., 55(1993)54
199. H.O. Lee, T.S. Jiang and K.S. Avramidis, J. Colloid Interface Sci., 146(1991)90
200. J. Krägel, S. Siegel, R. Miller, M. Born and K.H. Schano, Colloids and Surfaces A, 91(1994)169
201. G. Kretzschmar and K. Lunkenheimer, Ber. Bunsenges. Phys. Chem., 74(1970)1064
202. H. Fruhner and K. Wantke, Colloid and Surfaces A, 114(1996)53
203. K.S. Avramidis and T.S. Jiang, J. Colloid Interface Sci., 147]1991)262
204. H. Fruhner, K. Lunkenheimer and R. Miller, In: Materials and fluids under low gravity, L. Ratke, H. Walter and B. Feuerbach (eds.), Springer-Verlag, Berlin 1995, pp. 40-50
205. F. Stainsby, Nature, 169(1952)662
206. N. Ise, Angew. Chem., 1986, 98, 323.
207. N. Ise, ; T. Okubo; K. Yamamoto; H. Matsouka;H. Kawai; T. Hashimoto; M. Fujimura, J. Chem. Phys., 1983, 78, 541.
208. M. Djabourov, J. Leblond and P. Papon, J. Phys. (France), 49(1988)333
209. A. Scheludko, Kolloid-Z., 39(1957)155
210. T.O. Maternaghan and R.H. Ottewill, J. Photogr. Sci., 22(1974)289
211. A.I. Kudish, Ph.D. thesis, Polytechnic Inst. Of Brooklyn, 1970
212. A. Silberberg, J. Colloid Interface Sci., 38(1972)217
213. D.E. Graham and M.C. Phillips, J. Colloid Interface Sci., 70(1979)403
214. V.V. Rodin and V.N. lzmailova, Colloids and Surfaces A, 106(1996)95
215. G. Hearing and P.L. Luisi, J. Phys. Chem., 90(1986)5892
216. P.L. Luisi and L.J. Magid, CRC Crit. Rev. Biochem., 20(1986)409
217. C. Quellet and H.F. Eicke, Chimia, 40(1986)233
218. B.H. Robinson, P.J. Aktinson, A.M. Howe and R.K. Heeman, J. Chem. Soc. Fraday Trans., 87(1991)3389
219. S. Backlund, F. Eriksson, G. Hedstrom, A. Laine and M. Rantala, Colloid Polym. Sci., 274(1996)540
220. A. Seeboth, R. Wüstneck and J. Krägel, Colloid Polymer Sci., 272(1994)1151
221. E. Bianchi, R. Patrone, A. Tealdi and A. Ciferri, J. Biol. Chem., 245(1970)3341
222. A.I. Dolinny and V.N. Izmailova, Vysokomol. Soejed., 20B(1978)39.
223. E.P. Narishkina, V.N. Izmailova and A.I. Dolinny, Kolloid Zh. USSR, 45(1983)1106

224. M. Fujitsu, M. Hattori and T. Tamura, Colloid Polym. Sci., 275(1997)67
225. J.A. Mercier, W.A. Torpey and T.A. Russell, US-Patent 2761419, 1956
226. D.J. Hughes, US-Patent 3508947, 1970
227. H. Fruhner, J. Krägel and G. Kretzschmar, DD-Patent 214678, 1984
228. W. Schreiter, H. Fruhner, J. Krägel, G. Kretzschmar, D. Mähnert and W. Seibt, DD-Patent 217313, 1985
229. H. Fruhner, J. Krägel and G. Kretzschmar, DD-Patent 214304, 1984
230. H. Fruhner, J. Krägel, G. Kretzschmar and W. Schreiter, DD-Patent 281268, 1990
231. H. Fruhner, J. Krägel and G. Kretzschmar, DD-Patent 298676, 1992

8. LIST OF SYMBOLS

A	surface area
c	bulk concentration (of surfactant or protein)
CAC	critical aggregation concentration
CD	circular dichroism
CMC	critical micelle concentration
CTAB	hexadecyl trimethyl ammonium bromide
E	dilational elasticity
f	frequency
f^s	compliance limit
G^s	spontaneous elasticity modulus
IEP	isoelectric point
SDS	sodium dodecyl sulphate
λ	wave length
Θ	loss angle of viscoelasticity
$\Delta\sigma$	surface tension amplitudes
σ	surface tension
η^s	Newtonian viscosity

SUBJECT INDEX

FSC
www.fsc.org
MIX
Papier | Fördert gute Waldnutzung
FSC® C083411

Kolibri 360 GmbH
Ferdinand-Jühlke-Straße 7
99095 Erfurt, Deutschland
produktsicherheit@kolibri360.de